CW00432109

*C. N. R. Rao, A. Müller,*
*A. K. Cheetham (Eds.)*
**The Chemistry of**
**Nanomaterials**

## *Further Titles of Interest*

G. Schmid (Ed.)

**Nanoparticles**

**From Theory to Application**

2004

ISBN 3-527-30507-6

V. Balzani, A. Credi, M. Venturi

**Molecular Devices and Machines**

**A Journey into the Nanoworld**

2003

ISBN 3-527-30506-8

M. Driess, H. Nöth (Eds.)

**Molecular Clusters of the Main Group Elements**

2004

ISBN 3-527-30654-4

G. Hodes (Ed.)

**Electrochemistry of Nanomaterials**

2001

ISBN 3-527-29836-3

U. Schubert, N. Hüsing

**Synthesis of Inorganic Materials**

2000

ISBN 3-527-29550-X

*C. N. R. Rao, A. Müller, A. K. Cheetham (Eds.)*

# The Chemistry of Nanomaterials

Synthesis, Properties and Applications in 2 Volumes

Volume 1

WILEY-VCH Verlag GmbH & Co. KGaA

***Prof. Dr. C. N. R. Rao***
CSIR Centre of Excellence in Chemistry and Chemistry and Physics of Materials Unit
Jawaharlal Nehru Centre for Advanced Scientific Research
Jakkur P.O.
Bangalore – 560 064
India

***Prof. Dr. h.c. mult. Achim Müller***
Faculty of Chemistry
University of Bielefeld
Postfach 10 01 31
D-33501 Bielefeld
Germany

***Prof. Dr. A. K. Cheetham***
Director
Materials Research Laboratory
University of California, Santa Barbara
Santa Barbara, CA 93106
USA

First Edition 2004
First Reprint 2005
Second Reprint 2005
Third Reprint 2005
Fourth Reprint 2006



**Library of Congress Card No.: applied for**
A catalogue record for this book is available from the British Library.
Bibliographic information published by Die Deutsche Bibliothek
Die Deutsche Bibliothek lists this publication in the Deutsche Nationalbibliografie; detailed bibliographic data is available in the Internet at http://dnb.ddb.de



Printed in the Federal Republic of Germany.
Printed on acid-free paper.

**Composition** Asco Typesetters, Hong Kong
**Printing** betz-druck gmbh, Darmstadt
**Bookbinding** J. Schäffer GmbH & Co. KG, Grünstadt

**ISBN** 3-527-30686-2

# Contents

## Preface

Nanomaterials, characterized by at least one dimension in the nanometer range, can be considered to constitute a bridge between single molecules and infinite bulk systems. Besides individual nanostructures involving clusters, nanoparticles, quantum dots, nanowires and nanotubes, collections of these nanostructures in the form of arrays and superlattices are of vital interest to the science and technology of nanomaterials. The structure and properties of nanomaterials differ significantly from those of atoms and molecules as well as those of bulk materials. Synthesis, structure, energetics, response, dynamics and a variety of other properties and related applications form the theme of the emerging area of nanoscience, and there is a large chemical component in each of these aspects. Chemistry plays a particularly important role in the synthesis and characterization of nanobuilding units such as nanocrystals of metals, oxides and semiconductors, nanoparticles and composites involving ceramics, nanotubes of carbon and inorganics, nanowires of various materials and polymers involving dendrimers and block copolymers. Assembling these units into arrays also involves chemistry. In addition, new chemistry making use of these nanounits is making great progress. Electrochemistry and photochemistry using nanoparticles and nanowires, and nanocatalysis are examples of such new chemistry. Nanoporous solids have been attracting increasing attention in the last few years. Although the area of nanoscience is young, it seems likely that new devices and technologies will emerge in the near future. This book is intended to bring together the various experimental aspects of nanoscience of interest to chemists and to show how the subject works.

The book starts with a brief introduction to nanomaterials followed by chapters dealing with the synthesis, structure and properties of various types of nanostructures. There are chapters devoted to oxomolybdates, porous silicon, polymers, electrochemistry, photochemistry, nanoporous solids and nanocatalysis. Nanomanipulation and lithography are covered in a separate chapter. In our attempt to make each contribution complete in itself, there is some unavoidable overlap amongst the chapters. Some chapters cover entire areas, while others expound on a single material or a technique. Our gratitude goes to S. Roy for his valuable support in preparing the index manuscript.

We trust that beginners, teachers and practitioners of the subject will find the book useful and instructive. The book could profitably be used as the basis of a university course in the subject.

C. N. R. Rao
A. Müller
A. K. Cheetham

# List of Contributors

**S. Abbet**
University of Ulm
Institute of Surface Science and Catalysis
Albert-Einstein-Alle 47
D-89069 Ulm
Germany

**R. E. Bailey**
Departments of Biomedical Engineering and Chemistry
Georgia Institute of Technology and Emory University
1639 Pierce Drive, Suite 2001
Atlanta, GA 30322
USA

**J. M. Buriak**
National Institute of Nanotechnology
University of Alberta
Edmonton, AB
T6G 2V4
Canada

**K. K. Caswell**
Department of Chemistry and Biochemistry
University of South Carolina
Columbia, SC 29208
USA

**A. K. Cheetham**
Materials Research Laboratory
University of California, Santa Barbara
CA 93106-5121
USA

**J. F. Corrigan**
Department of Chemistry
University of Western Ontario
London, Ontario
Canada

**S. A. Davis**
Department of Chemistry
University of Bristol
Bristol, BS8 1TS
UK

**M. W. DeGroot**
Department of Chemistry
University of Western Ontario
London, Ontario
Canada

**S. Devarajan**
Department of Inorganic and Physical Chemistry
Indian Institute of Science
Bangalore 560 012
India

**E. Dujardin**
Department of Chemistry
University of Bristol
Bristol, BS8 1TS
UK

**K. J. Edler**
Department of Chemistry
University of Bath
Bath BA2 7 AY
UK

**P. M. Forster**
Materials Research Laboratory
University of California, Santa Barbara
CA 93106-5121
USA

**L. A. Gearheart**
Department of Chemistry and Biochemistry
University of South Carolina
Columbia, SC 29208
USA

**A. Gedanken**
Department of Chemistry
Bar-Ilan University, Ramat-Gan
Israel, 52900

**A. Govindaraj**
Chemistry and Physics of Materials Unit and
CSIR Centre of Excellence in Chemistry
Jawaharlal Nehru Centre for Advanced
Scientific Research
Jakkur P.O.
Bangalore 560 064
India

**Y. L. Gu**
Department of Chemistry
University of Science and Technology of China
Hefei, Anhui 230026
P.R. China

**U. Heiz**
University of Ulm
Institute of Surface Science and Catalysis
Albert-Einstein-Alle 47
D-89069 Ulm
Germany

**N. R. Jana**
Department of Chemistry and Biochemistry
University of South Carolina
Columbia, SC 29208
USA

**C. J. Johnson**
Department of Chemistry
University of Bristol
Bristol BS8 1TS
UK

**P. V. Kamat**
Notre Dame Radiation Laboratory, Notre
Dame
Indiana 46556-0579
USA

**G. U. Kulkarni**
Chemistry and Physics of Materials Unit
Jawaharlal Nehru Centre for Advanced
Scientific Research
Jakkur P.O.
Bangalore 560 064
India

**S. T. Lee**
Center Of Super-Diamond and Advanced
Films (COSDAF) & Department of Physics
and Materials Science
City University of Hong Kong
Hong Kong SAR
China

**Y. Lifshitz**
Center Of Super-Diamond and Advanced
Films (COSDAF) & Department of Physics
and Materials Science
City University of Hong Kong
Hong Kong SAR
China

**J. Lu**
Department of Chemistry
University of Science and Technology of China
Hefei, Anhui 230026
P.R. China

**S. Mann**
Department of Chemistry
University of Bristol
Bristol, BS8 1TS
UK

**Y. Mastai**
Department of Chemistry
Bar-Ilan University, Ramat-Gan
Israel, 52900

**A. Müller**
Faculty of Chemistry
University of Bielefeld
Postfach 100131
D-33501 Bielefeld
Germany

**C. J. Murphy**
Department of Chemistry and Biochemistry
University of South Carolina
Columbia, SC 29208
USA

**S. Nie**
Departments of Biomedical Engineering and
Chemistry
Georgia Institute of Technology and Emory
University
1639 Pierce Drive, Suite 2001,
Atlanta, GA 30322
USA

**S. O. Obare**
Department of Chemistry and Biochemistry
University of South Carolina
Columbia, SC 29208
USA

**P. O'Brien**
The Manchester Materials Science Centre and
the Chemistry Department

The University of Manchester
Oxford Road
Manchester, M139PL
UK

**N. Pickett**
Nano Co Ltd.
48 Grafton Street
Manchester, M139XX
UK

**Y. T. Qian**
Structure Research Laboratory and
Department of Chemistry
University of Science and Technology of China
Hefei, Anhui 230026
P.R. China

**C. N. R. Rao**
Chemistry and Physics of Materials Unit and
CSIR Centre of Excellence in Chemistry
Jawaharlal Nehru Centre for Advanced
Scientific Research
Jakkur P.O.
Bangalore 560 064
India

**S. Ramakrishnan**
Department of Inorganic and Physical
Chemistry
Indian Institute of Science
Bangalore 560012
India

**A. K. Raychaudhuri**
Department of Physics
Indian Institute of Science
Bangalore-560012
India

**S. Roy**
Faculty of Chemistry
University of Bielefeld
Postfach 100131
D-33501 Bielefeld
Germany

**S. Sampath**
Department of Inorganic and Physical
Chemistry
Indian Institute of Science
Bangalore 560 012
India

**S. Sapra**
Solid State and Structural Chemistry Unit
Indian Institute of Science
Bangalore-560012
India

**D. D. Sarma**
Solid State and Structural Chemistry Unit
and Centre for Condensed Matter Theory,
Indian
Institute of Science Bangalore-560012
India
and
Jawaharlal Nehru Centre for Advanced
Scientific Research
Jakkur
Bangalore-560064
India

**M. Sastry**
Materials Chemistry Division
National Chemical Laboratory
Pune – 411 008
India

**J. M. Schmeltzer**
Department of Chemistry
Purdue University
560 Oval Drive
West Lafayette, IN 47907-2084
USA

**R. Seshadri**
Materials Department
University of California, Santa Barbara
CA 93106-5050
USA

**P. J. Thomas**
Chemistry and Physics of Materials Unit
Jawaharlal Nehru Centre for Advanced
Scientific Research
Jakkur P.O.
Bangalore 560 064
India

**R. Q. Zhang**
Center Of Super-Diamond and Advanced
Films (COSDAF) & Department of Physics
and Materials Science
City University of Hong Kong
Hong Kong SAR
China

# 1
# Nanomaterials – An Introduction

*C. N. R. Rao, A. Müller, and A. K. Cheetham*

The term nanotechnology is employed to describe the creation and exploitation of materials with structural features in between those of atoms and bulk materials, with at least one dimension in the nanometer range (1 nm = $10^{-9}$ m). In Table 1.1, we list typical nanomaterials of different dimensions. Properties of materials of nanometric dimensions are significantly different from those of atoms as well as those of bulk materials. Suitable control of the properties of nanometer-scale structures can lead to new science as well as new devices and technologies. The underlying theme of nanotechnology is miniaturization. The importance of nanotechnology was pointed out by Feynman as early as 1959, in his often-cited lecture entitled "There is plenty of room at the bottom". The challenge is to beat Moore's law, according to which the size of microelectronic devices shrinks by half every four years. This implies that by 2020, the size will be in the nm scale and we should be able to accommodate 1000 CDs in a wristwatch, as predicted by Whitesides.

There has been an explosive growth of nanoscience and technology in the last few years, primarily because of the availability of new strategies for the synthesis of nanomaterials and new tools for characterization and manipulation (Table 1.2). There are many examples to demonstrate the current achievements and paradigm shifts in this area. Scanning tunneling microscope (STM) images of quantum dots (e.g. germanium pyramid on a silicon surface) and of the quantum corral of 48 Fe atoms placed in a circle of 7.3 nm radius being familiar ones (Figure 1.1). Several methods of synthesizing nanoparticles, nanowires and nanotubes, and their assemblies, have been discovered. Thus, nanotubes and nanowires of a variety of inorganic materials have been discovered, besides those of carbon. Ordered arrays or superlattices of nanocrystals of metals and semiconductors have been prepared. Nanostructured polymers formed by the ordered self-assembly of triblock copolymers and nanostructured high-strength materials are other examples.

Besides the established techniques of electron microscopy, diffraction methods and spectroscopic tools, scanning probe microscopies have provided powerful means for studying nanostructures. Novel methods of fabrication of patterned nanostructures as well as new device and fabrication concepts are constantly being

*The Chemistry of Nanomaterials: Synthesis, Properties and Applications, Volume 1.* Edited by C. N. R. Rao, A. Müller, A. K. Cheetham

ISBN: 3-527-30686-2

**Tab. 1.1.** Examples of nanomaterials.

| | *Size (approx.)* | *Materials* |
|---|---|---|
| Nanocrystals and clusters (quantum dots) | diam. 1–10 nm | Metals, semiconductors, magnetic materials |
| Other nanoparticles | diam. 1–100 nm | Ceramic oxides |
| Nanowires | diam. 1–100 nm | Metals, semiconductors, oxides, sulfides, nitrides |
| Nanotubes | diam. 1–100 nm | Carbon, layered metal chalcogenides |
| Nanoporous solids | pore diam. 0.5–10 nm | Zeolites, phosphates etc. |
| 2-Dimensional arrays (of nano particles) | several nm2–μm2 | Metals, semiconductors, magnetic materials |
| Surfaces and thin films | thickness 1–1000 nm | A variety of materials |
| 3-Dimensional structures (superlattices) | Several nm in the three dimensions | Metals, semiconductors, magnetic materials |

discovered. Nanostructures are also ideal for computer simulation and modelling, their size being sufficiently small to accommodate considerable rigor in treatment. In computations related to nanomaterials, one deals with a spatial scaling from 1 Å to 1 μm and a temporal scaling from 1 fs to 1 s, the limit of accuracy going beyond 1 kcal $mol^{-1}$. Prototype circuits involving nanoparticles and nanotubes for nanoelectronic devices have been fabricated. Quantum computing has made a beginning and appropriate quantum algorithms are being developed.

Let us not forget that not everything in nanoscience is new. Many existing technologies employ nanoscale processes, catalysis and photography being well-known examples. Our capability to synthesize, organize and tailor-make materials at the nanoscale is, however, of recent origin. Novel chemistry has been generated by employing nanoparticles, nanowires and other nanostructures. This includes electrochemical, photochemical, catalytic and other aspects. The immediate objectives of the science and technology of nanomaterials are: (i) to fully master the synthesis of isolated nanostructures (building blocks) and their assemblies with the desired properties, (ii) to explore and establish nanodevice concepts and systems architectures, (iii) to generate new classes of high performance materials, (iv) to connect

**Tab. 1.2.** Methods of synthesis and investigation of nanomaterials.

| *Scale (approx.)* | *Synthetic Method* | *Structural Tool* | *Theory and simulation* |
|---|---|---|---|
| 0.1 to ~10 nm | Covalent synthesis | Vibrational spectroscopy<br>NMR<br>Diffraction methods | Electronic structure |
| <1 to ~100 nm | Techniques of self-assembly | Scanning probe microscopies | Molecular dynamics and mechanics |
| 100 nm to ~1 μm | Processing, modifications | SEM, TEM | Coarse-grained models etc. |

**Fig. 1.1.** STM image of a quantum corral of 48 Fe atoms placed in a circle of 7.3 nm [IBM Research].

nanoscience to molecular electronics and biology, and (v) to improve known tools while discovering better tools of investigation of nanostructures.

## 1.1
## Size Effects

Size effects constitute a fascinating aspect of nanomaterials. The effects determined by size pertain to the evolution of structural, thermodynamic, electronic, spectroscopic, electromagnetic and chemical features of these finite systems with increasing size. Size effects can be classified into two types, one dealing with specific size effects (e.g. magic numbers of atoms in metal clusters, quantum mechanical effects at small sizes) and the other involving size-scaling applicable to relatively larger nanostructures. The former includes the appearance of new features in the electronic structure. In Figure 1.2, we show how the electronic structures of metal and semiconductor nanocrystals differ from those of bulk materials and isolated atoms. In Figure 1.3, we show the size-dependence of the average energy level spacing of sodium in terms of the Kubo gap ($E_F/N$) in K. In this figure, we also show the effective percentage of surface atoms as a function of particle diameter. Note that at small size, we have a high percentage of surface atoms.

Size affects the structure of nanoparticles of materials such as CdS and CdSe, and also their properties such as the melting point and the electronic absorption spectra. In Figures 1.4 and 1.5, we show such size effects graphically. It should be noted that even metals show nonmetallic band gaps when the diameter of the nanocrystals is in the 1–2 nm range. Hg clusters show a nonmetallic band gap which shrinks with increase in cluster size. It appears that around 300 atoms are necessary to close the gap. It is also noteworthy that metal particles of 1–2 nm diameter also exhibit unexpected catalytic activity, as exemplified by nanocatalysis by gold particles.

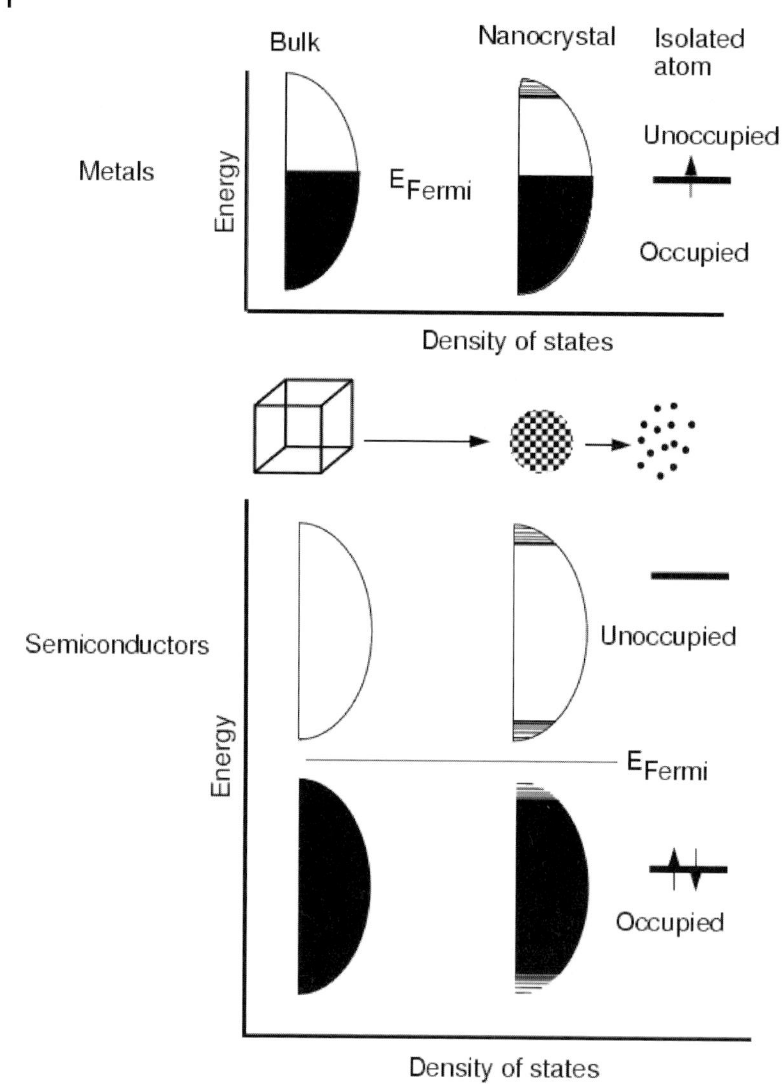


**Fig. 1.2.** Density of states for metal and semiconductor nanocrystals compared to those of the bulk and of isolated atoms [from C. N. R. Rao, G. U. Kulkarni, P. J. Thomas, P. P. Edwards, *Chem-Euro J.*, **2002**, *8*, 29.].

## 1.2 Synthesis and Assembly

The synthesis of nanomaterials and assembling the nanostructures into ordered arrays to render them functional and operational are crucial aspects of nanoscience. The materials/structures include nanoparticles, nanowires, nanotubes,

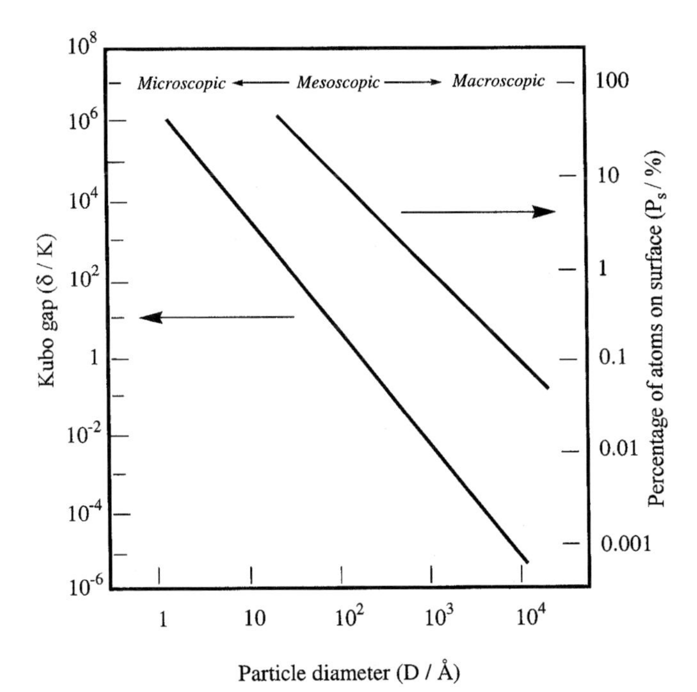


**Fig. 1.3.** A plot of the average electronic energy level spacing (Kubo gap, $\delta$) of sodium as a function of the particle diameter. Also shown is the percentage of sodium atoms at the surface as a function of particle diameter [From P. P. Edwards, R. L. Johnston and C. N. R. Rao, in *Metal Clusters in Chemistry*, ed. P. Braunstein et al., John Wiley, 1998.].

nanocapsules, nanostructured alloys and polymers, nanoporous solids and DNA chips. What is also noteworthy is that chemists have synthesized molecular entities of nanometric dimensions. In Figure 1.6, we show a two-dimensional crystalline array of thiolized metal nanocrystals to illustrate self-assembly.

## 1.3 Techniques

The emerging nanoworld encompasses entirely new and novel means of investigating structures and systems, besides exploiting the well known microscopic, diffraction and spectroscopic methods. Species as small as single atoms and molecules are manipulated and exploited as switches. Computer-controlled scanning probe microscopy enables a real-time, hands-on nanostructure manipulation. Nanomanipulators have also been designed to operate in scanning and transmission electron microscopes. A nanomanipulator gives virtual telepresence on the

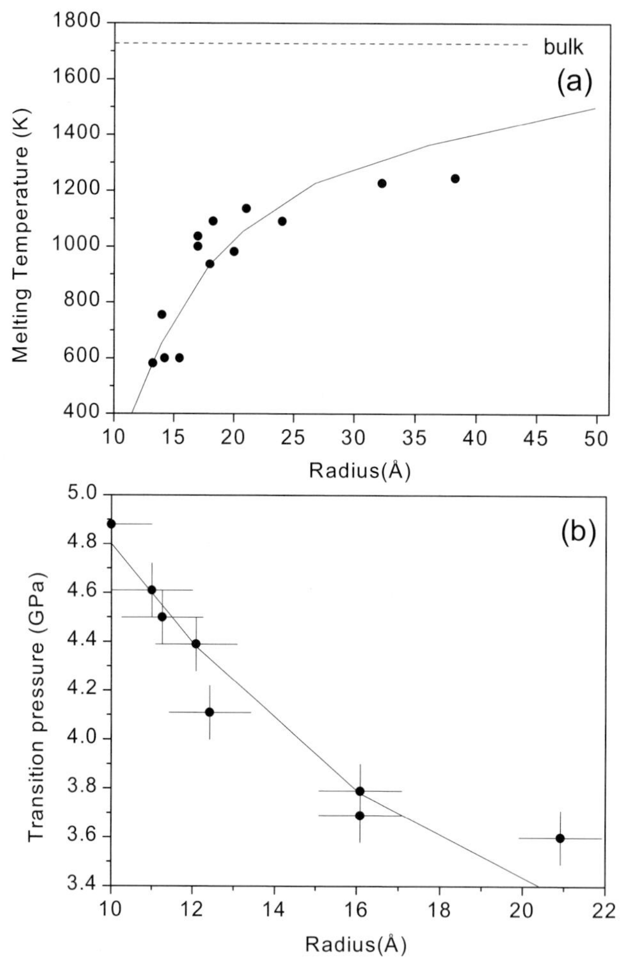


**Fig. 1.4.** (a) Size dependence of the melting temperature of CdS nanocrystals. (b) Size dependence of the pressure-induced wurtzite–rock salt transformation in CdSe nanocrystals [from A. P. Alivisatos, *J. Phys. Chem.*, **1996**, *100*, 13226.].

surface, with a scale factor of a million to one. Optical tweezers provide another approach to holding and moving nanometer structures, a capability especially useful in investigating the dynamics of molecules and particles. Questions such as, how does a polymer move, generate force, respond to an applied force and unfold, can be answered by the use of optical tweezers. It is noteworthy that the positioning of nanoparticles accurately and reliably on a surface by using the tip of an atomic force microscope as a robot has already been accomplished. Large-scale operations requiring parallel tip arrays are being explored in several laboratories.

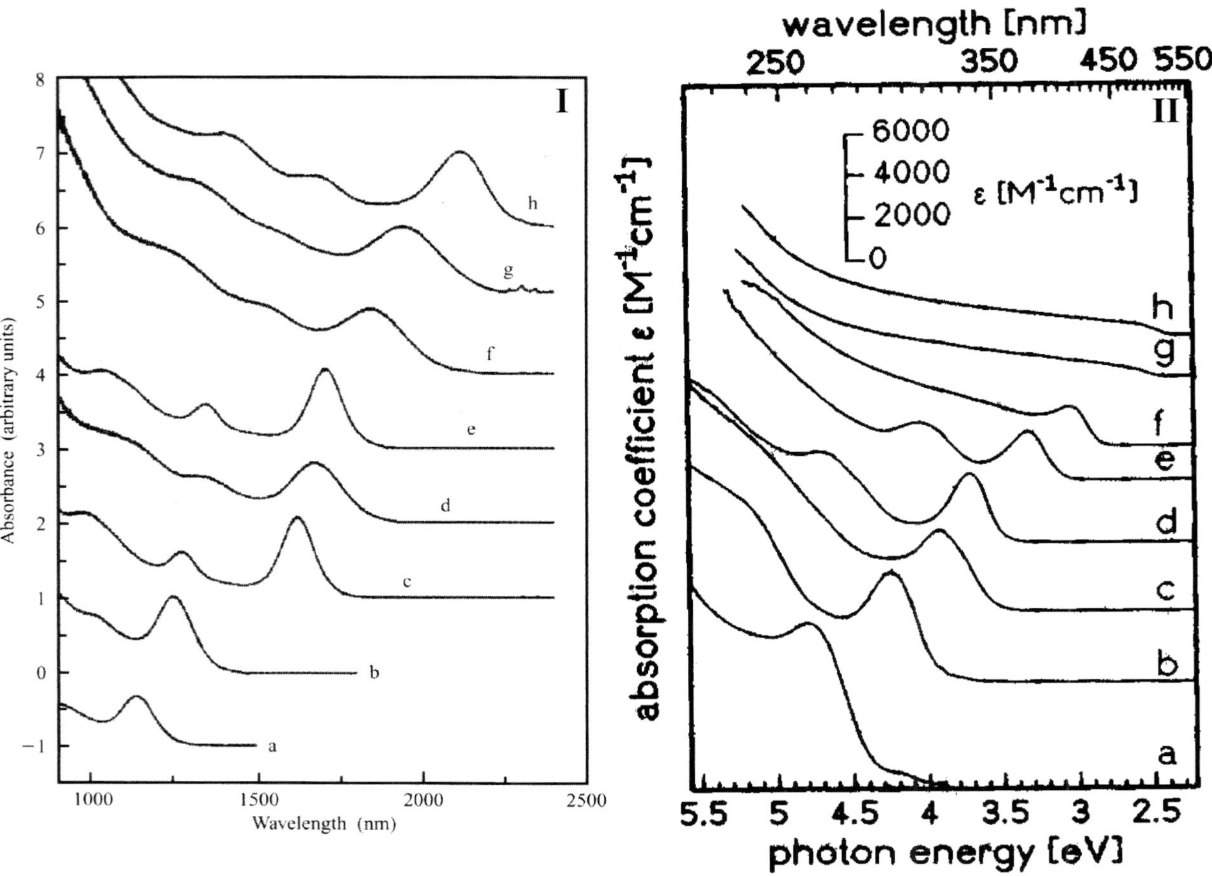


**Fig. 1.5.** Electronic spectra of (I) PbSe nanocrystals (a. 3.0 nm, b. 3.5 nm, c. 4.5 nm, d. 5.0 nm, e. 5.5 nm, f. 7 nm, g. 8 nm, h. 9 nm) and (II) of CdS nanocrystals (a. 0.64 nm, b. 0.72 nm, c. 0.8 nm, d. 0.93 nm, e. 1.94 nm, f. 2.8 nm, g. 4.8 nm) [from T. Vossmeyer et al. *J. Phys. Chem.*, **1994**, *98*, 7665 and R. W. Murray et al., *IBM J. Res. Dev.*, **2001**, *45*, 47.].

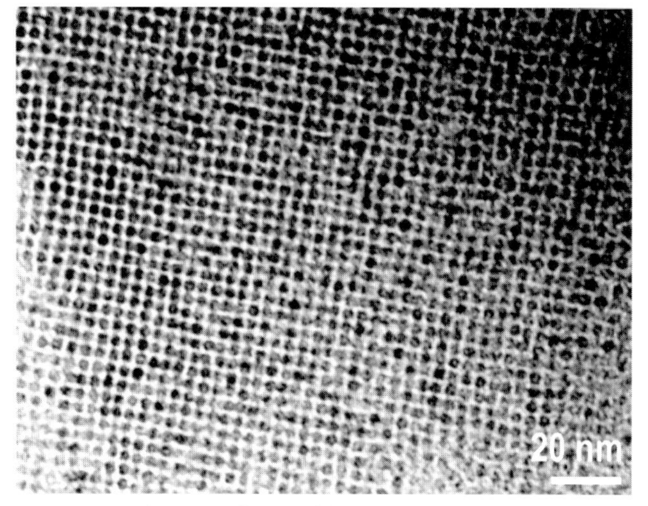


**Fig. 1.6.** Two-dimensional array of thiolized $Pd_{561}Ni_{561}$ nanocrystals [from P. J. Thomas, G. U. Kulkarni, C. N. R. Rao, *J. Nanosci. Nanotechnol.*, **2001**, *1*, 267.].

## 1.4 Applications and Technology Development

Some of the important applications and technologies based on nanomaterials are the following: (i) Production of nanopowders of ceramics and other materials, (ii) nanocomposites, (iii) development of nanolectrochemical systems (NEMS), (iv) Applications of nanotubes for hydrogen storage and other purposes, (v) DNA chips and chips for chemical/biochemical assays, (vi) gene targeting/drug targeting and (vii) nanoelectronics and nanodevices. The last one, which is probably the most challenging area, includes new lasers, nanosensors, nanocomputers (based on nanotubes and other materials), defect-free electronics for future molecular computers, resonant tunneling devices, spintronics and the linking of biological motors with inorganic nanodevices.

## 1.5 Nanoelectronics

The multidisciplinary area of nanoelectronics has two objectives: (i) utilization of a single nanostructure (e.g. nanocrystal, quantum dot, nanotube) for processing electrical, optical or chemical signals, and (ii) utilization of nanostructured materials involving assemblies of nanostructures for electronic, optoelectronic, chemical and other applications. While it is often difficult to make distinctions between the two, the first category is specifically intended to obtain single-electron devices and the second category is for the purpose of miniaturization in information storage etc. Typical examples of single nanostructure devices are those employing Coulomb blockade or a single-electron transistor. Arrays of quantum dots, scanning probe tips and nanotubes are examples of the second category. In Coulomb blockade, the addition of a single electron to a nanoparticle of radius $R$ gives rise to the charging energy, $W = W(\infty) + [b/R]$ where $W(\infty)$ relates to the charging energy of the bulk. The minimum voltage, $V_{min}$, required to inject an extra electron into the nanoparticle gives rise to the Coulomb staircase with voltage steps, $V_{min} = [W(\infty)/e] + [b/eR]$. The observation of the staircase provides a direct demonstration of the discrete electronic structure in such finite (nano) systems. In Figure 1.7, we show the $I$–$V$ characteristics of an isolated 3.3 nm Pd nanocrystal, exhibiting the Coulomb staircase phenomenon. The dependence of the charging energy on particle size is also shown in Figure 1.7.

## 1.6 Other Aspects

Consolidated nanostructures employing both ceramic and metallic materials are considered important in creating new generations of ultrahigh-strength, tough structural materials, new types of ferromagnets, strong and ductile cements, and new biomedical prosthetics. Typical of the nanostructured hard materials are Co/WC and Fe/TiC nanocomposites. Nanoparticle-reinforced polymers are being considered for automotive parts. Besides high strength materials, dispersions and powders as well as large bodies of novel morphologies are being produced. Coatings with highly improved features resulting from the incorporation of nanoparticles are being developed.

Nanoelectrochemical systems (NEMS) are likely to augment the already established micro analogue, MEMS. A related aspect pertains to molecular motors. Molecular motors are responsible for DNA transcription, cellular transport and muscle contraction. New fabrication tools enable us to understand and exploit these motors as actuators in nanoelectromechanical systems. These may lead to artificial biological devices that are powdered by ATP. Organic chemists are synthesizing molecules (e.g., rotaxanes) capable of various kinds of motions at the nanolevel. Using molecular motors as nanomachines and interfacing them with inorganic energy sources and other nanodevices would be of great interest.

DNA chips and microarrays represent a technology with applications in diagnostics and genetic research. DNA chips and arrays are devices wherein different DNA sequences are arrayed on a solid support, the arrays generally having 100 to 100,000 different pixels (DNA sites) on the chip surface. The chips will be useful in genomic research, drug discovery, forensics and different types of detection and diagnostics. Electronically active DNA microarrays and electronically directed DNA self-assembly technology could be of value in photonic and electronic devices and other areas. Appropriate nanoparticles containing DNA may indeed provide viable means of delivery in the near future. The gene gun is already being used to deliver genetic materials to transfect plant and animal cells.

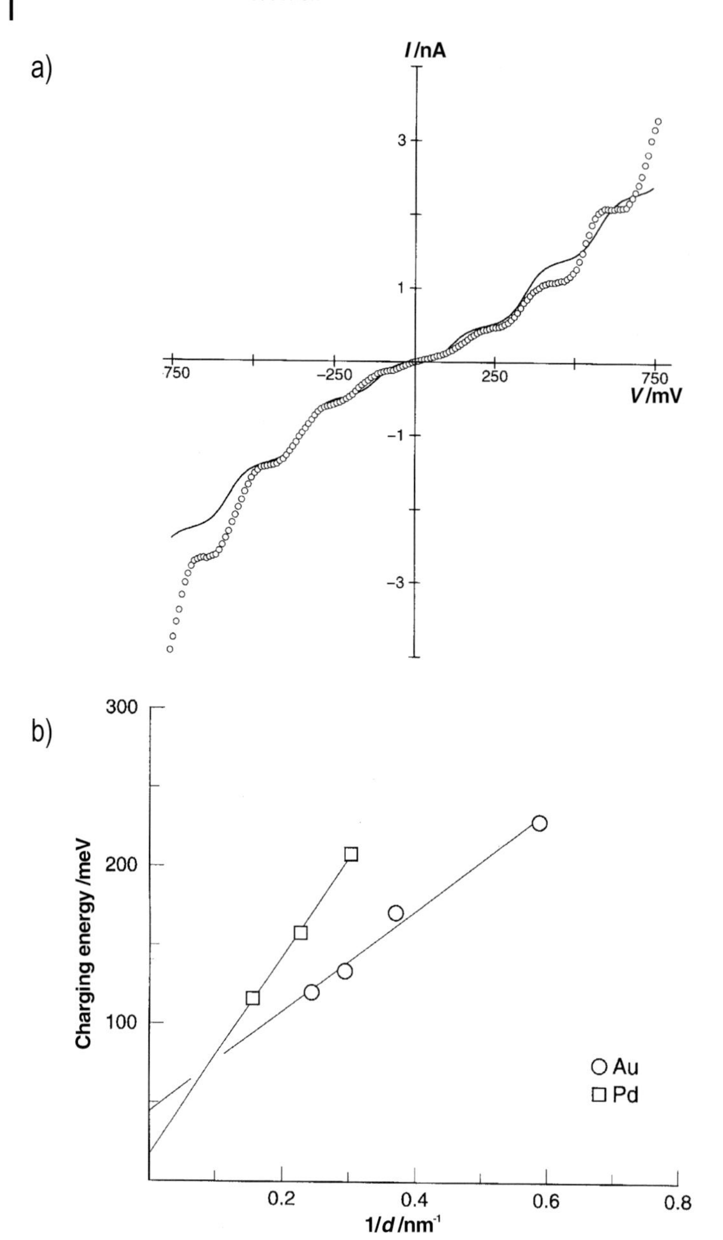


**Fig. 1.7.** (a) *I–V* characteristics of an isolated 3.3 nm Pd nanocrystal (dotted line) and theoretical fit (solid line) obtained at 300 K using a semiclassical model. (b) The size dependence of the charging energy [from P. J. Thomas, G. U. Kulkarni and C. N. R. Rao, *Chem. Phys. Lett.*, **2000**, *321*, 163.].

Semiconductor nanocrystals are being used as fluorescent biological labels. It is likely that sensors based on nanotechnology will revolutionize health care, climate control and detection of toxic substances. It is quite possible that we will have nanochips to carry out complete chemical analysis. Such nano-total analysis systems will have to employ new approaches to valves, pipes, pumps, separations and detection.

A knowledge of the processes related to nanoscale structures, natural as well as man-made, is useful in understanding transport and other aspects of these materials, and in developing technologies for preventing or minimizing harm to the environment. The use of homogeneous and heterogeneous catalysis (including nanocatalysis) for improving energy efficiency and reducing waste is well documented. The design of environmentally benign nanocomposites, the use of nanoparticles of $TiO_2$ and other nanomaterials for environmental cleansing processes and of nano-porous solids for sorption, are examples of the applications of nanotechnology for the protection and improvement of the environment. The use of nanoporous polymers for water purification and purification of liquids by photocatalysis of nanoparticles of $TiO_2$ are two other examples.

## 1.7 Concluding Remarks

The subject of nanomaterials is of great vitality and offers immense opportunities. It is truly interdisciplinary and encompasses chemistry, physics, biology, materials and engineering. Interaction amongst scientists with different backgrounds will undoubtedly create new science, and in particular new materials, with unforeseen technological possibilities. What is noteworthy is that nanotechnology is likely to benefit not only the electronics industry, but also the chemical and space industries, as well as medicine and health care. Chemistry has much to contribute to most aspects of the science and technology of nanomaterials.

## Bibliography

1 *Nanostructure Science and Technology*, eds. R. W. Seigel, E. Hu, M. C. Roco, National Science and Technology Council Report, Kluwer Academic Publishers, Boston 1999.

2 M. C. Roco, R. S. William, A. P. Alivisatos, *Nanotechnology Research Directions*, National Science and Technology Council Report, Kluwer Academic Publishers, Boston 2000.

3 Issues in Nanotechnology, *Science*, **2000**, *290*, 1523–1555.

4 Special topical issue on Nanostructural Systems, *Pure Appl. Chem.*, **2001**, *72*, **2002**, *74*.

5 C. N. R. Rao, A. K. Cheetham, *J. Mater. Chem.*, **2001**, *11*, 2887.

6 J. Jortner, C. N. R. Rao, *Pure Appl. Chem.*, **2002**, *74*, 1491.

7 *National Nanotechnology Initiative*, National Science and Technology Council, Washington D.C., June 2002.

# 2
# Strategies for the Scalable Synthesis of Quantum Dots and Related Nanodimensional Materials

*Paul O'Brien and N. Pickett*

## 2.1
## Introduction

At present there is a considerable interest in the potential for use of materials with dimensions that are best defined in terms of nanometers; especially so-called quantum dots as derived from bulk compound semiconducting materials [1–3]. To date almost all work on the synthesis of quantum dots has been carried out in the academic sector, with a strong emphasis on the synthesis of high quality material in small quantity. One product containing quantum dots has to date been marketed, a biological probe formed by binding particles of CdSe to form Qdot$^{TM}$ streptavidin conjugates [4]. However, there is potentially a wide range of applications for such materials which means that the production of quantities of material is becoming an issue of real interest.

The situation is analogous to the development of metal organic precursors for the deposition of compound semiconductors by metal organic chemical vapour deposition (MOCVD). These compounds underwent a period of intense research and development in the last 20 years of the 20$^{th}$ century [5]. Although work in this area continues it now underpins a mature technology, although the deposition of oxides is still undergoing active development. In its infancy this aspect of materials chemistry was crucial in delivering new functional devices especially in optoelectronics (e.g. high performance solid-state lasers and LEDs) many of which are in everyday use. It seems probable that the emergence of a materials chemistry for the scalable manufacture of nanodimensional materials will be of similar importance in the first half of the present century.

This chapter will address issues concerned with the synthesis of quantum dots in quantity. Many issues remain ill-defined in this area including the simple problem of defining what will form a proper specification for nanodimensional materials such as quantum dots. A typical material may involve two types of solid-state compound and an apparently ever increasingly complex organic coat used to enhance the stability of the material and the solvents in which it can be dispersed/processed. Again analogies with conventional semiconducting materials can be

*The Chemistry of Nanomaterials: Synthesis, Properties and Applications, Volume 1.* Edited by C. N. R. Rao, A. Müller, A. K. Cheetham

ISBN: 3-527-30686-2

made. In this case the specifications for this now singularly important class of materials emerged in the 1930s. It is important for the chemist to remember that the parameters defining the functional properties of such bulk materials: principally mobility and carrier type and concentration are defined by concentrations of impurities or dopants which are very small. The determination of bulk properties from such small concentrations made the initial proper definition of the intrinsic properties of silicon difficult, the history of this problem has been discussed recently [6].

This chapter will be structured as follows:

- The types of nanodimensional materials now routinely prepared, their structures and how these might be defined.
- The emergent uses for nanodimensional materials and how these lead to a market pull for volume manufacture.
- The general methods available for the synthesis of nanodimensional materials.
- The suitability of such methods for scaling.
- Conclusions and perspectives on the future.

In preparing this chapter we have tried to give a perspective on what is an emerging area. Due to constraints of space the referencing has been limited to key papers and indications of a number of major secondary sources are given. Examples have been drawn from the work of one author's own group at Imperial College and latterly in Manchester. The synthesis of nano-dispersed metals and oxides is not considered in detail, although some key references are provided. There are two reasons for omitting metals: Firstly, the problems in scale up are much greater for semiconductors than for metals and hence there is more scope for discussion. Secondly, gold in particular is readily available and widely used in nano-dispersed form [7]. For oxides in terms of well-defined processable nano-materials analogous of quantum dots there are relatively few reports, but for an outstanding example see [8], in contrast crude nanopowders of oxides are commonly available. A very useful review of nanodimensional magnetic materials has just appeared [9].

## 2.2 Defining Nanodimensional Materials

For the purposes of this article we will limit our discussion to particles defined by a minimum of two dimensions less than 100 nm but usually with 2-dimenions less than 10 nm. Current interest in these materials can principally be traced to work by Luis Brus in the mid-1980s in which he pointed out that the band gap of a simple direct band gap semiconductor such as CdS should be dependent on its size once its dimensions were smaller than the Bohr radius [10]. Experimental work confirmed this suggestion. Initial samples were prepared by low temperature aqueous precipitations but it was at first surprising to discover that the luminescence of such particles depended markedly on the surface coat. Particles treated with hydroxide were much more efficient than as-prepared particles. The importance of the surface in such materials reflects the relative importance of the surface sites for a typical nanodimension particle around 4 $nm^3$ e.g. a few percent of the atoms are found on the surface. Defects tend to anneal to the surface and hence the cores of such small particles are often 'relatively' defect free.

These observations and one further criterion start to define quantum dot systems. The second major possibility is the coating of the central particle with a second material e.g. CdS or ZnS on top of CdSe. When a wider band gap material is coated onto the outside of a narrow band gap material the confinement on the 'core' is enhanced, leading to enhanced optical properties, especially photoluminescence efficiencies.

The above enables us to define the nature of an isolated quantum dot. It will depend on

- The nature of the central materials, the core.
- The nature of any subsequent coating 'shell' layer.
- The nature of the final coat on the material, often an organic capping layer.

A typical such material is shown in Figure 2.1. In high quality materials the core of the dot will be a single crystal, the shell epitaxial or close to and the final coat will pacify defects and also confer solubility and or functionality for binding to a substrate or target molecule. However, what is shown in both these schematics and images is a single quantum dot. Each dot is a mesoscopic entity with individual properties, the properties of an ensemble of dots will additionally be determined by the particle size distribution and any differences in morphology within the ensemble. The best quality dots are themselves single crystals, as nicely illustrated by the PbS sample [11] shown in Figure 2.2.

There are as yet no standards defining the properties of such samples; these will clearly need to emerge as these materials become more generally utilised. Such standards will owe more to the definitions of materials and/or polymer science than to the world of the molecule and will include definitions probably statically derived from:

- Size, including aspect ratio.
- The distribution of particle sizes.
- The nature of the core.
- The nature of the shell.
- The nature of the final coat.

The above describe the physical composition of the particle; other properties may define its use, typically quantum efficiency in an optical material and other measures as appropriate, such as coercivity in a magnetic material.

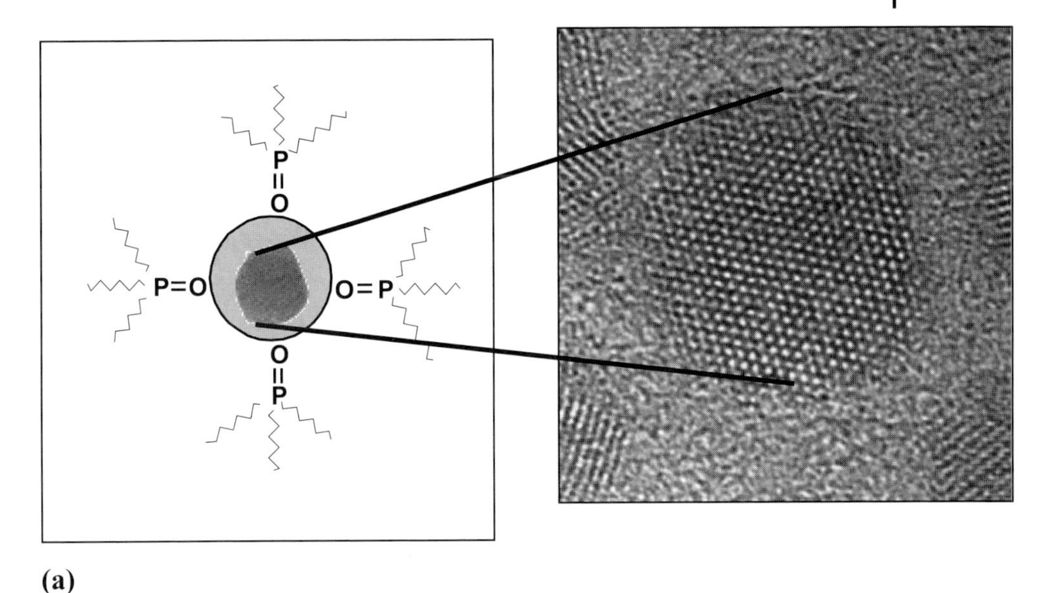


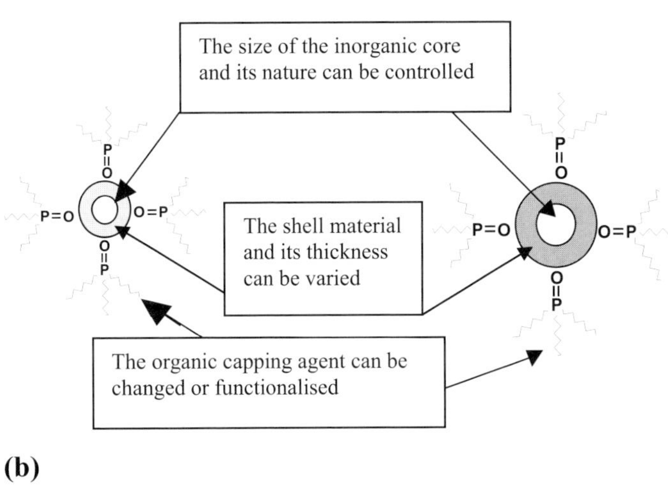


Fig. 2.1. (a) A ca. 4 nm core–shell dot of CdSe coated with CdS and a TOPO, and a schematic. (b) Anatomy of organically capped quantum dots.

## 2.3 Potential Uses for Nanodimensional Materials

In nanoparticles derived from semiconductors, once they are below a critical size, their electronic behaviour is governed by quantum physics (hence 'quantum dots'). The absorption of light at one wavelength leads to emission at slightly lower en-

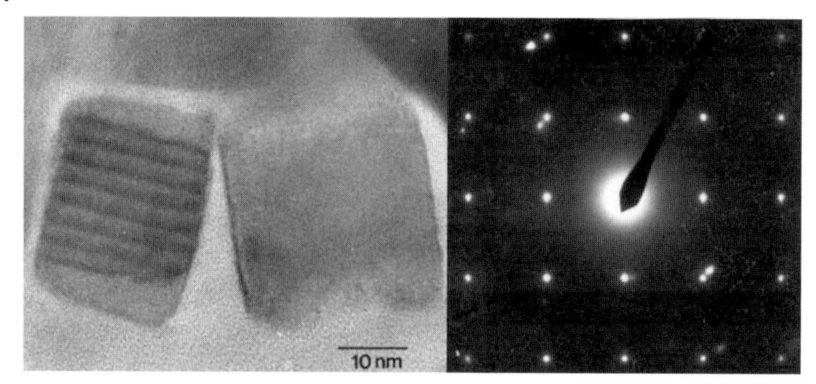


**Fig. 2.2.** Sample of PbS coated with TOPO.

ergy. Such behaviour is usually associated with extremely pure single crystals in bulk semiconductors such as the gallium arsenide or indium phosphide or related alloys used in solid-state lasers. The small shift in the emission is termed a 'Stokes shift'.

The characteristic wavelength of light at which these interactions occur is governed by the size of the particle. This effect could be exploited in:

- Inks and dyes (optical probes in general) with a unique color signature often only visible to the eye when exposed to UV light.
- Functional materials e.g. ones triggered by light to release payloads of proteins or DNA at an exact location in the human body.
- Solar cells probably hybrid organic/inorganic structures.

Leading to market applications in:

- **Security**: Authentication and anti-forgery, specifically the elimination of counterfeit currency, documents, brand name clothing, car parts, etc.
  Overt and covert anti-counterfeiting features – bank notes, paper documents, casino chips and brand protection, i.e. as replacement for specialty conventional (luminescent and fluorescent) dyes in security applications.
- **Life Sciences**: examples include: diagnostics; biological sensors; drug delivery; replacements for luminescent and fluorescent dyes in biological probes for high throughput screening applications.
- **Electronics**: data storage; LEDs; photovoltaics; flat panel displays. The materials are likely to be crucial in the evolution of an electronics industry based on soft materials such as plastics.

These possibilities are summarised schematically in Figure 2.3. Clearly there are a variety of markets in which quantum dots could be used if problems associated with their manufacture could be overcome. The striking sharp photoluminescence spectra of some quantum dots are shown in Figure 2.4 together with pictures of samples taken under UV irradiation.

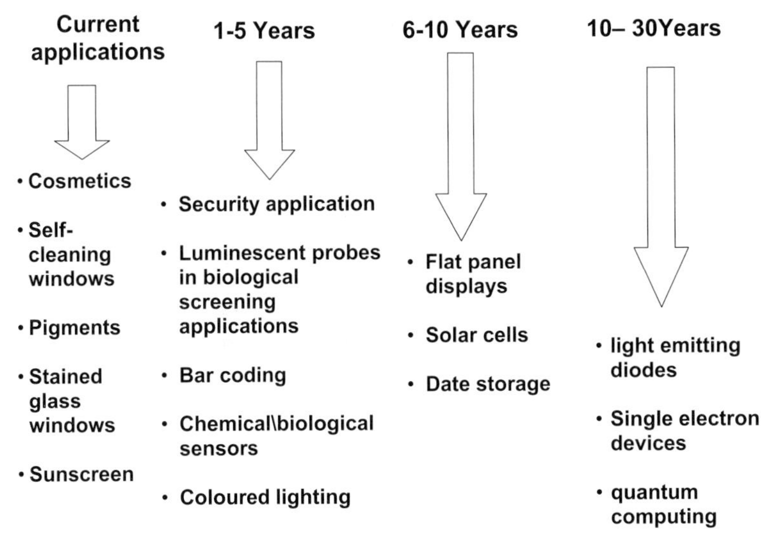


Fig. 2.3. Applications and future markets for nanoparticles and quantum dots.

## 2.4 The General Methods Available for the Synthesis of Nanodimensional Materials

There is much in common with the synthesis of colloidal materials in the preparation of dispersions of nanodimensional materials. Indeed it is a moot point as to whether such materials are large molecules or small colloidal particles. What probably distinguishes ensembles of quantum dots and related materials from

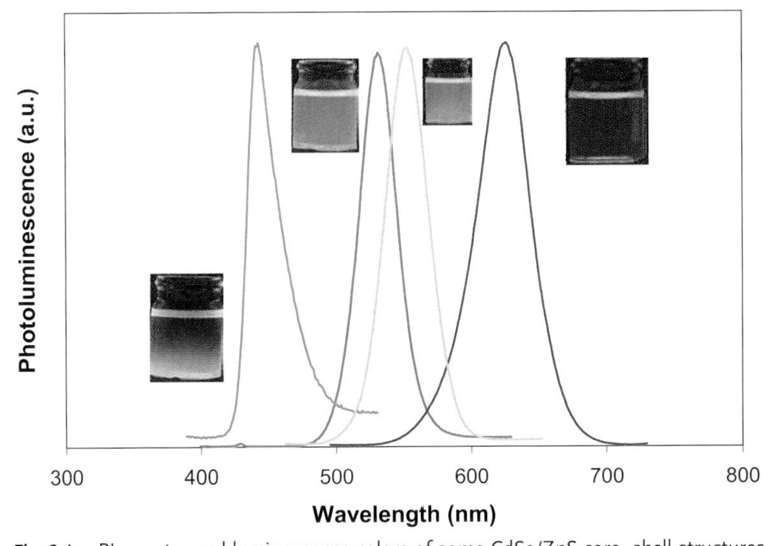


Fig. 2.4. PL spectra and luminescence colors of some CdSe/ZnS core–shell structures.

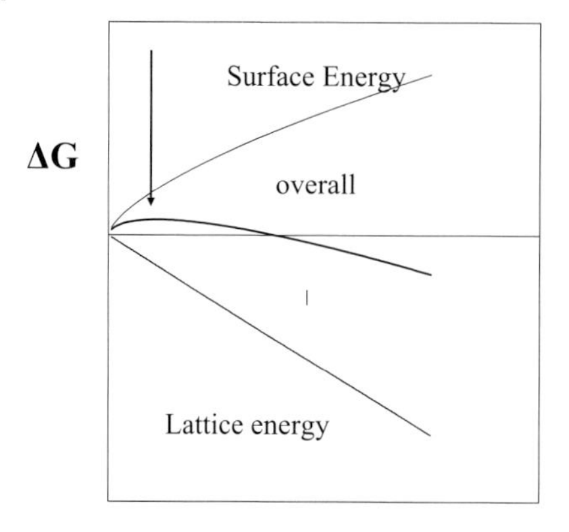


Fig. 2.5. Nucleation and growth.

collections of clusters is that each particle is, in principle, a mesoscopic object with its properties determined by its own nature, i.e. its precise morphology and the number of atomic units from which it is built. It inherently lacks the anonymous nature of a molecule in an assembly.

However there is much that can be learned from colloid chemistry when designing the synthesis of such materials, and it seems colloid chemistry is gaining a new lease of life from a better understanding of the assembly of nanoparticulates. Much of the understanding of crystallization is based on the ideas associated with supersaturation. In a typical system undergoing a crystallization process nuclei or small crystallites are generally held to have an unfavorable surface energy compensated by the evolving lattice energy of the solid, Figure 2.5. The situation is different from that in a bulk material as a relatively small percentage of atoms occupy surface sites in a macro or even conventionally termed microscopic particle. It is only when we reach the scale of nanometers that significant numbers of atoms reside at the surface. The majority of routes for the preparation of nanoparticulates avoid what might be termed "runaway" into the macroscopic world by using a molecule, ligand, which binds to the surface of the nanoparticle, the growth of particles is further inhibited by e.g. limiting the supply of the constituents forming the material, either by working at high dilution or by controlling their delivery by a chemical decomposition.

These above principles underlie most of the methods for the reproducible syntheses of nanomaterials, the main exception being those reactions which are physically constrained by being carried out in the pore of a solid state material or in a micelle or vesicle. There is a second ramification of the nucleation and growth process that is often used to effect in the reproducible synthesis of uniform assemblies of particles: the temporal separation of the nucleation and growth processes. If nucleation is induced by a perturbation of the system, e.g. a sharp elevation in temperature, and suddenly stopped; this effect is easily achieved by injecting a cold solution of a reactive precursor into a hot solution. Nucleation will occur but the cold solution injected will immediately cool the solution. If this

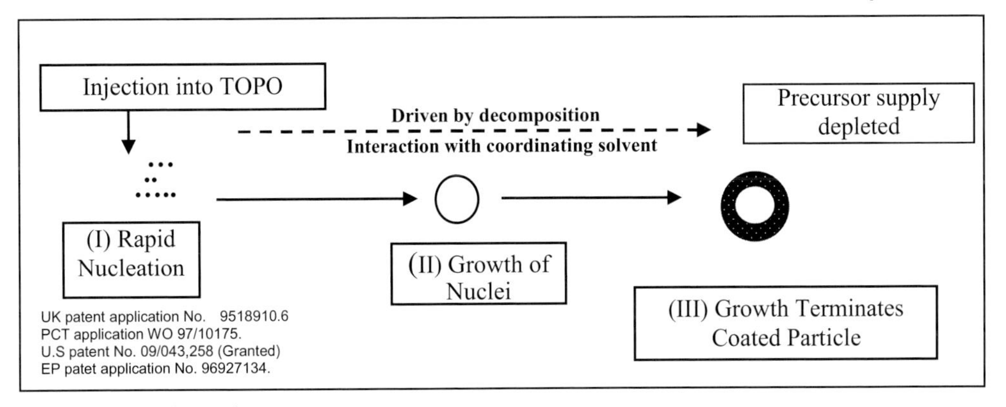


**Fig. 2.6.** A typical growth process.

cooling is significant, 10 °C or so, it is quite possible for little further nucleation to occur and for the growth process to be dominated by the originally created suite of nuclei growing in concert. This common type of process is generically illustrated in Figure 2.6.

Many specific methods have been reported for the preparation of quantum dots of compound semiconductors and include:

- Precipitative method, usually aqueous and at low temperature.
- Reactive methods in high boiling point solvents, often involving metal organic and or organometallic compounds.
- Hydrothermal and solvothermal methods.
- Vapour phase reactions.
- Reactions in confined solids or constrained on surfaces or involving micelles or confined reaction spaces.

The focus of this article is on the preparation of stable isolated particles and methods suitable for the preparation of such materials will be emphasised. A sound synthetic method will lead to crystalline nanoparticulates of high purity, with a narrow size distribution, that are suitably surface derivatized.

### 2.4.1 Precipitative Methods

This type of method was used by Brus and co-workers in studies on CdS and ZnS [12, 13], which have led to the explosion of interest in solid samples of semiconductors which show quantum confinement effects. They [12] prepared CdS nanoparticles by a process involving the controlled nucleation of CdS in dilute aqueous solutions of cadmium sulfate and ammonium sulfide. The dynamic equilibrium between solvated ions and solid CdS in acetonitrile, as a solvent in the presence of a styrene or maleic anhydride copolymer, allowed the preparation of stable CdS nanoparticles, with an average size of 3.4 nm or 4.3 nm, respectively [12] ZnS and CdS nanocrystallites have also been synthesised from aqueous and methanolic solutions without an organic capping agent [13].

Henglein and Weller made significant progress [14–16] using CdS colloids prepared by controlled precipitation methods [17–22]. However, to obtain highly monodispersed nanoparticles, post-preparative separation techniques such as size exclusion chromatography [20] and gel electrophoresis have been employed [21]. Gel electrophoresis was found to be superior to other separation techniques.

Weller et al. synthesised nanocrystallites of $Zn_3P_2$ and $Cd_3P_2$ by the injection of phosphine ($PH_3$) into solutions containing metal salts [18, 23]; control of particle size was achieved by varying the phosphine concentration and the temperature of the reaction. Samples of both $Zn_3P_2$ and $Cd_3P_2$ showed remarkable quantum size effects, as observed by changes in the color of the products. Bulk $Cd_3P_2$ is black whereas a solution containing small nanocrystallites (1.5 nm diameter) is colorless with a maximum in the electronic spectrum at around 310 nm. The color of such $Cd_3P_2$ changes from the black of macrocrystalline (band gap of 0.53 eV) through the visible to white for the 1.5 nm sample (with a band gap of ca. 0.8 eV). Hexametaphosphate was used as a stabiliser to prevent particle aggregation. The sample fluoresced when excited at 300 nm [23].

The relative solubility of inorganic salts can be used to prepare more complex structures by such methods and examples include: CdS/ZnS [24], CdSe/AgS [25] HgS/CdS [26], PbS/CdS [27, 28], CdS/HgS [29], ZnS/CdSe [30] and ZnSe/CdSe [31] particles. The main constraints on the production of such structures involve the relative solubility of the solids and lattice mismatches between the phases. The preparation of "quantum dot quantum well systems" such as CdS/HgS/CdS [32, 33], has also been reported, in which a HgS quantum well of 1–3 monolayers is capped by 1–5 monolayers of CdS. The synthesis grows less soluble HgS on CdS (5.2 nm) by ion-replacement. The solubility products of CdS and HgS are $5 \times 10^{-28}$ and $1.6 \times 10^{-52}$ respectively. The authors reported fluorescence measurements in which the band edge emission for CdS/HgS/CdS is shifted to lower energy values with increasing thickness of the HgS well [33].

Other examples of the preparation of semiconductor nanocrystallites by solution methods can be found in the literature [1–3]. Solution methods provide a cheap route to many nanoparticle materials. However, a lack of reaction control can be problematic when larger scale preparations are necessary. Also several important semiconductors are not easily obtained by this preparative method, with some being air and/or moisture sensitive, e.g. GaAs and InSb.

### 2.4.2
### Reactive Methods in High Boiling Point Solvents

The routine synthesis of well-defined semiconductor nanocrystallites was really opened up in a landmark paper by Murray, Norris and Bawendi [34]. They reacted solutions of dimethylcadmium $(CH_3)_2Cd$ (in tri-n-occtylphosphine TOP) and tri-n-octylphosphine selenide (TOPSe) in hot tri-n-octylphosphine oxide (TOPO) in the temperature range 120–300 °C. This reaction produced TOPO-capped nanocrystallites of CdSe. The size of the particles is controlled mainly by the temperature of reaction, with larger particles being obtained at higher temperatures. This TOPO method has advantages over previous synthetic methods, including, producing monodispersity ($\sigma \cong 5\%$) and the ability to produce hundreds of milligrams of materials in a single experiment. Alivisatos subsequently used higher temperatures for injection and growth to improve the quality of the material prepared [35]. In a series of recent papers interesting rod and tetrapodal structures have been grown especially in the CdSe system [36]. The method was readily adapted to the production of core–shell structures [37, 38] with materials having high quantum efficiencies being prepared.

In an early paper the use of a single-source precursor, oligomeric $Cd(Se(C_2H_5))_2$ [39] was used for the preparation of CdSe in 4-ethylpyridine. Such a compound obviates the need for the use of hazardous compounds such as toxic dimethylcadmium, $(CH_3)_2Cd$, which are especially toxic at high temperatures. The use of single-molecule precursors, i.e. a single compound containing all elements required within the nanocrystal, such as alkyldiseleno- or alkyldithio-carbamato complexes has been reported [40, 41]. An early example was the synthesis of cubic PbS [42] (see Figure 2.4) and PbSe [43], this work has been emulated and developed recently for the production of various morphologies [44]. Clusters of the general formula $[M_{10}Se_4(SPh)_{16}]^{4-}$ (M = Zn or Cd and SPh = phenyl thiolate) have been used to great effect in the synthesis of nanoparticulates of both the simple and core–shell type [45]; dialkyl-diseleo and dithio-carbamates have also been used to prepare core shell nanoparticles [46].

In devising safer syntheses for nanoparticulates based on the original TOPO type of methodologies it has occurred to more than one group that the use of metal alkyls may not be required [47–49]. In an early experiment cadmium chloride and TOPS were used to prepared CdS. More recently Peng has produced a series of papers in which cadmium salts, especially the acetate, have been used to prepare good quality CdSe [48, 49]. In the most recent report a blend of octadecene and oleic acid was used to produce CdS under conditions in which sulfur was the limiting reagent. Particle size control could be achieved by varying the amount of oleic acid in the reaction mixture. In related work lead-containing materials have been prepared. A recent review of the general area of the synthesis of nanodispersed semiconductors is available [50].

A TOPO-based method has been used [51] by Alivisatos et al. in the synthesis of InP nanocrystals (2–5 nm in diameter). The reaction used $InCl_3$ in TOPO followed by addition of $P(Si(CH_3)_3)_3$, with annealing of the resulting InP nanocrystals. The band gap for bulk InP is 1.35 eV whereas the InP nanocrystallites produced exhibit values ranging from 1.7 eV to 2.4 eV [51]. InAs has been prepared by a similar method by the dehalosilylation reaction between $As[Si(CH_3)_3]_3$ and $InCl_3$, surface oxidation did not change the properties of the resulting particles [52]. III/V semiconductors are less ionic in character than their II/VI analogs and thus do not crystallise as readily. Kaner et al. [53] used solid state metathesis involving the reaction of sodium pnictides with group III halides, at high temperatures, in a closed vessel, to produce III/V nanoparticles. A relatively low temperature method involving similar reactions in organic solvents has been reported for the preparation of GaP and GaAs nanoparticles, using gallium (III) halides and $(Na/K)_3E$ (E = P, As) [54]. This method avoids the use of hazardous phosphines or arsines. Nanocrystallites of InAs and InP were also synthesised [55] from the reaction of $InX_3$ (X = Cl, Br, I) with $As(SiMe_3)_3$ or $P(SiMe_3)_3$, respectively. Other analogous chemical routes to indium pnictides have also appeared in the literature [55].

GaAs nanocrystallites have been prepared by reacting $GaCl_3$ with $As(SiMe_3)_3$ in boiling quinoline, however, as yet unidentified species were found to mask the optical properties of the resulting particles and hence quantum size effects could not be properly determined [56, 57]. The alcoholysis or thermolysis of silylated single molecule precursors is one of several processes used for nanoparticle synthesis of III/V and II/V semiconductor materials. Another preparative method for the syntheses of GaAs [58] and InP [59] is by the methanolysis of organometallic compounds such as $[Cp^*(Cl)In(\mu\text{-}P(SiMe_3)_2]_2$. The chemical route led [60, 61] to bulk, amorphous, $Cd_3P_2$ after rapid flocculation of $Cd_3P_2$ nanoparticles. In order to control the kinetics a single molecule precursor with bulkier substituents was used, $Cd[P(SiPh_3)_2]_2$ [61, 62]. The methanolysis gave soluble nanoparticles of $Cd_3P_2$ with diameters ranging from 30–40 Å, however, the particles were not crystalline.

Other chemical routes to metal phosphides include phosphinolysis reactions and reactions between metallo-organics and $P(SiMe_3)_3$; all involve elimination and condensation processes. Buhro [61] reported the synthesis of several binary phosphide semiconductors along with ternary phases (e.g. $ZnGeP_2$), using solution-phase metallo-organic routes. A recent review of the preparation of II/V materials has appeared [63].

### 2.4.3
### Hydrothermal and Solvothermal Methods

An alternative approach to the use of high-temperature solvents, which can be both toxic and expensive, is to use more usual solvents conventionally limited by their rather low boiling points. However, solvents can be used well-above their boiling point at atmospheric pressure if heated in a sealed vessel (an autoclave or 'bomb') the autogenous pressure then far exceeds the ambient pressure raising the boiling point of the solvent. Such *solvothermal* reaction conditions are extensively used in the preparation of inorganic solids, especially zeolites [64]. For a comprehensive review of this method for nanoparticles see [65].

A few illustrative examples of this methodology follow. Qian and coworkers [66] have reported wurtzite ZnSe nanoparticles with sizes in the 18 nm range starting from the elements in ethylenediamine ($T = 120$ °C, $t = 6$ h.) This reaction yields a complex with the formula ZnSe(en) where en is ethylenediamine. Seshadri has used toluene as the solvent in a neat preparation of CdSe from cadmium stearate and Se powder [67, 68], tetralin (tetrahydronaphthalene) was added as a reducing agent. Tetralin is aromatized to naphthalene in the presence of Se, producing $H_2Se$

in the process. Monodisperse quantum dots (the particle size distributions have a width that is about 5% of the mean) with good luminescence properties are produced, the only report to date of a solvothermal route to surfactant-capped semiconductor chalcogenide nanoparticles.

$Cu_{2x}Se$ particles were obtained by Qian and coworkers [69] starting from CuI, Se, and ethylenediamine, with $T = 90$ °C, $t = 4$ h. The particles are spherical and quite monodisperse. Qian and coworkers have also reported [70] a solvothermal preparation of $CuInSe_2$, obtaining 15 nm particles from $CuCl_2$, $InCl_3$ and Se in either ethylenediamine or diethylamine, at 180 °C for 15 h in ethylenediamine, and 36 h for diethylamine. $CuInSe_{2x}S_x$ [71] also has been prepared by Qian and coworkers using $InCl_3$, $CuCl_2$, S and Se with ethylenediamine as the solvent.

Recently microwave-solvothermal reactions and flow-solvothermal reactions have been reported. These methods may be of particular interest in devising scalable synthesis of nanoparticlates. Komarneni et al. [72] have prepared a number of oxide nanoparticles, including 5–20 nm $MnFe_2O_4$, $CoFe_2O_4$, $NiFe_2O_4$ and $ZnFe_2O_4$ ferrite particles, usually by the reaction of metal nitrates in suitable ammoniacal solution. These solutions are microwaved, typically for around 4 min. In flow-hydrothermal techniques, a preheated solvent is mixed with the reactants just prior to introduction into a heated chamber with a back-pressure regulator. The reactants are pumped using a standard HPLC pump through the system. In this way Poliakoff et al. [73] have prepared $CeO_2$–$ZrO_2$. A residence time of 9 s at a (regulated) pressure of 25 MPa was sufficient to yield the product, which comprised nanoparticles with sizes as small as 4 nm. These two methods may well have great promise, not only for scale-up, but also for good monodispersity.

### 2.4.4
### Gas-Phase Synthesis of Semiconductor Nanoparticles

Most gas-phase methods of semiconductor nanoparticle synthesis involve atmospheric or low pressure evaporation of either powders or the pre-formed semiconductor, or the co-evaporation of the two elemental components, for example zinc metal and sulfur [74]. However, the use of these techniques usually results in deposits of particles with larger size distributions, in some cases ranging from 10 to 200 nm. Sercel et al. have reported the synthesis of GaAs nanoparticles by using the organometallic precursor trimethylgallium, $GaMe_3$, which on mixing in a furnace flow reactor with arsine gas, $AsH_3$, gives crystalline GaAs particles [75]. These approaches suffer from the problem of particle aggregation due to the absence of a surface passivating (capping) agent. The only report of gas-phase semiconductor nanoparticle synthesis using a capping agent was by Salata et al. who produced PbS and CdS nanoparticles covered with a polymer layer by reacting a polyvinyl alcohol precursor containing $Pb(NO_3)_2$ or $Cd(NO_3)_2$ in the gas phase with $H_2S$ gas [76].

An investigation of the prereactions which occur between $H_2S/H_2Se$ and $Me_2Cd/Me_2Zn$ when growing II–VI semiconductor films by CVD techniques, showed that the gas-phase reactions result in the formation of chalcogenide deposits of ZnS, ZnSe, CdS and CdSe, with the deposits consisting of poorly formed hexagonal phase micro-crystalline material [77–79]. The addition, into the gas phase, of small amounts of pyridine greatly improved the crystallinity exhibited by the particles, while the addition of larger quantities of pyridine retained the improved crystal quality and also led to a decrease in particle size to the range 10–100 nm. Moreover, as the gas phase concentration of pyridine increased particle size decreased. In the case of CdSe a quantization effect was observed by a change in color of the deposit, when $Me_2Cd$ and $H_2Se$ are mixed in the gas phase in the absence of pyridine the deposit is black, when a high concentration of pyridine is present the deposit is a yellowish-red color. For ZnSe when using hydrogen as the carrier gas, instead of helium or argon, with a high gas-phase concentration of pyridine, a relatively narrow particle size distribution (1–5 nm) could be obtained.

### 2.4.5
### Synthesis in a Structured Medium

The use of a matrix to define the reaction space is an intrinsically attractive approach to the preparation of large amounts of material which could be deposited from solution or from the vapor phase. A number of matrices have been used including: zeolites [80], layered solids [81], molecular sieves [82–84], micelles/microemulsions [85–89], gels [90–92], polymers [93–97] and glasses [98]. The matrix provides a mesoscopic reaction chamber in which the crystal can only grow to a certain size.

The properties of the nanocrystallites are determined, not only by the confinements of the host material, but also by the properties of the system, which can include a range of factors including the internal/external surface properties e.g. of the zeolite or the lability of a micelle. The particle size is controlled by the system chosen e.g. in zeolites the nanocrystallite diameter is limited by the pore size of the zeolite (typically smaller than 2 nm).

Wang and Herron [80] have studied the optical properties of both CdS and PbS clusters encapsulated in zeolites. The nanocrystallites were prepared in two different zeolites; modernite (unidirectional channels of 7 Å diameter) and zeolite Y (13 Å diameter channels with cages of tetrahedral symmetry interconnected by 8 Å windows, and 5 Å cages interconnected by 3 Å windows). For the preparation of CdS in zeolite Y, the sodium cations in the zeolite were first ion-exchanged with cadmium cations, by treatment with aqueous $Cd(NO_3)_2$ at pH 5. This was followed by passing hydrogen sulfide ($H_2S$) gas over the sample [80]. Depending on the loading level of cadmium ions within the zeolite, different sizes of CdS clusters could be obtained. At low loading levels (1:1 metal/sulfide) CdS clusters with an average size less than 13 Å were obtained. These gave an absorption peak at around 280 nm in their optical spectra. When an excess of cadmium was used the individual clusters aggregated into an extended structure, modulated by the internal cavities of the zeolite. These produced optical spectra showing an excitonic shoulder near 350 nm corresponding to CdS clusters of approximately 28 Å in diameter. The small dimensions reported in this work are typical of nanoparticles obtained when zeolites are used as the host structure.

Stable, cubic phase, PbS nanoparticles were prepared in a polymeric matrix by exchanging $Pb^{2+}$ ions in an ethylene – 15% methacrylic acid copolymer followed by reaction with $H_2S$ [91]. The size of the PbS nanoparticles was dependent on the initial concentration of $Pb^{2+}$ ions with diameters ranging from 13 to 125 Å. The smallest particles (13 Å) are reported to be molecular in nature and exhibit discrete absorption bands in their optical spectra. Two theoretical models, which take into account the effect of nonparabolicity, were proposed in order to explain the observed size-dependent optical shifts for PbS nanocrystallites. The authors reported that the effective mass approximation fails for PbS nanocrystallites.

Steigerwald et al. prepared capped CdSe, ZnS, ZnS/CdSe and CdSe/ZnS nanocrystallites from inverse micellar solutions [30, 99]. Silylchalcogenide reagents were added to micro-emulsions containing the appropriate metal ions. The particle surfaces were subsequently capped; for example with phenyl groups or with other semiconductor materials such as ZnS. Silylorganochalcogenides react readily with metal salts or simple metal alkyls to form metal–chalcogenide bonds [100]. Micelle stabilised CdSe nanocrystallites, with $Cd^{2+}$ rich surfaces, react similarly with $R[(CH_3)_3Si]_2Se$ to give larger CdSe crystallites encapsulated by a layer of organic ligands (R). These surface passivated crystallites can be isolated as powders, which are soluble in organic solvents such as pyridine. $^{77}Se$ NMR spectra of three size distributions of organic-capped CdSe were reported, with each giving different spectra [101], consisting of broad lines corresponding to bulk material along with additional peaks appearing at higher field and becoming more intense with decreasing particle size.

Several types of nanoparticles prepared from synthesis involving biologically related processes, biomimetic, have been reported [102–104]. For example, using empty polypeptide cages found in the iron storage protein ferritin; bio-inorganic nanocomposites of CdS–ferritin can be synthesised [102]. Another approach to nanocrystallite synthesis in a matrix was developed by Choi and Shea [90, 91], who report using porous inorganic–organic xerogel (polysilsesquioxanes) to produce CdS (6 and 9 nm) [90] and chromium particles (1–10 nm) [91]. The chromium precursor used was a zero-valent arene tricarbonyl chromium complex, introduced as a component of the xerogel matrix, which after heating under vacuum produced chromium nanoparticles. By first doping CdS into the starting material two different phases of chromium and CdS are reported to be obtained. Perhaps the most important use of a biological approach to nanoparticle growth is that taken in the commercial sector by Nanomagnetics Ltd. They have used the 8 nm cavity of the iron storage protein ferritin to grow iron oxide for use in magnetic storage devices and hope to fully commercialise the process.

## 2.5 The Suitability of Such Methods for Scaling

The vast majority of semiconductor nanoparticles are produced, at present, by batch methods in research laboratories in processes which essentially depend on the separation of nucleation and growth. In the semiconductor area high quality

**Tab. 2.1.** Some typical batch conditions used in the synthesis of nanoparticles.

| *Material* | *Precursor (mM)* | *Solvent system* | *Approx. [Precursor]/M* | *[Metal]/M* | *Ref.* |
|---|---|---|---|---|---|
| CdSe | $Me_2Cd$ (13.35)<br>TOPSe (10.00) | TOP 50 g<br>TOPO 50 ml | ca. 0.13 | 0.13 | 34 |
| CdSe<br>or<br>CdS | $Cd(Se_2CNMeHex)_2$ (1.2)<br>or<br>$Cd(S_2CNMeHex)_2$ (1.0) | TOP 25 ml<br>TOPO 25 g | ca. 0.06 | 0.06 | 42 |
| CdSe<br>or<br>ZnSe | $(Li)_4[Cd_{10}Se_4(SPh)_{16}]$ (0.28)<br>or<br>$(TMA)_4[Zn_{10}Se_4(SPh)_{16}]$ (0.20) | HDA 55 g | ca. 0.005 | 0.02 | 45 |
| CdS | CdO (0.1)<br>Oleic acid (21.2)<br>Sulfur (0.05) | ODE 4 g | ca. 0.025 | 0.025 | 49 |

TOP: tri-n-octylphosphine
TOPO: tri-n-octylphosphine oxide
HDA: hexadecylamine
ODE: octadecene

materials are likely to be grown at elevated temperatures. The relatively high dilution used in most batch processes in coordinating solvents will make production expensive and of relatively high environmental impact. In Table 2.1 some of the leading methods are compared.

Flow methods have the advantage of the potential for continuous processing but are rare. It has been pointed out that cheap solvents can be used in solvothermal methods [65]. It is clear from the literature that the majority of semiconductor dots and the few being sold are today being made by batch methods based on metal organic or organometallic routes.

A related process in which large quantities of particulates are produced is the manufacture of photographic emulsions. The manufacture involves the reaction of large quantities of material with rapid stirring, this type of method is unlikely to be useful for quantum dots. However the method used does draw our attention to the careful engineering of the manufacturing process. There is little evidence, to date, that serious consideration has been given to the engineering process for the bulk manufacture of quantum dots. In scaling a process the expense of the reagents, their toxicity and environmental impact will all be important factors. In minimising these problems vapor phase synthesis, or spray drying methods, as pioneered by Dobson and co-workers, may well be worth revisiting [76].

## 2.6 Conclusions and Perspectives on the Future

The synthesis of nanodimensional powders on a large scale remains in its infancy. We can draw inspiration from some of the ingenious processes that have been developed in closely related fields, perhaps the photographic process is the best example. The need to prepare high quality materials throws into sharp contrast the processing conditions chosen in the electronics industry as compared to biology. The latter tends to work slowly, close to equilibrium and at low temperatures. The electronics industry tends to use high temperature processes close to equilibrium or processes limited by chemical reaction (CVD in its various forms). The reproducible synthesis of well defined nanomaterials spans these two approaches. It seems likely that a modestly high temperature process is most likely to lead to material of the required quality. However, we should not overlook the interesting use of ferritin to produce fully functional magnetic materials. It has been impossible to provide more than an outline of the methods in use in this short chapter. The reader is referred to the many reviews cited for a more comprehensive account. We have tried to give a clear picture of how such processes might be scaled.

What is clear is that there is a major new technology to be developed using such particulates; for example the UK Parliamentary Office of Science and Technology has estimated that in the early part of this century the global market for nanotechnology products will be in excess of £80 bn pa. Those interest in the potential of this area might like to consider the following sources of information: New Dimensions for Manufacturing: A UK Strategy for Nanotechnology: DT and OST June 2002; DTI Nanotechnology 2001, DTI "The International Technology Service Missions on Nanotechnology to Germany and the USA" March 2001, European Commission "Technology Roadmap for Nanoelectronics" November 2000, Luxcapital "Nanotechnology The Nanotech Report: 2001", Red Herring July 17 2000.

## Acknowledgements

Our own work on quantum dots was instigated by Tito Trindade (Universidade Aveiro), principally exploiting carbamato precursors initially developed by Azad Malik, this initial work was ably extended by Mark Green (Oxonica Ltd) and Neerish Revaprasasu (University of Zululand) and is now carried on by the team in the University of Manchester and NanoCo Ltd. Our work has been extensively supported by the EPSRC in the UK and our fruitful collaborations with South Africa have been made possible by the long term support of the Royal Society in London and the NRF in South Africa.

## References

1 Alivisatos, A. P., *J. Phys. Chem.*, **1996**, *100*, 13226.
2 Green, M., O'Brien, P., *Chem Commun.*, **1999**, 2235–2241; Pickett, N. L., P. O'Brien, *The Chemical Record*, **2001**, *1*, 467–479; Trindade, T., Pickett, N. L., O'Brien, P., *Chem. Mater.*, **2001**, *13*, 3843–3858.
3 Rogach, A. L., Talapin, D. V., Shevchnko, E. V. et al., *Adv. Funct. Mater.*, **2002**, *12*, 653.
4 see Q.Dot Corp catalogue available at

*http://www.qdots.com/new/products/products.html.*
5 Jones, A. C., O'Brien, P., *The CVD of Compound Semiconducting Materials*, Wiley VCH, Weinheim 1998.
6 Cahn, R. W., *The Coming of Materials Science*, Pergamon Materials Science, Oxford 2001, Ch. 7, p. 256 et seq.
7 for fine examples see the work of Brus and Schiffrin: Kanaras, A. G., Kamounah, F. S., Schaumburg, K. et al., *Chem. Commun.*, **2002**, 2294; Clarke, N. Z., Waters, C., Johnson, K. A. et al., *Langmuir*, **2001**, *17*, 6048.
8 O'Brien, S., Brus, L., Murray, C. B., *J. Am. Chem. Soc.*, **2001**, *123*, 12085–12086.
9 Hycon, T., *Chem. Commun.*, **2003**, 927.
10 Brus, L. E., *J. Chem. Phys.*, **1984**, *80*, 4403.
11 Tito Trindade Ph.D. Thesis, Imperial College of Science Technology and Medicine, 1998.
12 Rossetti, R., Ellison, J. L., Gibson, J. M. et al., *J. Chem. Phys.*, **1984**, *80*, 4464.
13 Rossetti, R., Hull, R., Gibson, J. M. et al., *J. Chem. Phys.*, **1985**, *82*, 552.
14 Henglein, A., *Chem Rev.*, **1989**, *89*, 1861.
15 Weller, H., *Angew. Chem. Int. Ed. Engl.*, **1993**, *32*, 41.
16 Weller, H., *Adv. Mater.*, **1993**, *5*, 88.
17 Spanhel, L., Haase, M., Weller, H. et al., *J. Am. Chem. Soc.*, **1987**, *109*, 5649.
18 Baral, S., Fojtik, A., Weller, H. et al., *J. Am. Chem. Soc.*, **1986**, *108*, 375.
19 Fischer, C. H., Lilie, J., Weller, H. et al., *Ber. Bunsen-Ges. Phys. Chem.*, **1989**, *93*, 61.
20 Fisher, C. H., Weller, H., Katsikas, L. et al., *Langmuir*, **1989**, *5*, 429.
21 Eychmuller, A., Katsikas, L., Weller, H., *Langmuir*, **1990**, *6*, 1605.
22 Fojtik, A., Weller, H., Koch, U. et al., *Ber. Bunsen-Ges. Phys. Chem.*, **1984**, *88*, 969.
23 Weller, H., Fojtik, A., Henglein, A., *Chem. Phys. Lett.*, **1985**, *117*, 485.
24 Weller, H., Koch, U., Gutierrez, M. et al., *Ber. Bunsen-Ges. Phys. Chem.*, **1987**, *91*, 88.
25 Talapin, D. V., Rogach, A. L., Kornowski, A. et al., *Nano. Lett.*, **2001**, *1*, 207.
26 Hässelbarth, A., Eychmuller, A., Eichberger et al., *J. Phys. Chem.*, **1993**, *97*, 5333.
27 Zhou, H. S., Honma, I., Komiyama, H. et al., *J. Phys. Chem.*, **1993**, *97*, 895.
28 Zhou, H. S., Sasahara, H., Honma et al., *Chem. Mater.*, **1994**, *6*, 1534.
29 Eychmüller, A., Hässelbarth, A., Weller, H., *J. Lumin.*, **1992**, *53*, 113.
30 Kortan, A. R., Hull, R., Opila, R. L. et al., *J. Am. Chem. Soc.*, **1990**, *112*, 1327.
31 Hoener, C. F., Allan, K. A., Bard, A. J. et al., *J. Phys. Chem.*, **1992**, *96*, 3812.
32 Eychmuller, A., Mews, A., Weller, H., *Chem. Phys. Lett.*, **1993**, *208*, 59.
33 Mews, A., Eychmuller, A., Giersig, M. et al., *J. Phys. Chem.*, **1994**, *98*, 934.
34 Murray, C. B., Norris, D. J., Bawendi, M. G., *J. Am. Chem. Soc.*, **1993**, *115*, 8706.
35 Aliv, H. T., Shiang, J. J., Kadavanich, A. V. et al., *J. Phys. Chem.*, **1995**, *99*, 1741.
36 Manna, L., Scher, E. C., Li, L.-S. et al., *J. Am. Chem. Soc.*, **2002**, *124*, 7136; Manna, L., Scher, E. C., Alivisatos, A. P., *J. Am. Chem. Soc.*, **2000**, *122*, 12700.
37 Dabbousi, B. O., Rodriguez-Viejo, J., Mikulec, F. V. et al., *J. Phys. Chem. B*, **1997**, *101*, 9463.
38 Peng, X., Schlamp, M. C., Kadavanich, A. V. et al., *J. Am. Chem. Soc.*, **1997**, *119*, 7019.
39 Brennan, J. G., Siegrist, T., Carrol, P. J. et al., *J. Am. Chem. Soc.*, **1989**, *111*, 4141.
40 Trindade, T., O'Brien, P., *Adv. Mater.*, **1996**, *8*, 161.
41 Trindade, T., O'Brien, P., Zhang, X., *Chem. Mater.*, **1997**, *9*, 523.
42 a) Trindade, T., O'Brien, P., Zhang, X. et al., *J. Mater. Chem.*, **1997**, *7*, 1011; b) Malik, M. A., Revaprasadu, N., O'Brien, P., *Chem. Mater.*, **2001**, *13*(3); 913–920.
43 Trindade, T., Monteiro, O. C., O'Brien, P. et al., *Polyhedron*, **1999**, *18*, 1171.

44 Lee, S.-M., Ju, Y.-w., Cho, S.-N. et al., *J. Am. Chem. Soc.*, **2002**, *124*, 11244.

45 Cumberland, S. L., Hanif, K. M., Javier, A. et al., *Chem. Mater.*, **2002**, *14*, 1576.

46 Malik, M. A., O'Brien, P., Revaprasadu, N., *Chem. Mater.*, **2002**, *14*, 2004.

47 Lazell, M. R., O'Brien, P., *J. Mater. Chem.*, **1999**, *9*, 1381.

48 Peng, Z. A., Peng, X., *J. Am. Chem. Soc.*, **2001**, *123*, 183; Qu, L., Peng, Z. A., Peng, X., *Nanoletters*, **2001**, 333.

49 Yu, W. W., Peng, X., *Angew. Chem. Int. Ed. Eng.*, **2002**, *41*, 2368.

50 Trindade, T., *Curr. Opin. Solid State Mater. Sci.*, **2002**, 347.

51 Guzelian, A. A., Katari, J. E. B., Kadavanich, A. V. et al., *J. Phys. Chem.*, **1996**, *100*, 7212.

52 Guzelian, A. A., Banin, U., Kadavanich, A. V. et al., *Appl. Phys. Lett.*, **1996**, *69*, 1432.

53 Treece, R. E., Macala, G. S., Rao, L. et al., *Inorg. Chem.*, **1993**, *32*, 2745.

54 Kher, S. S., Wells, R. L., *Chem. Mater.*, **1994**, *6*, 2056.

55 Wells, R. L., Aubuchon, S. R., Kher, S. S. et al., *Chem. Mater.*, **1995**, *7*, 793.

56 Olshavsky, M. A., Goldstein, A. N., Alivisatos, A. P., *J. Am. Chem. Soc.*, **1990**, *112*, 9438.

57 Uchida, H., Curtis, C. J., Nozik, A. J., *J. Phys. Chem.*, **1991**, *95*, 5382.

58 Byrne, E. K., Parkanyi, L. K., Theopold, H., *Science* **1988**, *241*, 332.

59 Douglas, T., Teopold, K. H., *Inorg. Chem.*, **1991**, *30*, 594.

60 Goel, S. C., Chiang, M. Y., Buhro, W. E., *J. Am. Chem. Soc.*, **1990**, *112*, 5636.

61 Buhro, W. E., *Polyhedron* **1994**, *13*, 1131.

62 Matchett, M. A., Viano, A. M., Adolphi, N. L. et al., *Chem. Mater.*, **1992**, *4*, 508.

63 Green, M., *Curr. Opin. Solid State Mater. Sci.*, **2002**, 355.

64 (eds). *Recent Advances in Zeolite Science*, eds. Klinowski, J., Barrie, P. J., Elsevier, Amsterdam 1990.

65 Rajamathi, M., Seshadri, R., *Curr. Opin. Solid State Mater. Sci.*, **2002**, *6*, 337.

66 Zhan, J. H., Yang, X. G., Zhang, W. X. et al., *J. Mater. Res.*, **2000**, *15*, 629.

67 Gautam, U. K., Rajamathi, M., Meldrum, F. et al., *J. Chem. Soc., Chem. Commun.*, **2001**, 629.

68 Mitchell, P. W. D., Morgan, P. E. D., *J. Am. Ceram. Soc.*, **1974**, *57*, 278.

69 Wang, W., Yan, P., Liu, F. et al., *J. Mater. Chem.*, **1998**, *8*, 2321.

70 Li, B., Xie, Y., Huang, H. et al., *Adv. Mater.*, **1999**, *11*, 1456.

71 Xiao, J., Xie, Y., Xiong, Y. et al., *J. Mater. Chem.*, **2001**, *11*, 1417.

72 Komarneni, S., D'Arigo, M. C., Leonelli, C. et al., *J. Am. Ceram. Soc.*, **1998**, *81*, 3041.

73 Cabañas, A., Darr, J. A., Lester, E. et al., *J. Chem. Soc., Chem. Commun.*, **2000**, 901–902; Cabañas, A., Darr, J. A., Lester, E. et al., *J. Mater. Chem.*, **2001**, *11*, 561–568.

74 a) Kaito, C., Fujita, K., Shiojiri, M., *J. Appl. Phys.*, **1996**, *47*, 5161; b) Kaito, C., Fujita, K., Shiojiri, M., *J. Crystal Growth*, **1998**, *62*, 375; c) Kaito, C., Saito, Y., *J. Crystal Growth*, **1990**, *99*, 743; d) Agata, M., Kurase, H., Hayashi, S. et al., *Solid State Commun.*, **1990**, *76*, 1061.

75 Sercel, P. C., Saunders, W. A., Atwater, H. A. et al., *Appl. Phys. Lett.*, **1992**, *61*, 696.

76 a) Salata, O. V., Dobson, P. J., Hull, P. J. et al., *Adv. Mater.*, **1994**, *6*, 772; b) Salata, O. V., Dobson, P. J., Hull, P. J. et al., *Thin Solid Films*, **1994**, *251*, 1.

77 Pickett, N. L., Foster, D. F., Cole-Hamilton, D. J., *J. Mater. Chem.*, **1996**, *6*, 507.

78 Pickett, N. L., Foster, D. F., Cole-Hamilton, D. J., *J. Cryst. Growth*, **1997**, *170*, 476.

79 Pickett, N. L., Lawson, S., Thomas, W. G. et al., *J. Mater. Chem.*, **1998**, *8*, 2769.

80 Wang, Y., Herron, N., *J. Phys. Chem.*, **1987**, *91*, 257.

81 Cassagneau, T., Hix, G. B., Jones, D. J. et al., *J. Mater. Chem.*, **1994**, *4*, 189.

82 Abe, T., Tachibana, Y., Uematsu, T. et al., *J. Chem. Soc., Chem. Commun.*, **1995**, 1617.

83 Brenchley, M. E., Weller, M. T.,

*Angew. Chem. Int. Ed. Engl.*, **1993**, *32*, 1663.

**84** Blasse, G., Dirksen, C. J., Brenchley, M. E. et al., *Chem. Phys. Lett.*, **1995**, *234*, 177.

**85** Watzke, H. J., Fendler, J. H., *J. Phys. Chem.*, **1987**, *91*, 854.

**86** Steigerwald, M. L., Alivisatos, A. P., Gibson, J. M. et al., *J. Am. Chem. Soc.*, **1988**, *110*, 3046.

**87** Khan-Lodhi, A., Robinson, B. H., Towey, T. et al., in *The Structure, Dynamics and Equilibrium Properties of Colloidal Systems*, eds. Bloor, D. M., Wyn-Jones, E., Kluwer Academic Publishers 1990, p. 373.

**88** Towey, T. F., Khan-Lodi, A., Robinson, B. H., *J. Chem. Soc., Faraday Trans.*, **1990**, *86*, 3757.

**89** Korgel, B. A., Monbouquette, G., *J. Phys. Chem.*, **1996**, *100*, 346.

**90** Choi, K. M., Shea, K. J., *J. Phys. Chem.*, **1994**, *98*, 3207.

**91** Choi, K. M., Shea, K. J., *J. Am. Chem. Soc.*, **1994**, *116*, 9052.

**92** Carpenter, J. C., Lukehart, C. M., Stock, S. R. et al., *Chem. Mater.*, **1995**, *7*, 201.

**93** Wang, Y., Suna, A., Mahler, W. et al., *J. Chem. Phys.*, **1987**, *87*, 7315.

**94** Gao, M., Yang, Y., Yang, B. et al., *J. Chem. Soc., Chem. Commun.*, **1994**, 2779.

**95** Nirmal, M., Murray, C. B., Bawendi, M. G., *Phys. Rev. B*, **1994**, *50*, 2293.

**96** Tassoni, R., Schrock, R. R., *Chem. Mater.*, **1994**, *6*, 744.

**97** Moffitt, M., Eisenberg, A., *Chem. Mater.*, **1995**, *7*, 1178.

**98** Shinojima, H., Yumoto, J., Uesugi, N. et al., *Appl. Phys. Lett.*, **1989**, *55*, 1519.

**99** Steigerwald, M. L., Alivisatos, A. P., Gibson, J. M. et al., *J. Am. Chem. Soc.*, **1988**, *110*, 3046.

**100** Stuczynski, S. M., Brennan, J. G., Steigerwald, M. L., *Inorg. Chem.*, **1989**, *28*, 4431.

**101** Thayer, A. M., Steigerwald, M. L., Duncan, T. M. et al., *Phys. Rev. Lett.*, **1988**, *60*, 2673.

**102** Wong, K. K. W., Mann, S., *Adv. Mater.*, **1996**, *8*, 928.

**103** Mackle, P., Charnock, J. M., Garner, C. D. et al., *J. Am. Chem. Soc.*, **1993**, *115*, 8471.

**104** Meldrum, F. C., Heywood, B. R., Mann, S., *Science* **1992**, *257*, 522.

# 3
# Moving Nanoparticles Around: Phase-Transfer Processes in Nanomaterials Synthesis

*M. Sastry*

## 3.1
## Introduction

We are witnessing impressive advances in understanding the unusual physicochemical and optoelectronic properties of nanomaterials, their synthesis, assembly and packaging for commercial application [1]. One important area of nanotechnology is concerned with the development of reliable processes for the synthesis of nanomaterials over a range of sizes (with good monodispersity) and chemical compositions. Realizing that shape anisotropy could lead to interesting variation in the electronic and catalytic properties of nanoparticles [2, 3], much current research is directed towards development of experimental methods for the synthesis of nanoparticles of varying shapes. Nanorods and nanowires of silver [4, 5]/gold [6–12]/Au-core–Ag-shell [13]/CdSe [14]/tungsten sulfide [15], nanoprisms of silver [16]/gold [17, 18] and CdS [19] are some of the exotic nanocrystalline shapes that may be routinely synthesized in the laboratory today. From a fundamental angle, the ability to control the shape of nanocrystals is particularly exciting and has led to the first observation of two distinct quadrupole plasmon resonance modes in silver nanoprisms [16]. The need to develop eco-friendly synthesis protocols that do away with the use of toxic chemicals has also fuelled research in this direction and bio-related processes that use microorganisms such as bacteria [20–23], fungi [24–27] and actinomycete [28] have been developed to grow nanocrystals of silver and gold both inside and outside the biomass.

Gold nanoparticles have, in particular, been the subject of considerable attention over the ages and enjoy an interesting history dating back to the pioneering work of Faraday on the synthesis of gold hydrosols (gold nanoparticles dispersed in water) [29]. Gold nanoparticles find application in a variety of fields such as catalysis [30], as electron microscopy markers [31] and in DNA sequence determination. It is of little surprise, therefore, that there are many recipes for the synthesis of gold nanoparticles over a range of sizes in an aqueous environment. The procedures for synthesis of gold hydrosols include, (1) reduction of aqueous chloroaurate ions by a variety of reducing agents such as citric acid [33], sodium borohydride [34], and alkaline tetrakis(hydroxymethyl)phosphonium chloride [35]; (2)

*The Chemistry of Nanomaterials: Synthesis, Properties and Applications, Volume 1.* Edited by C. N. R. Rao, A. Müller, A. K. Cheetham

ISBN: 3-527-30686-2

radiation-induced reduction of gold ions [36, 37] and (3) sonochemical reduction of gold ions [38], to name just a few. The interested reader is directed to a comprehensive review by Handley that lists at least one dozen protocols for the syntheses of gold hydrosols with particle sizes in the range 10–640 Å.

The synthesis of gold nanoparticles in non-polar organic media is a considerably newer area of research whose origin may be traced to the seminal work of Brust and co-workers [39]. In this report, the authors have demonstrated the phase transfer of chloroaurate ions into toluene using a phase transfer molecule such as tetra-alkylammonium bromide. Thereafter, the gold ions were reduced using sodium borohydride to yield gold nanoparticles of excellent monodispersity capped with alkanethiol molecules. Analogous to the formation of self-assembled monolayers (SAMs) of alkanethiols on gold thin films [40], Brust and co-workers used thiolate chemistry to cap the gold nanoparticles with alkanethiols present in the organic phase during phase transfer and reduction of the gold ions, thus rendering them hydrophobic and soluble in the organic phase [39]. Such surfactant-stabilized gold nanoparticles behave like new compounds and can be easily separated out of solution in the form of a powder and re-dissolved in different organic solvents without significant variation in the particle size distribution. The Brust report has been rapidly followed by publications on the self-assembly on gold and silver nanoparticle surfaces of alkanethiol [41, 42], aromatic thiol [43, 44], alkylamine [45, 46], dialkyl disulfide [47] and thiolated cyclodextrin [48, 49] molecules. Very recently, we have shown in my group that the Brust protocol for the synthesis of hydrophobic metal nanoparticles may be considerably simplified by using a multifunctional molecule, 4-hexadecylaniline [50]. This molecule plays the role of a phase-transfer molecule, reducing and capping agent and results in the one-step synthesis of hydrophobic gold [50] and platinum nanoparticles [51] in a variety of organic solvents. The surface properties of the gold colloids may be tailored by chemisorption of terminally functionalized thiol molecules resulting in a number of interesting applications. One of the exciting areas of research using functionalized colloidal gold particles is the study of the reactivity of monolayer protected colloidal particles (or MPCs as they are termed) [52, 53]. Using a simple place-exchange reaction strategy, Murray and co-workers have demonstrated that alkanethiol derivatized colloidal gold particles in an organic solvent can be poly-$\omega$-functionalized [54]. This led to the possibility of using the polyheterofunctionalized colloidal particles as "nanofactories" where the metal core scaffolds support complex organic ligand structures that may include polymeric and hyperbranched domains [54]. Polar terminal functional groups such as carboxylic acid, ammonium ions, sulfonic acid etc. in the monolayers chemisorbed onto the gold particle surface may be used to increase the solubility of the particles in polar solvents such as water, thus enabling the synthesis of water-soluble gold nanoparticles that, like their hydrophobic counterparts, may be dried in the form of a powder and re-dispersed in water and other polar solvents without significant degradation in the particle size and monodispersity [55–57].

The two main approaches for the synthesis of gold nanoparticles, viz. synthesis in water and in non-polar organic solvents, have certain pros and cons that are enumerated below.

## 3.2 Water-Based Gold Nanoparticle Synthesis

### 3.2.1 Advantages

1. Water is a good solvent for a number of metal ions as well as a variety of capping molecules. The synthesis involves preparation of an aqueous gold salt solution followed by reduction of the metal ions in a single step. It is therefore considerably simpler than the multi-step Brust protocol [39].
2. No additional stabilization against aggregation of the gold nanoparticles is required – surface bound ions (citrate ions, chloroaurate ions etc.) normally stabilize the nanoparticles electrostatically in solution.
3. Electrostatic layer-by-layer assembly involving, for example, oppositely charged polyelectrolytes/surfactants and nanoparticles may be readily accomplished on suitably functionalized surfaces [58–61].
4. Nanoparticle shape control can be easily effected by using self-assembled structures such as micelles (arising due to spontaneous assembly of suitable surfactants in water) as templates [4–6].
5. Perhaps the biggest advantage of a water-based synthesis procedure is that bioconjugation of the gold and other metal/semiconductor nanoparticles with DNA [32], enzymes [62], antibodies [63] etc. may be easily accomplished.

### 3.2.2 Disadvantages

1. Ionic interactions limit the concentration of metal/semiconductor nanoparticles in the aqueous phase to very dilute levels, a big drawback in biological labeling of the nanoparticles.
2. Control over the particle size and monodispersity in a particular reduction protocol is not very good.
3. The gold nanoparticles do not spontaneously assemble into a close-packed hexagonal arrangement on solvent evaporation.
4. The nanoparticles synthesized in water are not easily separated from solution in the form of a powder that would be readily re-dispersible in water after storage [55–57].

## 3.3 Organic Solution-Based Synthesis of Gold Nanoparticles

### 3.3.1 Advantages

1. High degree of control may be exercised over the gold nanoparticle size, monodispersity [34] and chemical nature of the nanoparticle surface (via capping with terminally functionalized thiols, amines, amino acids etc.) [39, 41–57].

2. High concentrations of the gold nanoparticles in solution may be easily prepared.
3. Functionalized gold nanoparticles may be stored as a powder without sintering and irreversible aggregation of the particles.
4. The nanoparticles spontaneously assemble into close-packed, hexagonal monolayers upon solvent evaporation [39, 41, 42, 45, 46]. The collective properties of the nanoparticle assembly may be controlled by varying the interparticle separation via capping with different chain length alkanethiols.

### 3.3.2 Disadvantages

1. The procedure is a multi-step one involving, independently, phase transfer of the gold ions followed by their reduction and capping.
2. While close-packed monolayers of the gold nanoparticles may be deposited by solvent evaporation, there is little control over the process of assembly. Furthermore, superlattices of the gold nanoparticles cannot be readily deposited, in contrast with the layer-by-layer assembly that is possible for electrostatically stabilized gold nanoparticles in water.
3. Formation of bioconjugates with the gold nanoparticles is not possible in an organic environment and consequently, biological application of gold nanoparticles becomes difficult.

It is clear that both methods for the synthesis of gold nanoparticles have characteristic advantages. Depending on the particular application of the nanoparticles, the ideal condition would be to somehow marry the two methods and thus maximize their advantages. This may conveniently be done by effecting a *phase transfer* of gold nanoparticles synthesized in one medium (water/organic solvent) to the second medium (organic solvent/water). In addition to maximizing the benefits accruing from a combination of the two syntheses methods, the ability to move nanoparticles across liquid interfaces into environments of specific physicochemical properties to probe, for example, variation in the optical properties of the nanoparticle solution [64] is an attractive feature of phase-transfer protocols. In the remaining part of this chapter, I discuss some of the methods developed to carry out the phase transfer of gold nanoparticles in both directions. The experimental methods are quite general and may be extended to other chemical compositions such as metal sulfide nanoparticles. Examples will also be given wherever possible.

## 3.4 Moving Gold Nanoparticles Around

### 3.4.1 Phase Transfer of Aqueous Gold Nanoparticles to Non-Polar Organic Solvents

The movement of aqueous gold nanoparticles into non-polar organic solvents requires hydrophobization of the nanoparticles. The many techniques developed to accomplish phase transfer into organic solvents essentially differ in the nature of the capping molecule employed to hydrophobize the nanoparticles. Possibly the first report on the phase transfer of gold nanoparticles from water to an organic solvent such as butyl acetate was that of Underwood and Mulvaney [64]. In this study, the authors showed that aqueous gold nanoparticles synthesized by the Turkevich method [33] could be quantitatively transferred into butyl acetate by complexation of the particles with a 'comb stabilizer' present in the organic phase [64]. The comb stabilizer used was a co-polymer consisting of a backbone of methyl methacrylate and glycidyl methacrylate with poly(12-hydroxystearic acid) as pendant side chains. Gentle shaking of a biphasic mixture of the gold hydrosol and comb stabilizer in butyl acetate resulted in the emulsification of the gold hydrosol–butyl acetate mixture, this process accelerating the complexation of the polymer with the gold nanoparticle surface. The gold nanoparticles were thereby hydrophobized and were phase transferred to the organic phase. Gold sols possess a lovely pink to ruby red color [64] and therefore, the phase transfer of the gold particles from one phase to another is seen as a dramatic transfer of color between phases.

The focus of the work of Underwood and Mulvaney was to study changes in the optical properties of gold colloids as the refractive index of the organic solution was varied. This was done by preparing the gold sols in solutions consisting of mixtures of butyl acetate and $CS_2$ to obtain solutions possessing refractive indices in the range 1.336 (water) to 1.583. Figure 3.1 shows pictures of gold nanoparticles in the different butyl acetate–$CS_2$ solutions [64]. A clear variation in color with refractive index of the solution is seen and is in quantitative agreement with the Mie theory that is used to understand the optical properties of colloidal solutions [64]. The authors did not perform a detailed characterization of the polymer–gold nanoparticle complex and therefore, little is known of the interaction of the polymer with the particle surface etc.

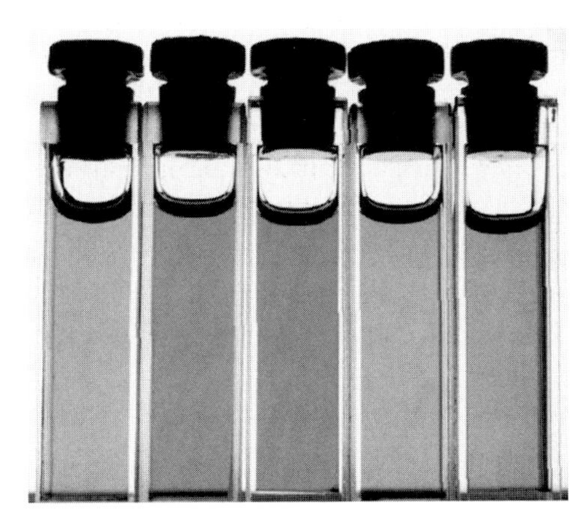

**Fig. 3.1.** Photograph of five sols of colloidal gold prepared in water and in mixtures of butyl acetate and $CS_2$. Refractive indices of the solutions at the absorption band maximum are 1.336, 1.407, 1.481, 1.525 and 1.583 ($\pm$0.004) respectively. (Reprinted with permission from [64], © 1994, American Chemical Society).

Recognizing that the strong interaction of alkanethiol molecules with gold nanoparticles may be used to hydrophobize gold nanoparticles at the liquid–liquid interface, Rao and co-workers have demonstrated the acid-facilitated phase transfer of aqueous gold and indeed, platinum and silver nanoparticles into organic solvents such as toluene [65, 66]. In a typical experiment, the authors took a mixture of a gold hydrosol and dodecanethiol in toluene. To this biphasic mixture, concentrated HCl was added under stirring. A swift movement of the gold nanoparticles into the organic phase containing the gold nanoparticles was observed, indicating capping of the metal nanoparticles with alkanethiol molecules [65, 66]. The alkanethiol-stabilized colloidal gold, silver and platinum particles transferred to toluene as described above could be self-assembled by solvent evaporation on different substrates yielding close-packed, hexagonal arrays of the nanoparticles [65, 66]. A concentrated solution of the different sized thiol stabilized-gold particles was placed on glass substrates and the particles assembled on the surface. Figure 3.2 shows X-ray diffraction patterns from the different films, the size of the particles is indicated next to the corresponding diffractogram.

The *d*-spacings obtained from the low angle peaks (indicated by arrows in Figure 3.2) are also listed in the figure. The low angle diffraction peaks arise from the arrangement of the gold particles in the array formed by solvent evaporation. It is observed that the separation between the clusters decreases as the size of the particles is reduced (Figure 3.2), in accordance with nanoparticle packing considerations. However, the *d*-spacing is smaller than that expected from the core + surfactant size considerations indicating some degree of interdigitation of the hydrocarbon chains from neighboring particles in the array [66]. The inset of Figure 3.2 shows the optical absorption spectra recorded from the colloidal gold particles in toluene for gold particles of different sizes. The surface plasmon resonance from the colloidal gold particles centered around 525 nm is clearly seen for the 4.2 nm and 2.1 nm sized particles and the intensity of the resonance is higher for the larger sized particles (Figure 3.2, inset). The resonance could not be detected for the smallest size particles (1.0 nm diameter) and this was attributed to the possibility of the gold particles in this size range being non-metallic [66].

Transmission electron microscopy images recorded from the self-assembled gold nanoparticle monolayer obtained by solvent evaporation are shown in Figure 3.3 [66]. It can be seen that the thiol-derivatized gold nanoparticles self-assemble into nanocrystalline arrays over tens of nanometer length scale. The spacing between the particles is highly regular and roughly 1 nm between the particles. While the larger particles of mean diameter 4.2 nm assembled into regular, close-packed domains, assemblies of smaller particles show a large fraction of voids within the domains [66].

It is known that alkylamine molecules also bind to gold nanoparticles quite strongly through a 'weak' covalent bond, as described by Leff, Brandt and Heath [45]. Using a process similar to that adopted for phase transfer of gold nanoparticles with alkanethiols, we have recently demonstrated that octadecylamine molecules present in the organic phase may be used to accomplish the phase transfer of aqueous gold nanoparticles into toluene [67]. Vigorous stirring of a bi-

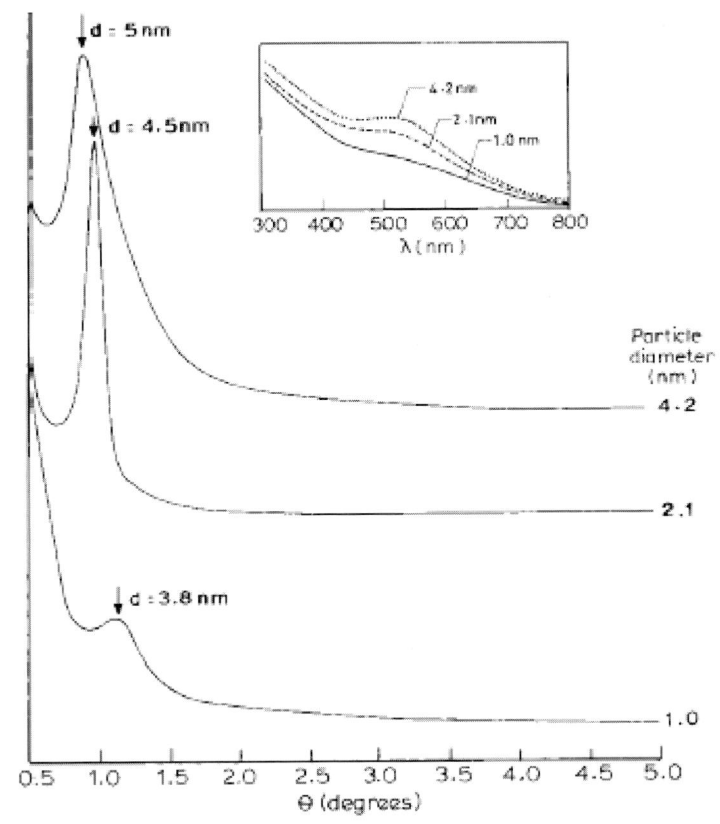


**Fig. 3.2.** XRD patterns from the nanocrystalline arrays of Au particles of different mean diameters (4.2, 2.1, and 1.0 nm). Inset of the figure displays the UV–vis spectra of these particles. (Reprinted with permission from [66], © 1997, American Chemical Society).

phasic mixture containing the gold hydrosol and octadecylamine (ODA) molecules in toluene resulted in the rapid transfer of gold nanoparticles into the organic phase without the use of acid. This is clearly demonstrated in Figure 3.4A that shows a picture of two test-tubes with the biphasic mixture before (test-tube to the right) and after shaking (test-tube to the left). The transfer of the red color from the lower, aqueous phase to the upper, toluene phase is clearly illustrated. UV–vis spectra recorded from the gold hydrosol before (curve 3) and after (curve 1) phase transfer are shown along with the toluene phase after phase transfer (curve 2 in Figure 3.4B). The almost complete disappearance of the gold nanoparticle plasmon resonance in curve 1 indicates facile phase transfer of the gold nanoparticles into toluene. The UV–vis spectrum recorded from the toluene phase after phase transfer of the gold nanoparticles (curve 2) shows a fairly sharp plasmon resonance from the nanoparticles, indicating little aggregation of the particles. Clearly, during

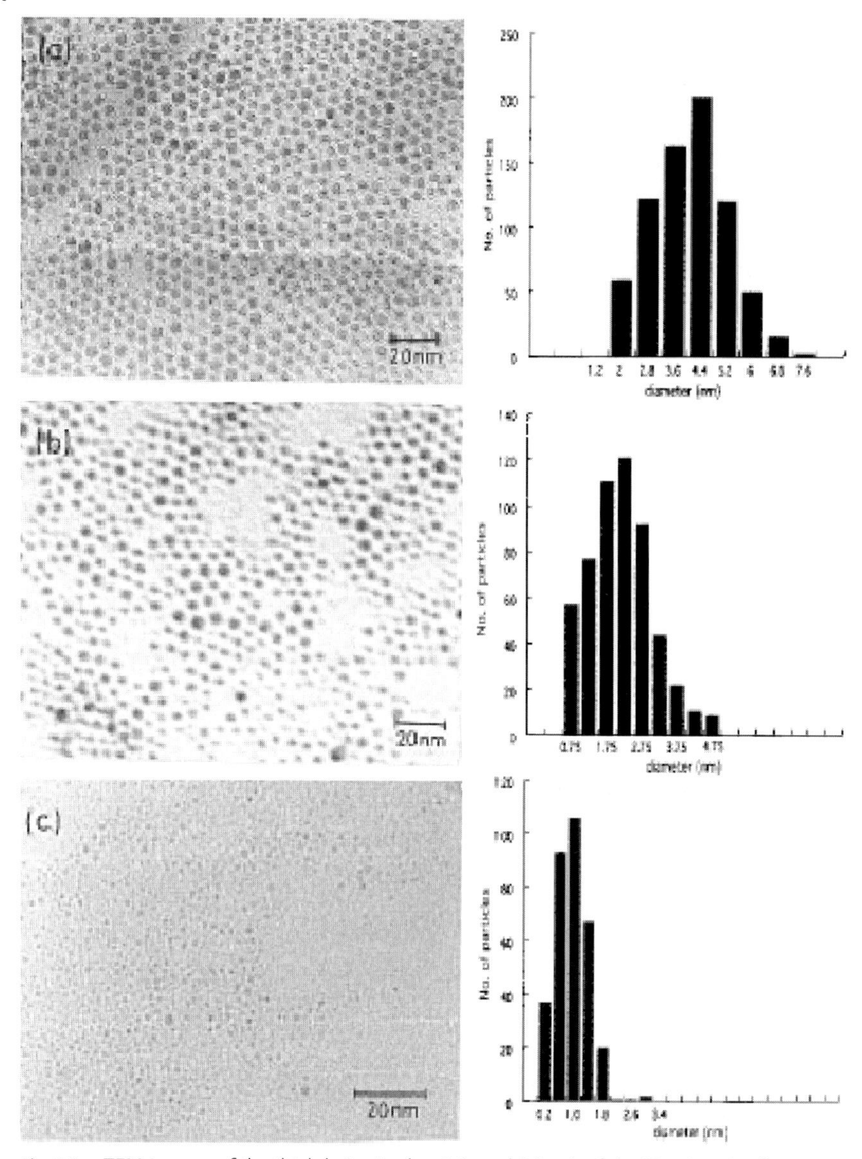


**Fig. 3.3.** TEM images of the thiol derivatized Au nanoparticles. The particle size distributions are shown in the form of histograms alongside the TEM images. The nanoparticles in (a), (b), and (c) were obtained by using 2.2, 2.0, and 1.8 mL of the 25 mM $HAuCl_4$ solution, respectively. (Reprinted with permission from [66], © 1997, American Chemical Society).

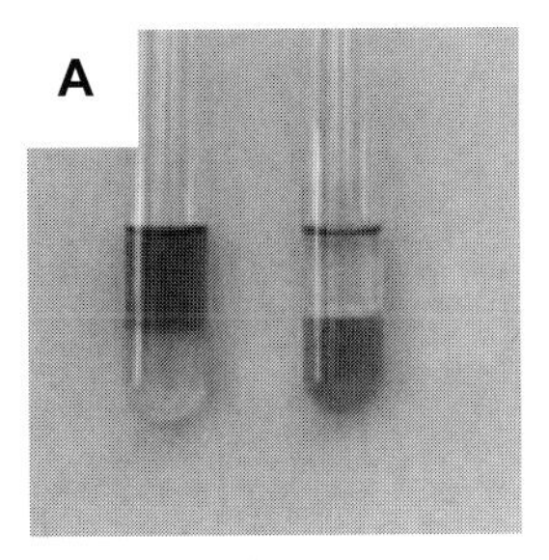


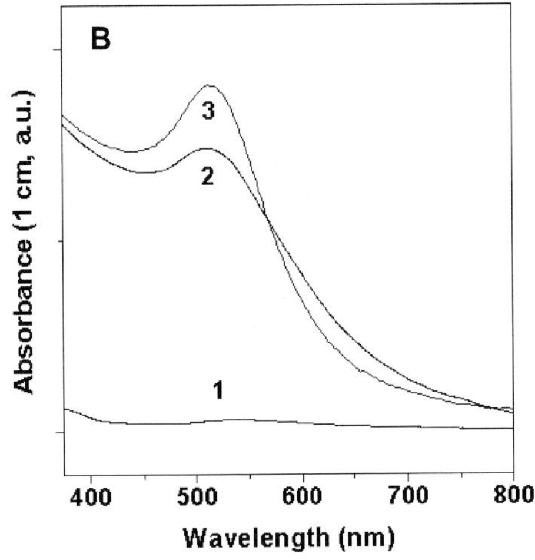


**Fig. 3.4.** (A) Picture showing the two-phase gold hydrosol–ODA containing toluene layers before (test tube on the right) and after (test tube on the left) phase transfer of the gold particles into toluene. (B) UV–vis spectra recorded from the toluene phase (curve 2) and the aqueous phase before (curve 3) and after (curve 1) phase transfer of the gold colloidal particles into toluene (Figure 3.4A, solutions taken from the test tube on the right). (Reprinted with permission from [67], © 2001, Elsevier Science).

stirring of the biphasic mixture, the ODA molecules bind to the gold nanoparticles, thereby rendering them hydrophobic and dispersible in the organic phase. The ODA-capped gold nanoparticles could be separated in the form of a dry powder by rotary evaporation of the toluene phase and could be readily redispersed in organic solvents such as chloroform, benzene, carbon tetrachloride etc. [67].

A drop of the toluene phase with the ODA-capped gold nanoparticles was placed on a TEM grid and analyzed after solvent evaporation. Figure 3.5 shows the image obtained wherein a well ordered hexagonal, close-packed configuration of gold nanoparticles can be seen. The ordered domains are not particularly large due to the fact that the monodispersity of the gold particles in the aqueous phase was ca. 15%. It is interesting to observe the separation of the gold nanoparticles into domains based on their size (Figure 3.5).

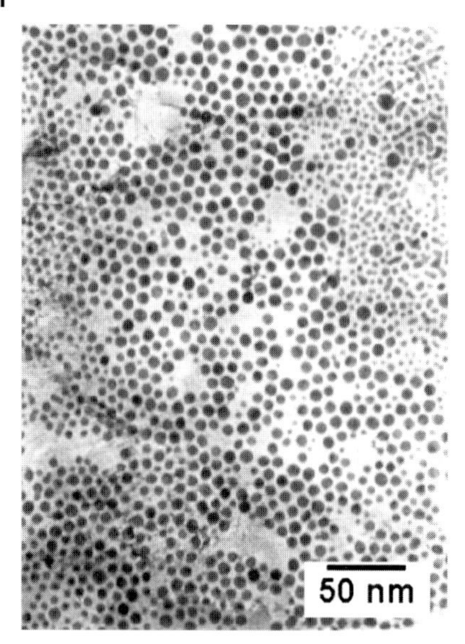


**Fig. 3.5.** TEM micrograph of gold nanoparticles phase transferred into toluene by complexation with ODA molecules and self-assembled by solvent evaporation.

Cyclodextrins (CDs), which are cyclic oligosaccharides consisting of six, seven or eight glucopyranose units ($\alpha$-, $\beta$- and $\gamma$-CDs respectively), are versatile host molecules for 'guests' such as alkanethiols, forming what are known as 'inclusion complexes'. Inclusion complexes of CD and alkanethiols in water have been self-assembled on gold thin films [68] as well as on gold nanoparticles [69]. We have shown recently that gold nanoparticles may be capped with octadecanethiol (ODT) molecules threaded with cyclodextrin and, furthermore, that during vigorous shaking of a biphasic mixture of this hydrosol with chloroform, the gold nanoparticles were rapidly transferred to the organic phase [70]. The inset of Figure 3.6 shows the test tubes before (test tube to the left) and after (test tube to the right) phase transfer of gold nanoparticles capped with inclusion complexes of ODT and CD into chloroform. The process of phase transfer proceeds via detachment of the CD molecules from the alkanethiol molecules capping the gold nanoparticles during stirring of the biphasic mixture. The gold nanoparticles become sufficiently hydrophobic at this stage and are phase transferred into chloroform. The schematics accompanying the picture illustrate the changes occurring in the monolayer of ODT on the surface of the gold nanoparticles during the shaking process leading to hydrophobization of the gold nanoparticles [70].

Just as in the case of gold nanoparticles, appropriate choice of the capping agent for modification of the surface of nanoparticles may be used to accomplish the phase transfer of aqueous silver nanoparticles. Wang, Efrima and Regev have shown that silver nanoparticles can be phase transferred from water to organic solvents by binding the nanoparticles to sodium oleate present in cyclohexane/

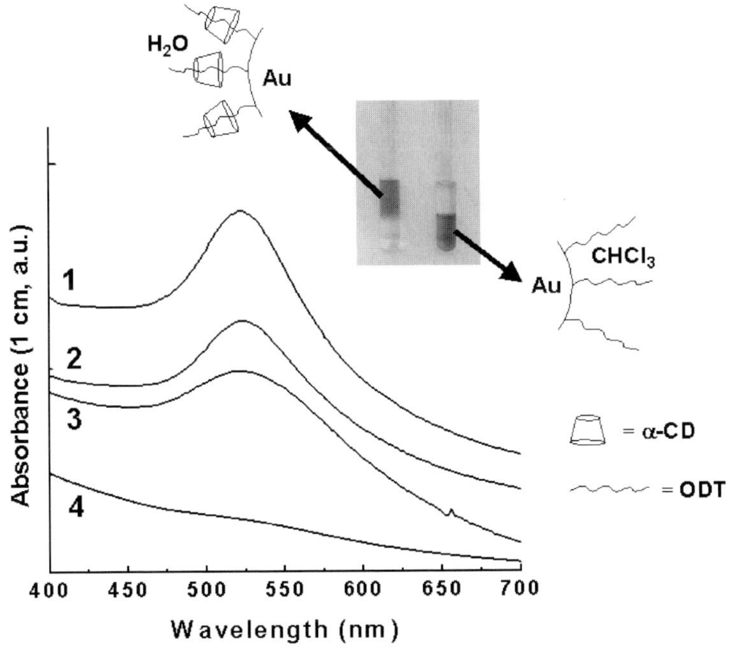


**Fig. 3.6.** UV–vis spectra recorded from the as-prepared gold colloidal solution (curve 1), the gold colloidal solution after capping with α-CD threaded ODT molecules (curve 2), the chloroform solution after phase transfer of the gold nanoparticles (curve 3), and the aqueous gold colloidal solution after phase transfer of the gold nanoparticles into chloroform (curve 4). The inset is a picture of test tubes containing solutions of chloroform and α-CD threaded ODT-capped gold hydrosol before (test tube on the left) and after phase transfer of the gold nanoparticles into chloroform (test tube on the right). The cartoons illustrate the nature of surface modification of the gold nanoparticles in the aqueous phase and in the organic phase. (Reprinted with permission from [70], © 2001, American Chemical Society).

dodecane [71]. Similar to the acid-facilitated phase transfer of gold nanoparticles using alkanethiols [65, 66], the phase transfer of silver particles occurs when a small amount of orthophosphoric/perchloric acid is added to the reaction medium [71]. As observed by other workers, the silver particles assembled into large domains of hexagonally packed nanoparticles [71]. What is interesting about this study is the change in conformation of the surface-bound oleate ions on transfer between the two phases. From FTIR studies, the authors inferred the presence of carboxylate ions on the surface of the silver particles in water that, upon phase transfer into the organic solvent, reversed direction to expose the hydrophobic tails of the oleate molecules towards the solvent. Detailed studies of the position of the double bond in the capping molecule in relation to the carboxylate ions indicated that this factor was crucial for efficient phase transfer of the silver nanoparticles [71].

There is much interest in the synthesis and electronic application of semiconductor nanoparticles, or quantum dots as they are more popularly known [72]. It is well known that thiols bind to quantum dots of CdS and, therefore, it should be possible to phase transfer aqueous CdS nanoparticles into organic solutions by complexation with alkanethiols and this has been demonstrated by us [73]. The experimental conditions for phase transfer of aqueous CdS into toluene containing octadecanethiol were slightly more stringent than those used for phase transfer of gold nanoparticles. It was observed that pre-formed CdS nanoparticles upon complexation with octadecanethiol molecules at the liquid–liquid interface assembled at the interface and were not transferred to the organic phase [73]. Bubbling $H_2S$ gas in a biphasic mixture of aqueous $CdCl_2$ solution and petroleum ether containing octadecanethiol during vigorous stirring resulted in the formation of nanoparticles of CdS capped with the thiol molecules that were rapidly transferred to the organic phase (Figure 3.7A) [73]. The CdS nanoparticle powder was extremely stable and could be readily redispersed in a number of organic solvents such as

A

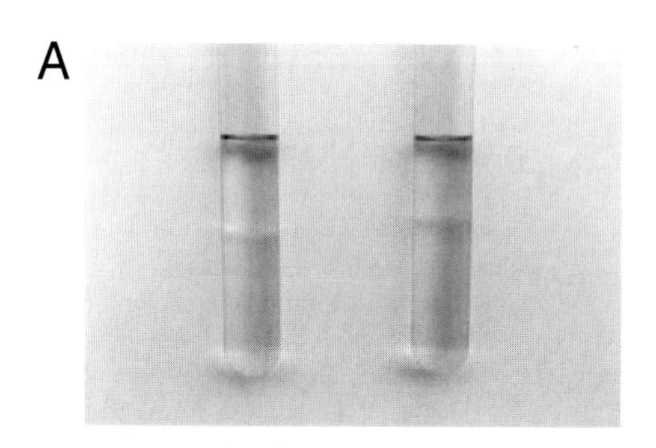

B

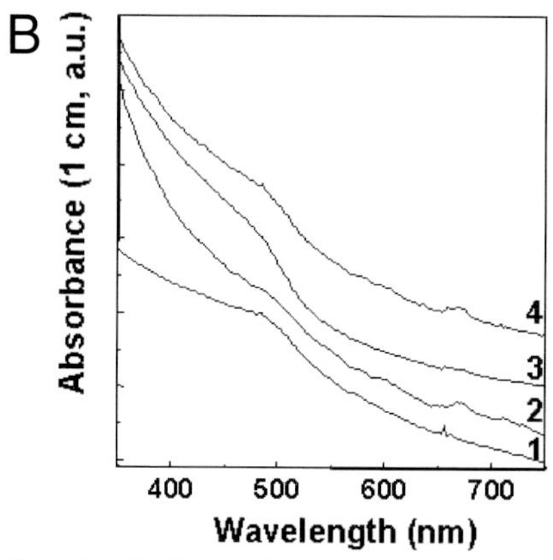


**Fig. 3.7.** (A) Picture showing test tubes containing the biphasic mixture of aqueous CdS and ODT in petroleum ether before (test tube on the left) and after (test tube on the right) phase transfer of the CdS nanoparticles into the organic phase. (B) UV–vis spectra recorded from CdS nanoparticles dispersed in different solvents. Curve 1: as-prepared CdS nanoparticle solution in water; curve 2: ODT-stabilized CdS nanoparticles in toluene; curve 3: ODT-stabilized CdS in chloroform and curve 4: ODT-stabilized CdS in benzene. (Reprinted with permission from [73], © 2001, American Chemical Society).

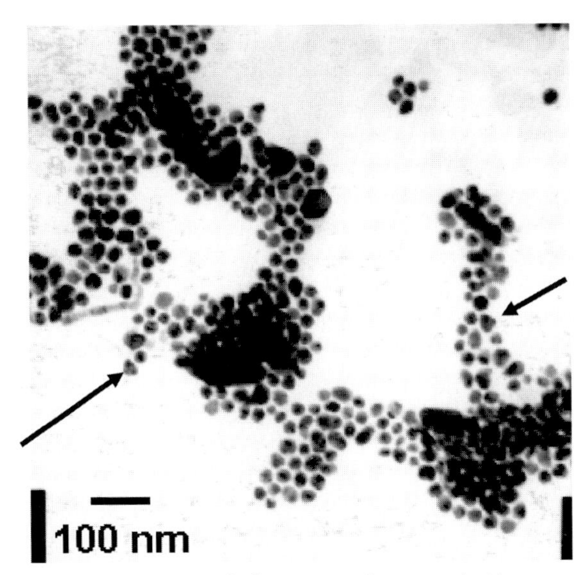


**Fig. 3.8.** TEM picture of CdS quantum dots capped with octadecanethiol molecules and phase transferred into petroleum ether. The arrows in the figure identify some triangular shaped CdS nanoparticles.

chloroform, benzene and toluene without any indication of sintering of the particles (Figure 3.7B) [73]. In the UV–vis spectra of hydrophobized CdS in the different solvents, the onset of absorption occurs at 475 nm and corresponds to a particle size of ca. 6 nm. This size is in good agreement with that determined from TEM studies of CdS nanoparticles transferred into petroleum ether (Figure 3.8). A fairly regular close-packed assembly of the CdS quantum dots can be observed in the TEM micrograph. A number of triangular CdS nanoparticles can also be seen in the figure (identified by arrows).

### 3.4.2
### Transfer of Organically Soluble Gold Nanoparticles to Water

While many different procedures have been developed for the transfer of aqueous gold nanoparticles into organic solvents, the number of reports on the phase transfer of gold nanoparticles in the reverse direction is much smaller. The key step is to replace the hydrophobic groups bound to the gold nanoparticle surface with polar functional groups, thus rendering the nanoparticles water-soluble. Rotello and co-workers used a place exchange mechanism to functionalize alkanethiol-capped organically soluble gold nanoparticles with carboxylic acid groups [74]. The process is illustrated in Figure 3.9 which depicts replacement of octanethiol molecules bound to the surface of 2 nm diameter gold nanoparticles by 11-thioundecanoic acid. Under the experimental conditions adopted by Rotello et al.,

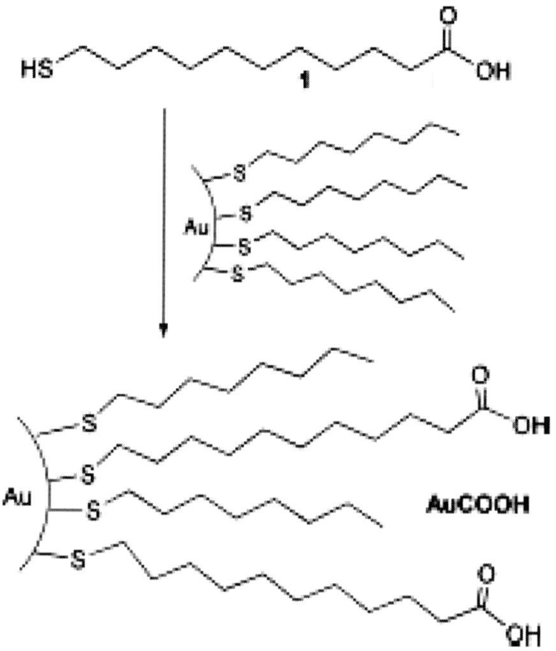

**Fig. 3.9.** Schematic showing the place exchange of octanethiol molecules on the surface of gold nanoparticles by $\omega$-thiol carboxylic acid molecules. (Reprinted with permission from [74], © 2000, Royal Society of Chemistry).

the ratio of $\omega$-thiol carboxylic acid:octanethiol on the gold nanoparticle surface after place exchange was found to be 1:1. The carboxylic acid derivatized gold nanoparticles were washed with dichloromethane and were found to be soluble in water. One advantage of using carboxylic acid functionality to render the gold nanoparticles water-soluble is that the charge on the nanoparticle surface can be modulated by varying the pH of the gold colloidal solution. Interesting variation in the optical properties of carboxylic acid derivatized gold [44] and silver nanoparticles [75] as a function of solution pH have been studied by us earlier. Rotello and co-workers have shown that as the pH of the gold colloidal solution was reduced below 7, the particles aggregated into close-packed assemblies, the size of the aggregates being largest at the lowest pH value (Figure 3.10).

A more direct method for the phase transfer of gold nanoparticles into water along the lines discussed earlier for phase transfer of aqueous particles into organic solvents involving a biphasic mixture has been demonstrated by Gittins and Caruso [76]. More specifically, Gittins and Caruso showed that gold and palladium nanoparticles synthesized in toluene by the Brust method and stabilized by tetra-alkyl ammonium salts could be rapidly and completely phase transferred to water by addition of an aqueous 0.1 M 4-dimethlyaminopyridine (DMAP) solution to aliquots of the gold nanoparticles in toluene. As in previous studies (Figures

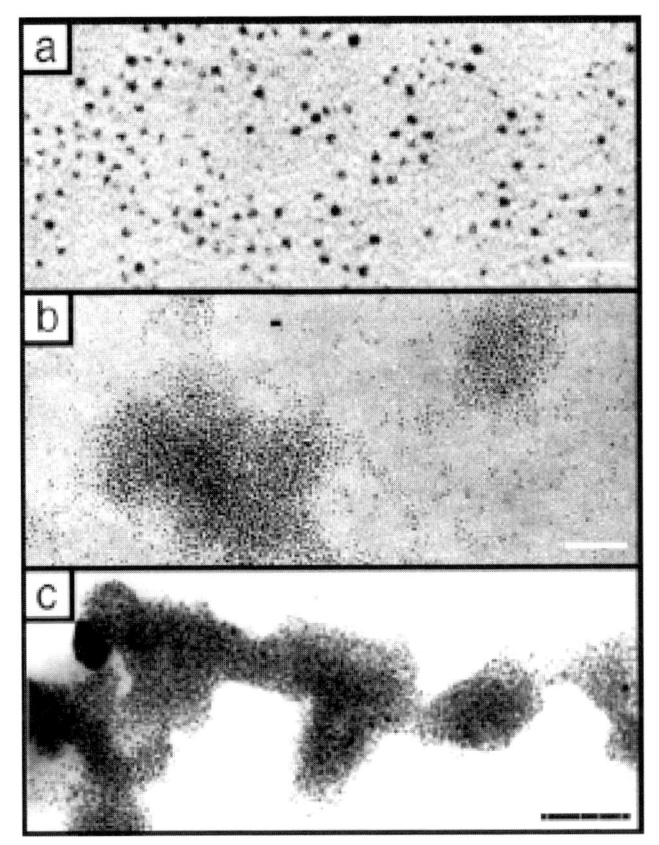


**Fig. 3.10.** TEM micrographs of $\omega$-thiol carboxylic acid functionalized gold colloid at (a) pH 10, (b) pH 7, and (c) pH 4. Scale bars represent 10 nm, 25 nm and 50 nm, respectively. (Reprinted with permission from [74], © 2000, Royal Society of Chemistry).

3.4 and 3.6), the process of phase transfer is seen as a transfer of color from the organic phase to the aqueous phase. That the movement of the gold particles between phases was almost 100% can be observed from the UV–vis spectra recorded from toluene before phase transfer (full line, Figure 3.11) and water after phase transfer (dashed line, Figure 3.11). The UV–vis spectra recorded from the two phases are almost identical. There is little broadening of the surface plasmon resonance after phase transfer, indicating no aggregation of the nanoparticles consequent to their movement [76]. The authors have shown that the aqueous gold and palladium solutions were extremely stable with no sign of degradation even after storage for several months [76]. This is graphically illustrated in the TEM micrographs recorded from the gold nanoparticles in the toluene phase (Figure 3.12A) and the aqueous phase one month after phase transfer of the gold nanoparticles

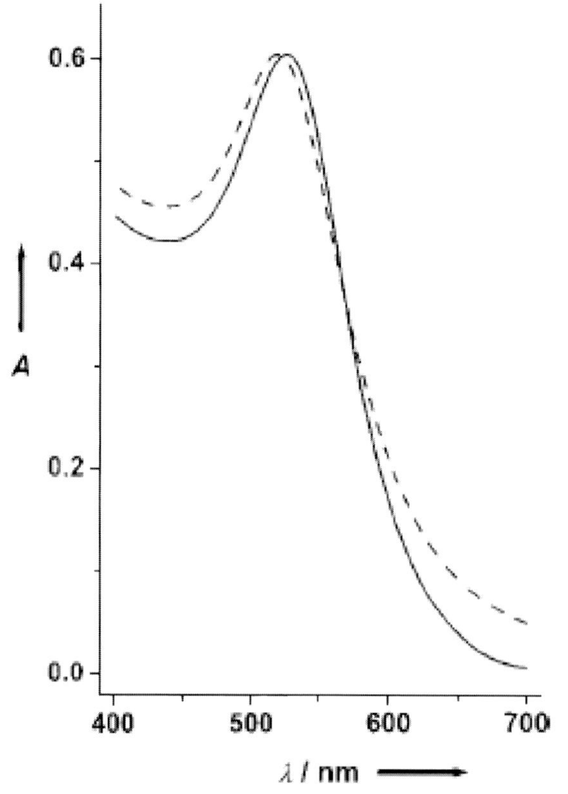


**Fig. 3.11.** UV–vis spectra of a diluted solution of gold nanoparticles in toluene (solid line) and the same sample transferred into an equal volume of 0.1 M DMAP solution at pH 10.5 (dashed line). (Reprinted with permission from [76], © 2001, WILEY-VCH Verlag GmbH).

(Figure 3.12B). A comparison of the two TEM pictures shows that little sintering/aggregation of the gold nanoparticles had occurred after one month of storage.

Gittins and Caruso have speculated on the mode of binding of the DMAP molecules on the gold nanoparticle surface during the phase-transfer process [76]. Based on a series of control experiments, they have proposed a mechanism for the phase transfer as illustrated in Figure 3.13.

Addition of an aqueous DMAP solution to the nanoparticle dispersion in toluene leads to partitioning of the DMAP molecules across the toluene/water boundary as shown in Figure 3.13. The DMAP molecules replace the tetraalkyl ammonium salts and form a labile donor–acceptor complex with the gold atoms on the surface of the nanoparticles through the endocyclic nitrogen atoms. The surface charge on the gold nanoparticles that stabilizes the particles in water arises due to partial

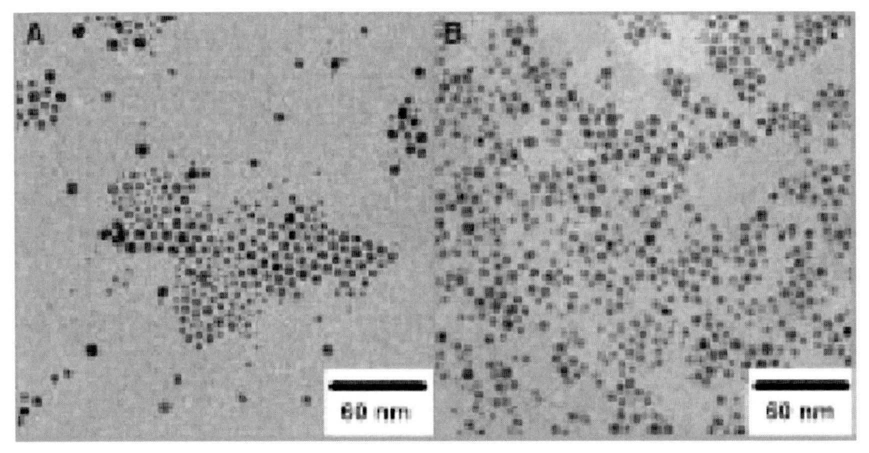


**Fig. 3.12.** Transmission electron micrographs of gold nanoparticles synthesized in toluene (A) and the same sample 1 month after being transferred into water by the addition of DMAP (B). (Reprinted with permission from [76], © 2001, WILEY-VCH Verlag GmbH).

protonation of the exocyclic nitrogens that extend away from the nanoparticle surface towards the solvent (Figure 3.13).

Gittins and Caruso have developed their approach and recently demonstrated the phase transfer of silver, gold, platinum and palladium nanoparticles using a number of exchanging ligands such as mercaptoundecanoic acid (MUA), mecaptosuccinic acid etc. [77]. A highlight of the work was the non-specific bioconjugation of the protein, bovine serum albumin (BSA) with MUA-functionalized gold nanoparticles, possibly through electrostatic and hydrogen bonding interactions between the protein and the ionized carboxylate ions on the nanoparticle surface [77].

In a completely different approach, interdigitated bilayers have been used to transfer gold nanoparticles present in non-polar organic solvents to water [78]. Dodecylamine-capped gold nanoparticles dispersed in chloroform were vigorously stirred with an aqueous solution containing the water-soluble surfactant cetyltrimethylammonium bromide (CTAB). During the stirring process, a secondary interdigitated monolayer of CTAB forms on the existing dodecylamine monolayer in contact with the gold nanoparticle surface. This results in significant hydrophilicity of the gold nanoparticles and a consequent phase transfer to the aqueous phase [78].

In conclusion, I have tried to outline the advantages of carrying out the phase transfer of inorganic nanoparticles (gold nanoparticles in particular) from aqueous to non-polar organic environments and vice versa. Various methods in the literature, including work from my group, in this fascinating area have been covered and if there are any omissions, it is unintentional. Future challenges include de-

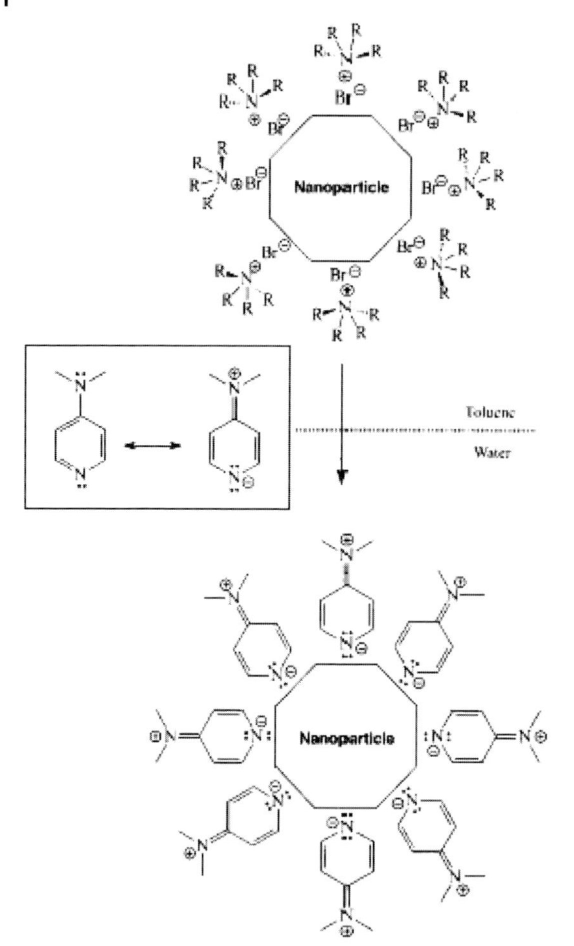


**Fig. 3.13.** Proposed mechanism for the spontaneous phase transfer of gold and palladium nanoparticles from an organic reaction medium (toluene) to water by the addition of DMAP. $R = C_8H_{17}$. (Reprinted with permission from [76], © 2001, WILEY-VCH Verlag GmbH).

velopment of strategies to accomplish the phase transfer of nanoparticles with high shape anisotropy and obtaining core–shell structures that can be moved across phase boundaries.

## Acknowledgments

The author wishes to thank his doctoral student, Mr. Ashavani Kumar and his post-doctoral fellow, Dr. Neeta Lala, who carried out much of the experimental work on nanoparticle phase transfer. Generous funding from the Department of Science and Technology (DST), Government of India, is gratefully acknowledged.

## References


1 The September, 2001 issue of *Scientific American* discusses exciting developments in the field of nanotechnology and current/futuristic applications envisaged for nanomaterials.
2 T. S. Ahmadi, Z. L. Wang, T. C. Green et al., *Science*, **1996**, *272*, 1924.
3 M. A. El-Sayed, *Acc. Chem. Res.*, **2001**, *34*, 257.
4 N. R. Jana, L. Gearhart, C. J. Murphy, *Chem. Commun.*, **2001**, 617.
5 S.-W. Chung, G. Markovich, J. R. Heath, *J. Phys. Chem. B*, **1998**, *102*, 6685.
6 K. Esumi, K. Matsuhisa, K. Torigoe, *Langmuir*, **1995**, *11*, 3285.
7 K. R. Brown, D. G. Walter, M. J. Natan, *Chem. Mater.*, **2000**, *12*, 306.
8 C. J. Johnson, E. Dujardin, S. A. Davis et al., *J. Mater. Chem.*, **2002**, *12*, 1765.
9 N. R. Jana, L. Gearheart, C. J. Murphy, *J. Phys. Chem. B*, **2001**, *105*, 4065.
10 S. O. Obare, N. R. Jana, C. J. Murphy, *Nano Lett.*, **2001**, *1*, 601.
11 R. Djalali, Y. Chen, H. Matsui, *J. Am. Chem. Soc.*, **2002**, *124*, 13660.
12 T. O. Hutchinson, Y.-P. Liu, C. Kiely et al., *Adv. Mater.*, **2001**, *13*, 1800.
13 C. S. Ah, S. D. Hong, D.-J. Jang, *J. Phys. Chem. B*, **2001**, *105*, 7871.
14 X. Peng, L. Manna, W. Yang et al., *Nature*, **2000**, *404*, 59.
15 S. I. Nikitenko, Y. Koltypin, Y. Mastai et al., *J. Mater. Chem.*, **2002**, *12*, 1450.
16 R. Jin, Y. Cao, C. A. Mirkin, K. L. Kelly et al., *Science*, **2001**, *294*, 1901.
17 N. Malikova, I. Pastoriza-Santos, M. Schierhorn et al., *Langmuir*, **2002**, *18*, 3694.
18 Y. Zhou, C. Y. Wang, Y. R. Zhu et al., *Chem. Mater.*, **1999**, *11*, 2310.
19 N. Pinna, K. Weiss, J. Urban et al., *Adv. Mater.*, **2001**, *13*, 261.
20 G. Southam, T. J. Beveridge, *Geochim. Cosmochim. Acta*, **1996**, *60*, 4369.
21 T. Klaus-Joerger, R. Joerger, E. Olsson et al., *Trends Biotech.*, **2001**, *19*, 15.
22 B. Nair, T. Pradeep, *Cryst. Growth Des.*, **2002**, *2*, 293.
23 T. J. Beveridge, R. G. E. Murray, *J. Bacteriol.*, **1980**, *141*, 876.
24 P. Mukherjee, A. Ahmad, D. Mandal et al., *Angew. Chem. Int. Ed. Engl.*, **2001**, *40*, 3585.
25 P. Mukherjee, A. Ahmad, D. Mandal et al., *Nano Lett.*, **2001**, *1*, 515.
26 P. Mukherjee, S. Senapati, D. Mandal et al., *ChemBioChem*, **2002**, *3*, 461.
27 A. Ahmad, P. Mukherjee, S. Senapati et al., *Colloid. Surf. B*, **2003**, *28*, 313.
28 A. Ahmad, S. Senapati, M. I. Khan et al., *Langmuir*, **2003**, *19*, 3550.
29 M. Faraday, *Philos. Trans. R. Soc. London*, **1857**, *147*, 145.
30 M. Haruta, T. Kobayashi, H. Sano et al., *Chem. Lett.*, **1987**, 405.
31 W. Baschong, N. G. Wrigley, *J. Electron. Microsc. Technique*, **1990**, *14*, 313.
32 R. Elghanian, J. J. Storhoff, R. C. Mucic et al., *Science*, **1997**, *277*, 1078.
33 J. Turkevich, G. Garton, P. C. Stevenson, *J. Colloid Sci.*, **1954**, *9*, 26.
34 D. A. Handley, *Colloidal Gold: Principles, Methods and Applications*, ed. M. A. Hayat, Academic Press, San Diego, 1989, Vol. 1, Ch. 2.
35 D. G. Duff, A. Baiker, P. P. Edwards, *Langmuir*, **1993**, *9*, 2301.
36 A. Henglein, *Langmuir*, **1999**, *15*, 6738.
37 E. Gachard, H. Remita, J. Khatouri et al., *New. J. Chem.*, **1998**, 1257.
38 Y. Mizukoshi, T. Fujimoto, Y. Nagata et al., *J. Phys. Chem. B*, **2000**, *104*, 6028.

39 M. Brust, M. Walker, D. Bethell et al., *J. Chem. Soc., Chem. Commun.*, **1994**, 801.
40 R. G. Nuzzo, D. L. Allara, *J. Am. Chem. Soc.*, **1993**, *105*, 4481.
41 D. V. Leff, P. C. Ohara, J. C. Heath et al., *J. Phys. Chem.*, **1995**, *99*, 7036.
42 R. L. Whetten, J. T. Khoury, M. M. Alvarez et al., *Adv. Mater.*, **1996**, *8*, 428.
43 S. R. Johnson, S. D. Evans, S. W. Mahon et al., *Langmuir*, **1997**, *13*, 51.
44 K. S. Mayya, V. Patil, M. Sastry, *Langmuir*, **1997**, *13*, 3944.
45 D. V. Leff, L. Brandt, J. R. Heath, *Langmuir*, **1996**, *12*, 4723.
46 L. O. Brown, J. E. Hutchison, *J. Phys. Chem. B*, **2001**, *105*, 8911.
47 L. A. Porter, D. Ji, S. L. Westcott et al., *Langmuir*, **1998**, *14*, 7378.
48 J. Liu, R. Xu, A. E. Kaifer, *Langmuir*, **1998**, *14*, 7337.
49 J. Liu, S. Mendoza, E. Roman et al., *J. Am. Chem. Soc.*, **1999**, *121*, 4304.
50 PR. Selvakannan, S. Mandal, R. Pasricha et al., *Chem. Commun.*, **2002**, 1334.
51 S. Mandal, PR. Selvakannan, D. Roy et al., *Chem. Commun.*, **2002**, 3002.
52 A. C. Templeton, M. J. Hostetler, C. T. Kraft et al., *J. Am. Chem. Soc.*, **1998**, *120*, 1906.
53 A. C. Templeton, M. J. Hostetler, E. K. Warmoth et al., *J. Am. Chem. Soc.*, **1998**, *120*, 4845.
54 R. S. Ingram, M. J. Hostetler, R. W. Murray, *J. Am. Chem. Soc.*, **1997**, *119*, 9175.
55 D. E. Cliffel, F. P. Zamborini, S. M. Gross et al., *Langmuir*, **2000**, *16*, 9699.
56 Y.-S. Shon, W. P. Wuelfling, R. W. Murray, *Langmuir*, **2001**, *17*, 1255.
57 PR. Selvakannan, S. Mandal, S. Phadtare et al., *Langmuir*, **2003**, *19*, 3459.
58 R. K. Iler, *J. Colloid Interface Sci.*, **1966**, *21*, 569.
59 T. Cassagneau, J. H. Fendler, *J. Phys. Chem. B*, **1999**, *103*, 1789.
60 M. Sastry, M. Rao, K. N. Ganesh, *Acc. Chem. Res.*, **2002**, *35*, 847.
61 A. Kumar, A. B. Mandale, M. Sastry, *Langmuir*, **2000**, *16*, 6921.
62 A. Gole, C. Dash, V. Ramachandran, A. B. Mandale et al., *Langmuir*, **2001**, *17*, 1674.
63 W. C. W. Chen, S. Nie, *Science*, **1998**, *281*, 2016.
64 S. Underwood, P. Mulvaney, *Langmuir*, **1994**, *10*, 3427.
65 K. V. Sarathy, G. U. Kulkarni, C. N. R. Rao, *Chem. Commun.*, **1997**, 537.
66 K. V. Sarathy, G. Raina, R. T. Yadav et al., *J. Phys. Chem. B*, **1997**, *101*, 9876.
67 M. Sastry, A. Kumar, P. Mukherjee, *Colloid. Surf. A*, **2001**, *181*, 255.
68 J. Yan, S. Dong, *Langmuir*, **1997**, *13*, 3251.
69 J. Liu, J. Alvarez, A. E. Kaifer, *Adv. Mater.*, **2000**, *12*, 1381.
70 N. Lala, S. P. Lalbegi, S. D. Adyanthaya et al., *Langmui*, **2001**, *17*, 3766.
71 W. Wang, S. Efrima, O. Regev, *Langmuir*, **1998**, *14*, 602.
72 A. P. Alivisatos, *Science*, **1996**, *271*, 933.
73 A. Kumar, A. B. Mandale, M. Sastry, *Langmuir*, **2000**, *16*, 9299.
74 J. Simard, C. Briggs, A. K. Boal et al., *Chem. Commun.*, **2000**, 1943.
75 M. Sastry, K. Bandyopadhyay, K. S. Mayya, *Colloid. Surf. A*, **1997**, *127*, 221.
76 D. J. Gittins, F. Caruso, *Angew. Chem. Int. Ed. Engl.*, **2001**, *40*, 3001.
77 D. J. Gittins, F. Caruso, *ChemPhysChem*, **2002**, *3*, 110.
78 A. Swami, A. Kumar, M. Sastry, *Langmuir*, **2003**, *19*, 1168.

# 4
# Mesoscopic Assembly and Other Properties of Metal and Semiconductor Nanocrystals

*G. U. Kulkarni, P. J. Thomas, and C. N. R. Rao*

**Abstract**

The properties of metal and semiconductor nanocrystals are briefly reviewed. The organization of metal and semiconductor nanocrystals into mesostructures is an important aspect of nanoscience. New methods of nanocrystal synthesis and functionalization relevant to mesoscopic assembly are described, with emphasis on procedures that deal with monodispersed nanocrystals. Programmed assemblies of nanocrystals in one, two and three dimensions are discussed together with attempts to measure the properties of such lattices. Future directions and potential applications of ordered assemblies are indicated.

## 4.1
## Introduction

A nanocrystal is a tiny chunk of the bulk measuring a few nanometers with a finite number of atoms in it. Nanocrystals in the size range 1–50 nm are considered important and are obtainable as sols: a dispersion of a solid in a liquid, also called a colloidal sol. Metal sols possess fascinating colors and have long been used as dyes. That such dyes indeed consist of tiny metal chunks was established as early as 1857 by Faraday [1], but a similar realization in the case of semiconducting nanocrystals had to wait for over a century [2, 3]. Modern techniques of synthesis enable one to obtain sols of metals and semiconductors that can be dried and re-dissolved like water colors. The colloidal particles display a remarkable tendency to remain single-crystalline and are hence named as nanocrystals. Nanocrystals posses high surface area, a large fraction of the atoms in a nanocrystal are on its surface. A small nanocrystal of 1 nm diameter will have as much as 30% of its atoms on the surface, while a larger nanocrystal of 10 nm ($\sim$1000 atoms) will have around 15% of its atoms on the surface [4].

An added dimension to research on nanocrystals is their size-dependent properties. The electronic, magnetic and optical properties of a nanocrystal depend on its size [4]. In small nanocrystals, the electronic energy levels are not continuous as

*The Chemistry of Nanomaterials: Synthesis, Properties and Applications, Volume 1.* Edited by C. N. R. Rao, A. Müller, A. K. Cheetham

ISBN: 3-527-30686-2

in the bulk but are discrete, due to the confinement of the electron wavefunction to the physical dimensions of the particles [5]. This phenomenon is called quantum confinement and, therefore, nanocrystals are also known as quantum dots. In other words, a small nanocrystal could be a very bad conductor although it is a tiny silhouette of the conducting bulk. Likewise, a tiny nanocrystal of a ferromagnet can be paramagnetic in nature. In the case of semiconductors, besides discretization of levels, the band gap increases due to quantum confinement. In several respects, small nanocrystals behave like molecules. The nanocrystals can be discretely charged with electrons with characteristic charging energies. This means that a nanocrystal carrying an extra electron can exhibit properties different from a neutral species.

The electronic absorption spectrum of metal nanocrystals in the visible region is dominated by the plasmon band. This absorption is due to the collective excitation of the itinerant electron gas on the particle surface and is characteristic of a nanocrystal of a given size. In metal colloids, surface plasmon excitations impart characteristic colors to the metal sols, the beautiful wine-red color of gold sols being well-known [6–8]. The dependence of the plasmon peak on the dielectric constant of the surrounding medium and the diameter of the nanocrystal was predicted theoretically by Mie and others at the turn of the last century [9–12]. The dependence of the absorption band of thiol-capped Au nanocrystals on solvent refractive index was recently verified by Templeton et al. [13]. Link et al. found that the absorption band splits into longitudinal and transverse bands in Au nanorods [6, 7].

In contrast to metals, exciton peaks dominate the absorption of semiconductor nanocrystals. Thus, yellowish CdS, exhibits an excitonic absorption around 600 nm, which gradually shifts into the UV region as the nanocrystal diameters are varied below 10 nm (see Chapter 1). The absorption band can be systematically varied across ranges of a few 100 nm by changing the size of the semiconductor nanocrystal [14–17]. Brus and others proposed an independent theory to describe size quantization effects in semiconductor nanocrystals, based on the effective mass approximation [18, 19] after recognizing the failure of Mie's theory. Since then, theories have grown in sophistication and rigour to include key effects like surface structure and coupling of electronic states [20–22]. However, our understanding of the optical properties of semiconductor nanocrystals is still incomplete and careful experiments on monodisperse nanocrytals are currently being pursued to unravel the mystery [23]. In addition to interesting absorption properties, the semiconductor nanocrystals also exhibit luminescent behaviour [24–27]. The emission from mono-disperse semiconductor nanocrystals such as CdSe is intense, narrow and can by brought about by excitation in a broad range of wavelengths [27]. The emission can be tuned by altering the diameter of the nanocrystal (see Figure 4.1). Further, control over the emission can be exercised by varying the surface structure and controlling the diameter distribution. The above factors have led to the exploration of a wide range of applications for luminescent semiconductor nanocrystals.

The shrinking dimensions of the current microelectronic devices and the realization that current lithographic processes cannot extend to the nanoworld [28] have lent tremendous thrust to research aimed at ordering nanocrystals into functional networks [29–34]. The nanocrystals akin to covalent systems, self-assemble

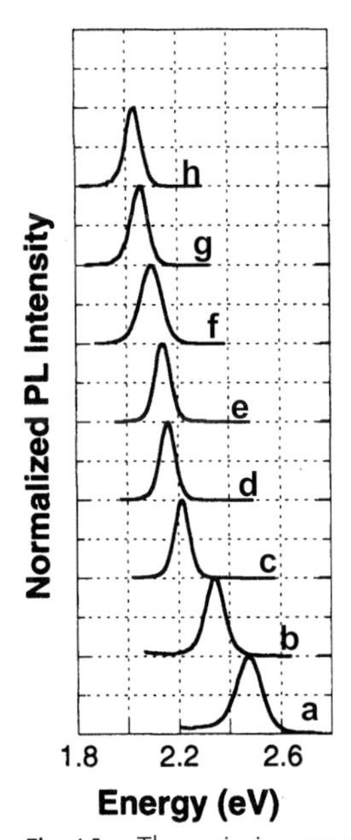


**Fig. 4.1.** The emission spectra of CdSe nanocrystals of different sizes (a) 2.4 nm; (b) 2.5 nm; (c) 2.9 nm; (d) 3.3 nm (e) 3.9 nm; (f) 4.1 nm; (g) 4.2 nm; (h) 4.4 nm. The change in the emission width is due to decrease in the nanocrystals diameter distribution with increase in diameter (reproduced with permission from [27]).

into ordered arrays in one, two and three dimensions under the right conditions. Lattices of nanocrystals consist of interacting nanocrystals and may exhibit novel properties arising out of such interactions. The ability to engineer such assemblies thus extends the reach of current lithographic techniques and holds promise for a new generation of electronics of the nanoworld [29]. In this context, synthesis and programmed assembly of nanocrystals assumes significance.

In this chapter, we discuss the structure and stability of mesoscopic organizations of nanocrystals in one, two and three dimensions, obtained by using a variety of surfactants. We also examine certain unusual organizations such as clusters of nanocrystals and microcolloidal crystals. Collective properties of nanocrystal organization are presented.

## 4.2
## Synthetic Strategies

### 4.2.1
### General Methods

Chemical synthesis of sols of metals and semiconductors results in nanoparticles embedded in a layer of ligands or stabilizing agents, that prevent the aggregation of particles. The stabilizing agents employed include surfactants such as long chain thiols or amines or polymeric ligands such as polyvinylpyrrolidone (PVP). Reduction of metal salts dissolved in appropriate solvents produces small metal particles of varying size distributions [35–38]. A variety of reducing agents have been employed for the reduction. These include electrides, alcohols, glycols, metal borohydrides and certain specialized reagents such as tetrakis(hydroxymethyl) phosphonium chloride. Si and Ge nanocrystals can be obtained by reduction of $GeCl_4$ or silanes with strong reducing agents such as lithium or sodium naphthalide [39–41].

Kinetic control of precipitation (arrested precipitation) is generally used to obtain semiconductor nanocrystals such as CdS [42], CdSe [43], ZnS, HgTe, PbS, CuS, $Cu_2S$, AgI, ZnO, AgI and $TiO_2$ [44–46]. The nanoparticles so obtained possess a broad distribution in diameter. Thermolysis methods involving the decomposition of organometallic precursors in high boiling organic solvents are used to prepare CdSe [47], CdS [47], PbSe [16], InP [17], ZnSe, GaAs, InSb, GaP nanocrystals [45–50]. Reverse micellar methods exploit the "water pools" in water-in-oil mixtures to synthesize nanocrystals and have been successfully utilized in the preparation of Ag, Au, Co, Pt, Co, CdS, CdTe, AgS nanocrystals [51, 52]. The synthesis of nanocrystals at the air–water interface, as in Langmuir–Blodgett films, or at a liquid–liquid interface, is currently attracting wide attention [30, 53, 54]. CdS, PbS and MgS nanocrystals have been prepared by exposing Langmuir–Blodgett films of fatty acids to $H_2S$ [55]. It has been shown recently that films of metal and semiconductor nanocrystals can be prepared using a water–toluene interface [56]. A typical film of Au nanocrystals is shown in Figure 4.2. Traditionally, clusters of controlled sizes have been generated by abalation of a metal target in vacuum followed by mass selection of the plume to yield cluster beams [57, 58]. Such cluster beams could be subjected to in situ studies or be directed on to solid substrates. In order to obtain nanocrystals in solution, Harfenist et al. [59] steered a mass-selected Ag cluster beam through a toluene solution of thiol and capped the vacuum prepared particles.

Colloids of alloys have been made by the chemical reduction of the appropriate salt mixture in the solution phase. In the case of semiconductor nanocrystals, a mixture of salts is subjected to controlled precipitation. Thus, Ag–Pd and Cu–Pd colloids of varying composition have been prepared by alcohol reduction of mixtures of silver nitrate or copper oxide with palladium oxide [60]. Fe–Pt alloy nanocrystals have been made by thermal decomposition of the Fe and Pt acetylacetonates in high boiling organic solvents [61]. Au–Ag alloy nanocrystals have been made by co-reduction of silver nitrate and chloroauric acid with sodium borohydride [62, 63]. Semiconductor nanocrystals of the form $Cd_xMn_{1-x}S$, $CdS_xSe_{1-x}$ have been obtained by the inverted micelle methods as well as in glasses by sol–gel methods [23, 64]. Alloys of controlled composition are also made by thermal decomposition of carefully chosen precursors, to achieve homogeneity. For example, $Mn_2(\mu\text{SeMe})_2(CO)_8$ was used as selenium source to obtain $Cd_{1-x}MnSe$ nanocrystals [65]. Au–Ag alloying and segregation has been brought about by the use of lasers on Au–Ag layered particles [66, 67].

**Fig. 4.2.** Nanocrystalline film of Au formed at the toluene–water interface (middle). Gold is introduced as a toluene solution of $Au(PPh_3)Cl$ while partially hydrolysed THPC (tetrakishydromethylphosphoniumchloride) in water acts as a reducing agent. The film is obtained when the two layers are allowed to stand for several hours. When dodecanethiol is added to the toluene layer, the film breaks up to form an organosol (left), while mercaptoundecanoic acid added to water produces a hydrosol (right). Shown below are the corresponding TEM images showing nanocrystals. Films of CdS nanocrystals could also be prepared by adopting the same methods. Scale bar 50 nm.

### 4.2.2 Size Control

The successful synthesis of nanocrystals involves three steps nucleation, growth and termination by the capping agent or ligand [35–37]. Though the reaction temperature and reagent concentrations provide a rudimentary control of the three steps, it is often impossible to independently control them and so the obtained nanocrystals usually exhibit a distribution in size. Typically, the distribution is log-normal with a standard deviation of 10% [37]. Given the fact that properties of the nanocrystals are size-dependent, it is significant to be able to synthesize nanocrystals of precise dimensions with minimal size-distributions. This can be accomplished to a limited extent by size selective precipitation, either by centrifugation or by use of a miscible solvent–non-solvent liquid mixture to precipitate the nanocrystals. However, single crystals of large clusters of semiconducting material such as $Cu_{147}Se_{73}(PEt_3)_{22}$ [68], $[Cd_{10}S_4(SPh)_{16}]^{4-}$ [69], $Cd_{32}S_{14}(SC_6H_5)_{36}$

$DMF_4$ [69], $Cd_{17}S_4(SCH_2CH_2OH)_{26}$ [69], $Cd_{32}S_{14}(SCH_2CH(CH_3)OH)_{36}$ [70], $Hg_{32}Se_{14}(SeC_6H_5)_{36}$ [71] have been obtained. Solutions of such clusters possess optical properties similar to those of the sols. Schmid [72] and Zamaraev [73] succeeded in preparing truly monodisperse nanocrystals which they called "cluster compounds". These cluster compounds are like macromolecules with a core containing metal–metal bonds yet are obtainable in definite stoichiometries, typical examples being $[Pt_{38}(CO)_{44}H_2]^{2-}$ and $Au_{55}(PPh_3)_{12}Cl_6$. The enhanced stability of $Au_{55}$ was recently demonstrated clearly by Boyen et al. [74] who exposed a series of $Au_n$ nanocrystals to oxidation. These nanocrystals have special stability because they consist of a 'magic number' of metal atoms which enables the complete closure of successive shells of atoms in a cubic close packed arrangement. The magic numbers 13, 55, 147, 309 and 561 correspond to the closure of 1, 2, 3, 4 and 5 shells respectively [75]. A schematic illustration of magic nuclearity nanocrystals is shown in Figure 4.3. Since the breakthrough, several magic nuclearity nanocrystals have been prepared including PVP-stabilized $Pd_{561}$ nanocrystals [76]. In Figure 4.4, are shown scanning tunnelling and transmission electron microscopic (TEM) images of polymer-protected $Pd_{561}$ nanocrystals.

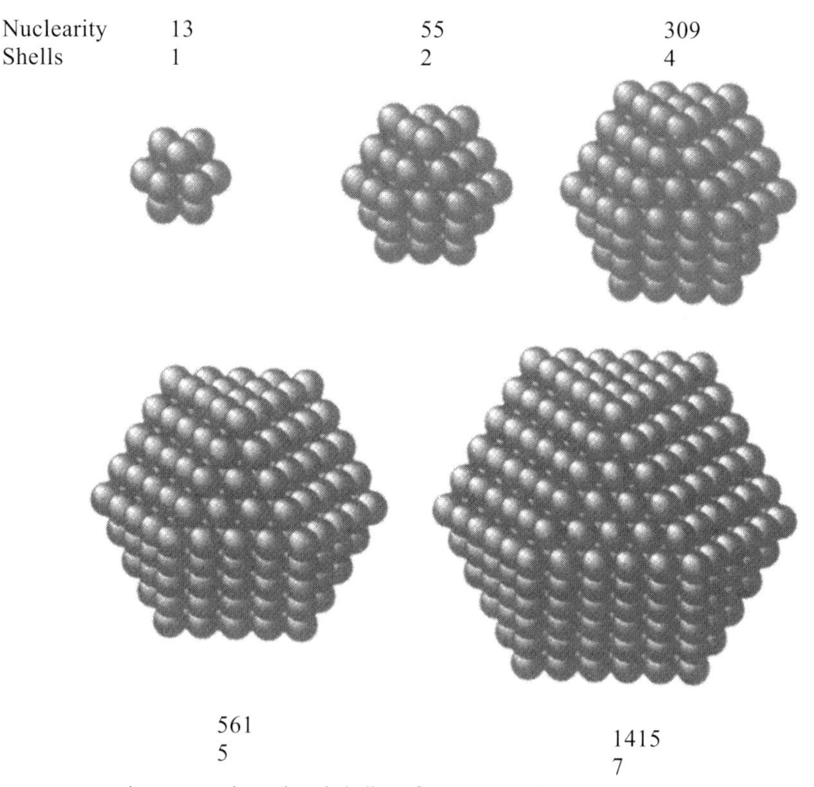


**Fig. 4.3.** Metal nanocrystals in closed-shell configurations with magic number of atoms.

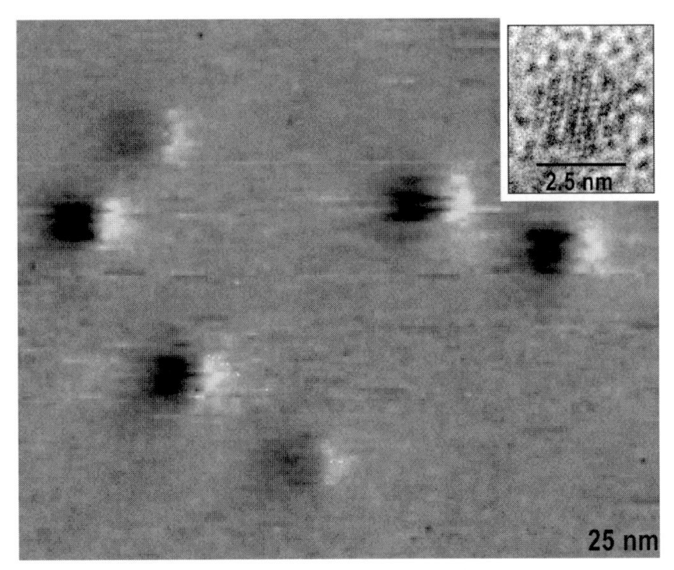


**Fig. 4.4.** Scanning tunneling microscopy image of polymer-coated $Pd_{561}$ nanocrystals. The nanocrystals are seen as fluffy balls against the plane background of the graphite substrate. The inset shows a high-resolution electron micrograph (HRTEM) of an individual nanocrystal. We see the characteristic 11 [111] fringes in the icosahedral shape measuring 2.5 nm. The diameter estimated from STM is ~3.4 nm, the difference being due to the ligand shell.

### 4.2.3 Shape Control

Since the properties of nanocrystals follow from the confinement of the electrons to the physical dimensions of the nanocrystals, it would be interesting to vary the shape of the nanocrystals and study the effect of confinement of electrons in such artificial shapes [77]. For example, it is predicted that light emitted from a nanorod would be linearly polarized along the growth-axis [23]. Such predictions have led to the revival of interest in synthetic strategies yielding non-spherical nanocrystals. Conventional methods such as those due to Turkevich [36] yield, in addition to spherical particles, a mixture of shapes: triangular, teardrop etc., which was then thought of as undesirable. Today, smarter synthetic schemes have been designed which selectively yield nanocrystals in the form of rods, elongated spheres, cubes and hexagons. CdSe nanocrystals in the form of rods, arrows, teardrops and tetrapods have been obtained by careful control of thermolysis conditions such as ratio of surfactants and injection volumes [23, 78]. In Figure 4.5 are shown, TEM micrographs of soluble CdSe nanorods of various aspect ratios. Triangular CdS nanocrystals have been obtained by inverse micelle methods [79]. Large tetrahedral Si nanocrystals as exclusive products have been obtained by careful control of the reducing conditions [80]. TEM images of triangular CdS and tetrahedral Si nano-

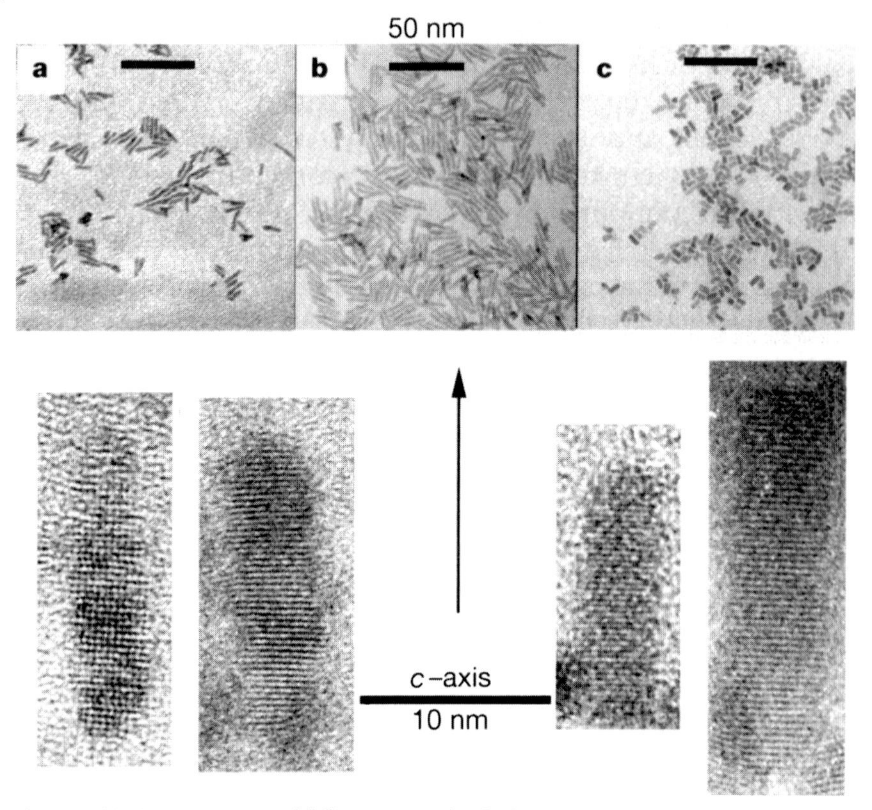


Fig. 4.5. (a)–(c) TEM images of different nanorods of CdSe with different sizes and aspect ratios, high resolution TEM images of four nanorods are shown below (reproduced with permission from [78]).

crystals are shown in Figure 4.6. Some shape control has also been demonstrated in the case of CdS and CdTe nanocrystals [81, 82].

### 4.2.4
### Tailoring the Ligand Shell

Nanocrystals in their native form are dominated by the surface species, the capping agents employed play a role in determining the property of the nanocrystals [83]. Hence, in addition to controlling the size and the shape of the nanocrystals it is also necessary to tailor its surface with the right capping agent. In addition to traditional capping agents that include ions, surfactants and polymers, a new breed of ligands: dendrimers, hydrogen bonding fragments of protein, DNA and dyes, with pendent thiol groups, as well as silica layers have been used as capping agents [84–86].

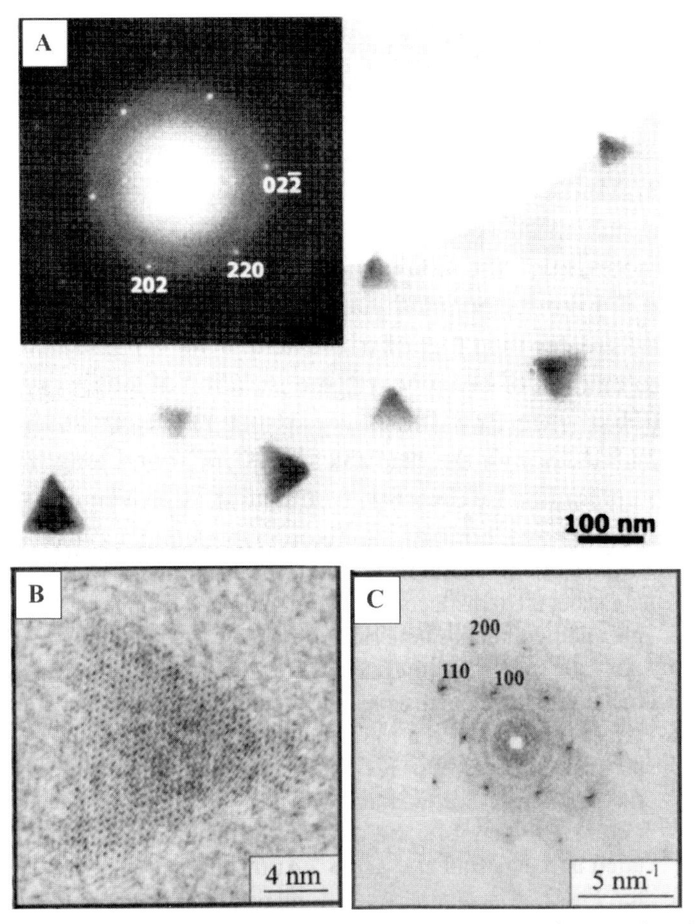


**Fig. 4.6.** (a) TEM micrograph showing tetrahedral Si nanocrystals. The inset shows a selected area electron diffraction pattern typical of a diamond structure. (b) A high resolution TEM image of a triangular CdS nanocrystal. The corresponding power spectrum is in (c). (reproduced with permission from [79, 80]).

In some cases, a layer of a noble metal is used as a buffer between the core nanocrystal and the ligand shell [87]. Thus, a layer of gold lends special stability to Fe nanocrystals and helps prevent oxidation and in preserving the magnetic properties of Fe (see Figure 4.7). A coating of a wider band gap material over a semiconductor nanocrystal aids in lifting the energy of the surface states from in between the highest occupied and the lowest unoccupied levels, thereby enhancing the luminescence efficiency [14, 24]. The lower band gap material acts as a seed for nucleation of the higher band bap material. Thus, core–shell nanocrystals such as CdSe–ZnS, Si–$SiO_2$, HgS–CdS, PbS–CdS [45, 46], ZnS–CdSe [88], ZnSe–CdSe [89], CdTe–HgTe nanocrystals have been obtained [90, 91].

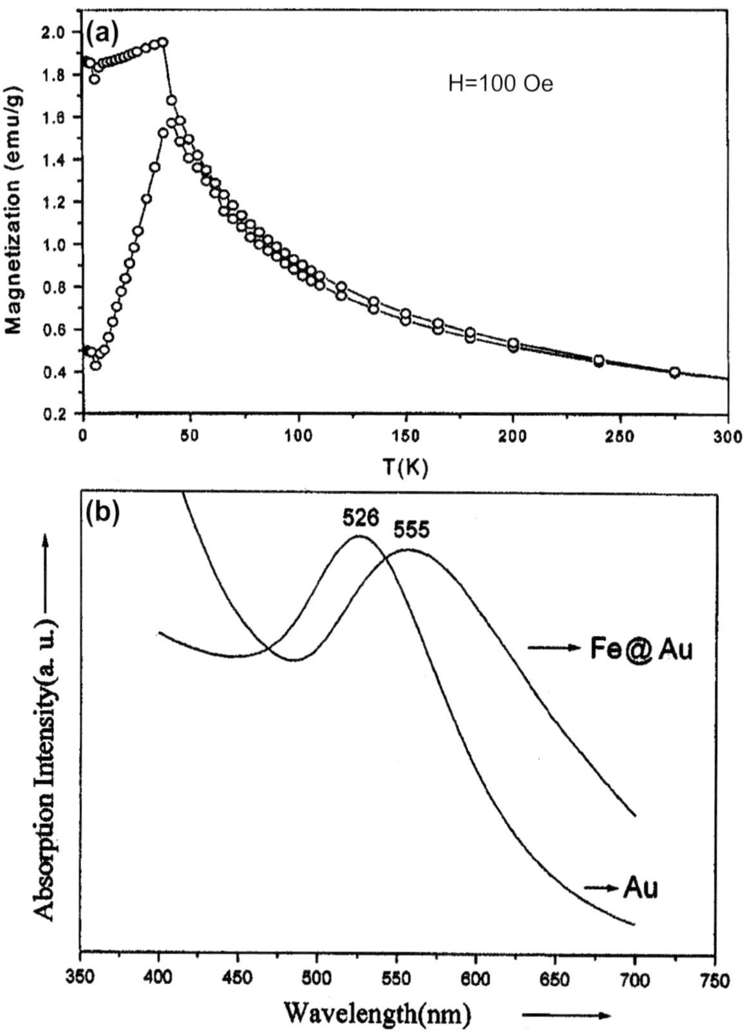


**Fig. 4.7.** (a) Zero field cooled and field cooled magnetization curves for Fe–Au core–shell nanocrystals. The blocking temperature is 42 K. (b) Absorption spectrum showing the Au surface plasmon, shifted due to capping by Fe. For comparison, the spectrum of Au hydrosol is also shown. (reproduced with permission from [87]).

Of special interest with regard to tailoring the ligand shells are reactions that enable the total replacement of one set of ligands with another [92–96]. These reactions also typically enable the transfer of nanocrystals from one phase to another. A novel method of thiol-derivatizing hydrosols of metal sols has been developed by Sarathy et al. [93, 94]. The procedure involves mixing vigorously a hydrosol containing metal particles of the desired size distribution with a toluene solution of an alkane thiol in the presence of a strong acid or reducing agent. The completion of the derivatization is marked by a vivid interchange of the colors from the aqueous layer to the hydrocarbon layer. The advantage of this method is that well-characterized metal particles can be easily thiol-derivatized in a nonaqueous medium. A variety of hydrosols of Au, Ag and Pt has been thiolized by this procedure. A simple modification of this technique is shown to be effective in the case of CdS nanocrystals [97].

## 4.3 Programmed Assemblies

Like molecular systems, nanocrystals capped with suitable ligands spontaneously assemble into ordered aggregates. That such self-assembly can occur through a variety of weak forces is being recognized. Cooperative assemblies of ligated metal and semiconductor as well as of colloidal polymer spheres seem to occur through the mediation of electrostatic and capillary forces [98–100]. The forces that govern the nanocrystal assembly, however, are different in many ways. Surface tension for example, plays an important role [37] because in a nanocrystal, a large fraction of atoms are present at the surface. Surfactant molecules which self-assemble on solid surfaces have proved to be the best means of obtaining ordered arrays of nanocrystals [100].

The way in which the nanocrystals organize themselves depends critically on the core diameter, the nature of the ligand, substrate and even the dispersive medium used [101]. Thiolized metal nanocrystals readily arrange into two-dimensional arrays on removal of the solvent [29]. Using suitable methods, they can also be put into one-dimensional organization in the form of strings or assembled in a stepwise fashion in a three-dimensional superlattice (see Figure 4.8).

### 4.3.1 One-Dimensional Arrangements

Hornayak and coworkers [102] used the ordered channels of porous alumina as templates to obtain linear arrangements of Au nanocrystals. By varying the pore size, the diameter of the nanowire could be controlled. A linear arrangement has also been obtained by coordinating Au particles ($\sim$1.4 nm) stabilized with phosphine ligands to single stranded DNA oligonucleotide of the desired length and specific sequence [103, 104]. Liquid crystalline phases of a genetically engineered virus–ZnS nanocrystal hybrid material was used as a template to obtain linear arrays of ZnS nanocrystal aggregates [105]. Similarly, Pt nanocrystals in the form of ribbons have been obtained using a cholesteric liquid crystalline template [106]. Organization of particles in a 1D lattice has met with limited success. Heath and coworkers [107] have fabricated wires of Ag nanocrystals by compressing a dispersion of Ag (4.5 nm) nanocrystals in toluene (Figure 4.9). The wires were one nanocrystal thick, a few nanocrystals wide and extended in length from 20–300 nm. The interwire separation distance and the alignment of the wires could

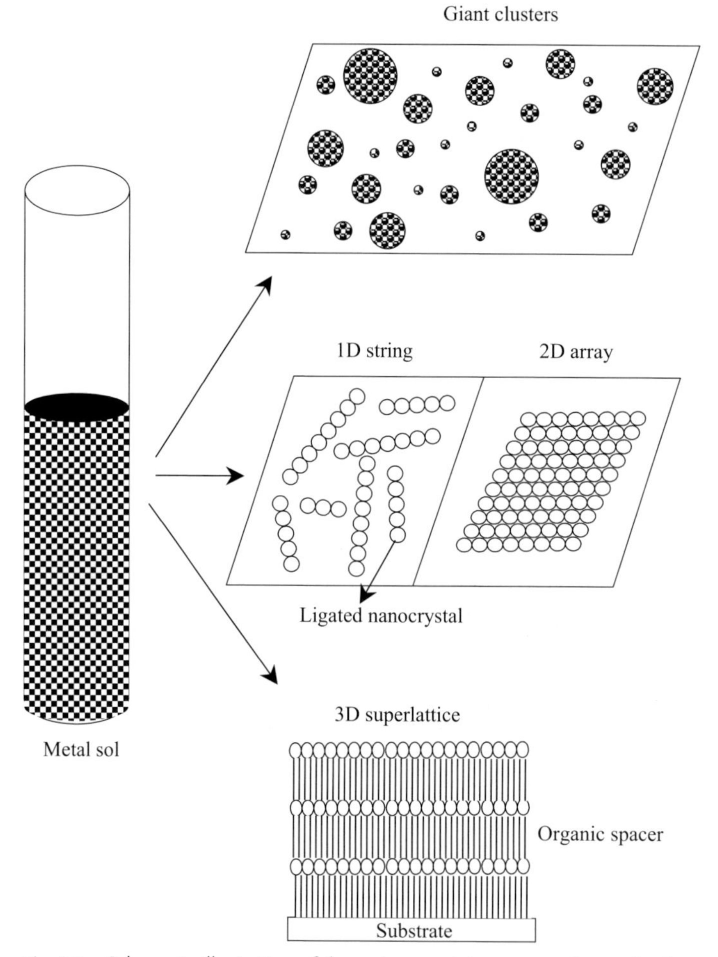


**Fig. 4.8.** Schematic illustration of the various metal nanocrystal organizations.

be controlled by compressing the film. Based on preliminary experimental observations, it has been suggested that tobacco mosaic virus tubules could serve as templates for the growth of 1D lattice of quantum dots [108].

### 4.3.2 Two-Dimensional Arrays

Ligands based on long chain thiols or phosphines have served as good candidates for assembling monodisperse nanocrystals on a flat substrate. Two-dimensional organizations of a variety of nanocrystals can be brought about by simply evaporating a drop of the sol on a flat substrate.

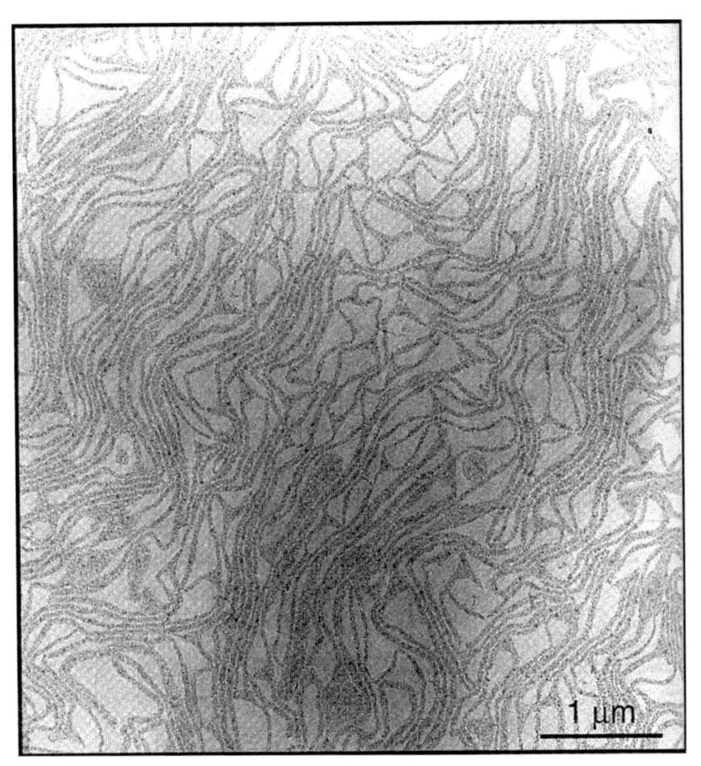


**Fig. 4.9.** TEM image of a continuous stratum structure of compressed LB film of Ag nanocrystals. One-dimensional strings of Ag nanocrystals are clearly seen (reproduced with permission from [107]).

#### 4.3.2.1 **Arrays of Metal Nanocrystals**

Gold organosols using alkane thiols as surfactants were first prepared by Schiffrin and co-workers [109] by phase transferring gold ions and carrying out reduction in the presence of thiols. Several workers have adopted this procedure to obtain thiolized metal nanocrystals [110–113].

Whetten et al. [111] centrifuged the organosol and separated out fractions containing nanocrystals of different mean sizes, to prepare well-ordered two-dimensional arrays of size-selected Au nanocrystals. Harfenist et al. [59] found that Ag nanocrystals prepared by using a cluster beam, were stable in air and formed extended two-dimensional arrays. In Figure 4.10 is shown a TEM image of a two-dimensional array of dodecanethiol covered Ag nanocrystals obtained by Fitzmaurice and coworkers [112]. The Ag nanocrystals were prepared following the method of Schiffrin and co-workers [109]. Well ordered arrays of magic nuclearity nanocrystals, $Pd_{561}$ and $Pd_{1415}$, have been successfully obtained (see Figure 4.11) after replacing their polymer coating by alkanethiols, following the phase transfer method discussed previously [114]. Long chain fatty acids have also been used

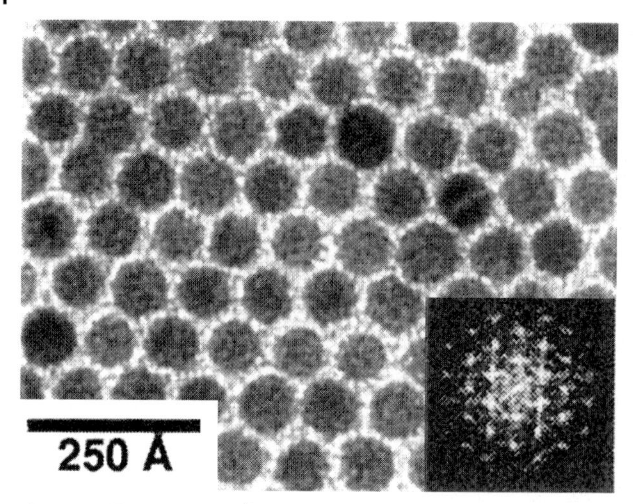


**Fig. 4.10.** Transmission electron micrograph showing hexagonal close-packed Ag nanocrystals (diameter, 7 nm) obtained by evaporating a chloroform dispersion on a carbon substrate. The average interparticle distance is 1.5 nm. Inset shows the 2D power spectrum of the image (reproduced with permission from [112]).

for ligating and assembling metal nanocrystals. Colloidal dispersion of Co nanocrystals capped with fatty acids were found to self-assemble to yield hexagonally ordered arrays similar to those obtained with alkanethiols [115, 116]. Similarly, Ag nanocrystals capped with fatty acids of appropriate lengths yield cubic or hexagonal close-packed structures [117, 118]. Schmid et al. [119] have reported an ordered two-dimensional array of small $Au_{55}$ nanocrystals (diameter $\sim$ 1.4 nm) on a polymer film (see Figure 4.12). At the other end of the size-regime, large Au nanocrystals of 15–90 nm dimensions have also been organized into two-dimensional arrays [120]. Arrays of Au–Ag [62, 63] and Fe–Pt alloy nanocrystals [61] have been obtained. Magic nuclearity $Pd_{561}$ nanocrystals, have been exploited to make Pd–Ni core–shell particles with variable Ni loadings [121]. The nanocrystals so obtained possess a core–shell structure, where a Ni layer covers a Pd seed. The magic nuclearity $Pd_{561}$ nanocrystals act as high quality seeds and promote the formation of monodisperse Pd–Ni core–shell nanocrystals.

Arrays of $Pd_{561}Ni_n$ ($n$ upto 10,000 atoms) have been prepared after thiolizing the core–shell nanocrystals as shown in Figure 4.13 [122]. By a simple extension of this technique, arrays of triple layer nanocrystals of the form $Pd_{561}Ni_{3000}Pd_{1500}$ were also obtained. Methods to organize non-spherical metal nanocrystals into two-dimensional arrays have met with very limited success. Thus, hexagonal Pt as well as elongated silver nanocrystals have been organized into ordered two-dimensional arrays [93, 123]. Interestingly, ordered two-dimensional lattices containing thiolized spherical Au particles of two different sizes have been reported by Kiely et al. (see Figure 4.14) [113], who found that the nanocrystals of different radii follow the

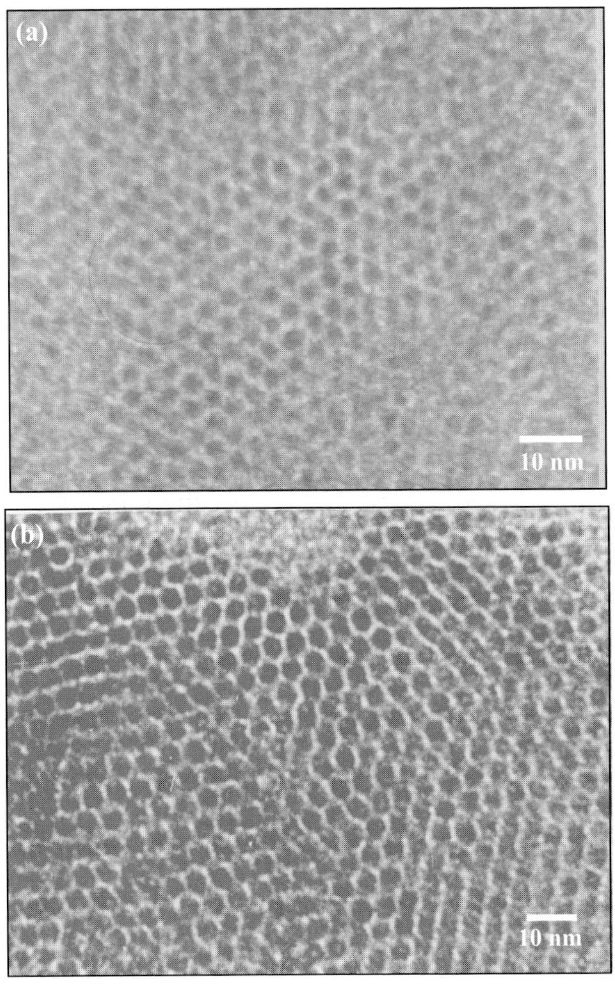


**Fig. 4.11.** TEM micrographs showing hexagonal arrays of thiolized Pd nanocrystals: (a) $Pd_{561}$ octanethiol (b) $Pd_{1415}$, octanethiol. Organized arrays of these nanocrystals extend to lengths over several microns.

radius ratio rules formulated for alloying of different metals. Alloy arrays consisting of Au and Ag nanocrystals of different sizes have been made [124].

#### 4.3.2.2 **Arrays of Semiconductor Nanocrystals**

Bawendi et al. first made monodisperse CdSe nanocrystals by rapid injection of a tri-n-octylphosphine (TOP) solution containing dimethylcadmium and tri-n-octylphosphine selenide into a hot solution of TOP and TOP oxide [47]. By a simple extension of this technique, CdTe and CdS nanocrystals can also be obtained

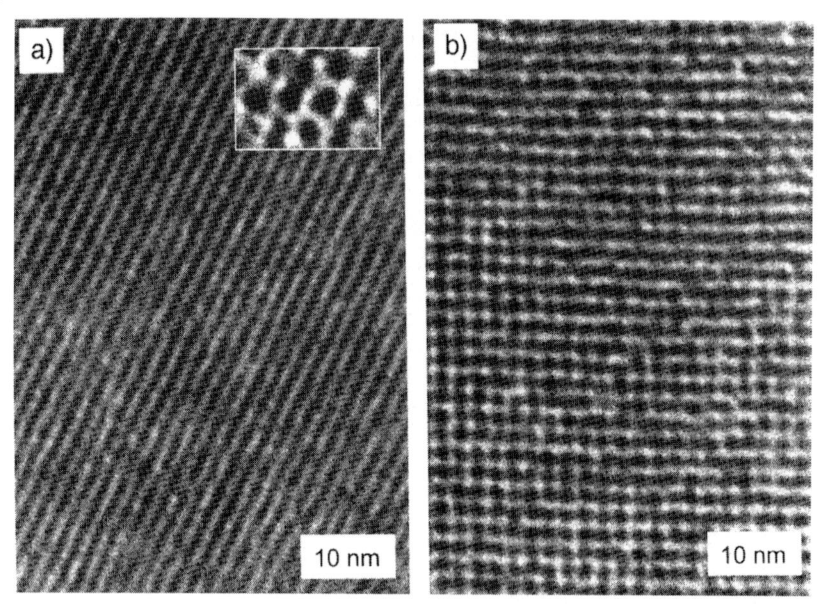


**Fig. 4.12.** TEM images of $Au_{55}$ monolayers showing a hexagonal (a) and a cubic (b) structure. The monolayers were prepared on a polyethyleneimine functionalized carbon grid. The magnified inset in (a) shows single clusters in the hexagonal form (reproduced with permission from [134]).

[47]. These nanocrystals could be size-selected to yield monodisperse CdSe nanocrystals [125]. Upon drying on a flat substrate, these CdSe nanocrystals assemble into superlattices that containing several layers of two-dimensionally ordered nanocrystals. TEM images showing the different facets of the two-dimensional layers in these superlattices are shown in Figure 4.15. Hexanethiol capped PbS nanocrystals, prepared by phase transferring Pb ions into an organic medium using hexanethiol, followed by reaction with $Na_2S$ were also organized into two-dimensional lattices (see Figure 4.16) [126, 127]. Motte et al. obtained a hexagonally ordered two-dimensional array of $Ag_2S$ nanocrystals synthesized by the reverse micelle method [64]. Ordered arrays of InP nanocrystals prepared by the thermolysis method have also been obtained [17].

#### 4.3.2.3 Arrays of Oxide Nanocrystals

The very first report of two-dimensional arrays was of $Fe_3O_4$ nanocrystals [128]. Bentzon et al. observed that the ferrofluid obtained by thermolysis of iron pentacarbonyl upon drying (over a period of several weeks) yielded well ordered two-dimensional arrays of $Fe_3O_4$ nanocrystals. Since then, easier methods have been devised to obtain arrays of $Fe_3O_4$ nanocrystals [129]. Two-dimensional arrays of amine-capped metal oxide nanocrystals such as $Co_3O_4$ have been obtained by start-

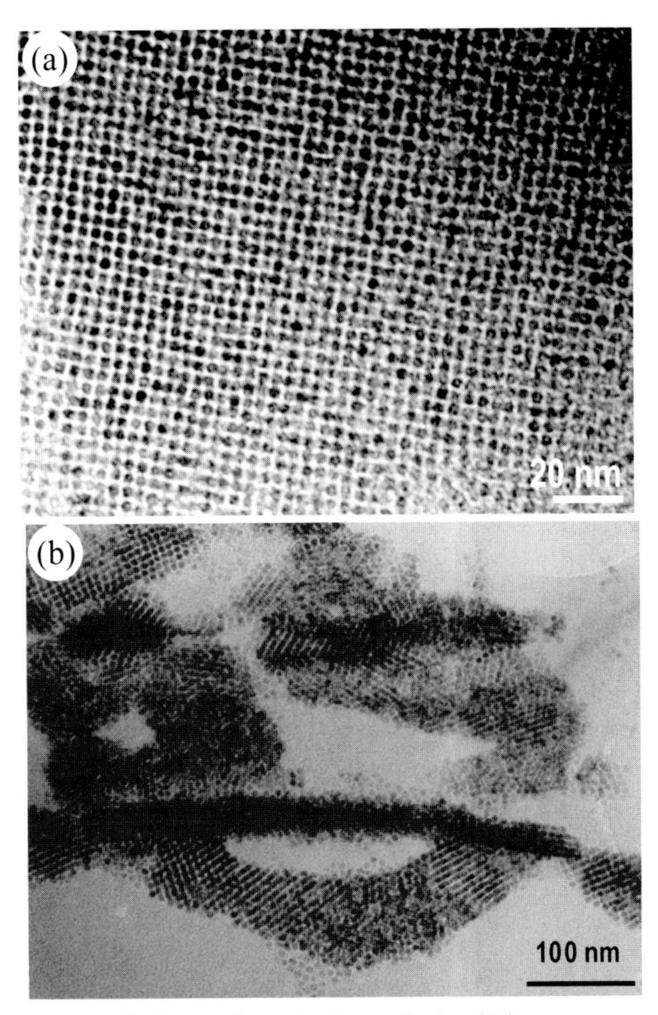


**Fig. 4.13.** TEM image of an ordered array of octanethiol capped (a) $Pd_{561}Ni_{561}$ and (b) $Pd_{561}Ni_{3000}$ nanocrystals. Ni was introduced in the form of its acetate during the reduction process. The nanocrystals were subsequently thiolized.

ing from metal oxide nanocrystals prepared by thermolysis of metal–cupferron complexes [130]. In contrast to metal nanocrystals, attempts to organize non-spherical oxide nanocrystals have met with reasonable success. Thus, tetrahedral CoO nanocrystals have been organized into extended two-dimensional arrays [131]. A rectangular superlattice made of prismatic $BaCrO_4$ has been observed [132]. More complex arrays such as those consisting of a mixture of nanocrystals of different sizes have been obtained using $Fe_3O_4$ and Fe–Pt nanocrystals [133]. The

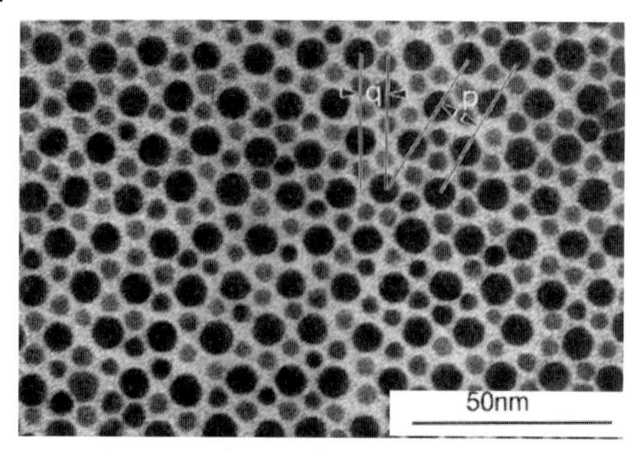


Fig. 4.14. A bimodal hexagonal array of Au nanocrystals. The radius ratio of the nanocrystals is 0.58 (reproduced with permission from [113]).

mixed arrays were obtained by evaporating a binary mixture of the metal oxide organosols.

#### 4.3.2.4 Other Two-Dimensional Arrangements

It is possible to obtain micrometer-sized rings of monodisperse nanocrystals, such as $Ag_2S$, Ag, instead of extended arrays, by varying the solvent evaporation rate and exploiting the resulting fluid instabilities (the Marangoni effect) that occurs during evaporation [134–138]. Neat Au nanocrystals as well as CdSe–ZnS core–shell nanocrystals of various sizes have been organized into two-dimensional lattices using a protein (chaperonin) template [139].

#### 4.3.2.5 Stability and Phase Behaviour of Two-Dimensional Arrays

The nanocrystal organizations mentioned above are mainly entropy-driven. The two lengths involved, the nanocrystal diameter ($d$) and the ligand chain length ($l$) play an important role in deciding the nature of the organization i.e., its orderliness. It has been observed experimentally that for a given diameter of the nanocrystal, the packing changes swiftly as the length of the thiol ligand is increased. The stability diagram in terms of $d$ and $l$ shown in Figure 4.17 illustrates that extended close-packed organizations of nanocrystals are found for $d/l$ values of $\sim 2$. Although entropy driven, the above cannot be treated as hard sphere organizations. Based on a study of the effect of the solvent polarity on the self-assembly of ligated metal nanocrystals, Korgel et al. [101, 112] proposed a soft sphere model taking the interparticle interaction into consideration. Accordingly, a ligated nanocrystal allows for penetration of the ligand shell up to its hard sphere limit. In this model, the total potential energy, $E$, is considered to be a result of two types of forces between the nanocrystals,

$$E = E_{\text{steric}} + E_{\text{vdW}} \tag{1}$$

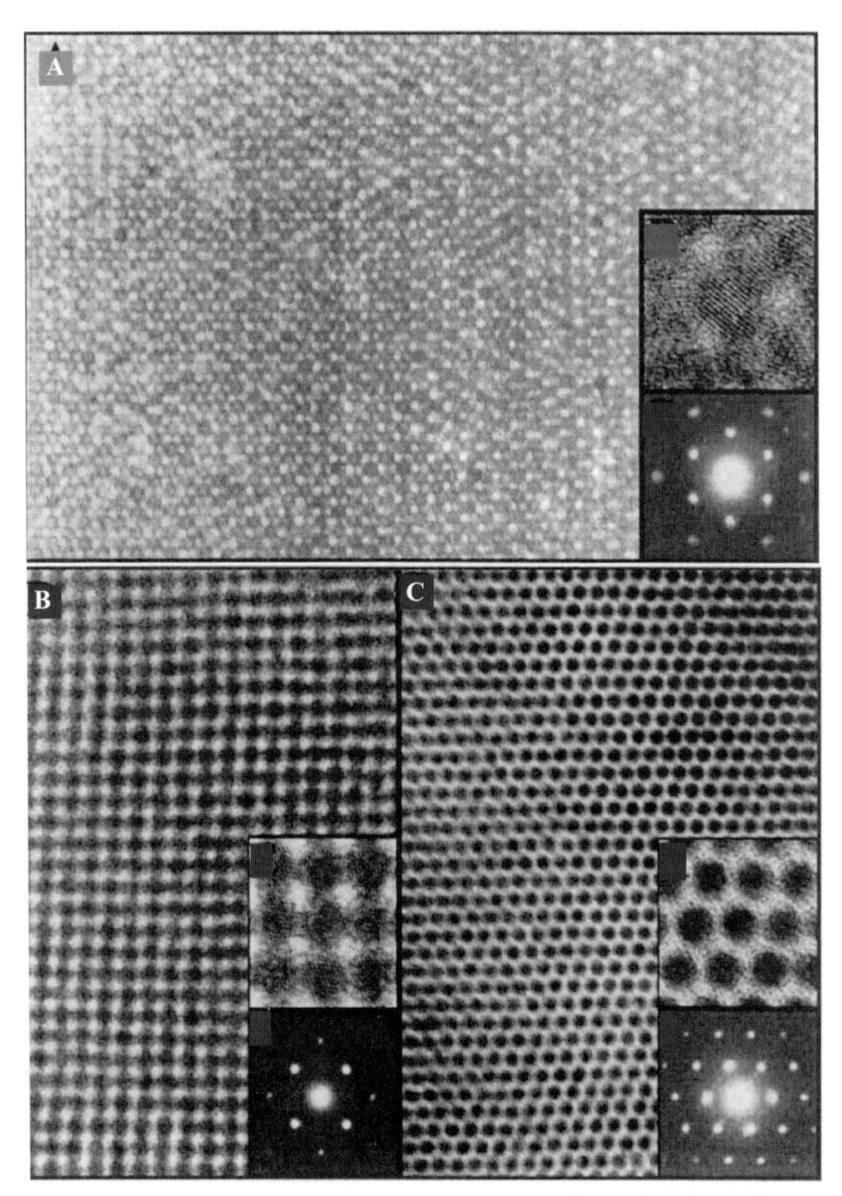


**Fig. 4.15.** Three-dimensional superlattices of 4.8 nm CdSe nanocrystals (a) along ⟨100⟩ orientation (b) along ⟨101⟩ orientation (c) along ⟨111⟩ orientation. High resolution images as well as a selected area electron diffraction pattern is shown alongside in each case (reproduced with permission from [125]).

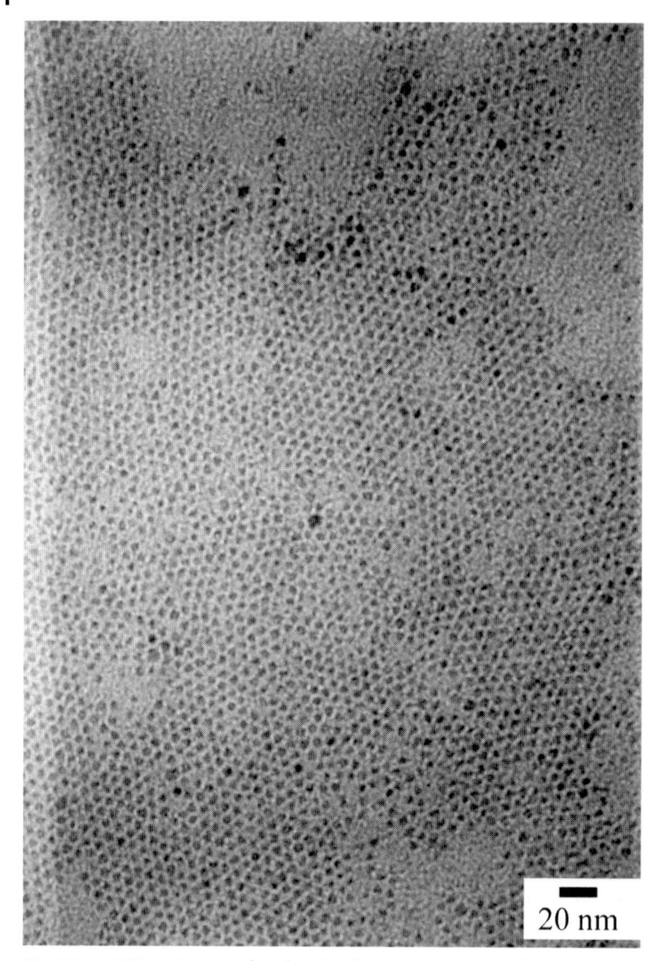


**Fig. 4.16.** TEM micrographs showing hexagonal arrays of hexanethiol capped PbS nanocrystals (unpublished results from our laboratory).

$$E_{\mathrm{vdW}} = \frac{A}{12}\left\{\frac{d^2}{\tau^2 - d^2} + \frac{d^2}{\tau^2} + 2\ln\left[\frac{\tau^2 - d^2}{\tau^2}\right]\right\} \tag{2}$$

$$E_{\mathrm{steric}} = \frac{50dl^2}{(\tau - d)\pi{\sigma_a}^3} kT e^{-\pi(\tau - d)} \tag{3}$$

The van der Waals interaction due to the polarization of the metal cores constitutes the attractive term and the steric interaction between the thiol molecules on the two surfaces forms the repulsive term, where $\tau$ is the interparticle distance. The Hamaker constant, $A$, for Pd nanocrystals, in toluene for instance, has been estimated to be 1.95 eV [140]. The calculated diameter of the area occupied by the thiol molecule (sa) on the particle surface is 4.3 Å [112]. The total energy is attractive

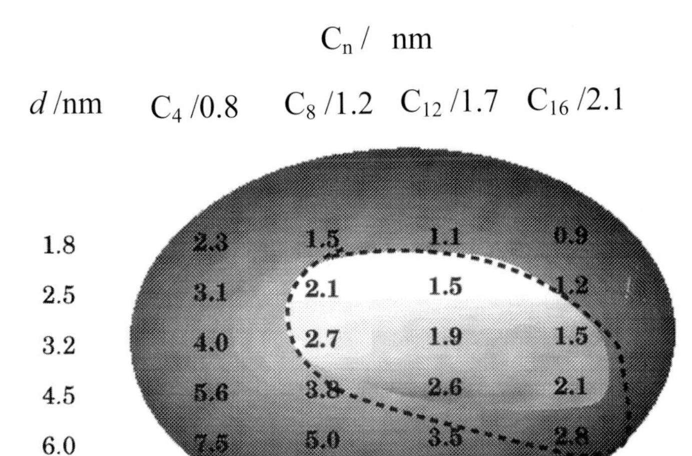


**Fig. 4.17.** The *d–l* phase diagram for Pd nanocrystals thiolized with different alkanethiols. The mean diameter, *d*, was obtained from the TEM measurements on as-prepared sols. The length of the thiol, *l*, is estimated by assuming an all-trans conformation of the alkane chain. The thiol is indicated by the number of carbon atoms, $C_n$. The bright area in the middle encompasses systems that form close-packed organizations of nanocrystals. The surrounding darker area includes disordered or low-order arrangements of nanocrystals. The area enclosed by the dashed line is derived from calculations from the soft sphere model (reproduced with permission from [114]).

over a range of interparticle distances, the magnitude increasing with fall in distance. There could be a range of interparticle distances where the attractive energy from the van der Waals term exceeds the repulsive energy due to the steric factor, giving rise to net stabilization of the two-particle system. This is illustrated in Figure 4.18 in the case of 4.5 nm Pd particles. Stabilization energies of 17 and 2 meV are obtained from the calculation for particles coated with octanethiol and dodecanethiol respectively. When the stabilization energies have moderate values, comparable to the thermal energy of the nanocrystals, ordered organizations can be expected (see the regime shown by dashed line in Figure 4.17). If the $d/l$ and hence the stabilization energy is not favorable, collapsed monolayers of nanocrystals or loosely packed structures are seen. Clearly, the interdigitation of thiol molecules plays a major role in attributing hardness to the ligated nanocrystal, which in turn decides the nature of the two-dimensional organization. A similar treatment should hold good for other metal and semiconductor nanocrystals.

### 4.3.3 Three-Dimensional Superlattices

Multilayer assemblies using monothiols such as those of CdSe (see Figure 4.15) are generally fragile and are not suited for use in functional devices. One of the

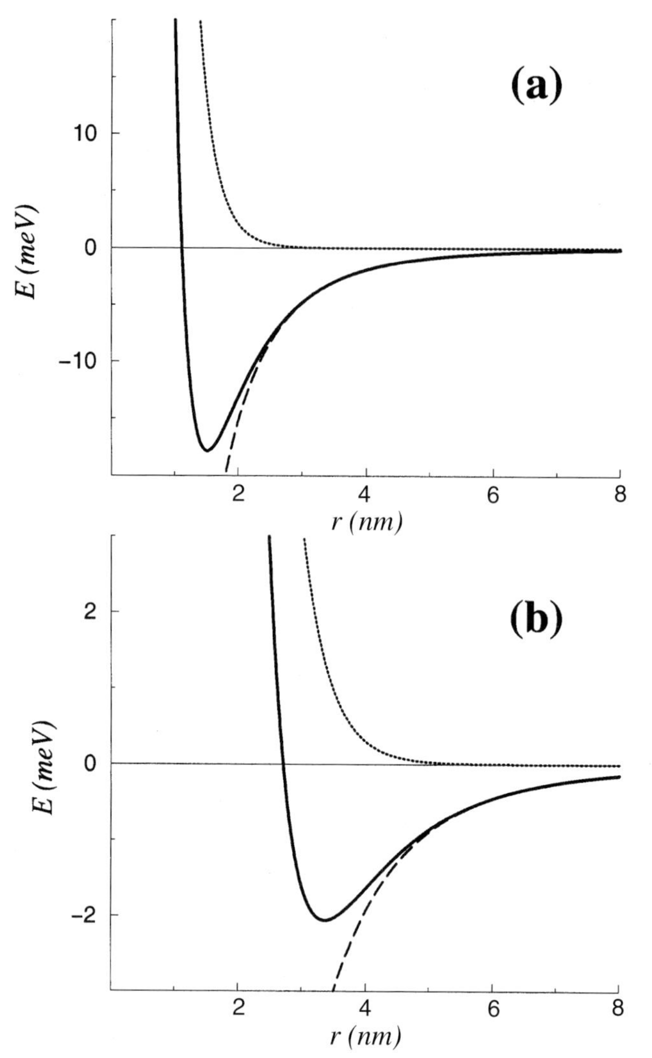


**Fig. 4.18.** Variation of the two components and the total potential energy versus the separation distance between two Pd nanocrystals of 4.5 nm diameter coated with (a) octanethiol (b) dodecanethiol (reproduced with permission from [114]).

means of obtaining robust structures involves multilayer deposition of nanocrystals and has been drawing a great deal of attention over the last few years, since it provides a convenient, low-cost means to prepare ultra-thin films of controlled thickness, suitable for device applications. In a typical experiment, one end of a monolayer forming bifunctional spacer, is tethered to a flat substrate such as gold, aluminum, indium tin oxide or glass, leaving the other end free to anchor nanocrystals [29, 30]. Subsequent layers can be introduced by dipping the substrate sequentially into the respective spacer molecule solution and the nanocrystal dispersion, with intermediate steps involving washing and drying. The formation of the multilayer assembly can be monitored using a variety of spectroscopy and microscopy tools as illustrated in Figure 4.19. Thus, employing Au substrates and dithiols as spacers, multilayer assemblies of several nanocrystals such as Au or CdS have been accomplished [141]. One such example of layer-by-layer deposition of Pt (5 nm) nanocrystals is shown in Figure 4.19. Brust et al. [142] have reported the formation of multilayers of Au nanoparticles using dithiols. These workers have confirmed the layer-by-layer deposition of particle arrays by employing UV–vis spectroscopy and ellipsometry. Multilayers with CdS nanocrystals prepared by the reverse micelle technique and spaced with alkane were made and measurable photocurrents were generated by these assemblies. Three-dimensional superlattices involving nanocrystals of different metals (e.g. Pt, Au) and of metals and semiconductors (e.g. Au, CdS) have also been prepared and characterized [141]. Such assemblies can be made with polyelectrolytes such as poly(diallyldimethylammonium chloride) (PDDA), Polyethyleneimine (PEI) [143, 144], Poly(allylamine hydrochloride) (PAH) and also polymers such as poly-phenylenevinylene (PPV) [145, 146]. Thus, mulilayers such as those of CdTe nanocrystals spaced with PDDA, CdSe spaced with PPV, have been prepared.

### 4.3.4
### Superclusters

It has been proposed that self-similarity in metal nanocrystal organization would manifest in the form of a giant cluster whose shape and size are direct consequences of the nanocrystals themselves [147]. The invariance of the shell effects in metal nanocrystals with scaling is shown schematically in Figure 4.20. Thus, $Pd_{561}$ nanocrystals would be expected to self-aggregate into a giant cluster of the type $(Pd_{561})_{561}$ under suitable conditions. The monodisperse nature of the nanocrystals is thought to be important in assisting the self-aggregation process. Formation of such clusters was observed in the mass spectra of magic nuclearity $Au_{55}$ nanocrystals. Secondary ion mass spectrometry indicated the presence of species with large $m/z$ values and these were attributed to $(Au_{13})_{55}$ giant clusters [148]. The giant clusters so obtained have, however, not been isolated or imaged. One such observation was made in the case of $Pd_{561}$ nanocrystals where the PVP covered nanocrystals aggregated to form giant clusters [149]. The TEM image in Figure 4.21 is revealing. There are regions where the nanocrystals are densely packed in the form of giant aggregates with estimated nanocrystal nuclearities corresponding to various magic numbers. It is possible that the formation of the giant clusters is facilitated by the polymer shell that encases them. Unlike in the case of Pd nanocrystals coated with alkanethiols, which self-assemble to form ordered arrays, the polymer shell effectively magnifies the facets of the metallic core thereby aiding a giant assembly of the nanocrystals.

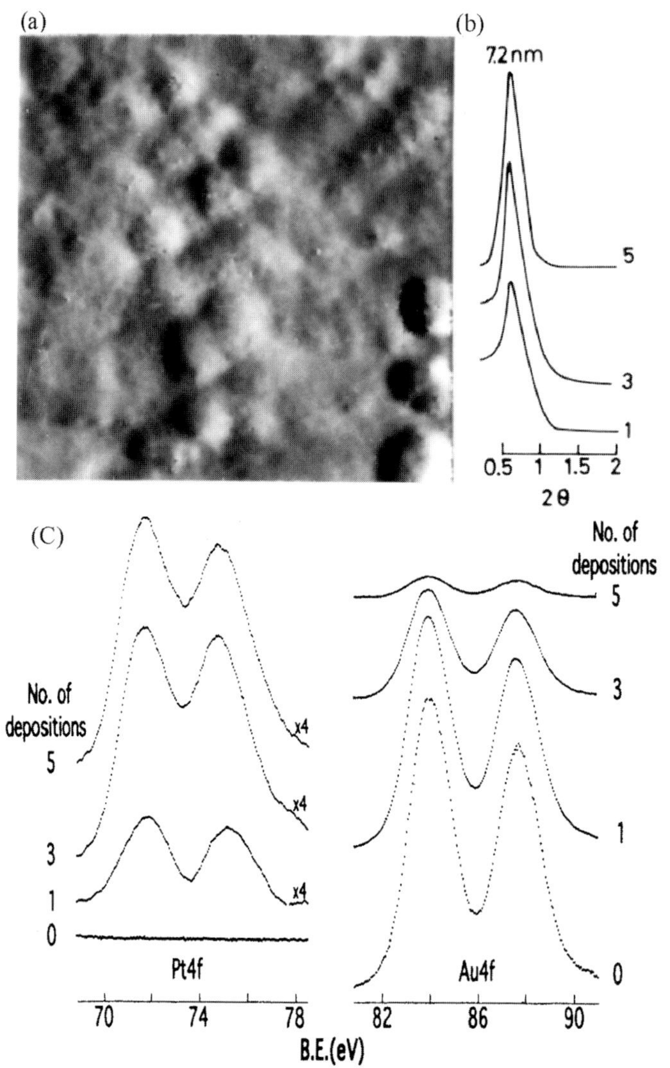


**Fig. 4.19.** Multilayer deposition of Pt (5 nm) nanocrystals on a polycrystalline Au substrate. After each deposition, the structure was characterized by STM, X-ray diffraction and XPS. (a) STM image obtained after the second deposition showing the presence of regular arrays of nanoparticles with an interparticle spacing of 2 nm, extending over 300 nm, corresponding to the size of a typical flat terrace on the substrate. (b) X-ray diffraction pattern of the arrays after the first, third and fifth depositions exhibiting a low-angle reflection with the *d*-spacings reflecting the particle diameter and the interparticle distance. (c) X-ray photoelectron spectra in the Pt(4f) and Au(4f) regions for the 5 nm Pt/Au system. The intensity of the Pt(4f) feature increases with the number of depositions, accompanied by a decreases in the Au(4f) intensity as the substrate gets increasingly shadowed due to the limited escape depth of the photoelectrons.

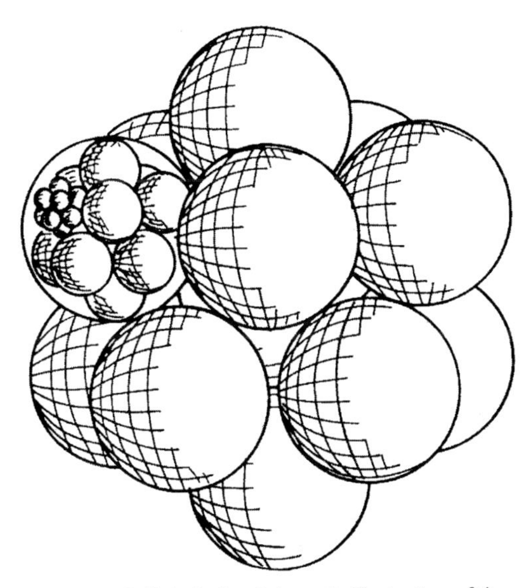

**Fig. 4.20.** Self-similarity: Schematic illustration of the formation of a cluster of metal nanocrystals (super cluster) and a cluster of superclusters. The size effects operating in nanocrystals could be invariant to scaling (reproduced with permission from [147]).

### 4.3.5 Colloidal Crystals

The tendency of monodisperse nanocrystals to arrange into ordered three-dimensional arrays extending to a few microns, has been noticed [150]. Careful tuning of crystallization conditions has yielded crystallites of micrometer dimensions consisting of $Au_{55}$ nanocrystals and Fe–Pt alloy nanocrystals (4.5 nm) as shown in Figure 4.22 [151, 152]. Micron-sized crystals consisting of TOPO capped CdSe nanocrystals have previously been obtained by the groups of Bawendi and Weller [125, 153]. However, it was observed that the nanocrystal arrangement in all the above crystallites was polymorphous. It is believed that such crystallites, consisting of ordered nanocrystals, could prove to be the best candidates to study the collective properties of an ensemble of nanocrystals.

### 4.3.6 Nanocrystal Patterning

Creating patterns of nanocrystals on surfaces has attracted wide attention. Such patterned substrates can act as templates to grow nanowires, etch masks to grow nanopillars and quantum dots [154–156]. Other than the layer-by-layer technique mentioned before, simple techniques such as spin coating have been employed to create a nanocrystalline pattern on surfaces [157].

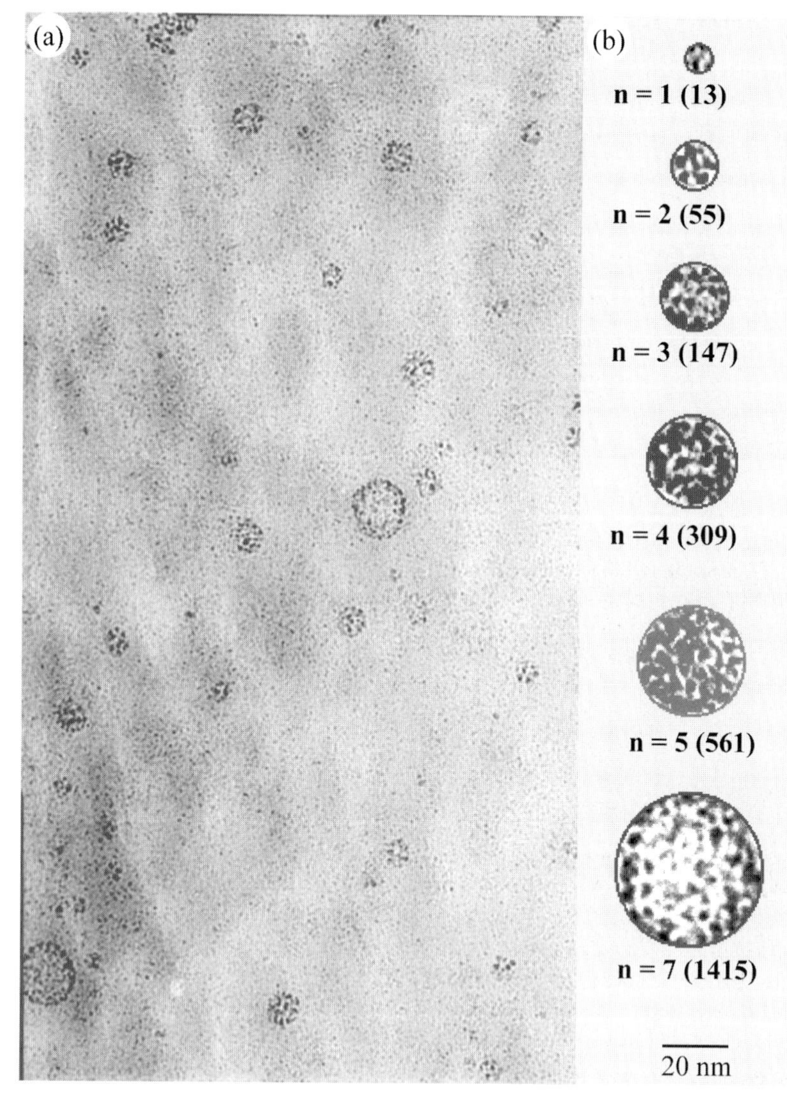


**Fig. 4.21.** TEM micrograph showing the giant clusters comprising $Pd_{561}$ nanocrystals. Sample for TEM was prepared by the slow evaporation of a PVP–$Pd_{561}$ hydrosol. Giant clusters are enclosed in circles whose diameters correspond to magic numbers. The *n* and the numbers in the parenthesis indicate the number of nanocrystals and closed-shells respectively.

In the example shown in Figure 4.23, a direct write lithographic technique, dip pen lithography [158, 159], which relies on a cantilever used for atomic force microscopy (AFM) to write on a substrate to create patterns of Au nanocrystals on mica substrates. Thus, nanocrystals of metals and semiconductors can be patterned into rectangles of varying aspect ratios.

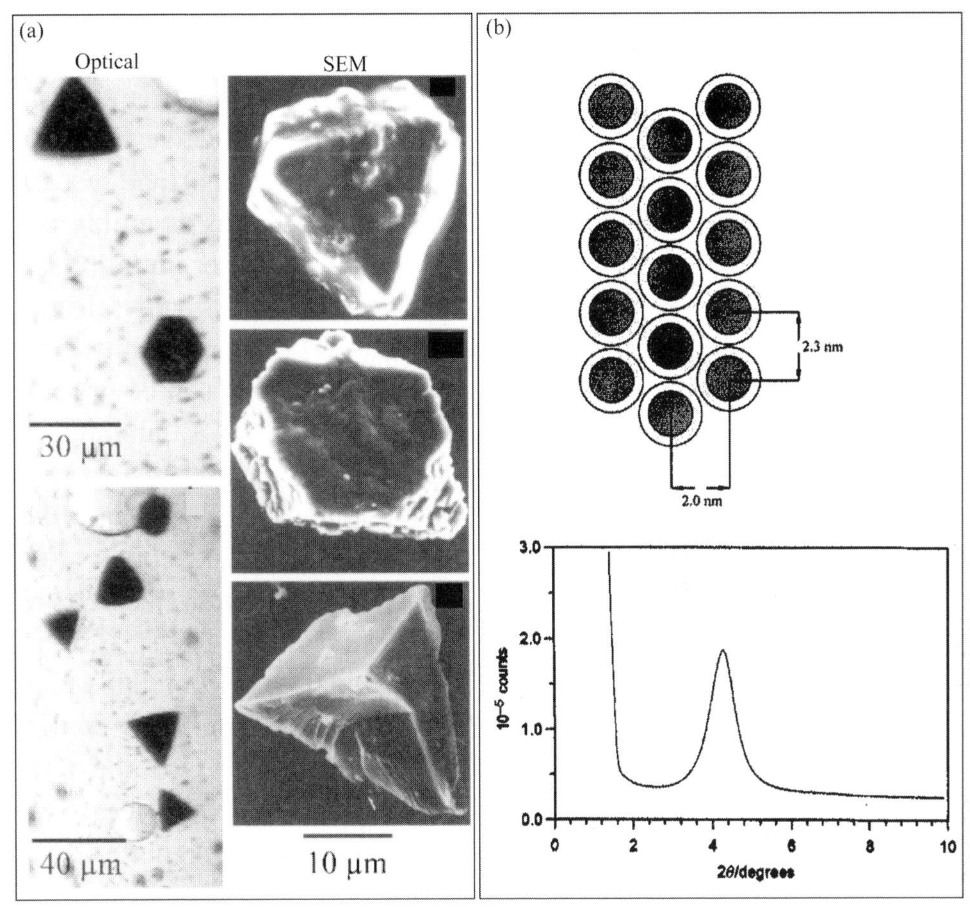


**Fig. 4.22.** Colloidal crystals: (a) from Fe–Pt nanocrystals (b) $Au_{55}$ nanocrystals. A schematic illustration of a $Au_{55}$ microcrystal along with the corresponding small angle X-ray diffraction pattern is shown alongside (reproduced with permission from [151] and [152]).

## 4.4 Emerging Applications

Several applications have been envisaged for nanocrystals, ranging from simple dyes to magnetic-resonance-imaging contrast agents [160], components of electronic circuitry [33, 161] and magnetic media [115], ingredients in catalyst and sensors, and so on. All the above applications seek to exploit the tunability provided by the size-dependent properties of the nanocrystals [4]. Before we dwell upon collective properties in mesostructures, the case of isolated nanocrystals is briefly discussed.

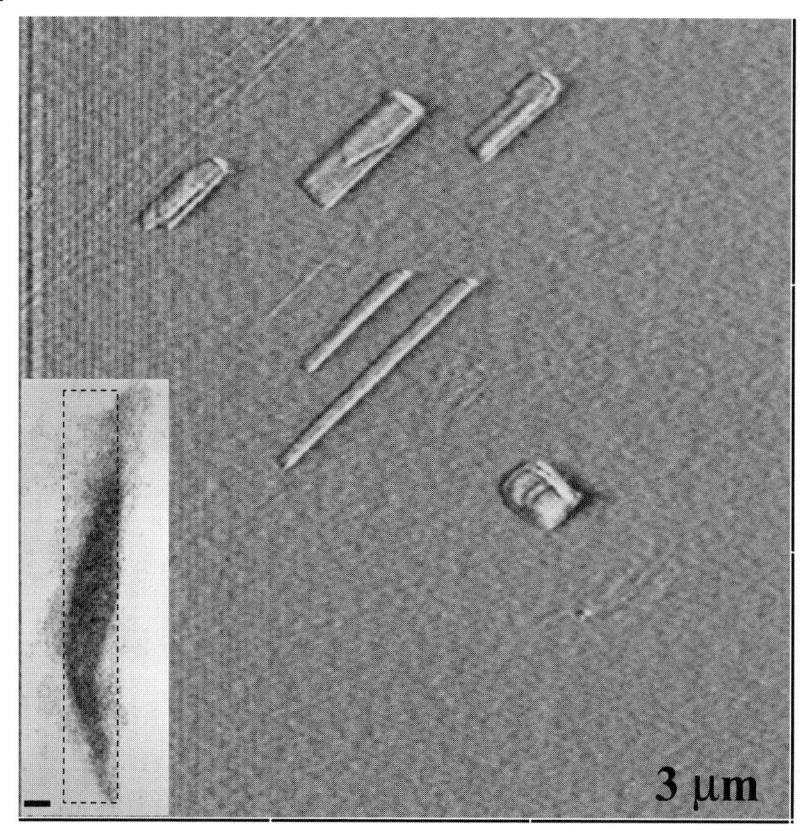


**Fig. 4.23.** Contact AFM scan of a $9\mu^2$ area on mica substrate showing rectangles of various aspect ratios filled with Au nanocrystals. The patterns were obtained by translating an AFM cantilever dipped in a sol across the surface. The inset shows a TEM image of a similar pattern on a holey carbon copper grid, the dotted line bounds the area sought to be filed. The scale bar in the inset corresponds to 50 nm.

### 4.4.1 Isolated Nanocrystals

A nanocrystal undergoes a size-induced metal-insulator transition when the diameter of the particles is decreased to below a few nanometers [5, 162, 163]. Scanning tunneling spectroscopy measurements of nanocrystals of various metals have revealed that the nanocrystals of dimensions ~1 nm exhibit a definitive band gap (up to 70 meV) that decreases gradually as the volume of the nanocrystal increases [162, 164, 165] (see Figure 4.24). Photoelectron spectroscopic measurements on mass selected $Hg_n$ nanoparticles ($n = 3$ to 250) in the gas phase reveal that the characteristic HOMO–LUMO (s–p) energy gap decreases gradually from ~3.5 eV for $n < 3$ to ~0.2 eV for $n < 250$, as shown in Figure 4.25. The band gap closure is predicted at $n \sim 400$. The change in the electronic structure of the nanocrystal manifests itself in many ways, one example being reactivity [166, 167].

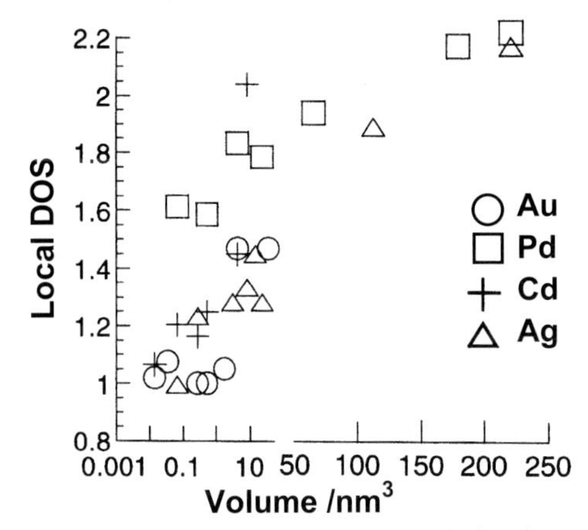


Fig. 4.24. (a) Variation of the nonmetallic band gap with nanocrystal size in metal nancrystals. The bandgaps were obtained based on scanning tunneling spectroscopic measurements (reproduced with permission from [162]).

The fluorescent properties of semiconductor nanocrystals have drawn wide attention because of their potential use as labels in fluorescence bio-assays [25, 26, 84, 168–170]. When compared to dyes currently in use, the emission from fluorescent nanocrystals is brighter and sharper. Further, the emission can be brought about by excitation over a broad range of wavelengths. It is therefore possible to excite nanocrystals of several different sizes simultaneously with a single source and obtain well resolved emission at different colours. In order that the nano-

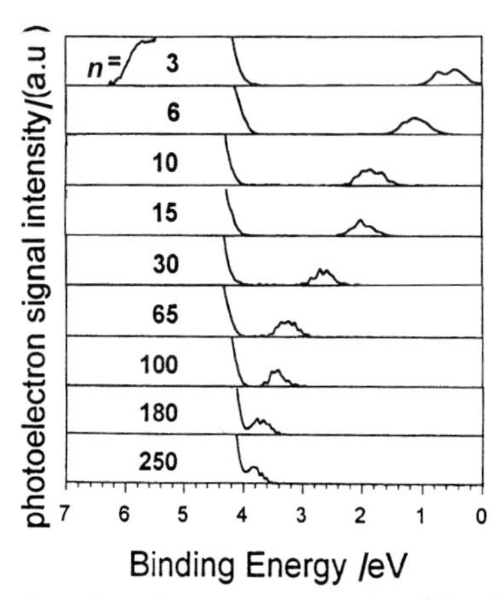


Fig. 4.25. Photoelectron spectra of Hg clusters of varying nuclearity. The 6p feature moves gradually towards the Fermi level, emphasizing that the band gap shrinks with increase in cluster size (reproduced with permission from [165]).

crystals are biocompatible, the nanocrystals need to be water soluble and possess pendent groups at the surface that bind to biomolecules like proteins. These changes can be brought about by tailoring the ligand shell with small DNA fragments or mercapto acids. Several in vivo and in vitro fluorescence biochemical assays have been carried out with nanocrystalline markers [25, 26, 84, 171]. A few studies have sought to exploit the dependence of the plasmon absorption band on the dielectric constant of the surrounding medium in metal nanocrystals to detect binding events taking place at the ligand shell. Thus, Au nanocrystals could colorimetrically determine the successful hybridization of oligonucleotide strands bound to its surface [99, 103]. It has been proposed that colorimetric sensing of heavy metal ions could be obtained by the use of carboxylic acid terminated bifunctional thiols bound to metal nanocrystals [172, 173]. The changes in the electronic absorption spectra of ~5 nm Ag nanocrystals capped with lipoic acid, following the addition of the heavy ions, $Cu^{2+}$ and $Fe^{2+}$ is shown in Figure 4.26. Such a dampening also brings about a change in color. It is apparent that $Cu^{2+}$ ions dampen the plasmon band more effectively than $Fe^{2+}$. It is hoped that mesoscalar organizations could provide useful substrates consisting of ordered nanocrystals that are required to carry out the above experiments in the solid state.

Nanocrystals are thought of as important in single electron devices, operating at room temperature, such as supersensitive electrometers and memory devices [161, 174]. Capped nanocrystals of both metal and semiconductors by virtue of their size, possess capacitance in the range of aF ($10^{-18}$ F). Charging a nanocrystal with an extra electron perturbs it to such an extent that the next electron requires an appreciable change in the charging potential. This is often seen as a 'Coulomb staircase' in the current–voltage tunneling spectra (see Chapter 1). Indeed, the charging energy varies linearly with the inverse of the diameter of the nanocrystal [175, 176]. The sensitivity of a nanocrystal to single electron charging makes it an ideal candidate for use in single electron transistors and memory devices.

It is well known that in the nanometric domain, the coercivity of magnetic nanocrystals tends to zero [115, 177]. Thus, the nanocrystals behave, as superparamagnets with no associated coercivity or retentivity. The blocking temperature which marks the onset of this superparamagnetism increases with the nanocrystal size. This scenario however changes in the case of interacting nanocrystals, where the interparticle interaction, and hence its magnetic properties, can be tuned by varying the interparticle distance. Thus, lattices of interacting magnetic nanocrystals are considered important in the future magnetic storage devices. Further, the magnetic moment per atom is seen to increase as the size of a particle decreases [178].

Several polymer/polyelectrolyte–nanocrystal hybrid devices have been fabricated seeking to exploit the electro and photoluminescent properties of such material [179–188]. Device fabrication in all these cases is by low-cost self-assembly based techniques. These devices utilize thin films of these hybrids obtained either by multilayer deposition or drop/spin casting methods. Thus, 'solar cells' have been made from poly(2-hexylthiophene)–CdSe nanorod multilayers, lasers from drop cast films of CdSe–titania composites and an infrared emitter from multilayers

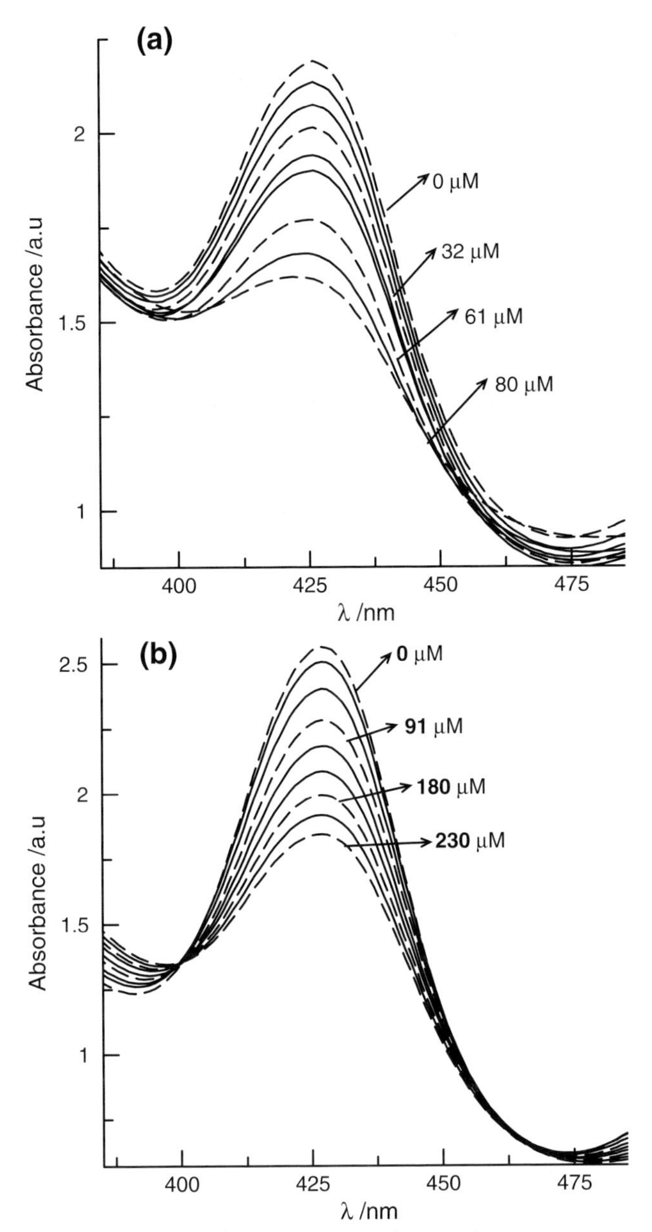


**Fig. 4.26.** Electronic absorption spectra of ~5 nm Ag nanoparticles showing changes accompanying the addition of (a) $Cu^{2+}$ and (b) $Fe^{2+}$ ions. The concentrations of the ions are indicated.

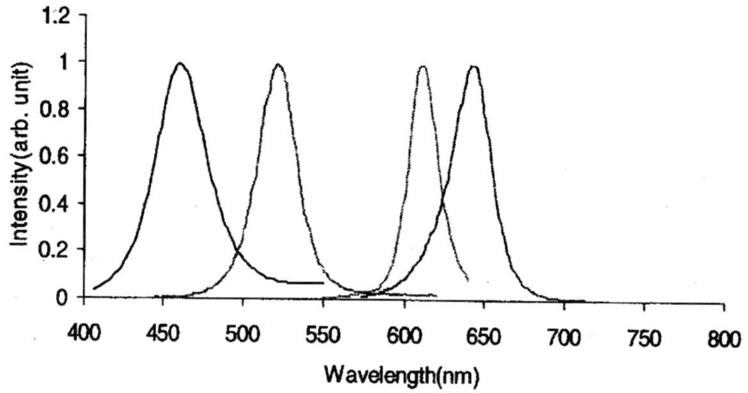


**Fig. 4.27.** Emission spectra of semiconductor nanocrystal polymer composites that were used to make a white light emitting film. (From left) CdS–ZnS core–shell nanocrystals (diameter 3 nm), CdSe–ZnS of diameters 1.3, 2.3, 2.8 nm respectively. The excitation was from a single source (mercury lamp with emission wavelength of 365 nm) (adapted from [187]).

of HgTe and PDDA. White light electroluminescence is seen in multilayers of CdSe–CdS–poly(phenylenevinylene), CdTe–PDDA. White light emission has also been obtained from drop cast films consisting of CdSe–ZnS, CdS–ZnS and poly-laurlymethacrylate (see Figure 4.27). The characteristics of all the above devices can be changed by altering the nanocrystal size.

### 4.4.2
### Collective Properties

The fact that the physical properties of nanocrystal organizations can be different from those of the isolated particles is being realized. Pellets of monodisperse nanocrystals, obtained by the use of a bifunctional ligand that binds to more than one nanocrystal or by applying pressure to dried nanocrystalline matter, have been used for electrical transport measurements [189–192]. Pellets made of small Au and Pd nanocrystals exhibit non-metallic behavior with specific conductivities in the range $10^6\ \Omega^{-1}\ cm^{-1}$ [189–191]. The conductivity however increases dramatically with an increase in the diameter of the nanocrystals. An insulator–metal transition has indeed been reported for pellets made of ~12.5 nm Au and Ag nanocrystals [192]. Electrical transport measurements on layer-by-layer assemblies of nanocrystals on conducting substrates have been carried out by adopting a sandwich configuration [193–195]. Nanocrystalline films with bulk metallic conductivity have been realized with Au nanocrystals of 5 and 11 nm diameter spaced with ionic and covalent spacers [194, 195]. The conductivity of a monolayered two-dimensional arrays of metal nanocrystals has been studied with patterned electrodes [196–201]. Structural disorder and interparticle separation distance are identified as key factors that determine the conductivity of such layers [196–199].

The conductivity of such layers can be enhanced by replacing alkane thiol with an aromatic thiol in situ [200, 201]. That the interaction energy of nanocrystals in such organizations can be continually varied by changing the interparticle distance was exploited by Heath and co-workers [202, 203], who prepared a monolayer of Ag (~3 nm) nanocrystals at the air–water interface in a LB trough and varied the interparticle distance by applying pressure. A host of measurements including reflectivity and non-linear optical spectroscopic techniques were carried out in situ. This study led to the observation of a reversible Mott–Hubbard metal–insulator transition in the nanocrystal ensemble wherein the coulomb gap closes at a critical distance between the particles. Tunnelling spectroscopic measurements on films of 2.6 nm Ag nanocrystals capped with decanethiol reveal a coulomb blockade behavior attributable to isolated nanocrystals [203]. On the other hand, nanocrystals capped with hexane and pentane thiol exhibit characteristics of strong interparticle quantum mechanical exchange (see Figure 4.28). Similar behavior was observed

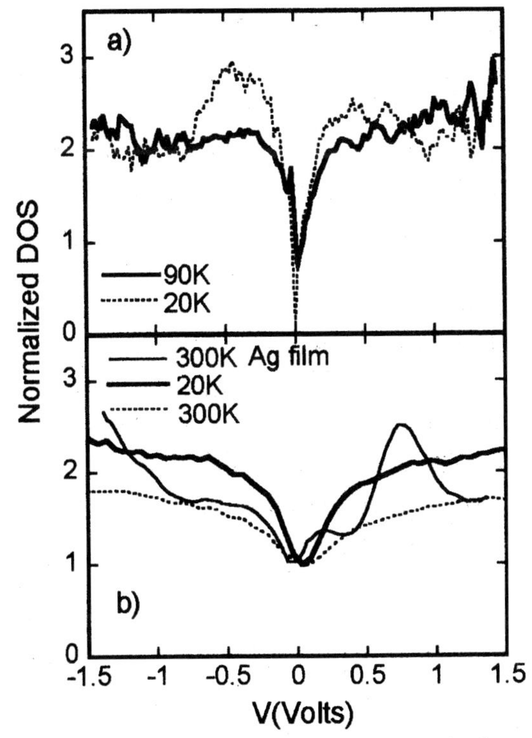


**Fig. 4.28.** Normalized density of states (DOS) measured from arrays of Ag nanocrystals of diameter ~2.6 nm capped with (a) decanethiol and (b) hexanethiol at various temperatures. The temperature dependence of DOS near 0 V for decanethiol-capped particles indicates that the films are non-metallic. In the case of hexanethiol-capped nanocrystals, the DOS around 0 V is temperature independent revealing the metallic nature of the film (reproduced with permission from [203]).

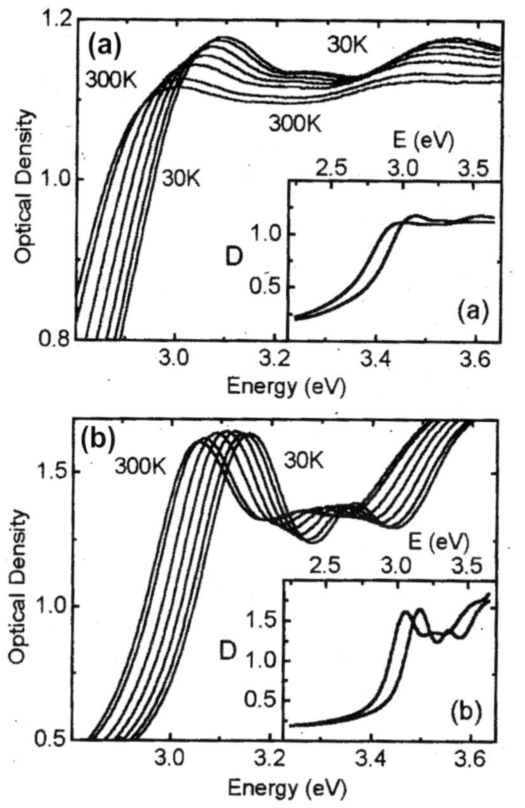


**Fig. 4.29.** Absorption spectra of thin films of close-packed (a) and isolated (b) CdSe nanocrystals at different temperatures (from right to left curves); 30 K, 80 K, 130 K, 180 K, 230 K, 280 K and 300 K. The insets show the full-range spectrum of optical density *D* for lowest and highest temperatures. A red shift and broadening of the peaks is seen in the case of close-packed films (reproduced with permission from [206]).

in the case of self-assembled two-dimensional arrays of Co nanocrystals and Au nanocrystals [204, 205].

The optical properties of a superlattice of semiconductor nanocrystals are different from those of the individual nanocrystals due to interparticle interactions [32]. Typical spectra showing such a change in the case of CdSe nanocrystals are shown in Figure 4.29. When present in close-packed organization, the absorption spectra of the nanocrystals is broadened and red shifted. This change has been attributed to interparticle dipolar interactions [206]. Bawendi and co-workers have studied such changes in an ensemble of CdSe nanocrystals of different diameters and have obtained evidence for long-range resonance transfer of electronic excitation from smaller to bigger nanocrystals due to dipolar interactions [207]. In a noteworthy experiment, Weller and co-workers, prepared drop cast films of giant CdS clusters of the form $Cd_{17}S_4(SCH_2CH_2OH)_{26}$ and $Cd_{32}S_{14}(SCH_2CH(CH_3)OH)_{36}$ with diameters of 1.4 and 1.8 nm respectively. Futher, an integrating sphere was used to collect absorption data, thereby virtually eliminating errors from inhomogenities and size distributions [208]. The experiments due to Weller also support the idea of dipolar interaction leading to the red shift and broadening. The signature of such interactions has also been seen in the case of CdS multilayer deposits [145]. The interparticle interactions, however, could range from weak dipolar interactions to strong exchange interactions based on the interparticle separation. Delocalization of the electronic states of nanocrystals in ensembles due to exchage interactions have been observed in experiments with CdSe nanocrystals. Gaponenko and coworkers have shown that the optical properties of an ensemble of small (~1.6 nm) CdSe nanocrystals are similar to those of bulk CdSe and are due to complete delocalization of the electronic states of individual nanocrystals [209].

Nanocrystals of Co when organized into two-dimensional arrays exhibit a higher superparamagnetic blocking temperature than isolated nanocrystals, i.e., they display a higher resistance to thermal reversal of their spins than when they are isolated [210]. Sun et al. report a lattice of nanocrystals each consisting of a Fe core and a Pt shell prepared by heating Fe–Pt alloy nanocrystals [61]. Following phase segregation, the interaction between the nanocrystals increases, leading to a ferromagnetic film capable of supporting high density magnetization reversal transitions (see Figure 4.30). Exchange spring magnets, nanocomposites that consist

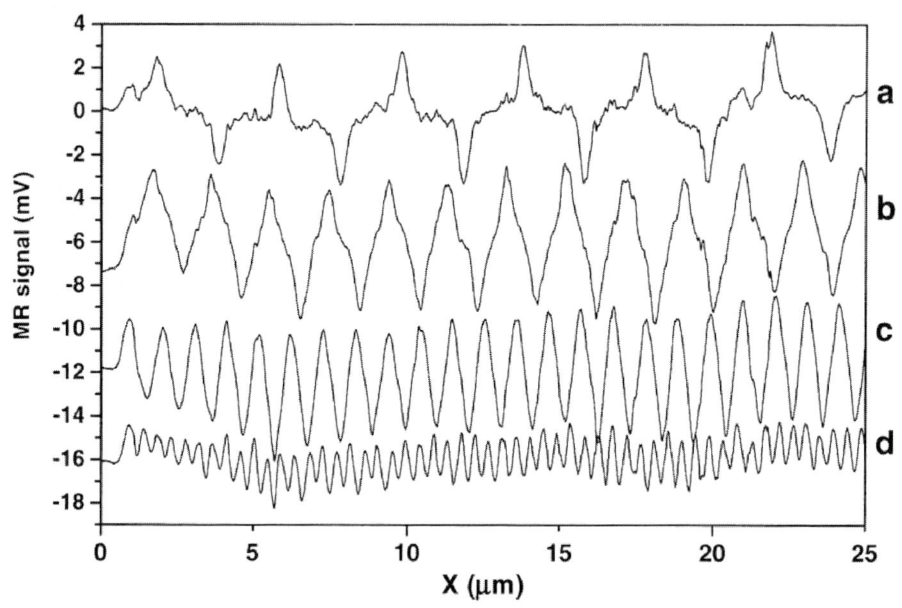


**Fig. 4.30.** Magnetoresistive (MR) read-back signals from written bit transitions in a array of 4 nm diameter $Fe_{48}Pt_{52}$ nanocrystals. The line scans reveal magnetization reversal transitions at linear densities of (a) 500, (b) 1040, (c) 2140, and (d) 5000 flux changes $mm^{-1}$.

of magnetically hard and soft phases interacting via magnetic exchange coupling, have been made by carefully annealing the mixed nanocrystal array consisting of Fe–Pt and $Fe_3O_4$ [133]. The easy magnetic axis of nanocrystals can be aligned by applying a magnetic field during evaporation of the colloid on a substrate to obtain films with high magnetic anisotropy [211–213].

Thus, ferromagnetic films with parallel anisotropy have been made of superparamagnetic $\gamma$-$Fe_2O_3$ nanocrystals [211]. By the use of substrate–nanocrystal interactions, films of the same $\gamma$-$Fe_2O_3$ nanocrystals can be made to exhibit perpendicular anisotropy [212] (see Figure 4.31). The film properties such as coercivity and anisotropy can be tuned by altering the size of the nanocrystals, the film thickness or by suitably doping the nanocrystals with magnetic ions [211–213].

### 4.4.3 Nanocomputing

Ordered arrays of nanocrystals, in principle, could be thought of as arrays of SETs, where the electrostatic interaction between neighboring SETs acts as a wireless communication means. It has been suggested by Korotkov [214] and Lent [215] that simple logical operations can be performed on a circuitry consisting of arrays of SETs in the form of chains or cells with suitable insulating spacers. An electric field applied in one direction polarizes the strings into either the 0 or the 1 state. Lent's scheme, called quantum cellular automata uses, instead, a square cell consisting of five nanocrystals to denote the state of polarization. Preliminary experiments to evaluate the schemes are currently being pursued.

The realization that a self-assembly driven fabrication process is not capable of producing defect-free structures, has fuelled a search for algorithms that can compute even with defective circuitry. Heath and co-workers [216] have developed Teramac, a computer that works despite a high concentration of defects in its bank of microprocessors. A more radical solution called amorphous computing aims to "engineer pre-specified, coherent behavior from the cooperation of large numbers of unreliable parts interconnected in unknown, irregular and time varying ways" [217–219].

## 4.5 Conclusions

Nanocrystals of metal and semiconductors with diameters in the range 1 to 50 nm form a class of materials with unusual properties which are size-dependent. Excellent electrical conductivity that primarily characterizes a metallic state, becomes a rare entity in small nanocrystals ($< 2$ nm) due to quantum confinement of the electronic states. Similarly, magnetic metals lose much of the coercivity with diminishing size. On the other hand, chemical properties such as reactivity may show up better at smaller sizes due to a greater number of surface bonding sites and other electronic effects. Considering the importance of nanocrystals in tech-

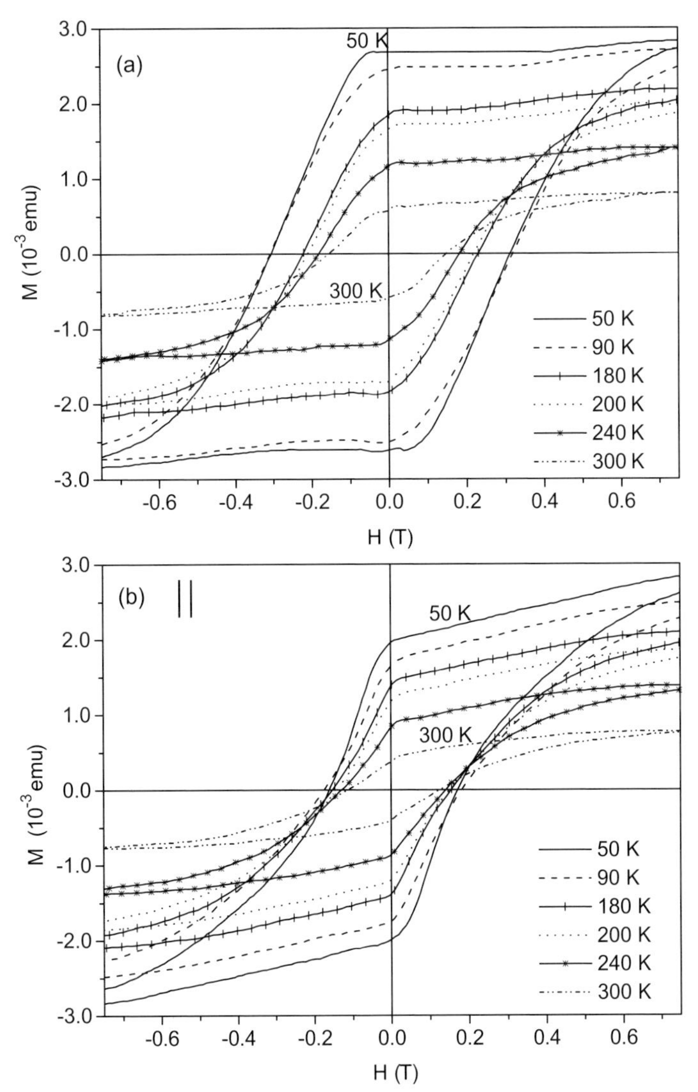


**Fig. 4.31.** Hysteresis loops from a film of $\gamma$-$Fe_2O_3$ deposited on Si(100) substrates at various temperatures with the substrate held (a) perpendicular and (b) parallel to the applied field direction. The increased coercivity along the perpendicular direction indicates perpendicular anisotropy caused by orientation of the easy magnetic axis of $\gamma$-$Fe_2O_3$ nanocrystals perpendicular to the substrate.

nological applications, a large number of synthesis methods have evolved in recent years which include reverse micelle and sonochemical methods as well as laser abalation. Control over size and shape as well as encasing the nanocrystals with ligands of specialized functionalities have become subjects of urgent enquiry. Semiconductor nanocrystals are becoming established as practical alternatives to flourescent dyes. Several exploratory devices, whose characteristics are changeable by varying the constituent nanocrystal diameter, have been made. While isolated nanocrystals are interesting by themselves, their organizations, of especially those which are capable of self-assembling into well-ordered arrays, have attracted greater attention. Nanocrystals anchored to fragments of DNA or similar molecules essentially form one-dimensional organizations. When coated with long-chain alkane thiols, nanocrystals exhibit a tendency to assemble into hexagonal arrays on flat substrates. The stability of such a two-dimensional organization depends on the diameter of the nanocrystals and the length of the ligand. Multilayers of nanocrystal arrays can also be made in a programmed way by selecting suitable spacer molecules. However, patterns of nanocrystals can be obtained using scanning probe techniques. Another mesoscalar aggregation known is the giant clusters of nanocrystals with definite nuclearities. It would be ideal to grow crystals of nanocrystals, but such efforts have met with only a limited success to date, giving micron-sized crystals. Nanocrystal organizations may exhibit properties very different from those of the individual nanocrystal. They are amenable to unprecedented control over the lattice, the size of the nanocrystal and the interparticle separation, being continuously variable over a range. Exploratory experiments for measuring such collective properties are currently underway in several laboratories around the globe.

## References


1 M. Faraday, *Philos. Trans. R. Soc. London*, *147*, 145, **1857**.
2 C. R. Berry, *Phys. Rev.*, *161*, 848, **1967**.
3 L. E. Brus, *J. Chem. Phys.*, *80*, 4403, **1984**.
4 C. N. R. Rao, G. U. Kulkarni, P. J. Thomas et al., *Chem. Eur. J.*, *29*, 27, **2002**.
5 P. P. Edwards, R. L. Johnston, C. N. R. Rao, in *Metal Clusters in Chemistry*, ed. P. Braunstein, G. Oro, P. R. Raithby, Wiley-VCH, Weinheim 1999.
6 S. Link, M. A. El-Sayed, *J. Phys. Chem. B*, *105*, 1, **2001**.
7 S. Link, M. A. El-Sayed, *Int. Rev. Phys. Chem.*, *19*, 409, **2001**.
8 P. Mulvaney, *Langmuir*, *12*, 788, **1996**.
9 G. Mie, *Ann. Phys.*, *25*, 377, **1908**.
10 G. C. Papavassilliou, *Prog. Solid State Chem.*, *12*, 185, **1980**.
11 R. Gans, *Ann. Phys.*, *31*, 881, **1911**.
12 R. Gans, *Ann. Phys.*, *47*, 270, **1915**.
13 A. C. Templeton, J. J. Pietron, R. W. Murray et al., *J. Phys. Chem. B*, *104*, 564, **2000**.
14 S. V. Gaponenko, *Optical Properties of Semiconductor Nanocrystals*, Cambridge University Press, Cambridge 1998.
15 T. Vossmeyer, L. Katsikas, M. Giersig et al., *J. Phys. Chem.*, *98*, 7665, **1994**.
16 C. B. Murray, S. Sun, W. Gaschler et al., *IBM J. Res. Dev.*, *45*, 47, **2001**.
17 O. I. Micic, K. M. Jones, A. Cahill et al., *J. Phys. Chem. B*, *102*, 9791, **1998**.
18 L. E. Brus, *J. Chem. Phys.*, *79*, 5566, **1983**.
19 L. E. Brus, *J. Chem. Phys.*, *80*, 4403, **1984**.
20 P. E. Lippens, M. Lannoo, *Phys. Rev. B*, *39*, 10935, **1989**.
21 M. V. R. Krishna, R. A. Friesner, *J. Chem. Phys.*, *95*, 8309, **1991**.
22 S. Sapra, N. Shanthi, D. D. Sharma, *Phys. Rev. B*, *66*, 205202, **2002**.
23 L. Manna, E. C. Scher, A. P. Alivisatos, *J. Am. Chem. Soc.*, *122*, 12700, **2000**.
24 T. Trindade, P. O'Brien, N. L. Pickett, *Chem. Mater.*, *13*, 3843, **2001**.

25 A. J. Sutherland, *Curr. Opin. Solid State Mater. Sci.*, *6*, 365, **2002**.
26 M. Bruchez, Jr., M. Moronne, P. Gin et al., *Science*, *281*, 2013, **1998**.
27 X. Peng, J. Wickham, A. P. Alivisatos, *J. Am. Chem. Soc.*, *120*, 5343, **1998**.
28 R. F. Service, *Science*, *274*, 1834, **1996**.
29 C. N. R. Rao, G. U. Kulkarni, P. J. Thomas et al., *Chem. Soc. Rev.*, *29*, 27, **2000**.
30 A. N. Shipway, E. Katz, I. Willner, *ChemPhysChem.*, *1*, 18, **2000**.
31 M. P. Pileni, *J. Phys. Chem. B*, *105*, 3358, **2001**.
32 C. B. Murray, C. R. Kagan, M. G. Bawendi, *Annu. Rev. Mater. Sci.*, *30*, 545, **2000**.
33 U. Simon, *Adv. Mater.*, *10*, 1487, **1998**.
34 G. Schmid, L. F. Chi, *Adv. Mater.*, *10*, 515, **1998**.
35 *Clusters and Colloids: From Theory to Applications*, ed. G. Schmid, VCH, Weinheim 1994.
36 J. Turkevich, P. C. Stevenson, J. Hillier, *Discuss. Faraday Soc.*, *11*, 55, **1951**.
37 A. I. Kirkland, D. E. Jefferson, D. G. Duff et al., *Proc. R. Soc. London Ser. A*, *440*, 589, **1993**.
38 *Physics and Chemistry of Metal Cluster Compounds*, ed. L. J. de Jongh, Kluwer, Dordrecht 1994.
39 R. K. Baldwin, K. A. Pettigrew, E. Ratai et al., *Chem. Commun.*, 1822, **2002**.
40 J. P. Wilcoxon, G. A. Samara, P. N. Provencio, *Phys. Rev. B*, *60*, 2704, **1999**.
41 A. Koronowski, M. Giersig, M. Vogel et al., *Adv. Mater.*, *5*, 634, **1993**.
42 A. L. Rogach, A. Kornowski, M. Gao et al., *J. Phys. Chem. B*, *103*, 3065, **1999**.
43 N. Herron, Y. Wang, H. Eckert, *J. Am. Chem. Soc.*, *112*, 1322, **1990**.
44 R. Rossetti, R. Hull, J. M. Gibson et al., *J. Chem. Phys.*, *82*, 552, **1985**.
45 A. C. C. Esteves, T. Trindade, *Curr. Opin. Solid State Mater. Sci.*, *6*, 347, **2002**.
46 M. Green, *Curr. Opin. Solid State Mater. Sci.*, *6*, 355, **2002**.
47 C. B. Murray, D. J. Norris, M. G. Bawendi, *J. Am. Chem. Soc.*, *115*, 8706, **1993**.
48 S. L. Cumberland, K. M. Hanif, A. Javier et al., *Chem. Mater.*, *14*, 1576, **2002**.
49 U. K. Gautam, M. Rajamathi, F. Meldrum et al., *Chem. Commun.*, 629, **2001**.
50 A. A. Guzelian, U. Banin, A. V. Kadavanich et al., *Appl. Phys. Lett.*, *69*, 1432, **1996**.
51 T. S. Ahmadi, L. Wang, A. Henglein et al., *Chem. Mater.*, *8*, 428, **1996**.
52 M. P. Pileni, *J. Phys. Chem.*, *97*, 6961, **1993**.
53 I. Moriguchi, F. Shibata, Y. Teraoka et al., *Chem. Lett.*, 761, **1995**.
54 M. Platt, R. A. W. Dryfe, E. P. L. Roberts, *Chem. Commun.*, 2324, **2002**.
55 E. S. Smotkin, C. Lee, A. J. Bard et al., *Chem. Phys. Lett.*, *152*, 265, **1998**.
56 Unpublished results from the authors laboratory.
57 K. Sattler, J. Mhlback, E. Recknagel, *Phys. Rev. Lett.*, *45*, 821, **1980**.
58 P. Milani, S. Iannotta, *Cluster Beam Synthesis of Nanostructured Materials*, Springer, Berlin 1999.
59 S. A. Harfenist, Z. L. Wang, R. L. Whetten et al., *Adv. Mater.*, *9*, 817, **1997**.
60 H. N. Vasan, C. N. R. Rao, *J. Mater. Chem.*, *5*, 1755, **1995**.
61 S. Sun, C. B. Murray, D. Weller et al., *Science*, *287*, 1989, **2000**.
62 N. Sandhyarani, M. R. Reshmi, R. Unnikrishnan et al., *Chem. Mater.*, *12*, 104, **2000**.
63 S. T. He, S. S. Xie, J. N. Yao et al., *Appl. Phys. Lett.*, *81*, 150, **2002**.
64 *Nanoparticles and Nanostructured Films*, ed. J. H. Fendler, Wiley-VCH, Weinheim 1998.
65 F. V. Mikulec, M. Kuno, M. Bennati et al., *J. Am. Chem. Soc.*, *122*, 2532, **2000**.
66 Y.-H. Chen, C.-S. Yeh, *Chem. Commun.*, 371, **2001**.
67 J.-P. Abid, H. H. Girault, P. F. Brevet, *Chem. Commun.*, 829, **2001**.
68 N. Herron, J. C. Calabrese, W. E. Farneth et al., *Science*, *259*, 1426, **1993**.

69 N. Zhu, D. Fenske, *J. Chem. Soc., Dalton Trans.*, 1067, **1999**; M. Bettenhausen, A. Eichhofer, D. Fenske et al., *Z. Anorg. Allg. Chem.*, *625*, 593, **1999**.

70 T. Vossmeyer, G. Reck, L. Katsikas et al., *Science*, *267*, 1476, **1995**.

71 S. Beherens, M. Bettenhausen, A. C. Deveson et al., *Angew. Chem. Int. Ed. Engl.*, *35*, 221, **1996**.

72 G. Schmid, *Inorg. Synth.*, *7*, 214, **1990**.

73 M. N. Vargaftik, V. P. Zagorodnikov, I. P. Stolyarov et al., *Chem. Commun.*, 937, **1985**.

74 H.-G. Boyen, G. Kastle, F. Weigl et al., *Science*, *297*, 1533, **2002**.

75 T. P. Martin, T. Bergmann, H. Ghlich et al., *J. Phys. Chem.*, *95*, 6421, **1991**.

76 T. Teranishi, M. Miyake, *Chem. Mater.*, *10*, 54, **1998**; T. Teranishi, H. Hori, M. Miyake, *J. Phys. Chem. B*, *101*, 5774, **1997**.

77 S. Link, M. B. Mohamed, M. A. El-Sayed, *J. Phys. Chem. B*, *103*, 3073, **1999**.

78 X. Peng, L. Manna, W. Yang et al., *Nature*, *404*, 59, **2000**.

79 N. Pinna, K. Weiss, H. Sach-Kongehl et al., *Langmuir*, *17*, 7982, **2001**.

80 R. K. Baldwin, K. A. Pettigrew, J. C. Garno et al., *J. Am. Chem. Soc.*, *124*, 1150, **2002**.

81 Y. Liu, J. Zhan, M. Ren et al., *Mater. Res. Bull.*, *36*, 1231, **2001**.

82 D. Ingert, M.-P. Pileni, *Adv. Mater.*, *11*, 136, **2001**.

83 P. Zhang, T. K. Sham, *Appl. Phys. Lett.*, *81*, 736, **2002**.

84 C. M. Niemeyer, *Angew. Chem. Int. Ed. Engl.*, *40*, 4128, **2001**.

85 V. Chechik, R. M. Crooks, *J. Am. Chem. Soc.*, *122*, 1243, **2000**.

86 P. Mulvaney, L. M. Liz-Marzan, M. Giersig et al., *J. Mater. Chem.*, *10*, 1259, **2002**.

87 J. Lin, W. Zhow, A. Kumbhar, J. Wiemann et al., *J. Solid State Chem.*, *159*, 26, **2001**.

88 A. Fojtik, A. Henglein, *Chem. Phys. Lett.*, *221*, 363, **1994**.

89 A. R. Kortan, R. Hull, R. L. Ophila et al., *J. Am. Chem. Soc.*, *112*, 1327, **1990**.

90 S. V. Kershaw, M. Burt, M. Harrison et al., *Appl. Phys. Lett.*, *75*, 1694, **1999**.

91 X. Peng, M. C. Schlamp, A. V. Kadavanich, A. P. Alivisatos, *J. Am. Chem. Soc.*, *119*, 7019, **1997**.

92 H. Harai, H. Aizawa, H. Shiozaki, *Chem. Lett.*, *8*, 1527, **1992**.

93 K. V. Sarathy, G. Raina, R. T. Yadav et al., *J. Phys. Chem. B*, *101*, 9876, **1997**.

94 K. V. Sarathy, G. U. Kulkarni, C. N. R. Rao, *Chem. Commun.*, 537, **1997**.

95 D. I. Gittins, F. Caruso, *Angew. Chem. Int. Ed. Engl.*, *40*, 3001, **2001**.

96 L. O. Brown, J. E. Hutchison, *J. Am. Chem. Soc.*, *121*, 882, **1999**.

97 A. Kumar, A. B. Mandale, M. Sastry, *Langmuir*, *16*, 9229, **2000**.

98 A. Terfort, N. Bowden, G. M. Whitesides, *Nature*, *386*, 162, **1997**.

99 C. A. Mirkin, R. L. Letsinger, R. C. Mucic et al., *Nature*, *382*, 607, **1996**.

100 C. P. Collier, T. Vossmeyer, J. R. Heath, *Annu. Rev. Phys. Chem.*, *49*, 371, **1998**.

101 B. A. Korgel, D. Fitzmaurice, *Phys. Rev. Lett.*, *80*, 3531, **1998**.

102 G. L. Hornayak, M. Krll, R. Pugin et al., *Eur. J. Chem.*, *3*, 195, **1997**.

103 A. P. Alivisatos, K. P. Johnsson, X. Peng et al., *Nature*, *382*, 60, **1996**.

104 A. Kumar, M. Pattarkine, M. Bhadbhade et al., *Adv. Mater.*, *13*, 341, **2001**.

105 M. Mitov, C. Portet, C. Bourgerette et al., *Nature Mater.*, *1*, 229, **2002**.

106 S.-W. Lee, C. Mao, C. E. Flynn et al., *Science*, *296*, 892, **2002**.

107 S. W. Chung, G. Markovich, J. R. Heath, *J. Phys. Chem. B*, *102*, 6685, **1998**.

108 E. Dujardin, C. Peet, G. Stubbs et al., *Nanoletters*, *3*, 413, **2003**.

109 M. Brust, M. Walker, D. Bethell et al., *Chem. Commun.*, 801, **1994**.

110 N. Sandhyarani, T. Pradeep, *Chem. Mater.*, *12*, 1755, **2000**.

111 R. L. Whetten, J. T. Khoury, M. M. Alvarez et al., *Adv. Mater.*, *8*, 428, **1996**.

112 B. A. Korgel, S. Fullam, S. Connolly et al., *J. Phys. Chem. B*, *102*, 8379, **1998**.
113 C. J. Kiely, J. Fink, M. Brust et al., *Nature*, *396*, 444, **1998**.
114 P. J. Thomas, G. U. Kulkarni, C. N. R. Rao, *J. Phys. Chem. B*, *104*, 8138, **2000**.
115 S. Sun, C. B. Murray, *J. Appl. Phys.*, *85*, 4325, **1999**.
116 C. Petit, A. Taleb, M. P. Pileni, *J. Phys. Chem. B*, *103*, 1805, **1999**.
117 M. P. Pileni, *New. J. Chem.*, 693, **1998**.
118 K. Abe, T. Hanada, Y. Yoshida et al., *Thin Solid Films*, *327–329*, 524, **1998**.
119 G. Schmid, M. Bäumle, N. Beyer, *Angew. Chem. Int. Ed. Engl.*, *1*, 39, **2000**.
120 B. Kim, S. L. Tripp, A. Wei, *J. Am. Chem. Soc.*, *123*, 7955, **2001**.
121 T. Teranishi, M. Miyake, *Chem. Mater.*, *11*, 3414, **1999**.
122 P. J. Thomas, G. U. Kulkarni, C. N. R. Rao, *J. Nanosci. Nanotechnol.*, *1*, 267, **2001**.
123 B. A. Korgel, D. Fitzmaurice, *Adv. Mater.*, *10*, 661, **1998**.
124 C. J. Kiely, J. Fink, J. G. Zheng et al., *Adv. Mater.*, *12*, 639, **2000**.
125 C. B. Murray, C. R. Kagan, M. G. Bawendi, *Science*, *270*, 1335, **1995**.
126 S. Chen, L. A. Truax, J. M. Sommers, *Chem. Mater.*, *12*, 3864, **2002**.
127 Unpublished results from our laboratory.
128 M. D. Bentzon, J. van Wonterghem, S. Morup et al., *Philos. Mag. B*, *60*, 169, **1989**.
129 S. Sun, H. Zeng, *J. Am. Chem. Soc.*, *124*, 8204, **2002**.
130 P. J. Thomas, P. Saravanan, G. U. Kulkarni et al., *Pramana*, *58*, 371, **2002**.
131 J. S. Yin, Z. L. Wang, *Phys. Rev. Lett.*, *79*, 2570, **1997**.
132 M. Li, H. Schnablegger, S. Mann, *Science*, *402*, 393, **1999**.
133 H. Zheng, J. Li, J. P. Llu et al., *Nature*, *420*, 395, **2002**.
134 M. Maillard, L. Motte, A. T. Ngo et al., *J. Phys. Chem. B*, *104*, 11871, **2000**.
135 M. Maillard, L. Motte, M. P. Pileni, *Adv. Mater.*, *16*, 200, **2001**.
136 P. C. Ohara, W. M. Gelbart, *Langmuir*, *14*, 3418, **1998**.
137 T. Vossmeyer, S.-W. Chung, W. M. Gelbart et al., *Adv. Mater.*, *10*, 351, **1998**.
138 C. Stowell, B. A. Korgel, *Nanoletters*, *1*, 595, **2001**.
139 R. A. McMillan, C. D. Paavola, J. Howard et al., *Nature Mater.*, *1*, 247, **2002**.
140 D. Bargeman, F. V. V. Vader, *J. Electroanal. Chem.*, *37*, 45, **1972**.
141 K. V. Sarathy, P. J. Thomas, G. U. Kulkarni et al., *J. Phys. Chem. B*, *103*, 399, **1999**.
142 M. Brust, D. Bethell, C. J. Kiely et al., *Langmuir*, *14*, 5425, **1998**.
143 R. Blonder, L. Sheeney, I. Willner, *Chem. Commun.*, 1393, **1998**.
144 Y. Liu, Y. Wany, R. O. Claus, *Chem. Phys. Lett.*, *298*, 315, **1998**.
145 A. Samokhvalov, M. Berfeld, M. Lahav et al., *J. Phys. Chem. B*, *104*, 8632, **2000**.
146 M. Gao, B. Richter, S. Kirstein, *Adv. Mater.*, 9, 802, **1997**.
147 H. G. Fritsche, H. Muller, B. Fehrensen, *Z. Phys. Chem.*, *199*, 87, **1997**.
148 H. Feld, A. Leute, D. Rading et al., *J. Am. Chem. Soc.*, *112*, 8166, **1990**.
149 P. J. Thomas, G. U. Kulkarni, C. N. R. Rao, *J. Phys. Chem. B*, *105*, 2515, **2001**.
150 M. Maillard, L. Motte, A. T. Ngo et al., *J. Phys. Chem. B*, *104*, 11871, **2000**.
151 G. Schmid, R. Pugin, T. Sawitowski et al., *Chem. Commun.*, 1303, **1999**.
152 E. Shevchenko, D. Talapin, A. Kornowski et al., *Adv. Mater.*, *14*, 287, **2002**.
153 D. V. Talapin, E. V. Shevchenko, A. Kornowski et al., *Adv. Mater.*, *13*, 1868, **2001**.
154 H. Ago, T. Komatsu, S. Ohshima et al., *Appl. Phys. Lett.*, *77*, 79, **2000**.
155 Y. Cui, L. J. Lauhon, M. S. Gudisksen et al., *Appl. Phys. Lett.*, *78*, 2214, **2001**.
156 P. A. Lewis, H. Ahamed, T. Sato, *J. Vac. Sci. Technol. B*, *16*, 2938, **1998**.
157 Y.-K. Hong, H. Kim, G. Lee et al., *Appl. Phys. Lett.*, *80*, 844, **2002**.

158 R. D. Piner, J. Zhu, F. Xu et al., *Science*, *283*, 661, **1999**.
159 C. A. Mirkin, S. Hong, L. Demers, *ChemPhysChem*, *2*, 37, **2001**.
160 C. R. Martin, D. T. Mitchell, *Anal. Chem.*, 322A, **1998**.
161 D. L. Feldheim, C. D. Keating, *Chem. Soc. Rev.*, *27*, 1, **1998**.
162 C. P. Vinod, G. U. Kulkarni, C. N. R. Rao, *Chem. Phys. Lett.*, *289*, 329, **1998**.
163 R. Busani, M. Folker, O. Chesnovsky, *Phys. Rev. Lett.*, *81*, 3836, **1998**.
164 M. Miyake, T. Torimoto, T. Sakata et al., *Langmuir*, *13*, 742, **1997**.
165 K. Rademann, O. D. Rademann, M. Schlauf et al., *Phys. Rev. Lett.*, *69*, 3208, **1992**.
166 M. Valden, X. Lai, D. W. Goodman, *Science*, *281*, 1647, **1998**.
167 U. Heiz, A. Sanchez, S. Abbet et al., *J. Am. Chem. Soc.*, *121*, 3214, **1999**.
168 H. Mattoussi, J. M. Mauro, E. R. Goldman et al., *J. Am. Chem. Soc.*, *122*, 12142, **2000**.
169 D. Gerion, F. Pinaud, S. C. Williams et al., *J. Phys. Chem. B*, *105*, 8861, **2001**.
170 W. J. Parak, D. Gerion, D. Zanchet et al., *Chem. Mater.*, *14*, 2113, **2002**.
171 W. C. W. Chan, S. Nie, *Science*, *281*, 2106, **1998**.
172 Y. Kim, R. C. Johnson, J. T. Hupp, *Nanoletters*, *1*, 165, **2001**.
173 S. Berchmans, P. J. Thomas, C. N. R. Rao, *J. Phys. Chem. B*, *106*, 4651, **2002**.
174 Single Charge Tunneling, Coulomb Blockade Phenomena in Nanostructures, ed. H. Graber, M. H. Devoret, Plenum, New York 1992, NATO ASI series B294.
175 P. J. Thomas, G. U. Kulkarni, C. N. R. Rao, *Chem. Phys. Lett.*, *321*, 163, **2000**.
176 J. Jortner, *Z. Phys. D*, *24*, 247, **1992**.
177 C. P. Bean, J. D. Livingston, *J. Appl. Phys.*, *30*, 1208, **1959**.
178 Van de Heer, P. Milani, A. Chatelain, *Z. Phys. D*, *19*, 241, **1991**.
179 J. Nanda, K. S. Narayan, B. A. Kuruvilla et al., *Appl. Phys. Lett.*, *72*, 1335, **1998**.
180 V. L. Colin, M. C. Schlamp, A. P. Alivisatos, *Nature*, *370*, 354, **1994**.
181 M. C. Schlamp, X. Peng, A. P. Alivisatos, *J. Appl. Phys.*, *82*, 5837, **1997**.
182 B. O. Dabbousi, M. G. Bawendi, O. Onitsuka et al., *Appl. Phys. Lett.*, *66*, 1316, **1995**.
183 K. S. Narayan, A. G. Manoj, J. Nanda et al., *Appl. Phys. Lett.*, *74*, 871, **1999**.
184 H.-J. Eisler, V. C. Sundar, M. G. Bawendi et al., *Appl. Phys. Lett.*, *80*, 4614, **2002**.
185 V. C. Sundar, H.-J. Eisler, M. G. Bawendi, *Adv. Mater.*, *14*, 739, **2002**.
186 W. U. Huynh, J. J. Dittmer, A. P. Alivisatos, *Science*, *295*, 2425, **2002**.
187 J. Lee, V. C. Sundar, J. R. Heine et al., *Adv. Mater.*, *12*, 1102, **2000**.
188 M. Gao, C. Lesser, S. Kirstein et al., *J. Appl. Phys.*, *87*, 2297, **2000**.
189 M. Brust, D. Bethell, D. J. Schiffrin et al., *Adv. Mater.*, *7*, 795, **1995**.
190 V. Torma, G. Schmid, U. Simon, *ChemPhysChem*, *1*, 321, **2001**.
191 U. Simon, R. Flesch, H. Wiggers et al., *J. Mater. Chem.*, *8*, 517, **1998**.
192 M. Aslam, I. S. Mulla, K. Vijayamohanan, *Appl. Phys. Lett.*, *79*, 689, **2001**.
193 R. H. Terrill, T. A. Postlewaite, C. Chen et al., *J. Am. Chem. Soc.*, *117*, 1237, **1995**.
194 M. D. Musick, C. D. Keating, M. H. Keefe et al., *Chem. Mater.*, *9*, 1499, **1997**.
195 Y. Liu, Y. Wang, R. O. Clauss, *Chem. Phys. Lett.*, *298*, 315, **1998**.
196 R. Parthasarathy, X.-M. Lin, H. A. Jaeger, *Phys. Rev. Lett.*, *87*, 186807, **2001**.
197 J. Schmelzer, Jr., S. A. Brown, A. Wurl et al., *Phys. Rev. Lett.*, *88*, 226802, **2002**.
198 R. C. Doty, H. Yu, C. K. Shih et al., *J. Phys. Chem. B*, *105*, 8291, **2001**.
199 T. Ogawa, K. Kobayashi, G. Masuda et al., *Thin Solid Films*, *393*, 374, **2001**.
200 R. G. Osifchin, W. J. Mahoney, J. D. Bielefeld et al., *Superlattices Microstruct.*, *18*, 283, **1995**.

201 R. G. Osifchin, W. J. Mahoney, J. D. Bielefeld et al., *Superlattices Microstruct.*, *18*, 275, **1995**.

202 G. Markovich, C. P. Collier, S. E. Hendricks et al., *Acc. Chem. Res.*, *32*, 415, **1999**.

203 G. Medeiros-Ribeiro, D. A. A. Ohlberg, R. S. Williams et al., *Phys. Rev. B*, *59*, 1633, **1999**.

204 A. Taleb, F. Silly, A. O. Gusev, F. Charra et al., *Adv. Mater.*, *12*, 633, **2000**.

205 T. P. Bigioni, L. E. Harrell, W. G. Cullen et al., *Eur. Phys. J.*, *D6*, 355, **1999**.

206 M. V. Artemyev, U. Woggon, H. Jaschinski et al., *J. Phys. Chem. B*, *104*, 11617, **1999**.

207 C. R. Kagan, C. B. Murray, M. G. Bawendi, *Phys. Rev. B*, *54*, 8633, **1996**.

208 H. Döllefeld, H. Weller, A. Eychmüller, *J. Phys. Chem. B*, *106*, 5604, **2002**.

209 M. V. Artemyev, A. I. Bibik, L. I. Gurinovich et al., *Phys. Rev. B*, *60*, 1504, **1999**.

210 V. Russier, C. Petit, J. Legrand et al., *Phys. Rev. B*, *62*, 3910, **2000**.

211 A. T. Ngo, M. P. Pileni, *J. Phys. Chem. B*, *105*, 53, **2001**.

212 Unpublished results from the authors laboratory.

213 A. T. Ngo, M. P. Pileni, *Adv. Mater.*, *12*, 276, **2000**.

214 *Molecular Electronics*, ed. J. Jortner, M. Ratner, Blackwell Scientific, London 1997, IUPAC A 'Chemistry for the 21st Century' monograph.

215 A. O. Orlov, I. Amlani, G. H. Berstein et al., *Science*, *277*, 928, **1997**.

216 J. R. Heath, P. J. Kuekes, G. S. Snider et al., *Science*, *280*, 1717, **1998**.

217 H. Abelson, D. Allen, D. Coore et al., *Technical Report A. I. Memo 1665*, Massachusetts Institute of Technology, Artificial Intelligence Laboratory, Aug. 1999.

218 D. Coore, R. Nagpal, R. Weiss, *Technical Report A. I. Memo 1614*, Massachusetts Institute of Technology, Artificial Intelligence Laboratory, Oct. 1997.

219 H. Abelson, D. Allen, D. Coore et al., *Commun. Assoc. Comp. Mach.*, May 2000.

# 5
# Oxide Nanoparticles

*R. Seshadri*


## Abstract

Recent developments in the preparation and use of nanoparticulate oxide materials, more specifically isolated nanoparticles of simple and compound oxides, are reviewed. While oxide nanoparticles have been known and studied for many decades, it is only in recent years that methods for their preparation have achieved the level of sophistication which permits monodisperse nanoparticles to be produced in quantity. In addition, it is only in recent years that the notion of capping of oxide nanoparticle surfaces by long-chain surfactants, with the corollary that they become soluble, has taken firm hold. The emphasis is on new routes for the preparation of oxide nanoparticles, and how these could be distinct from those used for metals or chalcogenides. Properties of oxide nanoparticles are discussed with special reference to how they are distinct from the bulk material. Present and possible applications are discussed in context.

Methods of preparation that are encouraging from the viewpoint of scale-up are particularly emphasized. Finally, some important issues in the field are outlined.


## 5.1
## Introduction

One of Michael Faraday's Bakerian Lectures, published as an article in Philosophical Transactions of the Royal Society in 1857 [1] reveals that already in the mid-19th century, a great deal of thought had been devoted to what happens to the properties of materials as their dimensions are made small. Among the materials investigated by Faraday are the now famous gold sols, which he prepared using a variety of means. From the colors that Faraday reports, including ruby red, one infers that he was able to prepare sub-100 nm particles. It would perhaps be proper to describe Faraday as among the first scientists to attempt a systematic study of what today would be called a nanomaterial. Nanomaterials have existed for much longer, an example being Maya blue paint [2], the bright and chemically stable

*The Chemistry of Nanomaterials: Synthesis, Properties and Applications, Volume 1.* Edited by C. N. R. Rao, A. Müller, A. K. Cheetham

ISBN: 3-527-30686-2

pigment whose color is believed to arise from the coexistence of metal nanoparticles, oxide superlattices and an organic pigment. Gold colloids find wide used in staining glass, and the nanoparticulate pigment purple of Cassius, obtained by reducing a gold salt using tin (II) chloride is named for the physician Andreas Cassius who died in 1673 [3].

What is then new in the area that it should receive such renewed attention. Firstly, a number of tools, both experimental (exemplified by high-resolution transmission electronic microscopy and scanning tunneling microscopy) and theoretical (exemplified by density functional theory), that permit a much better understanding of these systems have become available. For the first time, researchers in the area can "see" what they are doing. Secondly, in close conjunction with the improved experimental tools are synthetic skills that have been brought to bear upon this area – monodisperse nanoparticles (particularly of metals and chalcogenides), whose surfaces are stabilized by the use of capping agents, have become commonplace in recent years. For chalcogenide nanoparticles, key early studies by the Henglein group in Berlin [4], from Bell laboratories [5], and from the group of Bawendi in MIT [6] have played a role. In the case of metal nanoparticles, the two-phase synthesis of gold nanoparticles by Brust et al. [7] was perhaps the earliest report of well-capped particles. These authors made use of the well-known ability of long chain thiols to form close-packed monolayers on gold surfaces [8] to cap gold nanoparticles. Nanoparticles ensuing from these preparations, unlike the sols of Davy, Faraday and others, are distinct in that they can be precipitated (or collected through evaporation of solvent) and then redissolved. Colloidal sols such as clays in rivers, milk and India ink, once precipitated (or flocculated) cannot be redispersed without great effort [9]. Capped nanoparticles therefore resemble large molecules in their solubility behavior.

Solubility (in the molecular sense, rather than in the sense of forming dispersions and sols) opens up a number of possibilities. The first and perhaps most important, is that it allows size-selective precipitation [10], permitting monodisperse nanoparticles to be prepared. It is only when particles are monodisperse that their size-dependent physical properties can be studied in detail [6]. It is also possible to organize these monodisperse nanoparticles via slow evaporation to yield superlattices [11–13]. Superlattices of nanocrystals can rightly be described as a new class of materials, comprising crystals of crystals as opposed to most crystalline solids which are crystals of atoms [14]. In contrast, naturally occurring opals are crystals of amorphous silica spheres [15].

Oxides, particularly those of the transition metals, display the widest and most fascinating range of properties of any single class of materials. In recent years, oxide materials have been at the heart of many dramatic advances in condensed matter science – layered copper oxides exhibiting high-TC superconductivity readily spring to mind as a good example [16]. Another example is the finding that certain perovskite manganese oxides display dramatic (as large as twelve orders of magnitude) changes in their specific resistivities when subjected to magnetic fields, opening up the new sub-field of colossal magnetoresistance (CMR) [17].

In traditional electronics – LCR circuits, for example – the Ls and the Cs are invariably oxide materials. In the area of integrated semiconductor devices, gate dielectrics [18], dielectrics in dynamic random access memories [19], ferroelectrics in non-volatile memories [20], and decoupling capacitors [21] are all oxide materials. Oxides are also at the heart of many fuel cell [22] and secondary battery materials [23].

Given the importance of oxide materials, it is only natural to ask whether recent understanding regarding the synthesis, properties and applications of capped metal and semiconductor nanoparticles can be extended to oxides.

This review does not concern itself with the vast literature of nanophase materials, which are monolithic materials comprising crystalline grains whose dimensions are in the nanometer range. Such materials are typically prepared using sol–gel and related techniques. This is an older, more mature area that continues to attract a great deal of interest [24].

## 5.2 Magnetite Particles in Nature

While it was known that many organisms respond to the earth's magnetic field, Blakemore's demonstration in 1975 [25] that certain bacteria found in marine marsh muds tended to rapidly navigate along a specific direction – the local geomagnetic north – was seminal. In addition, Blakemore used transmission electron microscopy to demonstrate that these bacteria contained iron-rich crystals about 100 nm in diameter. Several crystals were found to align themselves into chains along their major axis [Figure 5.1]. Frankel, Blakemore and Wolfe [26] then found that an unclassified magnetotactic spirillum (denoted MS-1) when cultured in solutions containing ferric salts, precipitated uniform single crystals that were shown, by Mössbauer spectroscopy, to be mostly spinel $Fe_3O_4$. Since these early studies, the importance of magnetotaxis and magnetoreception in a number of organisms, from bacteria through higher vertebrates, continue to be the subject of investigation [27, 28]. It is interesting that the effects of magnetic fields on organisms – described as "a romping ground for quakes and charlatans, dating at least to the French Mesmerists in the late 18th century" [27] should finally find some legitimacy through the discovery that magnetic nanoparticles play a role in magnetotaxis.

Microbial magnetite (and the ferrimagnetic sulfide biomineral greigite $\gamma$-$Fe_3S_4$, found in magnetotactic bacteria that grow in marine, sulfidic environments) are fine examples of the control that Nature is able to exert over inorganic crystallizations. The magnetic particles are invariably at the optimal single domain size for the specific mineral. What this means is that each magnetic particle contains only a single magnetic domain. If the particle were smaller or larger, it would not be as effective a magnet. In addition, within the magnetosome, the magnetite (or greigite) crystals are aligned in such a manner that the assembly develops a per-

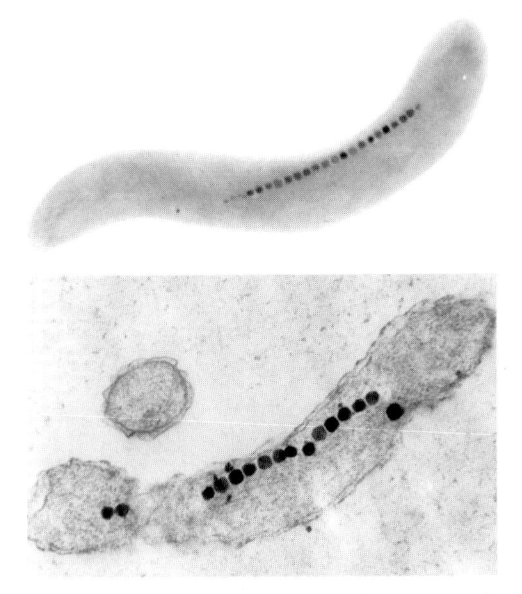

**Fig. 5.1.** Transmission micrographs of the magnetotactic bacteria *Magnetospirillium magnetotacticum* taken in two different directions. The precise control of both the magnetite nanocrystal shape and size, as well as the manner in which all the crystals are aligned, provide inspiration to practitioners of oxide nanoparticle preparations. Images kindly provided by Professor R. B. Frankel, California Polytechnic University, San Luis Obispo.

manent magnetic dipole making an effective compass. The natural remnant magnetization of the assembly approaches the saturation magnetization of magnetite [29]. A major concern in the preparation of oxide nanoparticles is to be able to control size and shape very precisely, and, in many cases, to assemble the nanoparticles in a templated manner. Biogenic magnetite provides great inspiration, suggesting what laboratory preparations can aspire to, and providing hints on how these goals could be achieved. Schüler and Frankel [30] have recently reviewed bacterial magnetosomes with reference to their microbiology, biomineralization and applications in biotechnology. Presently, a UK company is exploiting the principles of magnetite biomineralization by using apoferritin protein shells as nanoscale containers within which monodisperse nanoparticles, including nanoparticles of magnetite, can be grown in quantity [31].

Interestingly, the unusual and very specific control over magnetite morphology that Nature (as embodied in magnetotactic bacteria) exercises, has led a team of investigators to suggest that magnetite in certain calcitic globules from the Martian meteorite ALH84001 are actually magnetofossil records of life on Mars [32]. This claim has been disputed [33], based on evidence from electron holography of magnetite nanocrystals from magnetotactic bacteria, and their comparison with the meteoritic magnetite.

## 5.3
## Routes for the Preparation of Isolated Oxide Nanoparticles

### 5.3.1
### Hydrolysis

Traditionally, a number of oxide materials have been prepared from aqueous solution using hydrolytic methods. Many of these preparation techniques lend themselves to the preparation of nanoparticles. The term hydrolysis is used in a number of contexts, in most of which it involves breaking up of the water molecule [34]. It is useful to imagine that the hydrolysis of a hydrated trivalent metal ion into an oxide takes place in the following manner:

$$2M^{3+}(H_2O)_6 \rightarrow M_2O_3\downarrow + 6H^+ + 9H_2O$$

Such a scheme indicates that the water molecule is decomposed in the process and, in addition, that the formation of an oxide usually requires basic conditions. However, in the preparation of oxides of amphoteric species such as Al(III) or Ga(III), acid hydrolysis could be employed. For example:

$$2[Ga(OH)_4]^- \rightarrow Ga_2O_3\downarrow + 3H_2O + 2(OH)^-$$

An important aspect of such hydrolysis is that not all ions are amenable to it. The alkali, alkaline earth and rare-earth ions, for example, are too stable in solution, and too ionic for hydrolysis to take place. In fact, the hydration reaction is the one that would be favored. For $La^{3+}$ for example, the equilibrium:

$$La_2O_3\downarrow + 6H^+ + 9H_2O \leftrightarrow 2La^{3+}(H_2O)_6$$

can be shifted completely to the left only at temperatures as high as 1000 K.

Despite these limitations, hydrolytic routes have been the mainstay of many preparations of nanoparticulate oxide materials, particularly of the first row transition metals. It is of interest to note that much of the reported literature on hydrolytic routes concerns aqueous systems and it would not be incorrect to say that much needs to be done on the use of solvents other than water, particularly aprotic solvents, where hydrolysis may be more effective. The use of electrochemistry to assist in hydrolysis is a technique that has found much use in the preparation of transition metal oxide films [35], but it's use in the preparation of oxide nanoparticles is not widespread.

Magnetic oxide (typically ferrite spinels or $\gamma$-$Fe_2O_3$) nanoparticles, when dispersed at high concentrations in water or oil, form ferrofluids [36]. Particle sizes in these materials are usually in the 5–15 nm range – small enough that neither magnetic nor gravitational fields should cause their precipitation. The many uses of ferrofluids in magnetic seals, bearings, dampers etc. [36] have resulted in an extensive body of literature on the preparation and properties of dispersions of magnetic oxide nanoparticles. Many of the preparative routes involve hydrolysis. In many cases, the preparation simply involves raising the pH of a solution of metal ions, taken in the correct proportion, by addition of base. Sorenson, Klabunde and coworkers [37], have made detailed magnetic and Mössbauer studies of $MnFe_2O_4$ nanoparticles prepared by precipitation at high pH followed by digestion. Cabuil and coworkers [38] have shown in the case of the preparation of $CoFe_2O_4$ and $\gamma$-$Fe_2O_3$ particles that changing the temperature and the nature of the base allows particles of different sizes to be prepared. In addition, they have suggested methods of dispersing the particles in water, through controlling Coulomb repulsion between the particles. The particles can be dispersed in oils by modifying their surfaces with long chain surfactants. Cabuil et al. [39] have also shown that size-selection of particles is possible though control of a combination of surface charge and ionic strength of the dispersing medium. Morales et al. [40] have made a careful study of $\gamma$-$Fe_2O_3$ nanoparticles, with sizes ranging from 3 to 14 nm, prepared by hydrolytic means (and also by laser pyrolysis of $Fe(CO)_5$ in solution). They show that to account for the magnetic properties of the particles, it is necessary to use a model wherein the moments on the surface of the particle are canted.

The use of polar solvents other than water to perform hydrolysis is an exciting prospect. Ammar et al. [41] have demonstrated that glycols (specifically 1,2-propanediol) can be used as a solvent under reflux to hydrolyze a mixture of Co(II) and Fe(III) salts to obtain equiaxed particles of $CoFe_2O_4$ with an average diameter of 5.5 nm. A combination of Mössbauer spectroscopy, X-ray absorption near-edge structure (XANES) and magnetic measurements suggested that the particles were well-ordered both in terms of being crystalline and having all Co in the divalent state. Rajamathi et al. have been able to capitalize on the unusual properties of glycols – that they are sufficiently polar that they can dissolve metal salts and that they can support hydrolysis and yet they can dissolve long chain surfactants such as amines – in the preparation of n-octylamine-capped 5 nm $\gamma$-$Fe_2O_3$ nanoparticles [42]. The particles [Figure 5.2] can be dissolved in toluene when a little excess amine is added to the solvent. The particles can be precipitated through addition of a polar solvent such as 2-propanol, and then redissolved in toluene/n-octylamine. Diethylene glycol has also been used as a solvent by Carunto et al. [43] to prepare a number of transition metal ferrites capped by long chain carboxylic acids.

Pileni [44] has reviewed the extensive work from her group on the use of reverse micelles as nanoscale reaction chambers within which nanoparticles can be prepared. In a system of water and surfactant dispersed in oil, under suitable conditions, the water forms spherical droplets of radius $R_W$ given by:

$$R_w = \frac{3V_{aq}[H_2O]}{\sigma[S]}$$

where square brackets indicate concentration, S refers to surfactant, $\sigma$ is the area per head group of the surfactant molecule and $V_{aq}$ is the volume of a water molecule. By controlling the water and surfactant concentration, the diameter of the water droplet can therefore be controlled. If a nanoparticle is nucleated within the

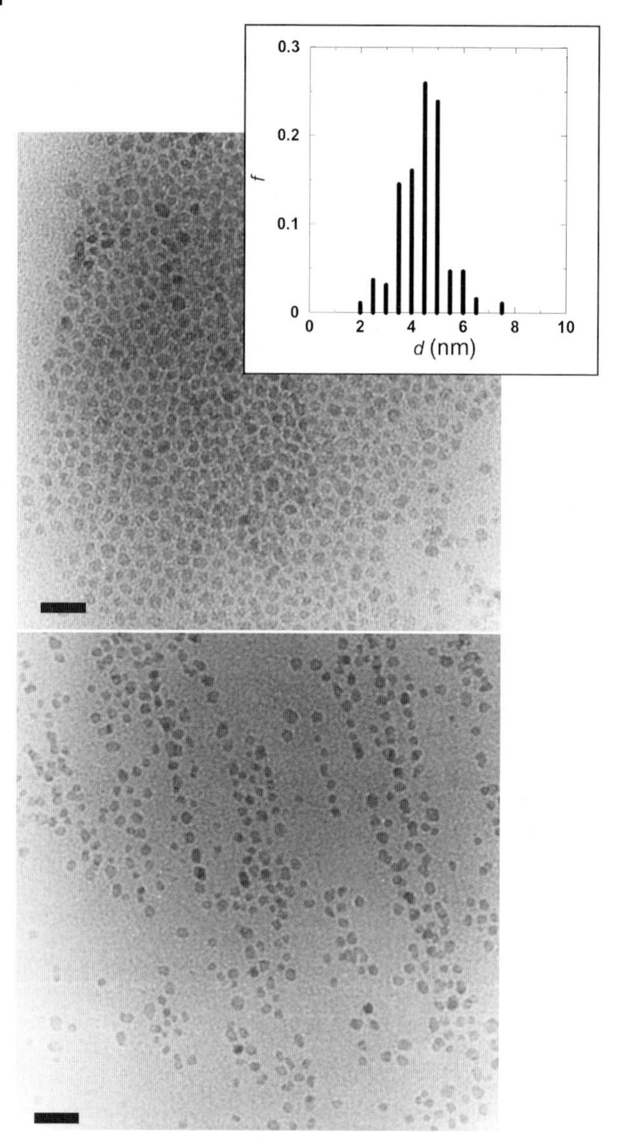


**Fig. 5.2.** Transmission electron micrographs of maghemite $\gamma$-$Fe_2O_3$ nanoparticles prepared by hydrolysis of Fe(III) salts in refluxing propylene glycol in the presence of amine capping agents. The nanoparticles are soluble in toluene solutions of n-octylamine and, as a result, display a tendency for superlattice self-assembly when concentrated (top) and to form aligned arrays in the direction of the earth's magnetic field when dilute (bottom). The bars are 20 nm. Reproduced from [42] with permission from the Royal Society of Chemistry.

water sphere, its growth is limited by the size constraint of the water droplet. Pillai and Shah [45] have utilized this route to prepare high-coercivity $CoFe_2O_4$ nanoparticles by mixing two water-in-oil microemulsions, one containing metal ions and the other containing base. These authors calcine the particles, so the resulting material is nanophase, rather than nanoparticulate. $MnFe_2O_4$ [46] and $CoCrFeO_4$ [47] have been prepared using such reverse-micelle routes by Zhang et al., and characterized by using a combination of magnetization studies and powder neutron diffraction.

Pileni has also pioneered the use of the surfactant-as-reactant approach in the preparation of nanoparticles. For example, in the preparation of $CoFe_2O_4$ nanoparticles with sizes between 2 and 5 nm, instead of preparing inverse-micellar dispersions of the Co and Fe salts, Moumen and Pileni [48] prepared the dodecylsulfonate (DS) analogs $Fe(DS)_2$ and $Co(DS)_2$. These were made to form micellar solutions, to raise the pH, aqueous methylamine solution was added. Stirring for 2 h resulted in a magnetic precipitate. Due to the low yield of Fe(II) to Fe(III) oxidation under these conditions, an excess of $Fe(DS)_2$ is required.

With an increase in concentration of the reactants, there is an increase in particle size. Zinc-doped cobalt ferrite nanoparticles [49] and zinc ferrite [50] have also been prepared using these methods. The 10 nm ferrite nanocrystal ferrofluids prepared using the $Fe(DS)_2$ route form deposits with different morphologies when evaporated on oriented graphite substrates [51]. The morphology can be strongly influenced by applying a magnetic field during the evaporation process. Thus magnetic properties of deposits prepared in the presence and absence of a field are quite different.

Hydrolysis can be assisted by using irradiation of the reacting bath with ultrasound. In these sonochemical preparations, acoustic cavitation results in the production of concentrated spots of extremely high temperatures. These "hot spots" accelerate the rate of metal ion hydrolysis. Gedanken and coworkers [52] have prepared ZnO, CuO, $Co_3O_4$ and $Fe_3O_4$ particles by subjecting solutions of the acetates to ultrasound irradiation using a high-intensity horn. Particle morphologies could be altered by using mixtures of water and dimethyl formamide instead of pure water as the solvent. Magnetite nanorods have been prepared by these authors [53] by ultrasonically irradiation of Fe(II) acetate in water in the presence of $\beta$-cyclodextrin. The authors suggest that cyclodextrin molecules are acting as size-stabilizing agents.

### 5.3.2 Oxidation

Oxides, unlike the other chalcogenides (sulfides, selenides, tellurides), are not associated with an oxide ion source. In other words, while the formation of a sulfide such as ZnS can be written:

$$Zn^{2+}\,(aq.) + S^{2-}\,(aq.) \rightarrow ZnS\downarrow$$

where the source of the sulfide ion in solution could be $Na_2S$ or thiourea or the like, it is not possible to frame a similar reaction scheme for oxides. On the other hand, it is possible to directly oxidize a metal source in solution. A popular route to this is to use zero-valent carbonyls, for example $Fe(CO)_5$. Decomposing these carbonyls in solvents results in very finely divided metal particles that are easily susceptible to oxidation. In fact, exposing them to an atmosphere of air is usually sufficient to convert the particles to oxides.

Bentzon et al. [11] have prepared iron oxide nanoparticles by decomposing $Fe(CO)_5$ in decalin in the presence of oleic acid as the stabilizing ligand. Aging the ferrofluid so formed in air for several weeks results in a mixture of hematite and spinel phases. The fascinating aspect of this work is that it is among the first reports of nanocrystal superlattice formations, and is still only one of few reports on highly coherent oxide nanocrystal superlattices. More recently, Hyeon et al. have prepared monodisperse $\gamma$-$Fe_2O_3$ nanoparticles in the size range 4–16 nm by decomposing $Fe(CO)_5$ complexes in octyl ether at 300 °C in the presence of oleic acid. Oxidation of bcc-Fe to $\gamma$-$Fe_2O_3$ was achieved by the addition of the organic oxidant $(CH_2)_2NO$. The resulting capped oxide nanoparticles can easily be redispersed in organic solvents such as hexane or toluene. Particle sizes are altered by varying the ratio of Fe to the capping agent (oleic or lauric acids). The remarkable feature of this preparation is that the as-prepared particles are sufficiently monodisperse that they form nanocrystal superlattices without the need for a size-selection process.

Wagner and coworkers have prepared yttrium oxide [55] and europium-doped yttrium oxide nanoparticles [56] by first reducing rare-earth salts in solution using alkalides (strong reducing agents that are complexes of crown ethers with alkali metals, and where the anion is a complexed electron) and then oxidizing the rare-earth metal nanoparticle so formed, in aerated water. The white powders were then annealed in air. Some more examples of oxide nanoparticles prepared through direct oxidation will be discussed in the solvothermal methods section.

### 5.3.3 Thermolysis

If one starts with a precursor complex wherein the ligands bind to metal ions through oxygen, it could be possible to envisage a decomposition reaction that would leave behind the metal oxide.

For a trivalent ion, such reactions could be generalized as:

$$[R{-}O]_3{-}M^{3+} \rightarrow M_2O_3 + \text{leaving groups}$$

Suitable design of the R group (stable leaving groups) would ensure that the reaction proceeds in a facile manner. It should then be possible to carry out such reactions in a suitable high temperature solvent under solvothermal conditions, possibly in the presence of a suitable capping agent.

Rockenberger, Scher and Alivisatos [57] have described the use of cupferron complexes as precursors to prepare transition metal oxide nanoparticles. Cupferron (*N*-phenyl, *N*-nitroso hydroxylamine) forms bidentate, univalent complexes with a number of different transition metals ions. These complexes easily decompose to give the oxide. The authors demonstrated the preparation of $\gamma$-$Fe_2O_3$, $Cu_2O$ and $Mn_3O_4$ nanoparticles prepared by injecting octylamine solutions of the corresponding cupferron precursors into refluxing trioctylamine. Size is controlled by controlling the temperature of the reaction. The particles so prepared form stable solutions in solvents such as toluene, from which they can be reprecipitated by the addition of methanol. This work is quite seminal in its generality, and particularly in the manner it which it suggests the search for suitable precursors for the preparation of oxide nanoparticles. Most importantly, it suggests thermolysis as an alternative to hydrolysis which, as pointed out earlier, is simply not viable for a number of metal oxides.

An important contribution to the surface chemistry of metal oxide nanoparticles has been made by Rotello and coworkers [58] who have prepared $\gamma$-$Fe_2O_3$ nanoparticles by the cupferron decomposition method, and compared the relative efficacies of different long chain surfactants as capping agents. The most stabilizing capping agent (as monitored by ease of redissolution and stability in solution) was obtained by using a two-tailed surfactant (with 12-carbon tails) with the polar part comprising a 1,3-diol.

The thermolysis of Fe(III) hydroxide caprylates in boiling tetralin under argon flow gives $\gamma$-$Fe_2O_4$ nanoparticles [59]. The surfaces of the nanoparticles could be modified by exchanging the capping caprylate groups with betaine, among other species [60]. Betaine-capped $\gamma$-$Fe_2O_3$ nanoparticles are reported to have high solubility in water. Using a combination of hydrolysis and oxidation, O'Brien, Brus and Murray [61] have shown that the treatment of a complex alkoxide containing $Ba^{2+}$ and $Ti^{4+}$, $BaTi(O_2CC_7H_{15})[OCH(CH_3)_2]_5$, an agent for the MOCVD growth of $BaTiO_3$, can be decomposed in diphenyl ether at 140 °C, in the presence of oleic acid as a capping agent, to give nanocrystalline, cubic $BaTiO_3$ [Figure 5.3]. After cooling to 100 °C, 30% $H_2O_2$ is added, and crystallization is induced over a 48 h period. Changing the ratio of capping agent to water, and the amount of peroxide added, permits the size of the particles to be varied. This is perhaps the only report of soluble perovskite oxide nanoparticles and, once again, is a route of great generality and interest. The authors also report using such routes to prepare $PbTiO_3$ and $TiO_2$ nanoparticles.

### 5.3.4 Metathesis

In a metathetic reaction, two compounds AB and CD, exchange species to give two new compounds AC and BD. Such routes have been explored in the preparation of nanoparticles. Arnal et al. [62] have reported two non-hydrolytic routes to sol–gel metal oxides, particularly of titanium. The first route involves the reaction of a metal halide with a metal alkoxide:

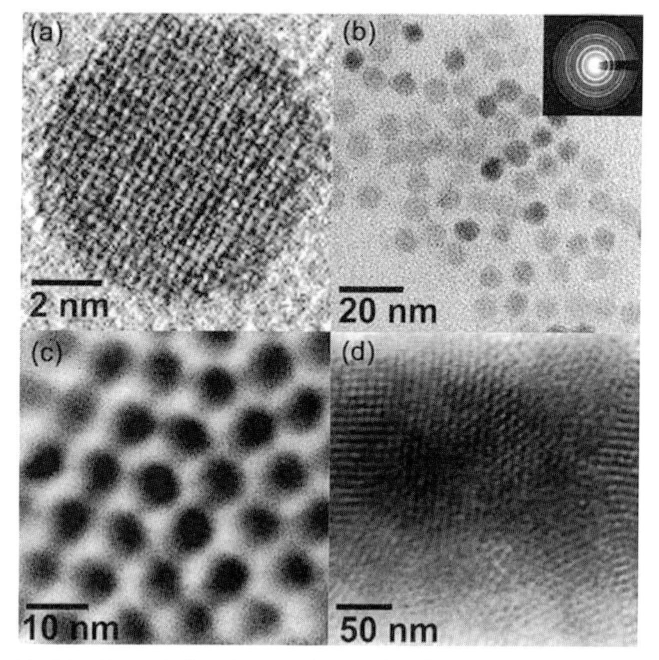


**Fig. 5.3.** Transmission electron micrographs of 8 nm $BaTiO_3$ nanoparticles prepared by the hydrolysis/oxidation of a Ba–Ti alkoxide precursor. Reproduced from [61] with permission from the American Chemical Society.

$$MCl_n + M(OR)_n \rightarrow 2MO_{n/2} + nRCl$$

and the second route involves the reaction of a metal halide with an ether:

$$MCl_n + (n/2)ROR \rightarrow MO_{n/2} + nRCl$$

In both reactions, some of the driving force comes from the removal of the volatile product RCl. This first of the two methods has been employed by Colvin and co-workers [63] to prepare capped $TiO_2$ nanoparticles in refluxing heptadecane with trioctyl phosphine oxide as the capping agent:

$$MCl_n + M(OR)_n \rightarrow 2MO_{n/2} + nRCl$$

The particles formed clear solutions in non-polar solvents, from which they could be precipitated through the addition of a polar solvent such as acetone. When trioctyl phosphine oxide was not taken with the starting materials, insoluble precipitates were obtained.

### 5.3.5 Solvothermal Methods

To prepare crystalline oxide materials, high temperatures, associated with the refluxing of reactants in high-boiling solvents, is often called for. There are many problems associated with such solvents, including possible toxicity, expense and their inability to dissolve simple salts. A simple method to circumvent this problem is to use solvothermal methods, employing solvents well above their boiling points in enclosed vessels that support high autogenous pressures. Solvothermal (called hydrothermal when the solvent is water) methods are widespread in their use, particularly in the preparation of crystalline solids, including silicate materials such as zeolites [64].

Recently, Rajamathi and Seshadri [65] have reviewed the uses of solvothermal methods for the preparation of oxide and chalcogenide nanoparticles. For oxide nanoparticles, these methods can involve hydrolysis, oxidation and thermolysis, all performed under hydrothermal or solvothermal conditions. Some of the more striking examples are provided here.

#### 5.3.5.1 Oxidation

An unusual reaction reported by Inoue et al. [66] is the direct oxidation of Ce metal in 2-methoxyethanol at temperatures between 200 °C and 250 °C. Most of the product obtained was bulk $CeO_2$ as a yellow solid, but in addition, they obtained a brown solution of 2 nm $CeO_2$ nanoparticles. The $CeO_2$ nanoparticles could be salted out by the addition of NaCl, and redispersed into solution at will. The solutions obeyed the Beer–Lambert law for the concentration dependence of the optical extinction, suggesting that the nanoparticulate dispersion was a genuine solution.

Starting with Zn powder and $GaCl_3$ in water, Qian and coworkers [67] have used the fact that the oxidation of Zn to $Zn^{2+}$ in water is associated with the production of hydroxy ions according to:

$$Zn + 2H_2O \rightarrow Zn^{2+} + 2OH^- + H_2\uparrow$$

The $OH^-$ ions help in the hydrolysis of $Ga^{3+}$, to eventually give, in the presence of $Zn^{2+}$, the spinel $ZnGa_2O_4$. The authors used a 10–20% over-stoichiometry of Zn and the reaction was carried out at 150 °C for 10 h. 10 nm $ZnGa_2O_4$ nanoparticles were obtained. Excess $ZnCl_2$ formed in the reaction could be washed away with water.

#### 5.3.5.2 Hydrolysis

Chemseddine and Moritz [68] have used the polycondensation of titanium alkoxide $Ti(OR)_4$ in the presence of tetramethylammonium hydroxide, to obtain highly crystalline anatase nanoparticles with different sizes and shapes [Figure 5.4]. A general, stepwise reaction for such polycondensation could be written:

$$2Ti(OR)_4 \rightarrow (OR)_3TiOTi(OR)_3 + ROR$$

$$(OR)_3TiOTi(OR)_3 + Ti(OR)_4 \rightarrow (OR)_3TiOTi(OR)_2OTi(OR)_3 + ROR$$

and so on to anatase, though it might not proceed in precisely this manner in the presence of base. The reactions were performed in two steps: First, alkoxide was added to the base at 0 °C in alcoholic solvents in a three-necked flask. The temperature was then raised to reflux. The second stage involved treating the product of the reflux in Ti autoclaves under a saturated vapor pressure of water (2500 kPa) at temperatures between 175 °C and 200 °C for 5 h. The nanocrystals obtained were sufficiently monodisperse that they formed coherent superlattices, as monitored by low-angle powder X-ray diffraction. By using other hydroxides with bulkier alkyl groups (for example, tetrapropylammonium hydroxide) some control over crystallite morphology became possible.

Wu et al. [69] have combined microemulsion and inverse-micelle techniques with hydrothermal techniques in the preparation of rutile and anatase $TiO_2$ nanoparticles. They used the system water–cyclohexane–Triton X-100 with n-hexanol as a second emulsifier. The water-filled micellar pockets were acidified (with HCl or $HNO_3$) $Ti(OR)_4$ (R = butyl). Treating the system at 120–200 °C for 12–144 h gave anatase particles. When high HCl concentrations were employed, rutile rods were obtained.

Hirano [70] has prepared spinel $ZnGa_2O_4$ nanoparticles by adjusting the pH of Zn and Ga sulfates with $NH_3$ to different initial values (varied from 2.5 through 10). The material was heat-treated hydrothermally at temperatures between 150 °C and 240 °C for times between 5 and 50 h. Particle sizes could be varied from 5 to 25 nm.

Cabañas et al. [71] have described the preparation of nanoparticulate $CeO_2$–$ZrO_2$ solid solutions under *flow-hydrothermal* conditions, wherein the reactants are taken to the final temperature very rapidly in a continuous process. The advantage of performing hydrothermal reactions in such a manner is, firstly, that a large amount of material can be processed, permitting simple scale up. Secondly, the nucleation step can be made very rapid, as a result of the rapid heating. This can help separate nucleation and growth, and can thereby satisfy the famous LaMer criterion [72] for obtaining monodisperse particles. Cabañas and Poliakoff [73] have also prepared a number of spinel ferrite samples in this way, starting from mixtures of different Fe(II) and M(II) acetates (M = Co, Ni, Zn and Co/Ni). Most of the preparations yielded a bimodal distribution of sample sizes, with the smaller samples being about 10 nm in diameter and the larger ones, about 100 nm.

#### 5.3.5.3 **Thermolysis**

Recently, Thimmaiah et al. [74] have extended the thermolytic route to oxide nanoparticles devised by Rockenberger, Scher and Alivisatos [57] to solvothermal conditions. Using solvothermal toluene (typically at 220 °C for 1 h), cupferron precursors of Fe, and Co and Fe are decomposed to obtain sub-12 nm maghemite $\gamma$-$Fe_2O_3$ nanoparticles and spinel $CoFe_2O_4$ nanoparticles. The reactions do not

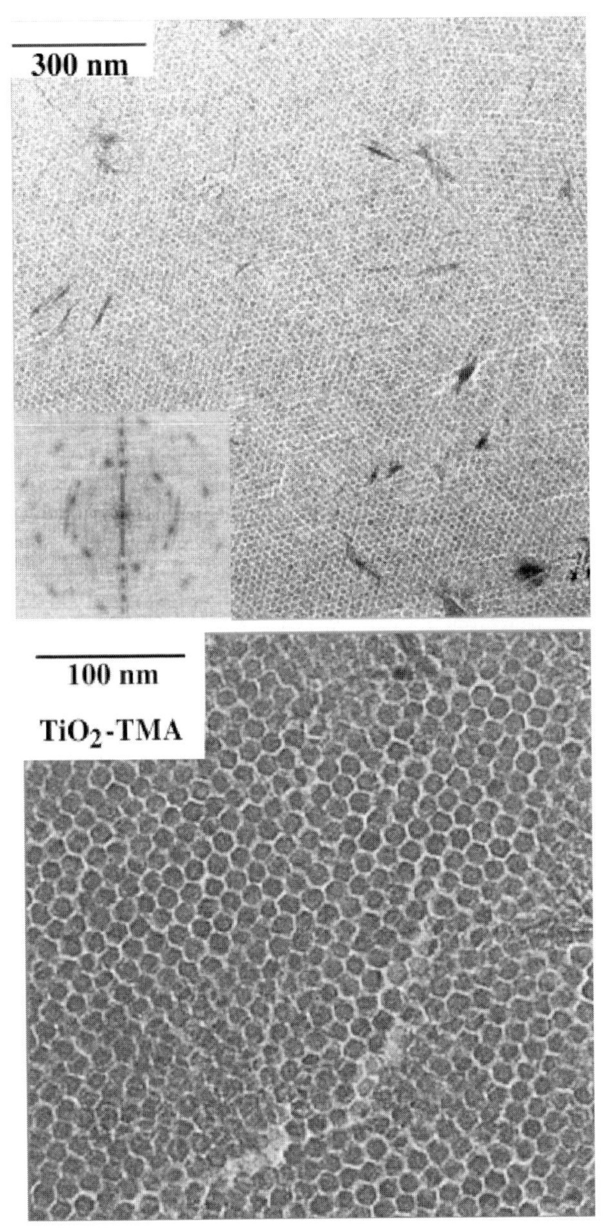


**Fig. 5.4.** Transmission electron micrographs of anatase $TiO_2$ nanoparticles prepared by solvothermal treatment of $Ti(OR)_4$ after hydrolysis with tetramethylammonium hydroxide. The inset in the upper micrograph displays a Fourier transform with discrete spots, attesting to the extent of superlattice coherence. The lower, high magnification image shows that the nanocrystals are well-faceted. Reproduced from [68] with permission from Wiley–VCH.

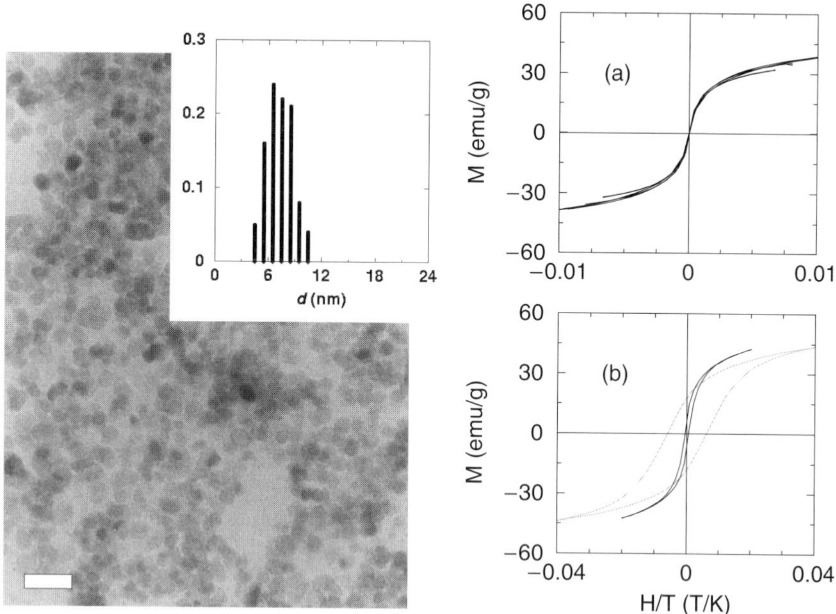


**Fig. 5.5.** Transmission electron micrograph of n-octylamine-capped spinel $CoFe_2O_4$ nanoparticles prepared by the thermolysis of Co(II) and Fe(III) cupferron precursors in solvothermal toluene. The panels on the right display the magnetic properties of these superparamagnetic particles. (a) is the *M* versus *H*/*T* plot for data on a pressed pellet of these nanoparticles taken at temperature of 300, 250, 200, 150 and 100 K, The collapse of all these traces onto a single S-shaped curve indicates superparamagnetic behavior. Only at 50 K and 5 K (black and gray traces in (b)) are the samples "blocked" and hysteretic behavior manifests. Reproduced from [74] with permission from the Royal Society of Chemistry.

work in the absence of long chain amines (n-octylamine and n-dodecylamine) which are added as capping agents. A complete characterization of the magnetic properties of these nanoparticles has been presented by the authors [Figure 5.5]. This is the only report to date of capped oxide nanoparticles prepared by a solvothermal route.

## 5.4 Prospects

As has been emphasized, research on the preparation of oxide nanoparticles is very much an active area that is being pursued across the world. While certain oxide structural classes (such as transition metal spinels) seem easily amenable to being prepared in nanoparticulate form, a number of other oxide materials are not. In particular, few attempts have been made to prepare capped perovskite oxide nanoparticles [61]. While some of the properties of oxide nanoparticles have already been mentioned alongside their syntheses, some unusual properties and applications call for special mention.

Individual magnetic nanoparticles, apart from displaying superparamagnetic behavior, possess other unusual qualities. Bonet et al. [75] have shown that small ferrite (in the magnetoplumbite structure) nanoparticles can have their magnetic polarizations completely reversed through uniform rotation of their magnetization. The spin precession of the whole particle magnetic moment of 4 nm $\gamma$-$Fe_2O_3$ nanoparticles has been monitored, using neutron scattering, by Lefmann et al. [76]. These authors find that the precession is coherent since it is associated with the excitation of ferrimagnetism in the nanoparticle.

Oxide nanoparticles, unlike nanoparticles of metals, display an expansion in their lattice parameters in comparison with the bulk. Tsunekawa et al. have examined sub-10 nm $CeO_{2-x}$ and sub-100 nm $BaTiO_3$ nanoparticles using a combination of electron diffraction, X-ray photoelectron spectroscopy and ab initio computer simulations. They find that in the $CeO_{2-x}$ system, the lattice expansion arises from a decrease in Ce valence, whilst in the $BaTiO_3$ system, the decreasing Ti–O covalency with decreasing particle size results in the expanded lattice.

There is continued interest in magnetic oxide nanoparticle assemblies for magnetic data storage. Recent advances in this area have come from the group of Sun, who have prepared magnetite nanoparticles [77] from Fe(II) acetylacetonate in refluxing diphenyl ether in the presence of oleic acid and oleylamine. Precipitation by addition of ethanol gave 4 nm $Fe_3O_4$ nanoparticles which could be used to seed the growth of larger $Fe_3O_4$ particles (reaching 16 nm). The larger $Fe_3O_4$ nanoparticles can be oxidized to $\gamma$-$Fe_2O_3$ with a distinct powder XRD pattern. The same group [78] have then self-assembled these nanoparticles as well as performed coupled self-assemblies of $Fe_3O_4$ nanoparticles with Fe–Pt nanoparticles and then annealed the films to obtain exchange-coupled nanocomposite assemblies.

An interesting and novel application for magnetic nanoparticles is their use as intracellular magnetic labels in nuclear magnetic esonance imaging. The presence of magnetic nanoparticles results in protons in their vicinity relaxing at a much faster rate. Since NMR imaging is based on the rate of magnetic relaxation, there is distinct contrast introduced. Biocompatible dextran-coated nanoparticles have been conjugated with peptides to improve their uptake into target cells. These have then been found to significantly aid the NMR imaging of the cells [79]. Weissleder et al. have also conjugated specific oligonucleotide sequences to magnetic oxide nanoparticles [80]. This results in their being able to employ NMR methods in the rapid detection of DNA sequences that are complementary to the oligonucleotide that is bound to the nanoparticle. In functional magnetic resonance imaging (fMRI) experiments, primates or humans are exposed to some external stimulus while their brains are imaged. This technique has proved invaluable in mapping the brain, in terms of associating region with function. Magnetic nanoparticles (dextran-coated magnetite) have been used to enhance fMRI imaging contrast in the Rhesus brain in experiments that measured photic response [81].

Magnetic oxide nanoparticles are also finding a number of other uses, including in magnetic drug delivery and in hyperthermic cancer therapy. In the former, the pharmacoactive molecule is bound to a small magnetic particle, and the particle is guided to its in vivo target through the use of magnetic fields. In the latter, cancerous tumors are destroyed by injecting magnetic particles into them and then coupling the particles to a strong radio-frequency magnetic field, resulting in large amounts of heats being delivered locally to, and eventually destroying, the tumor.

While heterogeneous catalysis is a traditional area of chemistry and is not usually classified as an application of nanoparticles, a recent development in cleaner-burning fuels is worth mentioning here. Two major manufacturers of diesel automobiles in Europe, Peugeot–Citröen and Ford, inject sub-10 nm $CeO_2$ nanoparticles (called by the trade name EOLYS TM, manufactured by the French company Rhodia) into diesel fuel before it is burnt in the engine. This results in an in situ catalyst that not only improves fuel efficiency but also greatly reduces the emission of particulates. Were the particles not sub-10 nm, it would not be possible to make homogeneous dispersions in the diesel fuel. It is expected that more than a million diesel automobiles will incorporate such systems by the year 2005. Another area where large quantities of nanoparticulate oxide materials are expected to be consumed is in the preparation of transparent sunscreens for topical application. Traditionally, sunscreens are applied as opaque creams. The active principle is a ZnO powder that absorbs most of the UV radiation incident upon the skin. There is great interest however, in material that can be sprayed on, and that is invisible. For this, the ZnO particles have to be sub-100 nm.

## Acknowledgments

I thank the office of the Dean of Engineering, UCSB for a seed grant. In writing this review, I have benefited greatly from discussions with Paul O'Brien and his group in Manchester during a recent visit made possible by a grant from the Royal Society of Chemistry. Susanne Stemmer is thanked for inputs on bulk oxide materials. My many coauthors in the area are acknowledged in the papers listed; Michael Rajamathi and Moumita Ghosh are thanked for their help in collating references. Larken Euliss is thanked for critically reading the manuscript.

## References


1 M. Faraday, *Philos. Trans. R. Soc. London*, **1857**, *147*, 145–181.

2 M. José-Yacamán, L. Rendón, J. Arenas et al., *Science*, **1996**, *273*, 223–225.

3 Oxford English Dictionary at http://dictionary.oed.com/

4 A. Fojtik, H. Weller, U. Koch et al., *Ber. Bunsen-Ges. Phys. Chem.*, **1984**, *88*, 969–977.

5 M. L. Steigerwald, A. P. Alivisatos, J. M. Gibson et al., *J. Am. Chem. Soc.*, **1988**, *110*, 3046–3050.

6 C. B. Murray, D. J. Norris, M. G. Bawendi, *J. Am. Chem. Soc.*, **1993**, *115*, 8706–8715.

7 M. Brust, M. Walker, D. Bethell et al., *J. Chem. Soc., Chem. Commun.*, **1994**, 801–802.

8 R. G. Nuzzo, D. L. Allara, *J. Am. Chem. Soc.*, **1983**, *105*, 4481–4483.

9 J. Prost, F. Rondelez, Suppl. to *Nature*, **1991**, *350*, 11–23.
10 A. Chemseddine, H. Weller, *Ber. Bunsen-Ges. Phys. Chem.*, **1993**, *97*, 636–637.
11 M. D. Bentzon, J. van Wonterghem, S. Mørup et al., *Philos. Mag. B*, **1989**, *60*, 169–178.
12 C. B. Murray, C. R. Kagan, M. G. Bawendi, *Science*, **1995**, *270*, 1335–1338.
13 R. L. Whetten, J. T. Khoury, M. M. Alvarez et al., *Adv. Mater.*, **1996**, *8*, 428–433.
14 C. P. Collier, T. Vossmeyer, J. R. Heath, *Annu. Rev. Phys. Chem.*, **1998**, *49*, 371–404; R. L. Whetten, M. N. Shafigullin, J. T. Khoury et al., *Acc. Chem. Res.*, **1999**, *32*, 397–406; C. B. Murray, C. R. Kagan, M. G. Bawendi, *Annu. Rev. Mater. Sci.*, **2000**, *30*, 545–610.
15 J. V. Sanders, *Acta Crystallogr. Sect. A*, **1968**, *24*, 427–434.
16 See for example, R. J. Cava, *J. Am. Ceram. Soc.*, **2000**, *83*, 5–28.
17 See for example, J. Z. Sun and A. Gupta, *Annu. Rev. Mater. Sci.*, **1998**, *28*, 45–78.
18 A. I. Kingon, J.-P. Maria, S. K. Streiffer, *Nature*, **2000**, *406*, 1032–1038.
19 J. F. Scott, *Annu. Rev. Mater. Sci.*, **1998**, *28*, 79–100.
20 O. Auciello, J. F. Scott, R. Ramesh, *Physics Today*, July **1998**, 22–27.
21 D. Dimos, C. H. Mueller, *Annu. Rev. Mater. Sci.*, **1998**, *28*, 397–419.
22 B. C. H. Steele, A. K. Heinzel, *Nature*, **2001**, *414*, 345–352.
23 J.-M. Tarascon, M. Armand, *Nature*, **2001**, *414*, 359–367.
24 H. Gleiter, *Acta Mater.*, **2000**, *48*, 1–29.
25 R. Blakemore, *Science*, **1975**, *190*, 377–379.
26 R. B. Frankel, R. P. Blakemore, R. S. Wolfe, *Science*, **1979**, *203*, 1355–1356.
27 J. L. Kirschvink, M. M. Walker, C. E. Diebel, *Curr. Opin. Neurobiol.*, **2001**, *11*, 462–467.
28 M. M. Walker, T. E. Dennis, J. L. Kirschvink, *Curr. Opin. Neurobiol.*, **2002**, *12*, 735–744.
29 R. E. Dunin-Borkowski, M. R. McCartney, R. B. Frankel et al., *Science*, **1998**, *282*, 1868.
30 D. Schüller, R. B. Frankel, *Appl. Microbial Biotechnol.*, **1999**, *52*, 464–473.
31 http://www.nanomagnetics.co.uk
32 K. L. Thomas-Keprta, D. A. Bazylinski, J. L. Kirschvink et al., *Geochim. Cosmochim. Acta*, **2000**, *64*, 4049–4081.
33 P. R. Buseck, R. E. Dunin-Borkowski, B. Devouard et al., *Proc. Natl. Acad. Sci. U.S.A.*, **2001**, *98*, 13490–13495.
34 D. F. Shriver, P. W. Atkins, *Inorganic Chemistry*, 3rd edition., W. H. Freeman and Company, New York.
35 G. H. A. Therese, P. Vishnu Kamath, *Chem. Mater.*, **2000**, *12*, 1195–1204.
36 V. Cabuil, *Curr. Opin. Colloid Interface Sci.*, **2000**, *5*, 44–48.
37 J. P. Chen, C. M. Sorenson, K. J. Klabunde et al., *Phys. Rev. B*, **1996**, *54*, 9288–9296.
38 S. Lefebure, E. Dubois, V. Cabuil et al., *J. Mater. Res.*, **1998**, 2975–2981.
39 R. Massart, E. Dubois, V. Cabuil et al., *J. Magn. Magn. Mater.*, **1995**, *149*, 1–5.
40 M. P. Morales, S. Veintemillas-Verdaguer, M. I. Montero et al., *Chem. Mater.*, **1999**, *11*, 3035–3064.
41 S. Ammar, A. Helfen, N. Jouini et al., *J. Mater. Chem.*, **2001**, *11*, 186–192.
42 M. Rajamathi, M. Ghosh, R. Seshadri, *J. Chem. Soc., Chem. Commun.*, **2002**, 1152–1153.
43 D. Carunto, Y. Remond, N. H. Chou et al., *Inorg. Chem.*, **2002**, *41*, 6137–6146.
44 M. P. Pileni, *J. Phys. Chem.*, **1993**, *97*, 6961–6973; *Langmuir*, **1997**, *13*, 3266–3276.
45 V. Pillai, D. O. Shah, *J. Magn. Magn. Mater.*, **1996**, *163*, 243–248.
46 C. Liu, Z. J. Zhang, *Chem. Mater.*, **2001**, *13*, 2092–2096.
47 C. Vestal, Z. J. Zhang, *Chem. Mater.*, **2002**, *14*, 3817–3822.
48 M. Moumen, M. P. Pileni, *Chem. Mater.*, **1996**, *8*, 1128–1134; *J. Phys. Chem.*, **1996**, *100*, 1867–1873; M.

MOUMEN, P. BONNEVILLE, M. P. PILENI, *J. Phys. Chem.*, **1996**, *100*, 14410–14416.

49 J. F. HOCHEPIED, Ph. SAINCTAVIT, M. P. PILENI, *J. Magn. Magn. Mater.*, **2001**, *231*, 315–322.

50 J. F. HOCHEPIED, P. BONNEVILLE, M. P. PILENI, *J. Phys. Chem.*, **2000**, *104*, 905–912.

51 A. T. NGO, M. P. PILENI, *J. Phys. Chem.*, **2001**, *105*, 53–58.

52 R. VIJAYA KUMAR, Y. DIAMANT, A. GEDANKEN, *Chem. Mater.*, **2000**, *12*, 2301–2305.

53 R. VIJAYA KUMAR, Y. KOLTYPIN, X. N. XU et al., *J. Appl. Phys.*, **2001**, *89*, 6324–6328.

54 T. HYEON, S. S. LEE, J. PARK et al., *J. Am. Chem. Soc.*, **2001**, *123*, 12798–12801.

55 J. A. NELSON, M. J. WAGNER, *Chem. Mater.*, **2002**, *14*, 915–917.

56 J. A. NELSON, E. L. BRANT, M. J. WAGNER, *Chem. Mater.*, **2003**, *15*, 688–693.

57 J. ROCKENBERGER, R. C. SCHER, A. P. ALIVISATOS, *J. Am. Chem. Soc.*, **1999**, *121*, 11595–11596.

58 A. K. BOAL, K. DAS, M. GRAY, V. M. ROTELLO, *Chem. Mater.*, **2002**, *14*, 2628–2636.

59 A. B. BOURLINOS, A. SIMOPOULOS, D. PETRIDES, *Chem. Mater.*, **2002**, *14*, 899–903.

60 A. B. BOURLINOS, A. BAKANDRITSOS, V. GEORGAKILAS et al., *Chem. Mater.*, **2002**, *14*, 3226–3228.

61 S. O'BRIEN, L. BRUS, C. B. MURRAY, *J. Am. Chem. Soc.*, **2001**, 123, 12085–12086.

62 P. ARNAL, R. J. P. CORRIU, D. LECLERCQ et al., *Chem. Mater.*, **1997**, *9*, 694–698.

63 T. J. TRENTLER, T. E. DENLER, J. F. BERTONE et al., *J. Am. Chem. Soc.*, **1999**, *121*, 1613–1614.

64 K. BYRAPPA, M. YOSHIMURA, *Handbook of Hydrothermal Technology*, William Andrew Publishing, Norwich, New York 2001.

65 M. RAJAMATHI, R. SESHADRI, *Curr. Opin. Solid State Mater. Sci.*, **2002**, *6*, 337–345.

66 M. INOUE, M. KIMURA, T. INUI, *J. Chem. Soc., Chem. Commun.*, **1999**, 957–958.

67 Y. LI, X. DUAN, H. LIAO, Y. QIAN, *Chem. Mater.*, **1998**, *10*, 17–18.

68 A. CHEMSEDDINE, T. MORITZ, *Eur. J. Inorg. Chem.*, **1999**, 235–245.

69 M. WU, J. LONG, A. HUANG, Y. LUO et al., *Langmuir*, **1999**, *15*, 8822–8825.

70 M. HIRANO, *J. Mater. Chem.*, **2000**, *10*, 469–472.

71 A. CABAÑAS, J. A. DARR, E. LESTER et al., *J. Chem. Soc., Chem. Commun.*, **2000**, 901–902; *J. Mater. Chem.*, **2001**, *11*, 561–568.

72 V. K. LAMER, R. H. DINEGAR, *J. Am. Chem. Soc.*, **1950**, *72*, 4847.

73 A. CABAÑAS, M. POLIAKOFF, *J. Mater. Chem.*, **2001**, *11*, 1408–1416.

74 S. THIMMAIAH, M. RAJAMATHI, N. SINGH et al., *J. Mater. Chem.*, **2001**, *11*, 3215–3220.

75 E. BONET, W. WERNSDORFER, B. BARBARA et al., *Phys. Rev. Lett.*, **1999**, *83*, 4188–4191.

76 K. LEFMANN, F. BØDKER, S. N. KLAUSEN et al., *Europhys. Lett.*, **2001**, *54*, 526–532.

77 S. SUN, H. ZENG, *J. Am. Chem. Soc.*, **2002**, *124*, 8204–8205.

78 H. ZENG, J. LI, Z. L. WANG et al., *Nature*, **2002**, *420*, 395–398.

79 L. JOSEPHSON, C.-H. TUNG, A. MOORE et al., *Bioconjugate Chem.*, **1999**, *10*, 186–191.

80 L. JOSEPHSON, J. M. PEREZ, R. WEISSLEDER, *Angew. Chem. Intl. Ed. Engl.*, **2001**, *40*, 3204–3206.

81 D. J. DUBOWITZ, K. A. BERNHEIM, D.-Y. CHEN et al., *NeuroReport*, **2001**, *12*, 2335–2340.

# 6
# Sonochemistry and Other Novel Methods Developed for the Synthesis of Nanoparticles

*Y. Mastai and A. Gedanken*

## Abstract

In this chapter the three synthetic methods that have been developed for the fabrication of nanomaterials in our laboratory will be described. The techniques are sonochemistry, sonoelectrochemistry, and microwave heating. In each category a short introduction to the technique will be presented in which an attempt will be made to interpret why the products produced by this method yield small size particles. In addition, we will report on the various nanosized particles that have been prepared by these methods, and other activities in these three fields related to nanochemistry. It should be mentioned that in 1996 [1a] Suslick and coworkers published an early review on the nanostructured materials generated by ultrasound radiation. Suslick and Price [1b] have also reviewed the application of ultrasound to materials science. Their review [1b] dealt with nanomaterials, but was not directed specifically to this topic. However, the review concentrated only on the sonochemistry of transition metal carbonyls, and catalytic reactions that involve the nanoparticles resulting from their sonochemical decomposition. Grieser and Asokkumar [2] have also written a review on a similar topic. A former coworker, J. J. Zhu, has recently submitted for publication a review article [3], entitled "*Novel Methods for Chemical Preparation of Metal Chalcogenide Nanoparticles*" in which he reviews the same above-mentioned three synthetic methods and their application in the synthesis of nanosized metal chalcogenides. It is for this reason that chalcogenide nanoparticles, their preparation and properties, will not be reviewed in this chapter. We attempt to review all the published literature related to these three synthetic methods published before the beginning of 2003.

## 6.1
## Sonochemistry

We are all aware of the use of ultrasound radiation in medicine, where it is being used mostly for diagnosis. More recently, however, focused ultrasound radiation is being used to burn cancer cells. Less is known regarding its application in chem-

*The Chemistry of Nanomaterials: Synthesis, Properties and Applications, Volume 1.* Edited by C. N. R. Rao, A. Müller, A. K. Cheetham

ISBN: 3-527-30686-2

istry, despite the fact that it has applications across almost the whole breadth of chemistry. The main advantage in conducting sonochemical experiments is that it is cheap to get started in the field, for example, in Romania after Ceaucescu's reign scientists carried out organic reactions using cheap ultrasonic baths as their source of radiation.

Sonochemistry is the research area in which molecules undergo chemical reaction due to the application of powerful ultrasound radiation (20 KHz–10 MHz) [4]. The physical phenomenon responsible for the sonochemical process is acoustic cavitation. Let us first address the question of how 20 kHz radiation can rupture chemical bonds (the question is also related to 1 MHz radiation), and try to explain the role of a few parameters in determining the yield of a sonochemical reaction, and then describe the unique products obtained when ultrasound radiation is used in materials science.

A number of theories have been developed in order to explain how 20 kHz sonic radiation can break chemical bonds. They all agree that the main event in sonochemistry is the creation, growth, and collapse of a bubble that is formed in the liquid.

The first question is how such a bubble can be formed, considering the fact that the forces required to separate water molecules to a distance of two van-der Waals radii, would require a power of $10^5$ W $cm^{-1}$ [4]. On the other hand, it is well known that in a sonication bath, with a power of 0.3 W $cm^{-1}$ [4] water is already converted into hydrogen peroxide. Different explanations have been offered; they are all based on the existence of unseen particles, or gas bubbles, that decrease the intermolecular forces, enabling the creation of the bubble. The experimental evidence for the importance of unseen particles in sonochemistry is that when the solution undergoes ultrafiltration, before the application of the ultrasonic power, there is no chemical reaction and chemical bonds are not ruptured.

The second stage is the growth of the bubble, which occurs through the diffusion of solute vapor into the volume of the bubble. The third stage is the collapse of the bubble, which occurs when the bubble size reaches its maximum value.

From here on we will adopt the hot spot mechanism, one of the theories that explain why, upon the collapse of a bubble, chemical bonds are broken. The theory claims that very high temperatures (5,000–25,000 K) [5] are obtained upon the collapse of the bubble. Since this collapse occurs in less than a nanosecond [5, 6], very high cooling rates, in excess of $10^{11}$ K $s^{-1}$, are obtained. This high cooling rate hinders the organization and crystallization of the products. For this reason, in all cases dealing with volatile precursors, where gas phase reactions are predominant, amorphous nanoparticles are obtained. While the explanation for the creation of amorphous products is well understood, the reason for the nanostructured products is not clear. One explanation is that the fast kinetics do not permit the growth of the nuclei. If, on the other hand, the precursor is a non-volatile compound, the reaction occurs in a 200 nm ring surrounding the collapsing bubble [7]. In this case, the sonochemical reaction occurs in the liquid phase. The products are sometimes nanoamorphous particles and, in other cases, nanocrystalline. This depends on the temperature in the ring region where the reaction takes place. The temperature in this ring is lower than inside the collapsing bubble, but higher than the temperature of the bulk. Suslick has estimated the temperature in the ring region as 1900 °C [7]. In short, in almost all the sonochemical reactions leading to inorganic products, nanomaterials were obtained. They varied in size, shape, structure, and in their solid phase (amorphous or crystalline), but they were always of nanometer size.

We cannot mention here how the various parameters: frequencies, power, gas under which the sonication takes place, pressure of the gas, etc. affect the sonochemical yield and rate. We will just mention one important parameter, the temperature. The equation of an adiabatic implosion is

$$T_{max} = T_0\{P_{ex}(\gamma - 1)/P_{bub}\}, \quad (1)$$

where $T_{max}$ is the temperature reached after the collapse of the bubble, $T_0$ is the temperature of the sonication bath, $\gamma = C_p/C_v$, $P_{ex}$ is the external pressure equal to the sum of the hydrostatic and acoustic pressure, and $P_{bub}$ is the pressure of the gas inside the cavity, at the radius at which it collapses. The choice of a non-volatile solvent (such as decalin, hexadecane, isodurene, etc.) guarantees that only the vapors of the solute can be found inside the cavitating bubble. Thus, $P_{bub}$ is practically the vapor pressure of the solute, and since it is found in the denominator, lower $P_{bub}$ results in higher temperatures and faster reaction rates. The conclusion is that the temperature affects the sonochemical reaction rate in two ways. On the one hand, lower temperatures cause a higher viscosity, which makes the formation of the bubble more difficult, and, on the other hand, the dominant effect is that at lower temperatures, higher rates will be achieved in sonochemical processes. This is why the sonic reaction involving volatile precursors is run at lower temperatures. Apparent negative activation energies were measured for sonochemical reactions.

Our first activity in the field combining sonochemistry and materials science was related to control over the particle size. We have demonstrated in three cases that for gas phase, as well as liquid phase reactions, diluting the precursor's solution decreases the particle size of the product [8–10]. This was illustrated for nanosized Fe [8], $Fe_2O_3$ [9], and for GaO(OH) [10]. In the last case, this conclusion was reached by measuring the particle's size from the transmission electron microscopy (TEM) micrograph. For the Fe, and $Fe_2O_3$ nanoparticles it was concluded from indirect magnetic, EPR, and differential scanning calorimetry (DSC) measurements. This is due to the high degree of aggregation that is observed in the TEM picture of these magnetic particles resulting from their strong magnetic interactions.

In the following section we will present the various inorganic systems that have been synthesized in the last few years, and we will then try to emphasize the unique features of sonochemistry, or what can be described by the famous song "anything you can do I (sonochemistry) can do better". In this section metals will serve as a demonstration of what can be done sonochemically. We will discuss the synthesis of nanometals, colloidal metallic solutions, formation of alloys, the coating of metallic nanoparticles on surfaces, SAM (self-assembled monolayer coating) on nanosized amorphous metals, and the formation of metal–polymer composites. The next section will be devoted to the sonochemical synthesis of mesoporous materials, and the use of ultrasound for the insertion of nanomaterials into the pores. In the final section we will discuss other important nanomaterials that have been prepared by the sonochemical technique (apart from chalcogenides).

## 6.1.1 Sonochemical Fabrication of Nanometals

### 6.1.1.1 Sonochemical Synthesis of Powders of Metallic Nanoparticles

In addition to the synthesis of the transition metals produced from the corresponding carbonyls (Fe from $Fe(CO)_5$ [6, 11], Ni from $Ni(CO)_4$ [12], and Co from $Co(CO)_3NO$ [13], other metals have also been synthesized sonochemically. Sonication of aqueous $Co^{+2}$ and hydrazine resulted [14] in the formation of anisometric (disk-shaped) cobalt nanoclusters that averaged about 100 nm in width and 15 nm in thickness. Electron diffraction from single particles revealed that they were oriented (001) crystals that conformed to a trigonal or hexagonal unit cell four times the size of the cell adopted by bulk alpha-cobalt. Nanophased particles of metallic copper [15] were formed by the sonochemical reduction of copper (II) hydrazine carboxylate ($Cu(N_2H_3COO)_2 \cdot 2H_2O$) in an aqueous solution. When the sonication was carried out under argon a mixture of $Cu_2O$ and metallic copper was obtained. However, sonicating the precursor solution under a mixture of hydrogen and argon yielded pure copper. The particles were obtained as porous materials with diameter 50 nm, smaller than those obtained by the thermal decomposition of the same precursor. A mechanism involving hydrogen radicals as the reducing agent is proposed [15].

$$Cu^{+2}_{(aq)} + 2H^{\cdot} \rightarrow Cu^{0}_{(s)} + 2H^{+}_{(aq)} \tag{2}$$

In a separate study, nanoparticles of palladium metallic clusters were prepared at room temperature by sonochemical reduction of a 1:2 molar mixture of $Pd(O_2CCH_3)_2$ and myristyltrimethylammonium bromide, $CH_3(CH_2)_{12}N(CH_3) \cdot BrNR_4X$, in THF or MeOH [16]. Apart from its stabilizing effect, $NR_4X$ acts as a reducing agent, probably due to the decomposition that occurs in the liquid-phase region immediately surrounding the collapsing cavity and provides reducing radicals. It is noteworthy that nanosized amorphous Pd is obtained in THF and a crystalline metal in MeOH.

A pioneer in the application of ultrasound to the formation of nanoparticles of noble metals is Y. Maeda. In an earlier study his group [17] synthesized sonochemically metallic nanoparticles of metals such as Ag, Pd, Au, Pt and Rh with a fairly narrow distribution (e.g., about 5 nm for Pd particles obtained from a 1.0 mM Pd(II) solution in polyethylene glycol monostearate solution). They suggested three different reduction pathways under sonication: (i) reduction by H atoms, (ii) reduction by secondary reducing radicals formed by hydrogen abstraction from organic additives with OH radicals and H atoms, and (iii) reduction by radicals formed from pyrolysis of the additives at the interfacial region between cavitation bubbles and the bulk solution. The reduction of Ag (I) and Pt (II) mainly proceeds through reaction pathway (ii). For Pd (II) and Au (III), the reductions mainly proceed through reaction pathway (iii). The reduction of Rh (III) was not achieved under the same conditions; however, on addition of sodium formate, reduction occurred and the preparation of Rh particles was successful. In another study [18] platinum nanoparticles were prepared sonochemically in an aqueous system in the presence of a surfactant (sodium dodecyl sulfate, SDS). The particles were stable, homogeneously spherical, and relatively monodispersed with an average diameter of 2.6 nm. Reducing species generated near and/or in the hot bubbles would react with the $PtCl_4{}^{-2}$ complexes to form the platinum nanoparticles. Three kinds of reducing species were proposed. Amorphous silver nanoparticles of 20 nm size have been prepared [19] by the sonochemical reduction of an aqueous silver nitrate solution in an atmosphere of argon–hydrogen. The sonochemical reduction occurs through the generation of hydrogen radicals during the sonication process.

The role of the surfactant was further studied in sonochemical reduction [20, 21] processes of Pt (IV) ions in water. It was investigated in the presence of various kinds of surfactants such as sodium dodecylsulfate (SDS) and sodium dodecylbenzenesulfonate (DBS) as anionic surfactants, polyethylene glycol monostearate (PEG-MS) as non-ionic, and dodecyltrimethylammonium chloride (DTAC) and bromide (DTAB) as cationic surfactants. An improved colorimetric determination reveals that Pt (IV) ion is reduced to zero-valent metal in two steps: step (1) Pt (IV) ion to Pt (II) ion, and step (2) Pt (II) ion to Pt (0), and after the completion of step (1), step (2) sets in.

Another Japanese group [22] reported on the preparation of Pd and Pt by sonochemical reduction of solutions containing $H_2PtCl_6$ or $K_2PdCl_4$. The effect of atmospheric gas on the particle size distribution was investigated. Average diameters and standard deviations of the Pd particles prepared under Ar (Pd/Ar) and $N_2$ (Pd/$N_2$) were found to be $3.6 \pm 0.7$ nm (Pd/Ar) and $2.0 \pm 0.3$ nm (Pd/$N_2$). A smaller and sharper distribution of the particle size was observed for the Pd particles formed under a $N_2$ atmosphere. In the case of Pt, a smaller and sharper distribution of the particle size was observed for the particles prepared under a Xe atmosphere. The importance of Xe can be explained in terms of a hot-spot temperature created by acoustic cavitation. Nanosized gold particles were also prepared from gold (III) (tetrachloroaurate (III)) [23]. The Au (III) was reduced in an aqueous solution containing only a small amount of 2-propanol to form colloidal gold nanoparticles. The rates of gold (III) reduction and the sizes of the gold particles formed could be sonochemically controlled by controlling the irradiation parameters such as the temperature of the solution, the intensity of the ultrasound, and the positioning of the reactor. The size of the gold particles depended strongly on the rate of gold (III) reduction, suggesting that this rate affects the initial nucleation of the gold particles. A similar study was published recently by Okitsu et al. [24]. The rates of gold (III) reduction were strongly dependent on the atmosphere, the temperature of the bulk solution, the intensity of the ultrasound, and the distance of the reaction vessel from the oscillator. For example, the rates of reduction under several atmospheres were in the order: $CH_4 = CO_2 < N_2 < Ne < He < Ar < Kr$, where no reduction proceeded under the $CH_4$ and $CO_2$ atmospheres. It was clearly seen that the rates of reduction were influenced by the cavitation phenomenon. Upon irradiation, colloidal gold particles having a surface plasmon absorption were formed, although in the absence of any stabilizers for the gold particles. It was found by TEM observation that the average size of the formed gold particles varied from 30 to 120 nm dependent upon the irradiation parameters. The size of the gold particles correlated to the initial rate of gold (III) reduction; the higher the rate of reduction, the smaller the particles.

The magnetic metals were also prepared by a method [25] based on the rapid expansion of supercritical fluid solutions (RESS) coupled with chemical reduction to produce nickel, cobalt, iron, and iron oxide nanoparticles of reasonably narrow size distribution. Under the protection of a polymer stabilization agent, the largely amorphous metal nanoparticles form stable suspensions in room-temperature solvents.

A nonmetallic element, silicon, was prepared sonochemically by reducing tetraethyl orthosilicate (TEOS) with a colloidal solution of sodium. The product was obtained as 2–5 nm sized, highly aggregated particles. The silicon exhibited a luminescence similar to that of porous silicon. This procedure is suggested as a general sonochemical reduction leading to the formation of metallic nanoparticles [26].

#### 6.1.1.2 Sonochemical Synthesis of Metallic Colloids

Although a few of the above-mentioned metallic nanoparticles were formed as colloidal solutions, and since this review is centered on their synthesis, they were included in the previous section. This section is devoted to the unique properties of metallic colloidal solutions prepared sonochemically. A recent survey by Grieser deals with the sonochemical formation of metallic colloids [27]. In the above-mentioned examples noble metal nanoparticles (e.g., Au, Pd, Ag) are obtained by sonicating aqueous solutions of the corresponding salts in the presence of a surfactant, which largely stabilizes the naked colloid. Likewise, a colloidal solution of metallic iron particles (8 nm, average size) has been obtained by sonolysis of $Fe(CO)_5$ in the presence of oleic acid [28]. Smaller particles are obtained with poly(vinylpyrrolidine), although in both cases the iron was amorphous and exhibited high magnetism. The sonochemical preparation of magnetic fluids has also been described by other authors [29, 30]. The systems include: (i) a cobalt colloidal solution in decalin stabilized by oleic acid, (ii) a colloidal dispersion of amorphous metallic iron in a polymeric matrix, and (iii) a $Fe_2O_3$ colloidal solution in hexadecane stabilized by oleic acid. For the cobalt colloidal solution uniform acicular-shaped particles were further obtained by an aging process in air (Figure 6.1). The figure demonstrates the conversion of 5–10 nm sized Co particles in a 1 μm sized acicular particle after 1 month of aging under ambient conditions.

Elongated copper nanoparticles were prepared by sonicating the above-mentioned precursor, copper hydrazine carboxylate, in an aqueous solution con-

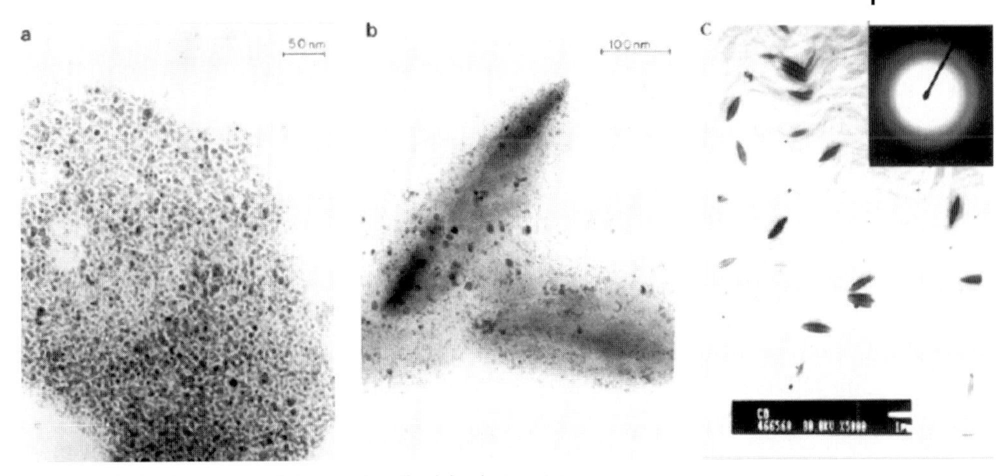


**Fig. 6.1.** TEM micrographs of the cobalt colloidal solution: (a) as prepared fresh, (b) after a week of aging, (c) after a month of aging (inset: ED pattern).

taining a zwitterionic surfactant [31]. In Figure 6.2 the TEM image of the elongated copper nanoparticles surrounded by surfactant molecules is presented. The dimensions of these copper nanoparticles were approximately 500 nm in length and 50 nm in width. In the absence of a surfactant, nanoparticles were spherical in

**Fig. 6.2.** Transmission electron micrograph of the elongated copper nanoparticles coated with CTAPTS.

shape, with a diameter of approximately 50 nm. A comparison of the XRD pattern with that obtained for the copper nanoparticles prepared without using a surfactant revealed an elongation taking place along the [111] and [200] planes of the copper nanocrystals.

A comprehensive study on the sonochemical synthesis of colloidal solutions of noble metals was conducted by Grieser and coworkers [32–34]. The 515 kHz ultrasound-initiated reduction of $AuCl_4^{-1}$ to Au (0) was examined as a function of the concentration of various surface-active solutes [32]. The amount of $AuCl_4^{-1}$ reduced in the presence of ethanol, 1-propanol, and 1-butanol was found to be dependent on the surface excess of the alcohol at the gas/solution interface, i.e., the relative concentration of the alcohol at the gas/solution interface compared to the bulk solution concentration. The efficiency of reduction of $AuCl_4^{-1}$ in the presence of the surfactants sodium dodecyl sulfate or octaethylene glycol monodecyl ether was found to be related to the monomer concentration of the surfactant in solution.

The ultrasound irradiation of aqueous solutions of $PtCl_6^{-2}$ was found to produce colloidal platinum of about 3 nm in diameter [33]. The presence of aliphatic alcohols significantly enhanced the reduction process. It is shown that the extent of reduction of $PtCl_6^{-2}$ in the presence of alcohol is directly related to the Gibbs surface excess of the alcohol at the air–water interface. The significance of this is discussed in relation to ultrasound-induced cavitation in solution. In a recent report Caruso et al. [34] reduced $AuCl_4^{-1}$ to colloidal gold in the presence of aliphatic alcohols and sodium dodecyl sulfate in aqueous solutions using 20 kHz ultrasound. The extent of reduction of $AuCl_4^{-1}$ was found to be directly dependent on the gas–solution interfacial activity of the solutes used. The diameters of the gold colloids produced were in the size range 9–25 nm. The particle size was dependent on the alcohol concentration in solution and on the hydrophobicity of the alcohol. For a particular alcohol concentration, the greater the hydrophobicity of the alcohol the smaller were the particles obtained. The authors present a detailed mechanism describing the overall sonochemically initiated reduction of $AuCl_4^{-1}$ leading to the formation of colloidal gold. Finally, Grieser has also published an early review on sonochemistry in colloidal systems [35].

#### 6.1.1.3 **Sonochemical Synthesis of Metallic Alloys**

Nanophased amorphous Fe/Co was obtained by sonicating a mixture of $Fe(CO)_5$ and $Co(CO)_3NO$ in decalin [36]. The Fe/Co, nanostructured alloys are formed and are non-crystalline by X-ray, neutron, and e-beam diffraction. In a further work, a Fe/Ni alloy was similarly obtained [37]. Remarkably, a 1:1 molar solution of Fe:Ni led to a $Fe_{20}Ni_{80}$ solid composition, which can be attributed to the ratio of the vapor pressure of the two carbonyls in the gas phase of the collapsing bubble. Magnetic susceptibilities of $Fe_{40}Ni_{60}$ and $Fe_{60}Ni_{40}$ indicated blocking temperatures of 35 K and a magnetic particle size of about 6 nm. Thermogravimetric measurements of $Fe_{20}Ni_{80}$ gave Curie temperatures of 322 °C for amorphous and 550 °C for crystallized forms. Differential scanning calorimetry exhibited an endothermic transition at 335 °C from a combination of the magnetic phase transition and alloy crystallization. The r Mössbauer spectrum of crystallized $Fe_{20}Ni_{80}$ shows a sextet pattern with a hyperfine field of 25.04 T. A similar study reports on the Co/Ni synthesis, characterization and properties [38].

M50 steel powder has been obtained by sonochemical decomposition of organometallic precursors, namely $Fe(CO)_5$, $(Et_xC_6H_{6-x})_2Cr$, $(Et_xC_6H_{6-x})_2Mo$, and $V(CO)_6$ in decalin [39]. The morphology of the amorphous (as evidenced by X-ray patterns) M50 powder was shown to be a porous coral-like microstructure. The consolidated iron pellet (the consolidation was carried out by vacuum hot press, conditions: 275 MPa, at 700 °C for 1 h) had a density of 100%. The iron sample had a high Rockwell C (RC) hardness of 37 as compared to 4–5 RC for conventional iron. The hardness of the M50 steel sample was 66.3 RC compared with 58–62 for conventional M50 steel after tempering. The authors attribute the extremely high hardness of the consolidated iron to the nanometer size particles, as well as to the low carbon and oxygen contamination.

In a different study [40], argon-saturated aqueous solutions of $NaAuCl_4$ and $PdCl_2$ or $K_2PtCl_4$ were reduced simultaneously by ultrasound irradiation to prepare noble metal alloy nanoparticles. The Au–Pd nanoparticles exhibited monodispersive distribution (8 nm), and consisted of a gold core and a palladium shell. Au–Pt alloy nanoparticles could not be produced from $NaAuCl_4$ and $K_2PtCl_4$ aqueous solutions by either simultaneous or successive reduction.

An aqueous solution of $AgNO_3$ in the presence of ammonia and $Fe(CO)_5$ was sonicated [41] under a $H_2$/Ar mixture, yielding a nanostructured homogeneous phase of Ag/$Fe_2O_3$. This composite material was further reduced at 300 °C under hydrogen to produce the nanophased Fe/Ag solid mixture. Finally, a ternary nanosized amorphous alloy, Fe/Ni/Co, was prepared by sonochemical decomposition of solutions of volatile organic precursors, $Fe(CO)_5$, $Ni(CO)_4$, and $Co(NO)(CO)_3$ in decalin, under an argon pressure of 100 to 150 kPa at 273 K [42]. Magnetic measurements indicated that the as-prepared amorphous Fe–Ni–Co alloy particles were super-paramagnetic. The observed magnetization measured up to a field of 1.5 kG of the annealed Fe–Ni–Co samples (75–87 emu $g^{-1}$) was significantly lower than that for the reported multidomain bulk particles (175 emu $g^{-1}$), reflecting the ultrafine nature of our sample.

#### 6.1.1.4 **Sonochemical Deposition of Nanoparticles on Spherical and Flat Surfaces**

Metallic nanoparticles were deposited on ceramic and polymeric particles using ultrasound radiation. A few papers report also on the deposition of nanomaterials produced sonochemically on flat surfaces. Our attention will be devoted to spheres. In a typical reaction, commercially available spheres of ceramic materials or polymers were introduced into a sonication bath and sonicated with the precursor of the metallic nanoparticles. In the first report Ramesh et al. [43] employed the Stober method [44] for the preparation of 250 nm silica spheres. These spheres were introduced into a sonication bath containing a decalin solution of $Ni(CO)_4$. The as-deposited amorphous clusters transform to polycrystalline, nanophasic, fcc nickel on heating in an inert atmosphere of argon at a temperature of 400 °C. Nitrogen adsorption measurements showed that the amorphous nickel with a high surface area undergoes a loss in surface area on crystallization.

The as-deposited amorphous nickel showed a super-paramagnetic behavior, while the polycrystalline nickel on silica was found to be ferromagnetic. FT-IR investigations showed a significant change in the surface silanol composition for the coated and uncoated silica. Ultrasound-driven cavitation desorbs the adsorbed water on silica, making the free silanols available for reaction with nickel species. A positively charged nickel species thus formed could constitute a nucleating site for further aggregation of nickel. An alternate mechanism for the interaction of nickel clusters with the silica surface is proposed, wherein ultrasound irradiation results in the dehydrative condensation of hydrogen-bonded silanols to form siloxane. This is followed by the formation of a bond between nickel and the bridging oxygen of the siloxane group. The surface topography and the adhesion of amorphous and polycrystalline nickel nanoparticles on the surface of silica submicrospheres (200–250 nm) were probed by atomic force microscopy (AFM), and reported in another publication [45]. Probe areas down to $50 \times 50$ nm in dimension were scanned on a single submicrosphere immobilized in a thermoplastic resin bed. Amorphous nickel particles formed by the aggregation of nickel clusters were soft, experienced poor adhesion to the silica, and caused huge tip-induced particle movements; but polycrystalline nickel nanoparticles in the size range 20–30 nm were hard and adhered strongly to the silica. The stronger adhesion of polycrystalline nickel is explained in terms of a silicate-type impurity phase formed in the nickel–silica interface during crystallization.

Ramesh has extended this study to another magnetic metal, Co. Ferromagnetic cobalt nanoparticles of a size similar to 10 nm well adhered to hard silica microspheres (225–250 nm) were synthesized by the sonochemical decomposition of a volatile organic precursor, cobalt nitrosyl carbonyl [$Co(CO)_3NO$], in a suspension of silica in decalin, followed by crystallization of the resultant amorphous product [46]. Silica spheres carrying ferromagnetic cobalt nanocrystals were deposited on a single crystalline silicon [100] substrate by spin coating. The two-dimensional organization of the magnetic microspheres on silicon and the adhesion of cobalt nanoparticles on the surface of microspherical silica have been examined by scanning electron microscopy and atomic force microscopy (AFM), respectively. Two-dimensional arrays of hard spherical particles carrying a nanoprobe hold potential as scanning tip arrays (STA) in force microscopy. Though rigid single molecules of proteins were originally envisioned as suitable probes, the authors suggested the possibility that magnetic nanoparticles could also fit the criteria.

Sonochemistry was also used to deposit Ni on amorphous and crystalline alumina [47]. The study concluded: (1) that amorphous alumina can provide a great number of active sites for reaction with nickel, and can yield a good coating effect in which most of nickel adheres tightly to the alumina surface, while in the case of the crystallized alumina as substrate, most of the nickel particles are distributed in the free space among the alumina submicrospheres. (2) As compared to the unadhered nickel, the adhered nickel has a strong interaction with the alumina core, which can retard the crystallization of elemental nickel and, conversely, promote the formation of the spinel phase $NiAl_2O_4$. (3) The first stage of the interaction between the nickel and the alumina may be through the isolated hydroxy groups to form a kind of interface Ni–O–Al bond. The connection positions further become nucleation centers for elemental nickel. After the sample is heated to high temperatures, DRS and IR results showed the diffusion of nickel ions into the vacant tetrahedral sites in alumina. At a higher temperature, an inversion process, the substitution of $Ni^{+2}$ ions for part of the $Al^{+3}$ at octahedral sites, takes place, resulting finally in the formation of the disordered spinel phase. (4) Magnetization measurements show that the as-prepared sonication products are superparamagnetic due to the ultrafine nature of nickel particles.

In a few studies sonochemistry was used to coat polymers with nanosized particles [48–50]. Of these three reports one [50] dealt with metals, more specifically with noble metals (Pt, Pd, and Au). In this research, metal colloids are adsorbed to the surface of neutral functionalized polystyrene microspheres, PSMS. The authors report on the synthesis and characterization of catalytically important noble monometallic colloids using various chemical and sonochemical methods. These metal colloids are then adsorbed onto suitably functionalized PSMS. The metal-immobilized microspheres are reacted with a linker such as 4-mercaptobutyl phosphonic acid and subsequently used to grow multilayers.

Lately, we have deposited silver nanoparticles with an average size of 5 nm on the surface of preformed silica submicrospheres with the aid of power ultrasound [51]. Ultrasound irradiation of a slurry of silica submicrospheres, silver nitrate, and ammonia in an aqueous medium for 90 min under an atmosphere of argon to hydrogen (95:5) yielded a silver–silica nanocomposite. By controlling the atmospheric and reaction conditions, we could achieve the deposition of metallic silver on the surface of the silica spheres. Figure 6.3 depicts a silica-coated sphere.

It can be seen that the level of deposition has been improved and we are currently able to create a very smooth coating of a monolayer of nanoparticles on the surface. This was achieved by a gradual reduction of the silver concentration, thus

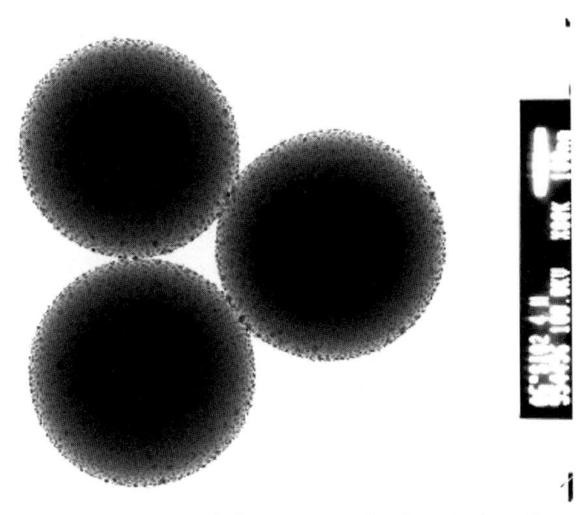

**Fig. 6.3.** TEM image of silver nanoparticles deposited on silica spheres.

reducing the thickness of the coated layer, as well as the amount of silver found between silica particles rather than on their surface.

Two questions were specifically investigated in these projects. First, the mechanism by which the deposited particles reach and are anchored to the host surface. Secondly, what is the mechanism by which the nanoparticles adhere to the surface, and are not removed, despite the severe stirring caused by the ultrasonic waves? Our answer to the first question involves the formation of shock waves and microjets created as after-effects to the collapse of the bubble [1a]. These effects always result when a bubble collapses near a solid surface. These jets, according to our interpretation, push the ultrafine particles towards the solid sphere at very high speeds, and are also known to cause the melting of colliding bodies and their sintering [52]. Once these nanoparticles collide with the surface, chemical bonds or weak interactions keep them on the surface in most cases. AFM studies have demonstrated [45] that the interaction becomes much stronger when the amorphous deposited particles are annealed at their crystallization temperature. The cantilever could scratch and move the particles only for the as-prepared products. Once they were annealed they were unaffected by the cantilever pushing power.

Ulman and coworkers have described in a few papers [53–55] a method by which sonochemically prepared nanoparticles are deposited on a flat surface, usually a silicon wafer. They describe their approach as a "plug and play", in which sonochemically synthesized amorphous $Fe_2O_3$ nanoparticles are incorporated onto device-quality Si wafers. After annealing the amorphous $Fe_2O_3$ nanoparticles they change their properties from super-paramagnetic to soft ferromagnetic. The samples exhibit multiple light emissions with wavelengths that are crucial for optical fiber communications.

Later, they demonstrate [55] that sonochemically synthesized $Fe_2O_3$ nanoparticles are introduced onto Si from an alcohol suspension. On annealing this sample in ultra-high vacuum, the oxygen atoms change the bonding partner from Fe to Si and desorb as SiO at 750 °C. This results in the formation of nanoparticles of Fe on the surface and exhibits ferromagnetic behavior. Deposition of a thin layer (2 nm) of Si onto the sample containing the metallic Fe nanoparticles followed by annealing at 560 °C leads to optically active Si.

Papadimitrakopoulos and coworkers reported on transparent Si/SiO$x$ nanocomposite films, spontaneously adsorbed on glass or quartz substrates from their colloidal suspensions via a sonication-assisted oxidation process [56]. Individual nanosilicon particles (ca. 20 nm) appear to cover a significant part of the substrate along with agglomerates of the order of 50–80 nm in thickness. Kinetic studies indicate a rapid initial adsorption that slows down significantly after 3 h.

#### 6.1.1.5 **Sonochemical Synthesis of a Polymer-Metal Composite**

The use of ultrasound radiation for polymerizing various monomers was reviewed in [1a]. Here we will discuss how ultrasound waves have been used successfully been to embed ultrafine metallic particles in a polymeric matrix. The first report was by Wizel and coworkers [57]. They used ultrasound radiation to prepare a composite material made of polymethylacrylate and amorphous iron nanoparticles.

Tab. 6.1. The concentration of iron in polymers prepared by the two methods.

| *Starting Materials* | *Fe in starting solution product (%)* | *Fe in final (%)* |
| --- | --- | --- |
| Methylacrylate $Fe(CO)_5$ | 0.5 | 0.56 |
| Methylacrylate (DMF) – 50 mg amorphous Fe | 0.3 | 5.4 |

Two preparation methods are described [57], in which the monomer, methylacrylate, is the starting material.

In the first, $Fe(CO)_5$ and distilled methylacrylate were sonicated as a mixture of neat liquids. During the sonication, the glass cell was wrapped in a dark cover to avoid photopolymerization. After 30 min of irradiation (avoiding degradation) at a dry ice–acetone temperature, the solution was treated with cold methanol, precipitating the polymeric product. In the second method, amorphous iron nanoparticles were prepared following Suslick's recipe [6]. The dried amorphous iron powder was introduced into the sonication cell without exposure to air and mixed with a solution of the methylacrylate monomer in *N,N′*-dimethylformamide (DMF). The concentration of the methylacrylate solution in DMF was 5.5 M. 35 ml of this solution was mixed with various amounts of amorphous iron nanoparticles. The amounts of the iron powder were changed from 50 mg to 200 mg. The chemical analysis of polymers prepared by the two methods is presented in Table 6.1.

The percentages presented in Table 6.1 are weight percentages of the iron in the mixture of $Fe(CO)_5$ and methylacrylate (first row, first column) in the starting solution. The second row presents the weight percent of iron in the starting solution (first column) and in the composite material (second column). The results according to Table 6.1 indicate that it is easier to introduce the iron by starting with amorphous iron as the precursor rather than $Fe(CO)_5$. Molecular weights of the polymers are presented in Table 6.2.

When the molecular weight ($M_w$) values of the polymeric composite material are compared with those of the polymeric product obtained from the irradiation of the monomer alone at the same irradiation time and concentration of the monomer, a 20% reduction is detected. This can be ascribed to the influence of the iron on the growth of the polymer. This would mean that the recombination of the iron

Tab. 6.2. Molecular weights of polymers obtained from various sonications.

| *Monomer* | *Sonication time (min)* | *$M_w$ (g $mol^{-1}$)* |
| --- | --- | --- |
| Methylacrylate | 30 | 134,000 |
| Methylacrylate in DMF | 90 | 1,760,000 |
| Methylacrylate in DMF containing 50 mg Fe | 90 | 142,000 |

pentacarbonyl dissociation products disturbs the growth of the polymeric radical. Thus, the dissociation products serve as a quenching agent for the creation of the polymer. The magnetic properties of the composite material are measured and reveal a super-paramagnetic behavior.

Wizel later extended her study and included another metallic nanoparticle, cobalt, and an additional polymer, poly(methylmethacrylate), in her metal–polymer composite research [58]. A significant difference in the solubility of the iron–poly(methylacrylate) and cobalt–poly(methylacrylate) in various solvents was observed. While the iron–poly(methylacrylate) composite (FePMA) and iron–poly(methylmethacrylate) composite (FePMMA) dissolved in chloroform, acetone, and toluene at room temperature, the corresponding cobalt–poly(methylacrylate) composite (CoPMA) was insoluble in these solvents at room temperature. At elevated temperatures (45 °C), dissolution of CoPMA in these solvents was observed. This difference is accounted for by the stronger interaction existing between the cobalt and the surrounding polymer. For iron–poly(methylacrylate) this interaction is weakened due to the formation of an iron complex. The $M_w$ of the various polymers and composites as a function of the metal-to-monomer weight ratio was measured and reported.

The general features observed are that the molecular weights of the CoPMA are always larger than those of the corresponding FePMA. This comparison is made for equal amounts of starting materials. The second observation is that the larger the amount of the metal, the smaller the molecular weight of the PMA. The opposite is observed for the FePMMA, where a larger molecular weight is obtained for higher amounts of iron clusters. This phenomenon is explained as due to the presence of the oxidized iron, especially $Fe^{+3}$. These ions accelerate the polymerization of the methylmethacrylate [59], and a higher amount of metal will favor a higher molecular weight polymer. The presence of oxidized iron is supported by XPS and X-ray absorption near-edge spectroscopy (XANES) measurements. Nano-sized amorphous iron was also imbedded in polystyrene [60] similarly.

A different approach was taken by Kumar and associates [61]. He also embedded metals in polymers, but used as his precursor the polymer and not the monomer. In his first study a composite material containing amorphous Cu nanoparticles and nanocrystalline $Cu_2O$ embedded in polyaniline matrices was prepared by a sonochemical method. These composite materials were obtained from the sonication of copper (II) acetate when aniline or 1% v/v aniline–water was used as the solvent. Mechanisms for the formation of these products are proposed and discussed. The physical and thermal properties of the as-prepared composite materials are presented. A band gap of 2.61 eV is estimated from optical measurements for the as-prepared $Cu_2O$ in polyaniline.

In a similar way, well-dispersed nickel nanoparticles in polystyrene were obtained by a sonochemical method [62]. The first step in this synthesis was the preparation of nickel formate, which was prepared according to a previously described method. The preparation of the nickel–polystyrene composite was carried out by the sonochemical method. Typically, 500 mg of nickel formate and 1 g of polystyrene (Aldrich; $M_w = 350{,}000$) are dissolved in 100 mL of *N*,*N*-dimethylformamide (DMF) and irradiated with a high-intensity ultrasonic horn under 1.5 atm of Ar–$H_2$ at room temperature for 3 h. The product is washed thoroughly with methanol in an inert glove box and dried overnight in a vacuum. The XRD diffraction indicates that the as-prepared composite material is crystalline. The XRD diffraction patterns match those of metallic nickel. The particle sizes measured from the TEM picture are about 5 nm in diameter and were very well dispersed in the polystyrene. The magnetization measurements established that the as-prepared nanocomposite materials are super paramagnetic due to their small size. The saturation magnetization (30.1 emu $g^{-1}$) and coercivity (5 Oe) of the materials were significantly smaller than those of the bulk nickel, reflecting the nanoparticle nature.

Kitamoto and Abe applied power ultrasonic waves (19.5 kHz, 600 W) to 300 ml of $FeCl_2$ aqueous solution (pH 7.0) at 70 °C, and succeeded in encapsulating polyacrylate spheres of 250 nm diameter with magnetite ferrite coatings [49]. From TEM observations of the cross sections it was seen that the polymer spheres were covered with uniform columnar crystallites of 30–40 nm in diameter at the bottom and 60–70 nm at the top. The ultrasound waves produce OH groups on the polymer surfaces which work as ferrite nucleation sites; this improves the quality of the ferrite coatings. The ferrite-encapsulated particles will greatly improve the performance of the enzyme immunoassay as a cancer test reagent. The above possible mechanism for the formation of the blue oxide is consistent with explanations in the literature for a sonochemical reaction.

Finally, it should be mentioned that Suslick sonicated the transition-metal carbonyl in a low volatility solvent in the presence of poly(vinylpyrrolidone) and obtained metallic colloids for Fe and Co [63, 64].

#### 6.1.1.6 Sonochemical Synthesis of Nanometals Encapsulated in a Carbon Matrix

The metal particles of an interstitial solid solution of palladium carbide, $PdC_x$ ($0 < x \leq 0.15$), were synthesized [65] at room temperature in an aqueous solution during the reduction of tetrachloropalladate (II) with sonochemically produced organic radicals. The sonochemical reduction was carried out using a 200 kHz ultrasonic generator operating at 200 W (6 W $cm^{-2}$, 65 mL). An aqueous solution of $PdCl_2 \cdot 2NaCl$ (1–10 mM, 60 mL) was placed in a cylindrical glass vessel (55 mL), which had a silicon rubber septum for gas bubbling or sample extraction, without exposing the sample to air. The vessel was fixed in a constant position and then irradiated for 1 h under argon at 20 °C. Under experimental conditions, the rate of formation of OH radicals and H atoms in the sonolysis of pure water was estimated to be 20 mM $min^{-1}$.

Organic compounds such as methanol, ethanol, hexanol, isopropanol, *tert*-butyl alcohol, and acetone, were injected into the solution using a microsyringe through the septum just before the irradiation, and acted as accelerators of the reduction of Pd (II).

The number of carbon atoms in the Pd particles was controlled by changing the concentration and the type of organic additives. The mechanism proposed for the PdC formation comprises the following steps: (i) an active Pd cluster is formed during the synthesis of Pd particles, (ii) organic additives are then adsorbed on the Pd cluster surface, and finally (iii) carbon atoms on the particle surface, which are formed from the catalytic dissociation of the additives, diffuse in the Pd metal lattice.

Similar studies in an organic solvent yielded almost the same product [66]. Nanostructured particles of amorphous carbon-activated palladium metallic clusters have been prepared (in situ) at room temperature by ultrasound irradiation of an organometallic precursor, tris-$\mu$-[dibenzylideneacetone]dipalladium [($\varphi$–CH=CH–CO–CH=CH–$\varphi$)$_3$Pd$_2$] in mesitylene. Characterization studies show that the product powder consists of nanosize particles, agglomerated in clusters of approximately 800 Å. Each particle is found to have a metallic core, covered by a carbonic shell that plays an important role in the stability of the nanoparticles. The catalytic activity in a Heck reaction, in the absence of phosphine ligands, has been demonstrated.

Gedanken and his group were searching to replace the $Ni(CO)_4$, which was the source for the preparation of nickel, and is known to be a hazardous material. They found [67] a new precursor for the sonochemical preparation of amorphous nickel, Ni(cyclooctadiene)$_2$, which yielded relatively large (200 nm) amorphous nanoparticles composed of nickel and carbon atoms. Small nickel particles were dispersed all over the particle. When these particles were heated slightly above their crystallization temperature, much smaller particles (5 nm) of encapsulated crystalline nickel in amorphous carbon were obtained. The XPS spectrum reveals that the crystallization process is also accompanied by the reduction of the surface $Ni^{+2}$ ions by the amorphous carbon atoms. The DSC measurements substantiate this assumption.

Walter and coworkers [68] intercalated small Pt nanoparticles into graphite using a sonochemical process. $H_2PtCl_6$ was intercalated into natural graphite by applying ultrasound to a mixture of graphite, $H_2PtCl_6$, $CCl_4$, and $SOCl_2$ for 3 days. X-ray diffraction data showed that the host lattice was partly intercalated by $H_2PtCl_6$. A mixture consisting of a third and fourth stage together with unreacted graphite was observed. The intercalation compound was suspended in acetone with hydrogen flowing through while the sonication took place for 2 days. Transmission electron microphotographs showed highly dispersed nanoparticles in a narrow size range inside the carbon lattice. X-ray photoelectron spectroscopy gave evidence that these particles are platinum metal (Pt(0)). Particle thickness estimated by X-ray diffraction indicated an average particle thickness of two layers. Selected-area electron diffraction microphotographs showed a pattern that could be hexagonally indexed. A ($2 \times$ alpha (graphite)) superstructure was observed for those quasi-two-dimensional aggregates formed by self-organization. This indicates a templating effect due to the carbon lattice.

Iijima and his group used ultrasound radiation to react carbon nanotubes [69]. They sonicated single-wall carbon nanotubes (SWNTs) in a monochlorobenzene (MCB) solution of poly(methyl methacrylate) (PMMA) and were able to react SWNTs with MCB or PMMA chemically. After the SWNTs reacted with these organic materials, they turned into ragged SWNTs (r-SWNTs) with many defects in the sidewall when burned in oxygen. They consider sonochemistry to be a simple method to functionalize SWNTs.

Nikitenko has succeeded recently in preparing iron nanoparticles that are air-stable [70]. Iron is easily oxidized in ambient conditions, but when it comes to iron nanoparticles they are pyrophoric and burn spontaneously in air. That is why obtaining iron nanoparticles that are air-stable is important. On the other hand, it is known that iron nanoparticles obtained sonochemically from alkane solutions are not stable in contact with air. In his study, Nikitenko prepared coated iron nanoparticles by the sonochemical decomposition of $[Fe(CO)_5]$ in diphenylmethane (DPhM), an aromatic solvent with physicochemical properties (m.p. 25 °C, b.p. 265 °C, vapor pressure 1 kPa at 77 °C) that are suitable for sonochemistry. The sonolysis of the neat DPhM forms a polymer-like solid product. It is proposed that small amounts of this product generated in situ would coat the surface of iron nanoparticles formed simultaneously from $Fe(CO)_5$. The as-prepared material contained 17.6 wt% C and 1.5 wt% H. The reaction yield was 53% with respect to iron. The presence of a significant amount of carbon and hydrogen indicates that the as-prepared material contains the sonolytic decomposition products of DPhM. At this stage the as-prepared product is annealed at 700 °C in argon. An air-stable, dark gray magnetic powder is formed as a result of the annealing. The annealed material contains 5.6 wt% C and 0.08 wt% H. The annealed product is air-stable. Its stability in air was checked once a month for the first 6 months after preparation by (1) Mössbauer spectroscopy, (2) XRD, and (3) magnetization measurements. No changes were observed in the measured parameters during this period. The annealed material is ferromagnetic, as follows from the magnetization curve. The saturation magnetization $M_s$ and coercive force $H_c$ are 212 emu $g^{-1}$ and 40 Oe, respectively. The $M_s$ value of the material is unexpectedly high and is close to that of bulk bcc Fe (222 emu $g^{-1}$). The air-stability of the product originates from a $Fe_3C/C$ layer surrounding the core of the iron nanoparticle.

### 6.1.2 Sonochemical Fabrication of Nano-Metallic Oxides

#### 6.1.2.1 Sonochemical Synthesis of Transition Metal Oxides from the Corresponding Carbonyls

Sonication of a decalin solution of $Fe(CO)_5$ in air yields amorphous nanoparticles of $Fe_2O_3$ [71]. This paper presents another advantage of using sonochemistry for preparation of an amorphous product. It is well known that, unlike iron, which can be obtained in the amorphous form by the cold quenching technique, amorphous iron oxide cannot be prepared in this way. Higher cooling rates are needed for iron oxides and other metal oxides. This is because the thermal conductivities of metal oxides are usually much lower than those of the metals. This is the reason that glass formers, whose purpose is to prevent crystallization, are added if the quenching method is applied. Amorphous metal oxides can be prepared by rapidly quenching the molten mixture of metal oxides only if a glass former, such as $P_2O_5$, $V_2O_5$, $Bi_2O_3$, $SiO_2$, and CaO, is added to the mixture [72–74]. As mentioned above, the cooling rates obtained during the cavitational collapse are estimated to be greater than $2 \times 10^9$ K $s^{-1}$. That is the reason why amorphous iron oxide, $Fe_2O_3$, can be prepared by sonicating $Fe(CO)_5$ in a decalin solution under air without adding any glass former.

The characterization of an amorphous material is difficult because it lacks characteristic XRD diffractions, so that Auger or Mössbauer spectroscopies are preferred, along with other conventional analytical assays such as spot test and iodometric titration [71]. The $Fe_2O_3$ nanoparticles are converted to crystalline $Fe_3O_4$ nanoparticles when heated to 420 °C under vacuum or when heated to the same temperature under a nitrogen atmosphere. The magnetization of pure amorphous $Fe_2O_3$ at room temperature is very low ($< 1.5$ emu $g^{-1}$) and it crystallizes at 268 °C.

Ultrafine powders of $Cr_2O_3$ and $Mn_2O_3$ have been prepared at room temperature by the sonochemical reduction of aqueous solutions containing $(NH_4)_2Cr_2O_7$ and $KMnO_4$, respectively [75]. The yield of the sonochemical reduction has been enhanced by raising the reaction temperature or by using a 0.1 M aqueous solution of ethanol. The amorphous powders are nanosized (50–200 nm), and the surface area varies from 35 to 48 $m^2$ $g^{-1}$. The crystallization of amorphous $Mn_2O_3$ and $Cr_2O_3$ could be achieved by heating them at 600 and 900 K, for 4 h, respectively.

Ultrasound irradiation of a slurry of $Mo(CO)_6$ in decalin for 3 h under ambient air produces blue-colored, $Mo_2O_5 \cdot 2H_2O$ [76]. FT-IR analysis of this material reveals the existence of Mo=O and Mo–O bonds as well as hydrogen-bonded and coordinated water molecules. The amount of water molecules was determined by thermogravimetric analysis. Characterization using powder X-ray diffraction (XRD) and transmission electron microscopy (TEM) with selected area electron diffraction (SAED) shows the amorphous nature of the blue product. The TEM picture shows that the blue oxide is composed of spongy platelet nanoparticles (20 nm in diameter). Heating the initial blue powder at 300 °C for 2 days under an oxygen, hydrogen, and nitrogen atmosphere yields X-ray crystalline $MoO_3$, $MoO_2$, and a mixture of $MoO_3$ and $MoO_2$, respectively. X-ray photoelectron spectroscopy (XPS), along with the potentiometric titration analysis of the blue oxide, confirms the formation of pentavalent molybdenum oxide. UV–visible absorption studies of the blue product demonstrate that the characteristic absorption of the Mo(V) (d1 – cation) oxide system and the Mo ions probably consist of two types of coordination symmetry (*Td* and *Oh*). Electron paramagnetic resonance (EPR) experimental results revealed an unusual doublet pattern, which is ascribed to superhyperfine coupling of pentavalent molybdenum with a proton of coordinated water. The nanostructured amorphous pentavalent molybdenum oxide (blue oxide) thus formed has also been successfully deposited ultrasonically on Stober's silica microspheres (250 nm). The TEM images of silica-supported blue oxide reveal uniform distribution and the strong adhering nature of the blue oxide. FT-IR spectroscopy illustrated the structural changes that occur when the amorphous $SiO_2$ is coated sonochemically with the blue oxide.

The sonochemical oxidation of molybdenum carbonyl would not occur in the bubble, due to its low vapor pressure. Therefore, the sonochemical reaction of

$Mo(CO)_6$ appears to take place in the interfacial region: Thus, the water molecule formed in the bubble may diffuse into the interfacial region or the water molecules that are available from the atmospheric air, precursor, and solvent, and stabilize the unusual pentavalent molybdenum oxide:

$$2Mo(CO)_6 + 17/2O_2 \rightarrow Mo_2O_5 + 12CO_2 \text{ (Sonolysis)} \quad (3)$$

$$Mo_2O_5 + 2H_2O \rightarrow Mo_2O_5{\cdot}2H_2O \quad (4)$$

The above possible mechanism for the formation of the blue oxide is consistent with explanations in the literature for a sonochemical reaction.

Magnetite nanorods have been prepared by the sonication of aqueous iron (II) acetate in the presence of $\beta$-cyclodextrin [77]. The as-prepared magnetite nanorods are ferromagnetic and their magnetization at room temperature is about 78 emu $g^{-1}$. The particle sizes measured from transmission electron micrographs are about 48/14 nm (L/W). A mechanism for the sonochemical formation of magnetite nanorods is discussed.

Only one report was found that discussed the effect of a magnetic field on sonication and sonication products [78]. The sonochemical irradiation of $Fe(CO)_5$ solution in decalin under argon has been carried out with and without an external magnetic field. The sonication cell placed between the poles of a magnetic field of 7 kG which was applied during the sonication. We have already pointed out that this reaction yields amorphous $Fe_2O_3$ [9]. Direct TEM measurements reveal that the sample obtained without a magnetic field consists of sponge-like particles with a mean size of about 25 nm, whereas the sample synthesized in a 7 kG magnetic field consists of highly acicular particles, 50 nm in length and 5 nm in width. Our finding sheds light on the process of particle nucleation during sonication, which cannot be a diffusion-assisted growth because of the very small time scale. We conclude, therefore, that particles are forced to form an acicular entity by direct magnetic interactions. The amorphous nature of the as-prepared substance was verified by X-ray diffraction, selected area electron diffraction, and differential scanning calorimetry. The magnetic moment vs. temperature measurements and Mössbauer spectroscopy reveal a large shift of the blocking temperature of about 70 K toward higher temperatures for the sample obtained in the magnetic field. We attribute the observed shift to the significant enhancement of the particle shape magnetic anisotropy.

#### 6.1.2.2 **Sonochemical Synthesis of Ferrites from the Corresponding Carbonyls**

It was just a reasonable extension that a mixture of carbonyls would be sonicated under air to yield the corresponding ferrites while the irradiation under argon yielded the metallic alloy [37]. Indeed, nanosized amorphous $NiFe_2O_4$ powder was prepared by sonochemical decomposition of solutions of volatile organic precursors, $Fe(CO)_5$ and $Ni(CO)_4$ in decalin at 273 K, under an oxygen pressure of 100–150 kPa [79]. The amorphous nature of these particles was confirmed by various techniques. Magnetic measurements, Mössbauer, and EPR spectral studies indicated that the as-prepared $NiFe_2O_4$ ferrite particles were super-paramagnetic. The Mössbauer spectrum of the crystallized sample showed a clear sextet pattern, with hyperfine field values of 500 and 508 kOe for A (tetrahedral) and B (octahedral) sublattices, respectively, of the inverse spinel $NiFe_2O_4$. Saturation magnetization of the annealed sample (25 emu $g^{-1}$) was significantly lower than that for the reported multidomain bulk particles (55 emu $g^{-1}$), reflecting the ultrafine nature of the sample. Thermogravimetric measurements with a permanent magnet gave Curie temperatures of 44 °C for amorphous and 560 °C for the crystallized forms.

Another ferrite exhibiting two unique properties was also synthesized sonochemically, $BaFe_{12}O_{19}$ [80]. A solution of $Fe(CO)_5$ and barium ethylhexanoate ($Ba–[OOCCH(C_2H_5)C_4H_9]_2$) in decane, in stoichiometric ratio, was decomposed by high-intensity ultrasonication. The as-prepared material was an amorphous $BaFe_{12}O_{19}$ precursor in colloidal suspension, where the particles are in the nanometer size regime and are homogeneously distributed. The precursor is extracted from the solution as powder by precipitation or by evaporation and then calcined at low temperature (600 °C) to obtain the final $BaFe_{12}O_{19}$ crystalline nanosized powders. The first unique property detected by Shafi was the formation of features such as the *Olympic Rings* on transmission electron micrographs of amorphous $BaFe_{12}O_{19}$ nanoparticles (Figure 6.4). Rings of smaller dimensions trapped inside

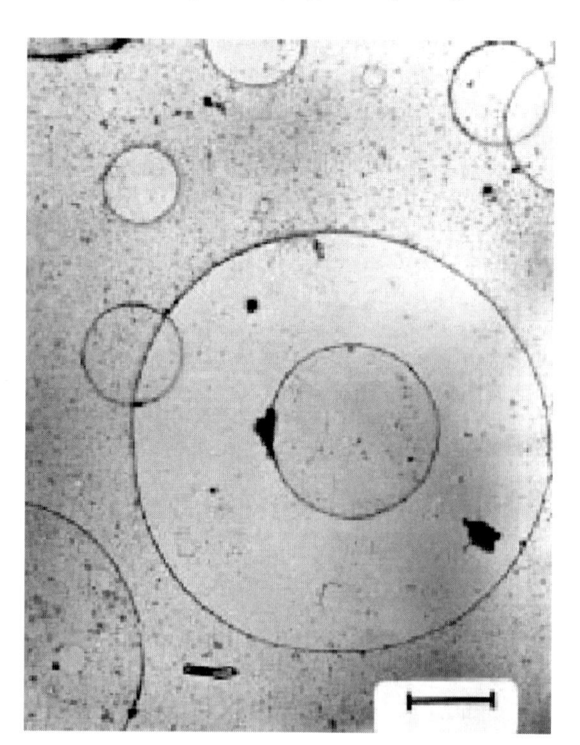

**Fig. 6.4.** TEM micrograph showing the self-organization of super-paramagnetic nanoparticles into submicron size rings the so-called *Olympic Rings* (scale bar is 0.7 μm).

the larger ones were another unique observation. The intersection of two rings is amazing, as this is in direct contradiction to the proposed mechanism for the ring formation, based on the dry hole formation on an evaporating thin film completely wetted to the substrate. The creation of this unique feature is attributed to the interplay of magnetic forces with the regular particle–substrate interactions. The second detected property was that barium hexaferrite was formed as a colloidal solution without the use of a surfactant.

#### 6.1.2.3 Sonochemical Preparation of Nanosized Rare-Earth Oxides

A long list of oxides was prepared sonochemically. Almost all the above-mentioned oxides were synthesized in organic solvents. The other oxides that will be discussed from here on were all prepared in aqueous solutions. Submicron size spheres of silica and alumina prepared by well-known methods were coated sonochemically by nanoparticles of oxides of europium and terbium using the same concentration of ions [81]. We have also used sonochemistry to prepare nanoparticles of silica and alumina doped with the same rare-earth ions for comparison. The highest luminescence intensities were observed for europium and terbium doped in nanoparticles of alumina of dimension 20–30 nm. The intensities are comparable or higher than in commercial phosphors.

The synthesis of the rare-earth oxides was as follows: europium oxide was dissolved in a minimum amount of nitric acid and evaporated to dryness. The dry nitrate was dissolved in 5 mL of water. The required amount of silica microspheres was put in a beaker with 30 mL of water and the europium nitrate solution was added. The sonication was carried out for 1 h in an open beaker kept in an ice bath. 5 mL of 25% aqueous ammonia was added dropwise during the sonication. The resulting product after sonication was washed extensively with water, centrifuged, and dried under vacuum.

The preparation of the europium oxide-doped silica nano-particles ($99.0SiO_2$–$1.0Eu_2O_3$ mol%) was different. It was also carried out by the hydrolysis of tetraethyl orthosilicate (TEOS) in the presence of $H_2O$, ethanol, and a europium nitrate solution. 25% aqueous ammonia was added dropwise during the sonication. In this way, europium oxide incorporated into silica particles was obtained.

Decay time measurements of $Eu_2O_3$, and $Tb_2O_3$ doped and coated on alumina were conducted [82]. The luminescence of the alumina substrate was found to be much shorter than that of the rare-earth oxides. Differences between the decay times of the deposited and doped materials are accounted for by the stronger guest–host interaction and the absence of concentration quenching in the doped material.

In a similar work, ultrasound radiation was used to prepare $Eu_2O_3$ doped in zirconia and yttrium-stabilized zirconium (YSZ) nanoparticles [83]. Europium oxide was also coated sonochemically on the surface of submicron spherical zirconia and YSZ, which were fabricated by wet chemical methods. Time decay measurements of the doped and coated materials were conducted using a pulsed laser source. Lifetimes $< 1.1$ ms radiative lifetime of the $Eu^{+3}$ ions were detected for the doped and coated as-prepared materials. When the doped and coated samples were annealed at 700 °C, longer lifetimes were measured. The shorter lifetimes were attributed to concentration quenching.

This series of experiments was continued and we recently deposited sonochemically a nanolayer of $Eu_2O_3$ on submicron-size titania [84]. Ultrasound irradiation of a slurry of titania, europium nitrate, and ammonia in an aqueous medium for 120 min yielded an $Eu_2O_3$–$TiO_2$ nanocomposite.

Finally, europium oxide nanorods have been prepared by the sonication of an aqueous solution of europium nitrate in the presence of ammonia [85]. The particle sizes measured from transmission electron micrographs and HRSEM are about 50 × 500 nm (W × L). Sonication of an aqueous solution of europium nitrate in the presence of ammonia results in the precipitation of europium hydroxide: The as-prepared material is europium hydroxide, as confirmed by TGA, DSC, XPS, and Mössbauer spectroscopy measurements, as well as by PXRD of the as-prepared sample assisted by microwave irradiation

$$Eu^{+3}_{(aq)} + 3H_2O \rightarrow Eu(OH)_{3(S)} + 3H^{+}_{(aq)} + NH^{+4}_{(aq)} \quad (5)$$

The presence of ammonium ion seems to be important for the formation of particles with morphology usually obtained in the absence of ammonia. In our case this was confirmed by the absence of precipitate after sonicating a solution of europium nitrate without addition of ammonia. Regarding the shape of the as-prepared product, careful examination of TEM, HRSEM, and HRTEM micrographs reveals that the rod-shaped particles of europium hydroxide consist of a few spherical particles aligned in one direction. The formation of europium hydroxide nanorods may be explained as follows. The $Eu(OH)_3$ particles formed upon collapse of the bubbles adsorb $NH_4{}^+$ ions or ammonia on their surfaces, thus forming a monolayer and fusing them by hydrogen bonding. In this way ammonium ions would be responsible for the observed morphology. This adsorption of $NH_4{}^+$ on the surface of the particles is supported by the results of Sherif et al. [86], who showed that ammonium nitrate ($NH_4NO_3$) is adsorbed on the surface of the powder after precipitation. We also evoke in the explanation the microjets mentioned above, and due to this very high velocity of liquid jets, the $Eu(OH)_3$, nanoparticles are pushed toward each other, forming the nanorods.

#### 6.1.2.4 The Sonohydrolysis of Group 3A Compounds

A detailed study of the sonohydrolysis of Ga, Al, In and Tl was conducted by sonicating the chloride aqueous solutions of these metals. Unlike the sonication of carbonyls, which resulted in nanophase amorphous products [6, 12], the sonication of inorganic salts such as $GaCl_3$, $InCl_3$, or $TlCl_3$ yields crystalline nanophase products. The sonochemical reaction of an aqueous solution of $GaCl_3$ led to the formation of GaO(OH) rolled up in a scroll-like layered structure to give cylinders of 80–120 nm diameter and 500–600 nm in length [89]. Small amounts of metallic Ga were incorporated with these tubes. The presence of zero-valent Ga (ca. 1%) was demonstrated by the DSC spectrum, which displayed an endothermic peak at 29.2 °C. The amount of metallic gallium is dependent on the irradiation time:

irradiating the starting solution for 90 min resulted in a DSC signal which was an order of magnitude smaller than that obtained after 6 h of sonication. In a parallel experiment, a 1 M aqueous solution of $GaCl_3$ was heated to 300 °C in a high-pressure cell, without sonication. No noticeable powder formation occurred after heating the solution for 4 h. The formation of GaO(OH) can occur by reaction (6);

$$Ga^{+3}_{(aq)} + 2H_2O_{(l)} \rightarrow CaO(OH)_{(s)} + 3H^{+3}_{(aq)} \qquad (6)$$

The observation that the pH of the solution changes from 2.37 prior to the sonication to 1.61 at the end of the reaction implies that reaction (6) is the dominant pathway. This reaction is endothermic [90] with an equilibrium constant at 50 °C of ca. $10^{-3}$. The high local temperature in the shell surrounding the collapsing bubble constitutes the driving force for the formation of the GaO(OH).

An explanation for the elongated layered structure of the GaO(OH) based on cavitation collapse occurring at the solid–liquid boundary has been suggested [89]. The particle size was found to decrease with decreasing concentration of $GaCl_3$ due to a reduced amount of reactant at the bubble–liquid interface.

We could not investigate and characterize the sonohydrolytic product of $AlCl_3$, because it is a very hygroscopic product. On the other hand, the products of the sonohydrolysis of $InCl_3$ [91], and $TlCl_3$ [92], were investigated in great detail, the latter producing inorganic fullerenes (IF) of $Tl_2O$.

$In(OH)_3$ nanopowder was prepared via the sonication of an aqueous solution of $InCl_3$ at room temperature and at 0 °C. At these temperatures, nonsonicated hydrolysis does not occur. The role of the ultrasound radiation and the mechanism of the reaction are discussed. The proposed mechanism is based on the sonohydrolysis of In (III) ions in the outer ring, and the liquid shell, of the collapsing bubble. The product, $In(OH)_3$, was obtained as needle-shaped particles.

The sonication of an aqueous solution of $TlCl_3$ under a flow of argon led to a precipitate, composed of two products. The major product is $Tl_2OCl_2$, while $Tl_2O$ is obtained in small quantities. The latter has the structure of a multi-shell closed compound. The identification of the inorganic fullerenes in the TEM picture as the thallium (I) oxide is based on SAED measurements. In Figure 6.5 we present a picture of the onion-like IF.

The three-dimensional structure of the IF was demonstrated when the sample was tilted under the electron beam and an identical picture was obtained at all tilting angles (see Figure 2 in [92]). The growth mechanism of both carbon and inorganic fullerenes is not yet fully understood. However, it is believed that the main stimulus for the formation of fullerene-like structures emanates from the large energy associated with the dangling covalent bonds at the edges of the layered structures. The growth conditions of fullerene-like structures, in most cases, are far from equilibrium. The large thermal energies during growth force the nanoclusters of layered materials. We suggest the following mechanism to explain the formation of the closed curved structures. In the first stage, $Tl_2O$ is formed via the sonochemical reduction of the $Tl^{+3}$. It is known that sonohydrolysis of aqueous solutions of metal ions is caused by the high local temper-

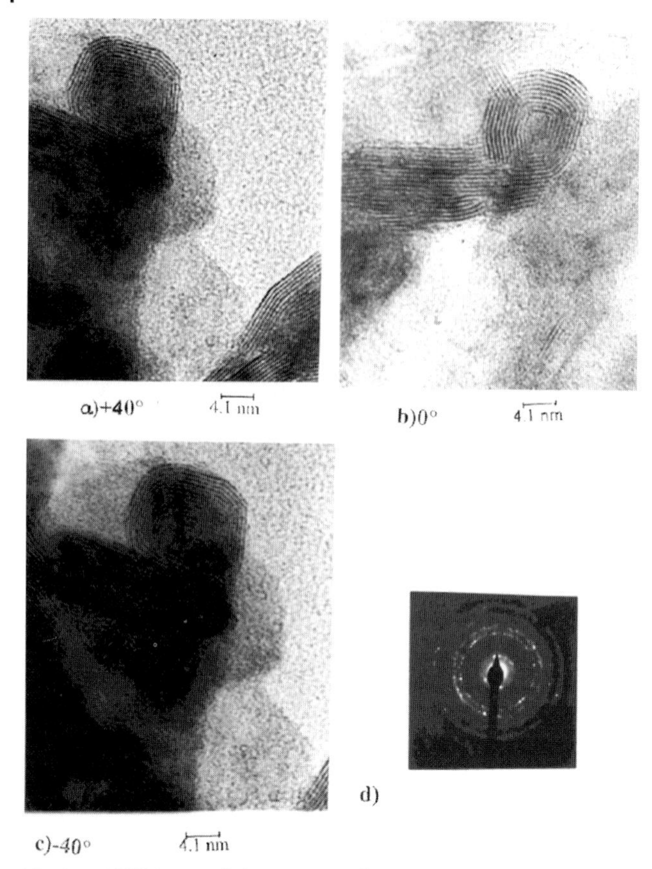


**Fig. 6.5.** TEM image of the as-prepared powder of fullerene-like inorganic $TlCl_3$, at different tilt angles. (a) +40°, (b) 0°, (c) −40°, (d) SAED inserted (Scale: 1 cm = 4.1 nm).

atures in or near the cavitation bubbles. If the original $Tl_2O$ particle is planar (two-dimensional), curvature and closure of the fullerene-like structures may occur around the collapsing bubble. The temperature gradient from the bubble surface into the solution should cause a temperature gradient across the particle that causes the curvature. This is helped by the energetics, which favors sheet closure due to bond energy released by eliminating reactive edges in the planar structures.

#### 6.1.2.5 The Sonochemical Synthesis of Nanostructured $SnO_2$ and SnO as their Use as Electrode Materials

In recent years there has been an increasing interest in the use of $SnO_2$ as an anode material for lithium batteries, starting with the announcement by Fuji Photo Film Celltee Co. Ltd., Japan, about their STALION lithium ion cell. These cells utilized an amorphous tin-based composite oxide as the anode (together with a

$Li_xCoO_2$ cathode). It appears that the basic anode process in these cells is lithium insertion into tin oxides, which forms Li–Sn alloys in a matrix of $Li_2O$. Hence, the reversible process that finally remains is lithium alloying with metallic tin. For this purpose we have prepared $SnO_2$, and later SnO semiconductor nanoparticles. They were synthesized by ultrasonic irradiation of an aqueous solution of $SnCl_4$ and azodicarbonamide under ambient air [93]. These nanoparticles are 3–5 nm in size, as calculated using the Debye–Scherrer formula, and as observed by TEM. Electrochemical tests were performed using the $SnO_2$ nanoparticles as the electrode materials in nonaqueous Li salt solutions. Amorphous, as well as crystalline, nanoparticles were prepared for this project. The performance of the anode was examined as a function of the crystalline state of the $SnO_2$, as well as the particle size.

These tests clearly indicated that the $SnO_2$ particles prepared by sonochemical synthesis have a promising capacity, reversibility, and cycle life in repeated lithium insertion and deinsertion processes. Heat treatment of the sonochemically prepared $SnO_2$ nanoparticles had a pronounced effect on their electrochemical behavior. The crystallization of the particles due to the heat treatment increased both the electrode's irreversible and reversible capacities measured during the first lithiation–delithiation cycle. However, while the reversible capacity retention upon cycling of electrodes made of as-prepared $SnO_2$ material was good, the capacity of electrodes made of the heat-treated materials degrades upon cycling.

SnO was prepared similarly to the preparation of $SnO_2$ [94]. Nanoparticles of SnO were synthesized sonochemically in mildly basic $SnCl_2$ solutions. The amorphous product thus obtained could be transformed to a nanocrystalline phase by heating to 200 °C. Composite electrodes comprised (by weight) of 80% SnO, 10% graphite flakes (conductive additive), and 10% polymeric binder (an optimal composition) were tested as anodes for rechargeable Li batteries. Both the amorphous and the nanocrystalline SnO are electrochemically active and can be reduced in nonaqueous Li salt solutions to matrixes of $Li_2O$ and $Li1_7Sn_4$. The nanocrystalline SnO was found to be much more effective and a superior anode material to the amorphous and microcrystalline phases, as an active material for electrodes. These electrodes could reach nearly their theoretical capacity (= 790 mA h $g^{-1}$, SnO) in electrochemical lithiation–delithiation processes versus a Li counter electrode in nonaqueous Li salt solutions. However, there is still a long way to go to the possible use of SnO as an anode material in practical batteries. This is due to its high irreversible capacity ($Li_2O$ formation and surface film precipitation due to reactions of lithium–tin compounds with solution species) and gradual capacity decrease during repeated charge–discharge cycling. Possible reasons for this capacity fading are discussed in [94].

#### 6.1.2.6 The Sonochemical Synthesis of Mesoporous Materials and the Insertion of Nanoparticles into the Mesopores by Ultrasound Radiation

In a recent review article the synthesis of mesoporous (MSP) silica is considered to be one of the four most important discoveries in solid-state and materials science in the last decade [95]. We have summarized our activities in this field in an article published in Chemistry – A European Journal [96]. We will therefore repeat briefly our early findings and introduce some new results.

The report demonstrates that ultrasound radiation can also be used for the preparation of MSP materials. MSP silica, MCM-41 [97], MSP titania [98], and YSZ (yittria stabilized zirconia) [99] were all prepared by this method. In addition, straight-extended layered mesostructures based on transition metal (Fe, Cr) and rare earth (Y, Ce, La, Sm, Er) oxides were also synthesized sonochemically [100]. The synthetic processes reproduced in most cases the already published sol–gel synthesis [101]. The main advantage of the sonication method is the short irradiation time. In most cases the reaction time was 3 h. The longest sonication period was 6 h. It was applied for the synthesis of MSP YSZ and caused the transformation of the product from a layered to a hexagonal mesostructure due to this prolonged irradiation time. A second advantage of the sonochemical synthesis is that the walls of the sonochemical product were thicker than those obtained conventionally (see Table 1, [97]). The thicker walls are responsible for the MCM-41 obtained sonochemically being more stable than MCM-41 prepared by conventional hydrothermal methods [102]. This was demonstrated when our product was treated with pure water, and its crystallinity changed only a little after heating under reflux for 6 h, and decreased by approximately 65% after heating under reflux for 12 h. In the literature [102] the MCM-41 prepared by using conventional hydrothermal methods became amorphous after refluxing for 12 h.

During the formation of the framework, despite the agitation of the ultrasound, which helps to disperse the small silica oligomers more homogeneously in the mixture, the formation of hot spots within the surfactant–silicate interface may accelerate the silica polymerization, which is slow and rate limiting under normal conditions. Thus, the fabrication of the meso-structure can be achieved more efficiently. On the one hand, acoustic cavitation etches the surfactant–silicate micelles on the surface; this results in a coarse outer surface or even the fragmentation of the micelles. An additional factor is that hot spots accelerate the condensation of surface silanol groups among micelles; in this way ultrasound radiation accelerates the formation of the MCM-41 framework and the growth of particles.

The discovery of MSP materials led immediately to the development of many experimental methods for the deposition of materials, especially catalysts, into the mesopores. We have deposited Mo oxide, and Co/Mo oxides into MCM-41 as well as into the pores of Al-MCM-41 [103]. We have also anchored $Fe_2O_3$ into the mesopores of titania [104]. A large variety of nanoparticles has been introduced into many MSP materials. This work, however, has not been published. In addition to the characterization studies of the composite catalyst-mesoporous product, catalytic studies have also been conducted.

The typical sonochemical reaction is performed as follows: a slurry of the MSP material, for example, Al-MCM-41, in decalin containing dissolved $Mo(CO)_6$ and/or $Co(CO)_3NO$, is sonicated. The sonication is carried out under ambient air at room temperature. The solid product is separated by centrifugation, thoroughly washed with dry pentane, and dried in vacuum at room temperature. The chemical composition of the solid phase was determined by EDAX to probe whether the transition metal is indeed found in the solid. The kinetics of the decomposition of the carbonyl in the presence of the MSP was faster than in its absence (7 times faster for Al-MCM-41). This is attributed to the large number of bubbles developed due to the solid surface. The amount of $MoO_x$ in the solid Al-MCM-41 could reach 67 wt%. However, it was necessary to find the location of the $MoO_x$ particles. In other words, to find out whether they have indeed entered the mesopores or can be found outside the pores. This location was determined by mapping the TEM grid employing TEM–EDAX measurements. The data clearly demonstrate [103] that the Mo oxide phase is located inside the support's pores and does not form separate particles up to an $MoO_3$ content of about 40–45 wt% upon ultrasonic deposition. High-resolution TEM (HRTEM) pictures [103] reveal that the $MoO_3$ deposition does not cause degradation of the Al-MCM-41 hexagonal pore structure. In addition to the EDAX mapping technique, four other methods are used to support these results. In the sample containing 67 wt% $MoO_x$, large amounts of the excess material are found outside the pores. Thus, a monolayer of $MoO_x$ particles is strongly anchored to the walls, while the rest are found outside. Considering the surface area of Al-MCM-41 used in the work and an Mo surface concentration of 5 Mo atoms $nm^{-2}$, the geometrical closed, packed monolayer capacity corresponds to 50 wt% $MoO_3$, which is in good agreement with our measurements. XPS measurements proved that chemical bonds are being formed between the silica and the molybdena, forming Si–O–Mo bonds.

The main advantage of using ultrasound for the insertion of nanoparticles into MSP materials is illustrated in Figure 6.6, which compares the normalized surface area (NSA) values obtained by the sonochemical methods with those of other methods such as impregnation and thermal spreading. The NSA is defined as

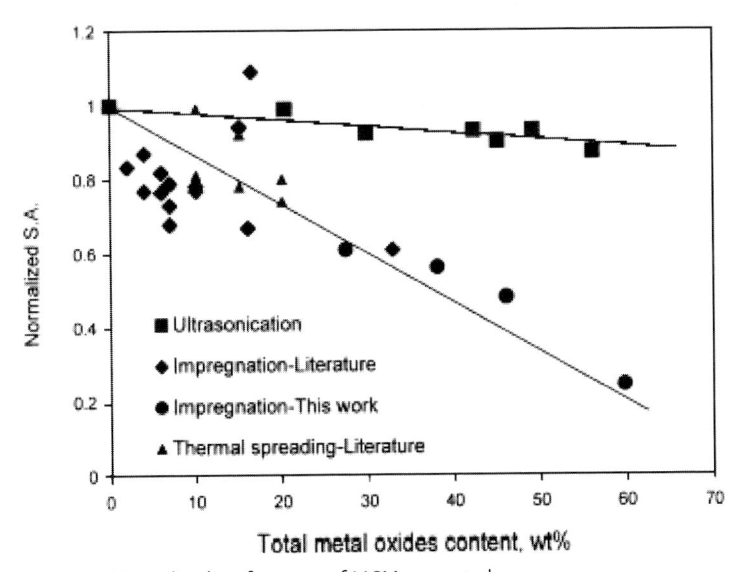


**Fig. 6.6.** Normalized surface area of MCM-supported catalysts: Mo–, Ni–, and Co(Ni)–Mo catalysts in oxide form.

$$NSA = SA_{catalyst}/(1 - y) \cdot 1/SA_{MCM} \quad (7)$$

where SA is the specific surface area of the parent Al-MCM-41 or the metal oxide/Al-MCM-41 composite, and $y$ is the weight fraction of metal oxides in the catalyst. In other words, NSA reflects the surface area per gram of MCM. If the active component is distributed at the support surface in the form of a close-packed monolayer of a given thickness (no pore blocking occurs), the NSA decreases only as a result of narrowing of the support pores. For example, for a 45 wt% $MoO_3$ phase loaded on a wide-pore Al-MCM-41 support, the reduction of pore diameter from 8.3 to 7.7 nm yields an NSA value of 0.93. The results shown in Figure 6.6 speak for themselves, showing that only a slight reduction in the NSA occurs when sonochemistry is used for the insertion of nanoparticles, or in other words, the loading of the MSP by large amounts of nanoparticles reduces the SA only slightly, whereas when other methods are used a considerable reduction in the SA is observed.

The remaining question is what is the role of the ultrasound radiation in the insertion of the nanoparticles into the mesopores? It is clear that the bubble cannot collapse inside the mesopores because the size that the bubble reaches before its collapse is estimated to be about 100 μm [105], while the pore diameter is less than 3 nm. Instead, we propose two possible mechanisms. The first is based on microjets and shock waves that result when a bubble collapses near a solid surface. As already explained, near a solid surface the collapse drives high-speed jets of liquid into the surface. Since most of the energy is transferred to the accelerating jet, the jet can reach velocities of hundreds of meters per second. In our case the small nanoparticles are pushed by these jets into the mesopores and, as a result of their reaction with the mesoporous support, are anchored to the inner surface of the mesoporous material. The other possibility is that the solution inside the pores undergoes a chemical reaction initiated by shock waves forming nanoparticles. The nanosized products interact with silanols, forming the chemical bonds described above. The second mechanism sounds a better explanation since it accounts for the homogeneous spreading of the nanoparticles in the pores.

Another active group in this field is that of Chen and Zhang. In their first paper [106] they reported on a sonochemical procedure at room temperature for the preparation of gold nanoparticles loaded in mesoporous silica. They show that the Au nanoparticles, with a mean size of 3–4 nm in diameter and a fairly narrow size distribution, are dispersed uniformly within the pores of the silica host. In a second paper they measured the optical absorption of the Au nanoparticles dispersed within ports of monolithic mesoporous silica after subsequent annealing treatment [107]. Charge transfer at the interfaces between Au nanoparticles and pore walls is introduced qualitatively to discuss the red-shift of Mie resonance absorption band with decreasing the Au particle size. A longer paper [108] gives more details on the sonochemical reduction of chloroauric acid ($HAuCl_4$) within the pores of silica. This reduction leads to the insertion of gold nanoparticles into mesoporous silica. The paper also reports on the optical measurements. In another publication related to gold inserted into MSP silica [109], they present an HR-TEM showing nearly spherical-shaped gold nanoparticles, with a mean diameter of 5.2 nm, located in the pores, most of which are less than 6 nm in diameter. The ultrasonic irradiation time dependence of optical absorption for the soaked porous solid sample, as measured by the variation in absorbance at 310 and 544 nm, indicated the reduction of Au(III) ions, and the nucleation and aggregation of gold nanoparticles within the pores of MSP silica. Additionally, the reaction rates estimated phenomenologically by the absorbance decay at 310 nm for both the porous sample and the corresponding soaking solution, showed the enhancement of the sonochemical reduction rate of Au(III) ions within the pores of mesoporous silica. It is assumed that the extensive liquid–solid interfacial zones in the pores, due to the high specific surface area and great porosity of the mesoporous solid, are the major regions where the efficient sonochemical reduction induced by the cavitation takes place.

In the latest of their publications related to the insertion of gold into MSP silica [110], the same authors discuss the structural changes that the silica skeleton undergoes upon sonication. The structure of mesoporous silica after sonochemical preparation of gold nanoparticles within its pores was studied by a nitrogen adsorption technique. It was shown that the structural parameters, such as specific surface area (SSA), porosity ($P$), and the mean pore diameter ($l$(p)) were increased significantly after ultrasonic irradiation. It is suggested that the collision of Au nanoparticles with the pore walls and localized erosion induced by the asymmetric implosive collapse of cavities on the extensive liquid–solid interface are responsible for the structural change in the mesoporous solid.

One paper by the same group reports on the sonochemical insertion of palladium nanoparticles loaded within mesoporous silica [111]. The formation of Pd nanoparticles (5–6 nm in diameter) was restricted by the coalescence of the sonochemically reduced Pd atoms inside the confined volumes of the porous solid.

One of the most exciting sonochemical preparation procedures of MSP titania did not use any surfactant [112]. Yu and his coworkers used monodispersed $TiO_2$ sol particles, which were formed initially by ultrasound-assisted hydrolysis of acetic acid-modified titanium isopropoxide. Then, the mesoporous spherical or globular particles, which have a narrow pore size distribution, were produced by controlled condensation and agglomeration of these sol nanoparticles under high intensity ultrasound irradiation. The mesoporous $TiO_2$ has a wormhole-like structure and a lack of long-range order. Nitrogen adsorption results indicate that the mesoporous $TiO_2$ retains mesoporosity with a narrow pore size distribution and high surface area to at least 673 K. The thermal stability of mesoporous $TiO_2$ is attributed to its thick inorganic walls, consisting of $TiO_2$ nanoparticles. A TGA study shows that this synthetic method is environmentally friendly. The photocatalytic activity of mesoporous $TiO_2$ for the oxidation of acetone in air was measured. As-prepared mesoporous $TiO_2$ has a negligible activity due to its amorphous structure. Calcined mesoporous $TiO_2$ shows a better activity than thecommercial photocatalyst P25. The reasons for the high activity of mesoporous $TiO_2$ are discussed.

In a continuing paper Yu and coworkers reported [113] on the preparation of mesoporous $TiO_2$ with a bicrystalline (anatase and brookite) framework, which was synthesized directly under high-intensity ultrasound irradiation. The synthesis was carried out separately, both with and without the use of a triblock copolymer. Without thermal treatment, mesoporous $TiO_2$ was formed by the agglomeration of monodispersed $TiO_2$ sol particles. The use of ultrasound irradiation assisted in the formation of the brookite phase. As the content of the brookite phase increased, the pore size and the crystalline sizes of anatase and brookite became larger when the triblock copolymer was used in the synthesis. Both as-prepared samples exhibited better activities than the commercial photocatalyst P25 in the degradation of n-pentane in air. The degradation rate of mesoporous $TiO_2$ synthesized in the presence of triblock copolymer was about two times greater than that of P25. The high activities of the mesoporous $TiO_2$ with a bicrystalline framework can be attributed to the combined effect of three factors: high brookite content, high surface area, and the existence of mesopores.

Two other MSP oxides were recently reported by Srivastava [114, 115]. Both preparations used the alkoxides of the metals as the inorganic precursor and CTAB as the template. The first report details the preparation of MSP $SnO_2$ [114]. The porous tin oxide prepared in this way was used in dye-sensitized solar cells.

The second paper [115] describes the synthesis of mesoporous iron oxide. Iron (III) ethoxide was used as an inorganic precursor and CTAB as an organic structure directing agent. After sonication, the surfactant was removed by calcination and solvent extraction methods. FTIR spectra demonstrated the removal of the surfactant from the pores of the mesoporous iron oxide. The surface area after solvent extraction is found to be 274 $m^2\ g^{-1}$. The as-prepared amorphous $Fe_2O_3$ shows paramagnetic behavior, but after calcination at 350 °C it changes to $\gamma$-$Fe_2O_3$ with good magnetic properties. The catalytic activity of mesoporous iron oxide was studied in the reaction of cyclohexane oxidation under mild conditions. The mesoporous $Fe_2O_3$ catalyst showed 36% conversion of cyclohexane into cyclohexanone and cyclohexanol, with a high selectivity. This is the highest conversion percentage ever reported for the oxidation of cyclohexane. The oxidation was conducted under 1 atm of $O_2$ at 70 °C.

MSP titania was also used to make electrodes, which were tested in a dye-sensitized solar cell [116]. The short-circuit photocurrent, open-circuit photovoltage and fill factor increased with increasing sintering temperature, having a performance threshold at 450 °C, showing that the more ordered structures are required for high solar cell conversion efficiencies.

Rana [117] has recently demonstrated that ultrasound radiation can be employed for the formation of vesicular mesoporous silica. The dimension of the vesicles ranged from 50–500 nm. If the synthesis is compared with a previous work on the synthesis of MSP silica vesicles [118], the advantages of the sonochemical synthesis are as follows: (1) It employs the commonly used CTAB as a surfactant, instead of Gemini surfactant, $C_nH_{2n+1}NH(CH_2)_2NH_2$; (2) the sonochemical reaction takes 1 h as compared with 48 h; (3) the reaction is conducted at 25–35 °C instead of 100 °C; and (4) a higher surface area is obtained, 940, as compared with 280–520 $m^2\ g^{-1}$. The special role of the bubbles in the formation of the vesicle is also explained.

#### 6.1.2.7 The Sonochemical Synthesis of Mixed Oxides

In addition to the above reported synthesis of ferrites our search has revealed that aluminates [119], nickelates [120], and manganates [121], have also been prepared by the sonochemical method. Nanosized nickel aluminate spinel particles have been synthesized [119] with the aid of ultrasound radiation by a precursor approach. Sonicating an aqueous solution of nickel nitrate, aluminum nitrate, and urea yields a precursor which, on heating at 950 °C for 14 h yields nanosized $NiAl_2O_4$ particles with a size of ca. 13 nm and with a surface area of about 108 $m^2\ g^{-1}$.

Nanostructured $LaNiO_3$ was prepared by co-precipitation under ultrasonic radiation [120]. Using various characterization methods and evaluation of catalytic activity, the effects of ultrasound on the structural properties and catalytic activity of $LaNiO_3$ were studied. The TEM showed that the ultrasound could cause a decrease in the particle size. The average particle size of $LaNiO_3$ prepared by sonochemistry is 20 nm. The specific surface area of $LaNiO_3$ is 11.27 $m^2\ g^{-1}$. Ultrasound could lead to increased surface oxide content and surface crystal oxygen vacancies. The TPR result showed that the $LaNiO_3$ prepared by sonochemistry has a lower reduction temperature and a higher ratio of surface oxygen to crystal oxygen. The evaluation of catalytic activity showed that ultrasound could increase the catalytic activity of $LaNiO_3$ for NO decomposition.

Lanthanum strontium manganate, known also as LSM, is known for its magnetoresistance properties, and is also used in solid oxide fuel cells (SOFC). Sonochemistry was used for its preparation [121]. Electron magnetic resonance (EMR) spectra of nanosized sonochemically sintered powders of $La_{0.7}Sr_{0.3}MnO_3$ (annealed, $T_C = 340$ K) were studied.

#### 6.1.2.8 The Sonochemical Synthesis of Nanosized Hydroxides

The synthesis of nanosized α-nickel hydroxide has been reported by two groups [122, 123]. The first reported reaction prepared nanosized α-nickel hydroxide with an interlayer spacing of 7.2 Å. The synthesized hydroxide was found to possess good stability in a KOH medium, and the material might be interesting from the application point of view in secondary alkaline batteries. The second synthesis prepared nanocrystalline α-$Ni(OH)_2$ by an ultrasonic precipitation/stirring method [123]. Compared with the sample prepared without ultrasonic stirring, the crystal structure of the α-phase sample has been changed from the β-phase. The crystalline size of the sample is about 20 nm, which is smaller than the sample produced without ultrasonic stirring (70 nm).

Cobalt hydroxide with an interlayer spacing of 7.53 Å and needle-like morphology has been synthesized with the aid of ultrasound radiation [124]. Characterization methods indicate the formation of α-cobalt hydroxide. Thermal decomposition of the hydroxide at 300 °C under air or argon yields nanometer-sized oxide particles of $Co_3O_4$ (ca. 9 nm) and CoO (ca. 6 nm), respectively.

In a different study [125], an α-cobalt hydroxide with an interlayer spacing of 12.65 Å was synthesized in sheet shapes with dimensions of 100–120 nm with the aid of sonication. Acetate anions are intercalated into the interlayer region of the as-prepared $\alpha$-cobalt hydroxide in the form of a free ion state. $\beta$-Cobalt hydroxide has also been prepared and formed as crystallized thin hexagonal platelets with a diameter of 100 nm. Pure cobalt oxyhydroxide with a particle size of 10–30 nm has also been obtained.

#### 6.1.2.9 Sonochemical Preparation of Nanosized Titania

Because of its many applications, a few attempts are known in which ultrasonic waves have been applied for the synthesis of nanosized titania. Yu and coworkers [126] discovered a novel method for preparing highly photoactive nano-sized $TiO_2$ photocatalysts with anatase and brookite phases. The method has been developed by hydrolysis of titanium tetraisopropoxide in pure water or a 1:1 $EtOH:H_2O$ solution under ultrasonic irradiation; the photocatalytic activity of $TiO_2$ particles prepared by this method exceeded that of Degussa P25.

Huang, in a similar reaction [127], found a way to selectively prepare anatase or rutile as well as their mixtures. This was achieved by using various precursors and employing ultrasound irradiation. The products, the particle sizes of which are nanometric (<9 nm), are dependent both upon the reaction temperature and the precursor used; a substantial reduction in reaction time, as well as reaction temperature is observed, as compared to the corresponding hydrothermal processes.

A novel sonochemical method for the direct preparation of anatase nanocrystalline $TiO_2$ was reported by Guo et al. [128]. Nanocrystalline $TiO_2$ was synthesized by the hydrolysis of titanium tetrabutyl in the presence of water and ethanol under high-intensity ultrasonic irradiation (20 kHz, 100 W $cm^{-2}$ at 363 K for 3 h.). The structure and particle sizes of the product were dependent upon the reaction temperature, the acidity of the medium, and the reaction time. The TEM images showed that the particles of $TiO_2$ were columnar in shape and the average sizes were ca. 3 nm × 7 nm. The sonochemical formation mechanism of nanocrystalline $TiO_2$ is as follows: The hydrolytic species of titanium tetrabutyl in water condensed to form a large number of tiny nuclei, which aggregated to form larger clusters. Ultrasound irradiation generated many local hot spots within the gel, and the crystal structural unit was formed near the hot spots with decrease in the gel nuclei, which led to the formation of nanocrystal particles.

A process has been developed for the formation of titania whiskers and nanotubes with the assistance of sonication [129]. Titanate whiskers are obtained as a slender sheet with length about 1 μm and width 60 nm; arrays of titania whiskers with diameter 5 nm are prepared from the titanate whiskers; titania nanotubes (see Figure 6.7) with diameter about 5 nm, and length 200–300 nm are also synthesized.

The application of ultrasound dramatically increases the rate of exfoliation of $H_xTi_{2-x/4}O_4{\cdot}yH_2O$ in the presence of aqueous tetrabutylammonium (TBA) hydroxide [130]. The effect of ultra sonication power and processing time on particle size distributions are evaluated. Applied powers of 60–300 W and reaction times of 2–30 min effectively reduce the H–Ti particle size to <100 nm. Both particle size distribution analysis and UV–Vis spectroscopy were used to study the effect of the

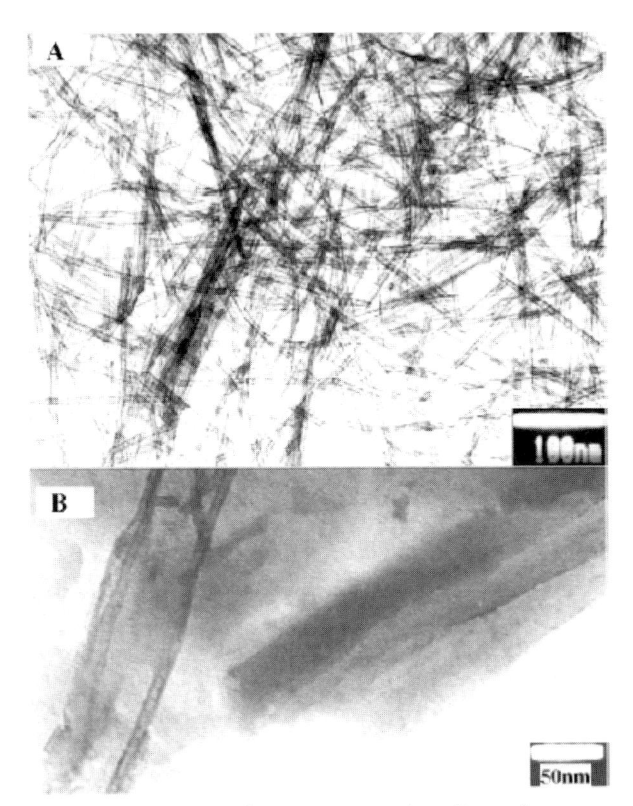


Fig. 6.7. TEM images of (a) titania nanotubes, (b) powders obtained by heat-treating sonicated products for 4 h at 110 °C followed by washing with water.

ratio of the TBA ion to exchangeable protons in H–Ti; a minimum ratio of TBA/H greater than or equal to 0.5 is required for rapid exfoliation.

Shafi et al. found that sonochemistry is a fast and efficient technique for coating octadecyltrihydrosilane ($CH_3(CH_2)_{17}SiH_3$) onto titania surfaces [131]. Infrared spectroscopy and thermal analysis confirm that complete coating is achieved after 30 min. Solid-state C-13 NMR spectroscopy establishes the bonding of trihydrosilane to the titania particles. Raman microscopy gives the expected rutile structure and further confirms the presence of an octadecyl monolayer. X-ray diffraction confirms that the rutile structure of the titania particles does not change during sonication. Anatase titania undergoes the same reaction when sonicated in the presence of octadecyltrihydrosilane.

#### 6.1.2.10 The Sonochemical Preparation of Other Oxides

Brij-35 [polyoxyethylene (23) lauryl ether] stabilized palladium nanoparticles, obtained on attempted sonochemical reduction of $PdCl_2$, by sodium sulfite in water under argon, instantaneously oxidized to PdO [132]. The particles obtained were stable and have a narrow size distribution with an average diameter of 10 nm. PdO nanoparticles were reduced to Pd nanoparticles in an autoclave by treatment with 50 bar hydrogen at 140 °C. The catalytic behavior of the Pd nanoparticles thus obtained is unusual in comparison with conventional Pd catalysts.

Two reports have been published on the sonochemical preparation of nanosized ceria, $CeO_2$ [133, 134]. Zhu and coworkers [133] synthesized nanocrystalline ceria ($CeO_2$) particles via sonochemical and microwave assisted heating routes from aqueous solutions containing $(NH4)_2Ce(NO3)_6$, hexamethylenetetramine and poly (ethylene glycol) (PEG $M_w = 19{,}000$). Analysis of the results showed that the products had a uniform shape, narrow size distribution, and displayed conspicuous quantum size effects.

In the second synthesis, cerium oxide ($CeO_2$) nanoparticles were prepared sonochemically by using cerium nitrate and azodicarbonamide as starting materials, and ethylenediamine or tetraalkylammonium hydroxide as additives [134]. The additives have a strong effect on the particle size and particle size distribution. $CeO_2$ nanoparticles with small particle size and narrow particle size distribution are obtained by the addition of additives; while highly agglomerated $CeO_2$ nanoparticles are obtained in the absence of additives. Monodispersed $CeO_2$ nanoparticles with a mean particle size of ca. 3.3 nm are obtained when tetramethylammonium hydroxide (TMAOH) is used as the additive and the molar ratio of cerium nitrate/azodicarbonamine/TMAOH is 1/1/1. Blue shifts of the absorption peak and the absorption edges of the products are observed in the UV–Vis absorption spectra as a result of the quantum size effect.

A composite material made of zinc oxide and polyvinyl alcohol was prepared by a sonochemical method [135]. Annealing of the composite under air removed the polymer, leaving porous spheres of ZnO. This change was accompanied by a change in the surface area from 2 to 34 $m^2\ g^{-1}$. The porous ZnO particles were used as the electrode material for dye-sensitized solar cells (DSSCs). They were tested by forming a film of the doped porous ZnO on a conductive glass support. The performance of the solar cell is reported.

A very unusual and previously unknown oxide of copper, the paramelaconite, $Cu_4O_3$, was synthesized sonochemically in a polyaniline matrix [136]. An aqueous solution of copper (II) acetate and aniline (1:10 molar ratio) is irradiated by ultrasound to produce nanophased $Cu_4O_3$ embedded in a polyaniline matrix. The as-prepared $Cu_4O_3$–polyaniline is characterized by X-ray diffraction (XRD), and other methods. The mechanism for the fabrication of $Cu_4O_3$–polyaniline is proposed and discussed. This method is general and works also for the production of nanocrystalline $Fe_3O_4$ and $Cu_2O$ embedded in polyaniline. This technique is also an easy route for the production of other metal oxides embedded in polyaniline.

Amorphous tungsten oxide has been prepared by ultrasound irradiation of a solution of tungsten hexacarbonyl $W(CO)_6$ in diphenylmethane (DPhM) in the presence of an Ar (80%) and $O_2$ (20%) gaseous mixture at 90 °C [137]. Heating this amorphous powder at 550 °C under Ar yields snowflake-like dendritic particles consisting of a mixture of monoclinic and orthorhombic $WO_2$ crystals. Annealing of the as-prepared product in Ar at 1000 °C leads to the formation of a $WO_2$–$WO_3$ mixture containing nanorods (around 50 nm in diameter) and packs of these nanorods. Heating the product in air for 3 h leads to triclinic $WO_3$ crystal formation, with a basic size of 50–70 nm.

Finally, Vijaykumar et al. reported on a general sonochemical reaction in which metal acetates can be converted to the corresponding metal oxides [138]. The acetates examined in this research were, Zn, Cu, Co, and Fe(II), yielding nanocrystalline CuO, ZnO, $Co_3O_4$, and $Fe_3O_4$. The solvents were water and a 10% water–DMF mixture. The diameters of the particles were: 20 nm (length ($L$)) and 2 nm (width ($W$)), (6), 340 nm (250), 30 nm (20), and 20 nm (8), respectively, when water (water–DMF) is used as the solvent. The results of DRS are analyzed in detail, and the band gap energies for CuO, ZnO, and $Co_3O_4$ are seen to be 2.18, 3.35, and 2.26 (3.40) eV, respectively.

#### 6.1.2.11 **Sonochemical Synthesis of Other Nanomaterials**

Metals, metal oxides, and chalcogenides constitute the main body of the sonochemical research. Only very few other groups of materials have been prepared by using power ultrasound. There may be two reasons for this: first, the difficulty in preparing these materials and, secondly, lack of interest. However, we believe that the first reason can explain why an important material such as GaN, for example, has not been prepared sonochemically.

In this section we will survey briefly other important groups of materials that have been reported as prepared using ultrasonic waves. Only one paper was found describing the preparation of nitrides [139]. This reports on a method for the preparation of nanoparticles of iron nitride powders. Iron nitride particles have been synthesized by two methods. In the first, $Fe(CO)_5$ was sonicated in a decane solution under a gaseous mixture of $NH_3$ and $H_2$ (3.5:1 molar ratio) at ca. 0 °C. The second method was based on nitriding the sonochemically prepared amorphous iron at ca. 400 °C for 4 h under a mixed stream of $NH_3$ and $H_2$ (3.5:1 molar ratio). Different products were obtained in the two cases. The product of the sonication of $Fe(CO)_5$ was amorphous $Fe_{2-3}N$ and a small quantity of iron oxide. The X-ray diffraction patterns in the second case showed $Fe_4N$ as the main product. The magnetic properties of both products were measured. The coercive force $H_c$ of the $Fe_4N$ is 190 Oe, and the saturation magnetization $\sigma(s)$, is 170 emu $g^{-1}$.

More publications were found related to carbides. First, Suslick's early report [64] that certain carbonyls sonicated in a decalin solvent under argon. For Fe and Co, nanostructured metals are formed; for Mo and W, metal carbides (e.g., $Mo_2C$) are produced. Molybdenum carbide was used later as a catalyst. The selectivity and catalytic activity of the Mo and W carbides was examined in the dehydrogenation of alkanes [140]. Another carbide that has already been mentioned is that of Pd [65], which was prepared by Maeda's group. Iron carbide was a byproduct that served as protective layer in Nikitenko's work on air-stable iron nanoparticles [70].

Ultrasonic irradiation (22 kHz, Ar atmosphere) of Th(IV) $\beta$-diketonates $Th(HFAA)_4$ and $Th(DBM)_4$, where HFAA and DBM are hexafluoroacetylacetone and dibenzoylmethane respectively, causes them to decompose in hexadecane solutions, forming solid thorium compounds [141]. The first-order rate constants for Th(IV) beta-diketonate degradation were found to be $(9.30–8) \times 10^{-3}$ for $Th(HFAA)_4$ and $(3.80–4) \times 10^{-3}$ $min^{-1}$ for $Th(DBM)_4$, ($T = 92$ °C, $I = 3$ W $cm^{-2}$). The rate of the sonochemical reaction increased with increasing $\beta$-diketonate volatility and decreased with increasing hydrocarbon solvent vapor pressure. Solid sonication products consisted of a mixture of thorium carbide $ThC_2$ and Th(IV) $\beta$-diketonate partial degradation products. The average $ThC_2$ particle size was estimated to be about 2 nm. $ThC_2$ formation was attributed to the high-temperature reaction occurring within the cavitating bubble. The thorium $\beta$-diketonate partial degradation products were formed in the liquid reaction zones surrounding the cavitating bubbles.

It is more difficult to prepare III–V semiconductors than the II–VI. Two sonochemical investigations reported on the preparation of these materials. The first paper details a safe method for the preparation of transition metal arsenides, FeAs, NiAs, and CoAs [142]. At room temperature, well-crystallized and monodispersed arsenide particles were successfully obtained under high-intensity ultrasonic irradiation for 4 h from the reaction of transition metal chlorides ($FeCl_3$, $NiCl_2$, and $CoCl_2$), arsenic (which is the least toxic arsenic feedstock) and zinc in ethanol. Different characterization techniques show that the product powders consist of nanosize particles. The ultrasonic irradiation and the solvent are both important in the formation of the product.

Another III–V semiconductor was prepared by Li and coworkers [143]. A room temperature sonochemical method for the preparation of GaSb nanoparticles using less hazardous Ga and antimony chloride ($SbCl_3$) as the precursors has been described. TEM and SAED results show that the as-prepared solid consists of nanosized GaSb crystals with sizes in the 20–30 nm range. The photoacoustic spectrum result reveals that the GaSb nanoparticles have a direct band gap of about 1.21 eV. On the basis of the control experiments and the extreme conditions produced by ultrasound, an ultrasound-assisted in situ reduction/combination mechanism has been proposed to explain the reaction.

We will conclude this survey of the synthesis of nanomaterials by sonochemical methods by mentioning that the most important material of the last decade, carbon nanotubes, were also synthesized by ultrasound radiation [144]. The carbon nanotube is produced by applying ultrasound to liquid chlorobenzene with $ZnCl_2$ particles and to *o*-dichlorobenzene with $ZnCl_2$ and Zn particles. It is considered that the polymer and the disordered carbon, which are formed by cavitational collapse in homogeneous liquid, are annealed by the inter-particle collision induced by the turbulent flow and shockwaves.

## 6.2 Sonoelectrochemistry

The effect of ultrasound in electrochemistry, i.e. that the application of ultrasonic energy can increase the rate of electrolytic water cleavage, was discovered as early as 1934. As for the case of sonochemistry, the use of ultrasound in electrochemistry went through a period of neglect until the early 1980s when there was again an upsurge of interest in the field. At present, ultrasound is used in a wide range of electrochemical processes such as: metal plating, deposition of polymers, electrogeneration of gases and solids, and electrochemical waste processing.

Ultrasound and electrochemistry provide a powerful combination for several reasons. Ultrasound is well known for its capacity to promote heterogeneous reactions, mainly through increased mass-transport, interfacial cleaning, and thermal effects. Effects of ultrasound in electrochemistry may be divided into several important branches: (1) Ultrasound greatly enhances mass transport, thereby altering the rate, and sometimes the mechanism, of the electrochemical reactions. (2) Ultrasound is known to affect surface morphology through cavitation jets at the electrode–electrolyte interface; it usually acts to increase the surface area. (3) Ultrasound reduces diffusion layer thickness and therefore ion depletion. A comprehensive review of the field has recently been given by Compton et al. [145].

However, it is only recently that the potential benefits of combining sonochemistry with electrochemistry have increasingly been studied. It should be noted that electrochemical methods, mainly electrodeposition, are well established for the preparation of metals and semiconductor nanomaterials (for a review see Mastai et al. [146]).

### 6.2.1
### Sonoelectrochemical Synthesis of Nanocrystalline Materials

Reisse and co-workers [147–149] were the first to describe a novel device for the production of metal powders using pulsed sonoelectrochemical reduction. This device exposes only the flat circular area at the end of the sonic tip to the electrodeposition solution. The exposed area acts as both cathode and ultrasound emitter, named by Reisse et al. as "sonoelectrode". A pulse of electric current produces a high density of fine metal nuclei. This is immediately followed by a burst of ultrasonic energy that removes the metal particles from the cathode, cleans the surface of the cathode, and replenishes the double layer with metal cations by stirring the solution. In [145], a list is given of chemically pure fine crystalline powders, mostly metals or metallic alloys, prepared by this method, with particle sizes varying between 10 and 1000 nm depending on deposition conditions.

Powder CdSe nanoparticles prepared by a pulsed sonoelectrochemical technique with a sonoelectrochemical device similar to that described by Reisse, namely the "sonoelectrode", were reported first by Mastai et al. [150]. In Figure 6.8 we present the experimental set-up for pulsed sonoelectrochemical deposition of CdSe nanoparticles, namely the "sonoelectrode", and schematics of the sonic and electrochemical waveforms.

In the "sonoelectrode" design, a titanium horn acts both as the cathode and the ultrasound emitter. The electroactive part of the sonoelectrode is the planar circular surface at the bottom of the horn. This sonoelectrode produces a sonic pulse that is triggered immediately following a current pulse. One pulse driver is used to con-

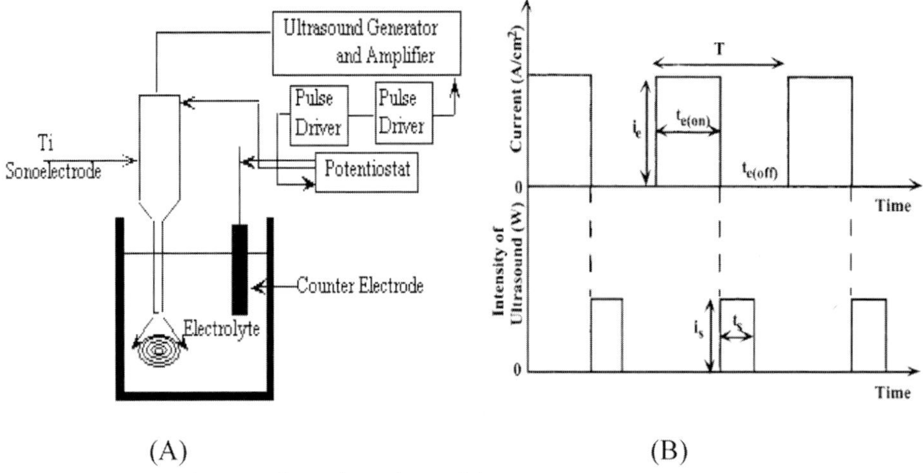


**Fig. 6.8.** Schematic of sonoelectrochemical deposition set-up and the sonic and electrochemical waveforms.

trol a potentiostat and a second controls the ultrasonic processor, which is adapted to work in the pulse mode.

The sonoelectrochemical deposition of CdSe nanoparticles is carried out from an aqueous solution of $CdSO_4$, complexed with potassium nitrilotriacetate ($N(CH_2CO_2K)_3$) NTA as a source for Cd, and a selenosulfate solution, ($Na_2SeSO_3$), as source for Se. The CdSe crystal size could be varied from X-ray amorphous up to 9 nm by controlling the various electrodeposition and sonic parameters. The effects of various electrodeposition and sonic parameters on the properties of CdSe nanoparticles are presented in Table 6.3.

**Tab. 6.3.** Effect of various deposition parameters on the CdSe nanoparticles properties in pulsed sonoelectrochemical synthesis.

| *Variable Parameter* | *Crystal Size (nm)* | *Bandgap $E_g$ (eV)* |
|---|---|---|
| Temperature (°C) | | |
| 5 | 4.5 | 2.10 |
| 25 | 5.5 | 2.10 |
| 75 | 6.5 | 1.93 |
| Ultrasound Intensity (W) | | |
| 0 | 9 | 1.90 |
| 10 | 10 | 1.85 |
| 60 | 4 | 2.15 |
| Continuous ultrasound | Amorphous | – |
| Current Pulse Width (sec) | | |
| 0.68 | 7 | 1.95 |
| 0.34 | 5 | 2.05 |
| 0.11 | 4 | 2.15 |

The fundamental basis of the sonoelectrochemical technique to form nanoparticles is massive nucleation using a high current density electrodeposition pulse (ca. 150–300 mA $cm^{-2}$), followed by removal of the deposit from the sonoelectrode by the sonic pulse. Removal of the particles from the electrode before the next current pulse prevents crystal growth. Overall there are many experimental variables involved in sonoelectrochemical deposition: electrolyte composition and temperature, electrodeposition conditions including current density ($I_e$), pulse-on time ($t_{e(on)}$) and ratio between pulse-on time and pulse-off time ($t_{e(off)}$) (the duty cycle); sonic probe conditions: sonic power ($I_s$), sonic pulse parameters, $t_{s(on)}$ and $t_{s(off)}$. The effects of the various sonoelectrochemical parameters on crystal size can be rationalized, in general, as follows:

- *Sonic intensity.* The greater the sonic intensity, the greater will be the efficiency of removal of the deposit and therefore the less chance there will be for crystal growth of existing nuclei. Typically above a certain intensity where all the deposit is removed, further increase in intensity is not expected to affect crystal growth much (in the case of CdSe this maximum sonic intensity is ca. 60 W $cm^{-2}$). Generally long sonic duration ($t_{s(on)}$) resulted in smaller crystal sizes, however, it should be noted that sonoelectrodeposition under continuous ultrasound wave leads to the formation of amorphous materials.
- *Deposition current pulse width.* Quite separately from the sonic wave effects, pulse electrodeposition is well known to result in a smaller crystal-sized deposit. This is particularly pronounced for high current densities, where a high rate of nucleation occurs during each pulse. The shorter the pulse duration ($t_{e(on)}$), the less chance there is of crystal growth occurring by deposition of new material on a previous nucleus. In normal pulse plating, crystal size may or may not increase with the number of pulses, depending on whether each new pulse forms a new nucleus or adds to pre-existing ones. In sonoelectrochemical deposition, where the deposit is removed during each sonic pulse, only new nuclei should be formed, and therefore the crystal size will be smaller with decreasing pulse duration ($t_{e(on)}$).
- *Temperature.* Temperature can affect crystal growth in several ways, all of them resulting in smaller crystal size at lower temperatures. The simplest is that crystal growth is slower at lower temperatures. Within the time between sonic pulses, growth can occur, either by coalescence during the deposition pulse or by migration on the substrate and coalescence at any time. Another effect of temperature is through the thermodynamic instability of very small nuclei below a certain critical size. These nuclei should re-dissolve, but may be stable for long enough to grow larger than the critical size, after which they are thermodynamically stable.

In a series of papers Gedanken and co-workers described the use of a pulse sonoelectrochemical technique for the preparation of nanocrystalline materials. In the first paper [151] in the series, the synthesis of silver nanoparticles of different shapes: spheres, rods, and dendrites, is described. The nanocrystalline Ag deposited from an aqueous solution of $AgNO_3$ using NTA as a complex agent. The shape of the Ag nanoparticles could be modified as a function of the concentration of $AgNO_3$ and NTA in the deposition solutions. In the second paper [152], X-ray amorphous silver nanoparticles were prepared by a pulse sonoelectrochemical method from an aqueous solution of AgBr in the presence of gelatin. Finally, it was shown that PbSe [153] nanoparticles, ca. 12 nm, could be prepared by a pulse sonoelectrochemical technique from an aqueous solution of sodium selenosulfate and lead acetate. Most recently, Zhu et al. [154] utilized a sonoelectrochemical route for the preparation of uniform silver nanowires of a single crystalline nature. Finally, in a very recent publication, silver nanoparticles were used as a model case to study the processes involved in sonoelectrochemical synthesis of nanoparticles [155]. The authors propose that sonoelectrochemical synthesis is highly affected by the formation of a suspensive electrode. This new model, the suspensive electrode, is based on the concept that the nanoparticles suspended in solution gain the sonoelectrode potential and thus act as part of the electrode. That is to say, during sonoelectrochemical synthesis, a suspension of charged nanosized particles is formed in solution and acts as part of the sonoelectrode.

Hodes et al. [156] demonstrated that the sonoelectrochemical technique could be used to synthesize closed fullerene-like structures of $MoS_2$ at room temperature by electrodeposition from a thiomolybdate solution. It should by noted that either electrodeposition or ultrasonic irradiation alone results in X-ray amorphous $MoS_2$ products, but the combination of both gives well-crystallized, closed structures of $MoS_2$ and $MoS_2$ nanotubes. A mechanistic study to explain the formation of the closed curved $MoS_2$ structures shows that in the first step amorphous $MoS_2$ is formed by electrodeposition onto the sonic probe cathode. At this point the deposit could be spheroidal or planar. Crystallization of the amorphous $MoS_2$, which normally requires high temperatures occurs in the collapse of the cavitation bubbles. The formation of $MoS_2$ nanotubes occurs if the bubble collapse occurs at the electrodeposit that is still on the electrode surface. In this case, the effect of the bubble collapse is asymmetric, leading to an asymmetric-shaped product, namely the nanotube.

In conclusion, the combination of electrochemical and sonic processes provides many experimental variables which should allow control of particle size and shape, and will probably be applicable to the formation of closed structures of other layered compounds that can be prepared by electrochemical (and quite likely also chemical) techniques.

## 6.3 Microwave Heating

Microwave heating (MWH) is the second method for the fabrication of nanomaterials that will be discussed in this chapter. Microwaves are electromagnetic radiation, whose wavelengths lie in the range 1 mm to 1 m (frequency range 0.3 to 300 GHz). A large part of the microwave spectrum is used for communication purposes and only narrow frequency windows centered at 900 MHz and 2.45 GHz are allowed for microwave heating purposes. Heating by microwave has been known since the early 1940s and has been used successfully in the food industry. Microwave irradiation as a heating method has found a number of applications in chemistry since 1986, especially in the work of Mingos [157]. Microwaves have been in use for accelerating organic reactions for a quite a while [158, 159]. Microwave synthesis is generally quite fast, simple, and very energy efficient. The exact nature of microwave interaction with reactants during the synthesis of materials is somewhat unclear and speculative [160]. However, transfer of energy from microwaves to the material is believed to occur either through resonance or relaxation, which results in rapid heating. This knowledge is widely used in the discussion of reaction mechanisms. Many successful examples of the applications of MW heating in organic chemistry have been reported [161, 162], although its expansion into the area of inorganic chemistry has been much slower. Only recently has it been noticed that metallic powders can be heated to considerably high temperatures in a microwave oven without arcing [161, 163]. Some metal chalcogenides have been prepared by microwave solid-state reaction [164]. These solid-state reactions do not produce nanoparticles. On the other hand, MW reactions in solutions do yield nanoparticles. The use of MWH in the preparation of solid state inorganic materials has been reviewed recently by the two main contributors to this field, Professor Mingos [159] and Professor K. J. Rao [160]. The preparation of nanoparticles by MWH started only recently. The aim of this section is to review the solution work that has been conducted using a regular domestic microwave oven. We will, therefore, not discuss the many reactions leading to the fabrication of nanomaterials conducted with a microwave plasma. An example of a domestic microwave oven that has been modified for the synthesis of nanomaterials is presented in Figure 6.9.

However, we should mention the pioneering work of Chou and Phillips, where metallic iron and iron oxide particles were produced by injecting ferrocene into the afterglow region of a low-pressure, low-power, plasma, generated using a microwave power source [160]. This gas phase reaction was carried out as part of an attempt to explore the feasibility of using flow-type microwave plasmas for the production of metal nanoparticles.

A short introduction to as to how the irradiated molecule is affected by microwaves is first presented. It is based on the reviews of Rao and Mingos [159, 160].

In general, materials fall into three categories, with respect to their interaction with microwaves [160]: (1) microwave reflectors, typified by bulk metals and alloys, such as brass, which are therefore used in making micro-waveguides; (2) microwave transmitters which are transparent to microwaves, typified by fused quartz, zircon, several types of glass, and ceramics (not containing any transition element), teflon etc.; they are therefore employed for cookware and containers for carrying out chemical reactions in microwaves; and (3) microwave absorbers, which constitute the most important class of materials for microwave synthesis; they take up the energy from the microwave field and heat up very rapidly.

A dielectric material is one that contains either permanent or induced dipoles

**Fig. 6.9.** A microwave oven with a reflux system.

which, when placed between two electrodes, act as a capacitor [160]. The polarization of dielectrics arises from the finite displacement of charges or rotation of dipoles in an electric field. At the molecular level, polarization involves either the distortion of the distribution of the electron cloud within a molecule, or the physical rotation of molecular dipoles. The latter are particularly significant in the context of microwave dielectric heating. The permittivity of a material, $\varepsilon$, is a property which describes the charge storing ability of that substance, irrespective of the sample's dimensions. The dielectric constant or relative permittivity is the permittivity of the material relative to that of free space. It is represented in [159] as $\varepsilon_S$. Compounds, which have large permanent dipole moments, also have large dielectric constants, because the dielectric polarization depends primarily on the ability of their dipoles to reorient in an applied electric field. In the gas and liquid phases the molecules rotate so rapidly that they are normally able to respond to field reverses occurring at $10^6$ times a second or higher. The permittivity of the material, is frequency-dependent, and, for a polar liquid, generally shows a marked decrease as the frequency of the electromagnetic radiation increases from $10^6$ (radio frequencies) to $10^{12}$ Hz (infrared frequencies). For polar molecules with molecular weights less than a few hundred, this relaxation process occurs in the microwave region, i.e., in the frequency range 300 MHz to 300 GHz. If the frequency of the electromagnetic radiation is in the microwave region ($\sim 10^9$ Hz) the rotations of the polar molecules in the liquid begin to lag behind the electric field oscillations. The resulting phase displacement $\delta$, acquires a component $I \sin \delta$ in phase with the electric field, and thus resistive heating occurs in the medium. This is described as dielectric loss and causes energy to be absorbed from the electric field. Since the dipoles are unable to follow the higher frequency electric field oscillations, the permittivity falls at the higher frequency and the substance behaves increasingly like a non-polar material.

$$\tan \delta = \varepsilon''/\varepsilon' \tag{8}$$

where $\varepsilon'$ is the real part of the dielectric constant and $\varepsilon''$ is the loss factor which reflects the conductance of the material. In his review, Mingos [159] introduces the Debye expression for the complex permittivity and shows how $\varepsilon''$ and $\varepsilon'$ depend on the relaxation time, which is itself dependent on the molecular size and the viscosity of the liquid. Mingos emphasizes that the interaction between the microwave radiation and the polar solvent, which occurs when the frequency of the radiation approximately matches the frequency of the rotational relaxation processes, is not a quantum mechanical resonance phenomenon. Transition between rotational energy levels is not involved and the energy transfer is not a property of a specific molecule, but the result of a collective phenomenon involving the bulk. In the Debye interpretation, the heat is generated by friction forces occurring between the polar molecules whose rotational velocity has been increased by the coupling with the microwave radiation and neighboring molecules. The dielectric heating is a broadband phenomenon and rapid energy transfer occurs even when the frequency of the microwaves and the relaxation frequency are not perfectly matched.

It is worth noting that much of the work for which domestic microwave ovens were used for fabricating nanomaterials concentrated on using ethylene glycol as the solvent. Table 6 in Mingos's review [159] shows that ethylene glycol indeed has a high loss tangent at 2.45 Ghz, and when this is coupled with its high boiling temperature (193 °C), this makes it an excellent candidate for dielectric heating and, hence, there is little penetration of microwaves. In large metal samples, as well as in metal films, large electric field gradients occur in the microwave cavity, giving rise to electric discharges. On the other hand, in metal powders no such discharge takes place and, due to eddy currents and localized plasma effects, very rapid heating takes place which may be as high as 100 K $s^{-1}$. Eddy current phenomena arise from the alternating magnetic field associated with microwaves.

### 6.3.1 Microwave Synthesis of Nanomaterials

#### 6.3.1.1 Microwave Synthesis of Nanometallic Particles

The first report of a microwave-assisted reaction leading to the formation of nanometallic particles was by Jiang and coworkers [167]. They prepared Au nanoparticles by a microwave high-pressure procedure with alcohol as the reducing agent. The color of colloidal Au nanoparticles is blue–violet. The maximum absorption spectrum of colloidal Au is at 580 nm, and the resonance scattering peak is at 580 nm. Using this method, the colloidal Au solution was very stable and the preparation is described as simple and quick.

Poly(vinylpyrrolidone) (PVP) is a polymer capable of complexing and stabilizing Ag and Au nanoparticles formed through the reduction of $Ag^+$ or $AuCl^-{}_4$ ions with *N,N*-dimethylformamide [168]. The reduction is efficiently performed both at reflux and under microwave irradiation, but each of these methods leads to different nanoparticle morphology and colloid stability. The use of microwave irradiation provides an extra degree of control of the reduction process. The use of PVP with different polymer chain lengths leads to particles with similar sizes, though with a different degree of stability. The colloids are also stable in ethanol for months, but only marginally stable in water. Alloys were also prepared by MWH thermal treatment of $(\eta\text{-}C_2H_4)(Cl)Pt(\mu\text{-}Cl)_2Ru(Cl)(\eta^3{:}\eta^3\text{-}2,7\text{-dimethyloctadienediyl})$/Vulcan carbon composites under appropriate oxidizing and reducing conditions. Using microwave dielectric loss heating affords PtRu/Vulcan carbon nanocomposites [169]. This composite consisted of PtRu alloy nanoparticles highly dispersed on a powdered carbon support. Two such nanocomposites containing 16 or 50 wt% total metal and alloy nanoclusters of 3.4 or 5.4 nm average diameter are formed within only 100 or 300 s of total microwave heating.

The polyol reaction was developed by Fivet's group. It was the most popular method for the fabrication of metallic nanoparticles using MWH. In the 1980s, Fievet et al. [170] used ethylene glycol as a solvent and reducing agent for the preparation of submicrometer particles of the transition metals. The mechanism of this reaction is still only poorly understood. It is, however, known that the reduction is based on the decomposition of the ethylene glycol and its conversion to diacetyl. Recently, Tarascon and co-workers [171] demonstrated that in these reactions the temperature is a dominant factor in affecting the reactivity. It is worth mentioning that the preparation of nanophased chalcogenides was also based on the polyol reaction [3].

The noble metals were the favorite metals for demonstrating the usefulness of the microwave operation in conducting the polyol reaction. Among the noble metals platinum was synthesized most frequently. Polymer-stabilized platinum colloids with nearly uniform spherical shape were prepared by Yu and coworkers by microwave dielectric heating [172]. The average diameters of the as-prepared platinum colloids were 2–4 nm with a narrow size distribution in regard to the preparation conditions.

The same group, using the polyol method, have prepared uniform and stable polymer-stabilized colloidal clusters of Pt, Ir, Rh, Pd, Au and Ru by microwave irradiation with a modified domestic microwave oven [173]. The as-synthesized colloidal clusters have small average diameters and narrow size distribution. The microwave method is characterized by rapid and homogeneous heating compared with conventional heating methods, although its thermal effects are similar to those of other heating methods.

Finally, using the polyol reaction and the same reactants as in [172] in a third paper by this group, they synthesize polymer-stabilized Pt colloids with small particle sizes and narrow size distributions by a continuous microwave synthesis [174]. This synthesis method has good reproducibility for synthesizing uniform metal colloids in bulk amounts.

Komarneni and coworkers [175] conducted a polyol reaction for the preparation of Pt and Ag nanoparticles. The synthesis of the metal nanoparticles [175] was conducted in a double-walled digestion vessel which has an inner liner and a cover made of Teflon PFA and an outer high-strength shell of Ultem polyetherimide. The starting precursors (chloroplatinic acid or silver nitrate) were mixed with ethylene glycol and poly(*N*-vinyl-2-pyrrolidone) of different molecular weights, with and without NaOH, and reacted at 150 °C for 15 min in a microwave system. Large Pt particles 100–400 and 50–150 nm were obtained with PVP molecular weights of 8,000 and 10,000, respectively. Non-uniform Pt particles of $<10$ nm were crystallized from PVP of 40,000 molecular weight. For the Ag particles strings were obtained. The growth of the Ag metal particles in the form of strings decreased as the molecular weight of the PVP increased. With a PVP molecular weight of 1,300,000 and without NaOH Ag particles of 100 nm were obtained.

In another attempt [176], silver nanoparticles, yellow in color, were prepared by a microwave high-pressure synthetic method, using polyacrylamide as the reducing agent and stabilizer. The maximum absorption peak of colloidal silver is at 421.6 nm, and its strongest resonance scattering peak is at 470 nm. The experimental results showed that the silver nanoparticles are well dispersed and the colloidal solution has good stability. The average diameter of silver nanoparticles is 66 nm. The procedure was rapid and convenient.

Spherical and uniform Pt nanoparticles, 3.5–4.0 nm supported on carbon, were prepared by microwave irradiation [177]. The products exhibited very high electrocatalytic activity in the room-temperature oxidation of liquid methanol. The preparation method was similar to that of Komarneni [175], namely, a polyol reduction. A 100 mL beaker, containing chloroplatinic acid dissolved in an ethylene glycol, KOH solution was added dropwise. Carbon XC-72 was uniformly dispersed in the mixed solution by ultrasound. The beaker was placed in the center of a microwave oven (National NN-S327WF, 2450 MHz, 700 W) and heated for 60 s.

Towards the end of 2002, a paper [178] appeared, claiming to have prepared polychrome silver nanoparticles by a soft solution approach under microwave irradiation from a solution of silver nitrate, $AgNO_3$, in the presence of poly (*N*-vinyl-2-pyrrolidone) PVP, without *any other reducing agent.* Different morphologies of silver colloids with lovely colors could be obtained using different solvents as the reaction medium. The influence of the solvent on the morphology of silver was investigated. In a typical procedure, the reaction solutions were prepared by dissolving PVP and $AgNO_3$ ethanol in a Pyrex flask to obtain a homogeneous reaction mixture, which was placed on the turntable of the microwave oven. The mixture was irradiated discontinuously at a power of 300 W.

#### 6.3.1.2 **The Synthesis of Nanoparticles of Metal Oxides by MWH**

Similar to sonochemical research, much attention has been devoted to the preparation of nanosized metallic oxides by MWH. When surveying the literature, we realized that iron oxide was the most popular material to be synthesized by this method.

The first report was by Komarneni and coworkers [179] using a microwave-hydrothermal process to catalyze the synthesis of crystalline oxides such as $TiO_2$, $ZrO_2$ and $Fe_2O_3$, and binary oxides such as $KNbO_3$ and $BaTiO_3$. The importance of this work was that this technique led to fine powders of these materials. The effect of different parameters, such as concentration of chemical species, time and temperature, on the crystallization kinetics of the above phases has been investigated under microwave-hydrothermal conditions using microwaves of 2.45 GHz frequency.

The second work was by Dong et al. [180] who report on the preparation of submicron uniform $\alpha$-$Fe_2O_3$ by microwave-induced hydrolysis of ferric salts with concurrent dissociation of urea. Although it was not conducted in a domestic MW oven, it is still on our list of references. A mixed solution of $FeC_3$ and $CO(NH_2)_2$ was placed in an Erlenmeyer flask equipped with a reflux condenser. This flask was inserted into a cylindrical resonant cavity. After microwave irradiation (2.4 GHZ, 500 W) for a certain period of time, the flask was moved to a homothermal water bath (at 94 °C) to be aged. Uniform particles of approximately spherical shape with mean diameter 75 nm were obtained. An almost identical paper was published a little later by Li and Wei [181]. $Fe(NO_3)_3$ was the iron source instead of the chloride [180]. They conducted the reaction at different pH values. In this way they obtained a larger variety of shapes, including acicular cubic and spherical particles, all in the nanometer range.

Palchik [182] obtained nanosized amorphous iron oxide ($Fe_2O_3$) by the pyrolysis of iron pentacarbonyl, $Fe(CO)_5$, in a modified domestic microwave oven in refluxing chlorobenzene as solvent under air. The reaction time was 20 min. Separate particles of iron oxide, 2–3 nm in diameter, were obtained together with aggregated spheres with a diameter of 25–40 nm. Differential scanning calorimetry measurements showed an amorphous/crystalline phase transition at about 250 °C.

A study of microwave effects on the formation of nanoparticles in the hydrolysis reaction of $FeCl_3$ with $NaH_2PO_4$ to get spindle-type colloidal hematite particles under microwave radiation, was reported by Han and coworkers [183]. They found that the reaction rate increased greatly, and the reaction conditions, for example, the acidity of the solution, concentration ratio of the components, and the microwave radiation time, had important effects on the nanoparticle formation and their morphology. They discussed the roles of microwaves and the concentration of $H_2PO_4^-$ during the hydrolysis process.

Nanometer-sized quasicubic and spindle $\alpha$-$Fe_2O_3$ particles were prepared by microwave heating from $Fe^{+3}$ salt solutions [184]. The obtained $\alpha$-$Fe_2O_3$ particles formed by MWH have smaller size and more uniform distribution than with conventional heating. Inorganic ions such as $H^+$, $OH^-$ and NaF, were found to affect the precipitating rate of spindle $\alpha$-$Fe_2O_3$ and accelerate the hydrolysis of ferric ions.

Another research project leading to the formation of amorphous nanoparticles was undertaken by Zhu's group [185]. The product was synthesized by microwave irradiation heating of an aqueous solution containing ferric chloride, polyethylene glycol-2000, and urea. Amorphous $Fe_2O_3$ nanoparticles of about 3–5 nm in size were obtained.

Komarneni returned to the synthesis of $\alpha$-$Fe_2O_3$ in 2001 [186]. This time he was interested in this product as a red pigment in porcelains. The methods of preparation were microwave-hydrothermal and conventional-hydrothermal reactions. The precursors were $FeCl_3{\cdot}6H_2O$ and HCl solutions. The reaction were carried out at 100–160 °C. Acicular and yellow $\beta$-FeOOH (akaganite) particles 300 nm in length and 40 nm in thickness were predominantly formed at 100 °C after 2–3 h, while spherical $\alpha$-$Fe_2O_3$ particles 100–180 nm in diameter were preferentially formed after 13 h using a conventional-hydrothermal reaction. However, a microwave-hydrothermal reaction at 100 °C led to monodispersed and red $\alpha$-$Fe_2O_3$ particles 30–66 nm in diameter after 2 h without the formation of $\beta$-FeOOH particles. They investigated the effect of microwave radiation during hydrothermal treatment at 100–160 °C on the formation, yield, kinetics, morphology phase type, and color of $\alpha$-$Fe_2O_3$.

Several oxides, including $\gamma$-$Fe_2O_3$, and $Fe_3O_4$, have been prepared recently by the interaction of electromagnetic radiation with a physical mixture of metal nitrates and amides/hydrazides [187]. A judicious choice of such redox mixtures undergoes exothermic reactions when they are coupled with microwave radiation. The coupling of electromagnetic radiation with metal salts and amides/hydrazides depends on the dielectric properties of the individual components in the reaction mixture. This approach has been used to prepare $\gamma$-$Fe_2O_3$, $Fe_3O_4$, $MgCr_2O_4$, $\alpha$-$CaCr_2O_4$, and $La_{0.7}Ba_{0.3}MnO_3$.

A microwave hydrothermal route was employed to synthesize various phases of iron oxide powders by using ferrous sulfate and sodium hydroxide as starting chemicals [188]. All the reactions were carried out under the identical microwave hydrothermal conditions of 190 °C, 154 psi, 30 min by varying the molar ratio (MR) of $FeSO_4$/NaOH (i.e., pH variation) from 0.133 to 4.00 in the solution. It was found that the variation of the molar ratio of $FeSO_4$/NaOH has a profound effect on the crystallization of various phases of iron oxides under identical processing conditions. The stoichiometric, submicron-sized (0.15–0.2 μm), spherical agglomerates of $Fe_3O_4$ powders were obtained if the MR of $FeSO_4$/NaOH was 0.133 ($pH \geq 10$) was maintained. On the other hand, non-stoichiometric $Fe_3O_4$ powders were obtained for all higher MR of $FeSO_4$/NaOH between 0.133 and 4.00 ($6.6 < pH < 10$). However, when the MR of $FeSO_4$/NaOH was 4.00 ($pH \approx 6.6$), a varied distribution of shape and size (1–5 μm) of agglomerates of $\alpha$-$Fe_2O_3$ powders was produced.

A paper reporting on the microwave-hydrothermal treatment of alcoholic solutions of ferrous chloride ($FeCl_2$) and sodium ethoxide (EtONa) solutions with a microwave autoclave designed by the authors (the RAMO system) has been published recently [189]. Depending on the initial concentrations, hematite ($\alpha$-$Fe_2O_3$), spinel phase ($Fe_{3-x}O_4$) or iron-magnetite (Fe(0)-$Fe_3O_4$) nanocomposites are obtained with a lower grain size than conventional composites. Indeed, X-ray diffraction analysis reveals grain sizes close to 20 nm for magnetite and 60 nm for metallic iron. However, the amount of metal is smaller (close to 11%). Furthermore, these particles are inert in the ambient atmosphere. Consequently, the RAMO (French acronym of Reactor Autoclave MicroOnde) system appears to provide an efficient source of energy for the rapid production of inert powders of iron, magnetite, and iron–magnetite composites.

Although most of the synthetic work related to iron oxide nanoparticles has concentrated, as we have demonstrated, on $Fe_2O_3$, and especially on the $\alpha$, hematite phase, a few papers also reported on the preparation of magnetite, $Fe_3O_4$. Submicron-sized (0.15–0.2 μm) spherical agglomerates of magnetite ($Fe_3O_4$) powders have been prepared successfully by microwave hydrothermal (MH) reaction of ferrous sulfate and sodium hydroxide in the temperature range 90–200 °C [190]. This work is closely related to the report of [190]. The Mössbauer spectra of these powders indicated that stoichiometric $Fe_3O_4$ particles are obtained only when the molar ratio of Fe/NaOH $\geq 0.133$ is maintained in the solution. It is observed that the Fe/NaOH ratio is an important parameter for the controlled oxidation of ferrous salts in alkaline media under MH conditions to produce stoichiometric $Fe_3O_4$. Furthermore, the kinetics of MH synthesis are one order faster than the reported conventional hydrothermal (CH) synthesis. The value of the saturation magnetization $M = 70$ emu g$^{-1}$ is obtained in the case of stoichiometric $Fe_3O_4$. However, when ferric salt is treated in an alkaline medium, single-phase $\alpha$-$Fe_2O_3$ is obtained under the MH conditions of 200 °C (194 psi).

Nanometric ferrites were also prepared by the MWH method. Komarneni [191] has reported on the synthesis of technologically important ferrites such as $ZnFe_2O_4$, $NiFe_2O_4$, $MnFe_2O_4$, and $CoFe_2O_4$ by using novel microwave-hydrothermal processing. The precursors, nitrates of zinc, nickel, manganese, or cobalt, were mixed with ferric nitrate and neutralized with ammonia to a specific pH. Nanophase ferrites with high surface areas, in the range of 72–247 m$^2$ g$^{-1}$, have been synthesized in a matter of a few minutes at temperatures as low as 164 °C. The rapid synthesis of nanophase ferrites via an acceleration of reaction rates under microwave-hydrothermal conditions is expected to lead to energy saving.

Other oxides have also been prepared by the MWH methods. Ceria, $CeO_2$, titania, $TiO_2$, and zirconia, $ZrO_2$, and the tin oxides were the most popular. Zhu's group [134] prepared nanometer-sized $CeO_2$ by two methods; sonochemical, and MWH. The precursors were aqueous solutions containing $(NH_4)_2Ce(NO_3)_6$, hexamethylenetetramine and poly (ethylene glycol) (PEG $M_w = 19{,}000$).

The influence of microwave power and reaction time on the formation of $CeO_2$ nanoparticles was investigated. When the microwave power is in the range 10% to 40%, the as-prepared $CeO_2$ nanoparticles are of similar size and morphology. Violent 'bump' boiling of the solvent occurs when the power is greater than 50%. When the reaction time was less than 5 min, no turbidity was observed and the solution remained transparent. After exposure to microwave irradiation for 8 min the solution became a turbid yellow, indicating the formation of the product. After 10 min, the reaction was complete, and the yield was as high as 90%. If the reaction time was prolonged to 30 min or even longer, the yield did not increase further, and the size and morphology of the $CeO_2$ nanoparticles remained almost unchanged [134]. Narrow size distribution is obtained with an average particle size of 2.8 nm. A communication [192] reporting only on the MWH part of these findings preceded this paper [134].

Nanosized zirconium oxide ($ZrO_2$) powders were prepared by adding NaOH to a zirconyl chloride aqueous solution under microwave-hydrothermal conditions [193]. The results showed that the tetragonal polymorph increased with increasing NaOH concentration in the starting solution and reached a maximum value by using 1 M $ZrOCl_2$. The authors emphasize the simplicity of the method and that it can lead to powders with desirable characteristics such as very fine size, narrow size distribution, and good chemical homogeneity. The microwave-hydrothermal treatments were conducted at 200 psi for 2 h. The time, pressure, and power were computer controlled. TEM analysis confirmed the effect of concentration on particle size. In particular, the calculated average particle size ranged from 16 ($\pm 3$) to 9 ($\pm 3$) nm, with the $ZrOCl_2$ concentration varying from 0.5 M to 1 M. TEM observation of the particles revealed spherical-shaped particles with no agglomeration.

Forced hydrolysis preparation of zirconia sols and powders by microwave heating of zirconium tetrachloride solutions at a temperature of 180 °C led in a few minutes to monodispersed nanoscale zirconia particles [194]. Synthesis was performed using the above-mentioned RAMO system [189]. This process combines the advantages of forced hydrolysis (homogeneous precipitation) and microwave heating (very fast heating rates). Sols are colloidally stable, which means that after 6 months no sedimentation is observed and the size distribution given by photon correlation spectroscopy (PCS) measurements does not change. For all synthesis conditions (with or without HCl, zirconium salt concentration, and synthesis time), zirconia polycrystalline particles were produced. According to the different analyses, these zirconia polycrystalline particles were aggregates of small primary clusters.

Zirconia, and polymer-stabilized tetragonal $ZrO_2$ nanopowders with an average size of ca. 2.0 nm have been prepared by microwave heating in an aqueous solution containing $Zr(NO3)_4 \cdot 5H_2O$, PVA, and NaOH [195]. The photoluminescence of the synthesized $ZrO_2$ fine particles has been investigated at two different excitations with an excitation wavelength of 254 nm; three fluorescence emissions at 402 nm, 420 nm, and 459 nm, respectively, could be observed. The PL spectrum obtained at 412 nm excitation exhibited a maximum at 608 nm, with a weak satellite peak at 530 nm. The emissions that appear at short wavelength excitation are ascribed to the near band-edge transitions.

Nanocrystalline $SnO_2$ powders of about 3 nm in size have been prepared by a microwave irradiation heating technique from an aqueous solution in the presence of $SnCl_4$ and urea [196]. A bandgap estimated to be 4.5 eV is obtained from the optical measurement of the nanoparticles. HRTEM pictures show that the as-prepared SnO powders are crystalline with ca. 3 nm particle size, and the particles are held together by an irregular network.

High-purity powders of SnO nanocrystallites with crystallite sizes less than 30 nm and surface areas up to 40 $m^2\ g^{-1}$ have been synthesized by a solution process in which an amorphous oxy–hydroxy precipitate of $Sn^{+2}$ is crystallized with microwave heating [197]. Microwave heating was found to have selectively accelerated SnO crystallization, but not the concurrent $Sn^{+2}$ to $Sn^{+4}$ oxidation, which otherwise prevails in the conventional thermal heating process. Control studies give a strong indication of a non-temperature effect of the microwave irradiation in the present process.

Microwave-hydrothermal synthesis of titanium dioxide under various reaction conditions was reported by Komarneni [198]. Crystallization of rutile from $TiOCl_2$ solutions was found to be extremely rapid. Titanium dioxide, particle size, morphology and polymorph can be controlled by changing various parameters, such as: concentration, pH, pressure (or temperature), time, and anionic species. The main advantages of microwave-hydrothermal processing of $TiO_2$ are: (1) rapid heating to required temperature and (2) extremely rapid kinetics of crystallization. Rutile was the only crystalline phase when various concentrations (3 M, and 2 M) of $TiOCl_2$ solutions were treated at a variety of pressures (190, 100, 50 and 25 psi) for 2 h. The yield of rutile was 95% at all pressures, which showed that the crystallization of this solution was practically complete. When the $TiOCl_2$ was further reduced, a mixture of anatase and rutile phases was obtained.

Another microwave study yielded only the anatase phase [199]. In this work $TiO_2$ was synthesized from the alkoxide by the polyol method using various polyols (1,4-butanediol, 1,5-pentanediol, or 1,6-hexanediol) under MW radiation. The authors demonstrated that the crystallite size, which was always less than 10 nm, could be controlled by the quantity of added water and by the nature of the polyol (see Table 1 in [199]).

A group of binary oxide nanophase (titanates and zirconates) materials were prepared using a microwave-assisted soft-chemical route [200]. $BaTiO_3$, $Ba_6Ti_{17}O_{40}$, $BaZrO_3$ and $PbTiO_3$ were prepared from $BaCl_2$ hydrate, $Pb(Ac)_2$, $Ti(OPri)_4$ and $ZrOCl_2$. All reactions were performed in ethylene glycol, which acted both as a solvent and as a growth regulating agent, under atmospheric pressure in a microwave reactor.

The Ni/NiO composite was prepared by using the fast method of microwave-assisted oxidation [201]. Amorphous Ni nanoparticles were used as a precursor, and the oxidizing agent was oxygen. By using vapors of $H_2O_2$, almost complete oxidation of nickel was achieved.

CuO nanoparticles with an average size of ca. 4 nm have been successfully prepared by microwave irradiation, using copper(II) acetate and sodium hydroxide as the starting materials and ethanol as the solvent [202]. The as-prepared CuO nanoparticles have regular shape, narrow size distribution and high purity. The band gap is estimated to be 2.43 eV according to the results of the optical measurements of the CuO nanoparticles.

Before summarizing this section and emphasizing the advantages of using microwave radiation for the synthesis of nanoparticles, we would like to remind the reader that perhaps the largest body of work in this field was conducted in the preparation of nanosized chalcogenides. However, following our own guidelines mentioned above, we have left it to another review [3]. Almost all the authors quoted in this review have mentioned the short reaction time in the microwave processes as perhaps the biggest advantage. The simplicity of the reaction, whether the reaction vessel is put directly in the microwave oven or hooked up to a reflux system, is that the reaction cell is still very simple and it is easy to assemble the experimental set-up. In almost all his papers Komarnani has pointed out the energy saving in conducting a microwave reaction. MWH is homogeneous: it starts from the inside and progresses towards the glass container, while the opposite is true for conventional heating. Temperature and concentration gradients are avoided in a MW reaction. Whether there is a specific microwave effect, namely a reaction carried out under microwave radiation leading to a different product, compared with the same reaction under *conventional heating, or not, is not clear.* There are a few well known examples showing results different from those obtained by conventional heating [203–205]. However, more detailed kinetic studies have established [206–208] that chemical reactions, which are carried out under MW radiation, are controlled by the same fundamental thermodynamics and kinetics as conventional reactions. The title of [206] is "Specific activation by MW-Myth or Reality".

In a MWH process one precursor can be heated at much higher heating rates and reach a higher temperature than its surroundings. In this respect it is similar to sonochemistry, where hot spots are formed in the liquid. In the polyol reactions where the first step was the formation of metallic fine particles, we could see the solution of ethylene glycol hot points reaching 600–700 °C. The difference between sonochemistry and MWH is that in the latter method there is no direct contact between the energy source and the solution, while in the sonochemistry the horn is dipped into the solution.

In MWH a reaction can be conducted at temperatures higher than the boiling point of the solvent while employing a simple apparatus. According to Mingos [159], it is possible to increase the temperature of a reaction in common organic solvents up to 100 °C above the conventional boiling point of the solvent.

## Acknowledgements

This work could not have been performed without the help of so many good friends from all around the world. I thank them all without trying to name all of them. I will name however, my coworkers in my laboratory at Bar-Ilan. I would like to thank Dr. X. Cao, Dr. Ziyi Zhong, Dr. Y. Zhao, Dr. S. Ramesh, Dr. M. Shafi, Dr. Arul Dhas, Prof. J. J. Zhu, Dr. Liu Suwen, Dr. R. A. Salkar, Dr. P. Jeevanandam, Prof. X. Tang, Dr. Yanqin Wang, Prof. W. Huang, Prof. Guansheng Peng, Prof. Ynigchun Zhu, Dr. R. Vijayakumar, Dr. Rohit Kumar Rana, Dr. Hong-Liang Li, Dr. Jinping Xiong, Dr. Qiaoling Li, Dr. D. N. Srivastava, Dr. V. Ganesh Kumar, Dr. M. Sivakumar, Dr. Gentao Zhou, Dr. S. Nikitenko, and Dr. Qiu Longhui, my foreign postdoctorate, for adapting rapidly to the Israeli system and enthusiastically conducting innovative research. I also thank my Israeli Ph. D. and M. Sc. students, Dr. Gina Katabi, Dr. Shlomit Wiizel, Dr. Oleg Palchik, Mrs. Tatiana Prozorov, Mr. Ronnen Polsky, Mr. Menachem Motiei, Mrs. Sigalit Avivi (Levi), Mr. Stanislav

Kishinevsky, Mrs. Ayelet Gamili, Haviv Grisaru, Vilas G. Pol, Riki Kerner (Harpeness), Mrs. Alexandra Gabashvili, and Mr. Yaakov David Solomon.

Special thanks are given to Professor Yuri Koltypin, and Dr. Elena Sominska, who helped me in the early stages in switching from spectroscopy to materials science, and liquidating the old equipment. They in turn were helped at the later stages by Dr. Nina Perkas.

This work could not have been accomplished without a few institutions who financed this research. I would like to thank the BSF (Bi-US-Israel Binational Science Foundation), the Israeli Ministry of Science, Culture, and Sport through the Indo-Israeli program in Materials Science (two grants), the Sino-Israeli program in Materials Science, the NEDO Organization, the Israeli Ministry of Science, Culture, and Sport for two Infrastructure Grants, The EC for 4 5$^{th}$ Program grants, and 3 INTAS grants, and the Israeli Science Foundation for two equipment grants. I would also like to thank the German BMBF for a DIP (Deutsche-Israel Projects) and a grant provided via the Energy Program.

This review could not have been written without an invitation from Professor S. T. Lee from the City University of Hong Kong for me to spend a short sabbatical at his COSDAF Institute. This invitation enabled this review, and I thank Professor Lee for extending this invitation to me.

## References

1 a) K. S. Suslick, T. W. Hyeon, M. W. Fang, *Chem. Mater.*, **1996**, *8*, 2172; b) K. S. Suslick, G. J. Price, *Annu. Rev. Mater. Sci.*, **1999**, *29*, 295.
2 M. Ashokkumar, F. Grieser, *Rev. Chem. Eng.*, **1999**, *15*, 41.
3 Jun-Jie Zhu, Hui Wang, Novel Methods for Chemical Preparation of Metal Chalcogenide Nanoparticles, in *Encyclopedia of Nanoscience and Nanotechnology*, ed. Hari Singh Nalwa, American Scientific Publishers, New York, 2003.
4 T. J. Mason, Sonochemistry, Royal Society of Chemistry, Cambridge, 1990.
5 K. S. Suslick, S. B. Choe, A. A. Cichowlas et al., *Nature*, **1991**, *353*, 414.
6 K. S. Suslick, R. E. Cline, D. A. Hammerton, *J. Am. Chem. Soc.*, **1986**, *108*, 5641.
7 K. S. Suslick, R. E. Cline, D. A. Hammerton, *J. Am. Chem. Soc.*, **1986**, *108*, 5641.
8 X. Cao, Yu. Koltypin, G. Kataby et al., *J. Mater. Res.*, **1995**, *10*, 2952.
9 X. Cao, Yu. Koltypin, R. Prozorov et al., *J. Mater. Chem.*, **1997**, *7*, 2447.
10 S. Avivi, Y. Mastai, G. Hodes et al., *J. Am. Chem. Soc.*, **1999**, *121*, 4196.
11 M. W. Grinstaff, A. A. Cichowlas, S. B. Choe et al., *Ultrasonics*, **1992**, *30*, 168.
12 Yu. Koltypin, G. Katabi, X. Cao et al., *J. Non Cryst Solids*, **1996**, *201*, 159.
13 K. S. Suslick, M. Fang, M. T. Hyeon et al., *Mater. Res. Soc. Symp. Proc.*, **1994**, *351*, 443.
14 C. P. Gibson, K. J. Putzer, *Science*, **1995**, *267*, 1338.
15 N. A. Dhas, C. P. Raj, A. Gedanken, *Chem. Mater.*, **1998**, *10*, 1446.
16 N. A. Dhas, A. Gedanken, *J. Mater. Chem.*, **1998**, *8*, 445.
17 K. Okitsu, Y. Mizukoshi, H. Bandow et al., *Ultrason. Sonochem.*, **1996**, *3*, S249.
18 Y. Mizukoshi, R. Oshima, Y. Maeda et al., *Langmuir*, **1999**, *15*, 2733.
19 R. A. Salkar, P. Jeevanandam, S. T. Aruna et al., *J. Mater. Chem.*, **1999**, *9*, 1333.

20 Y. Mizukoshi, E. Takagi, H. Okuno et al., *Ultrason. Sonochem.*, **2001**, *8*, 1.
21 T. Fujimoto, Y. Mizukoshi, Y. Nagata et al., *Scr. Mater.*, **2001**, *44*, 2183.
22 T. Fujimoto, S. Terauchi, H. Umehara et al., *Chem. Mater.*, **2001**, *13*, 1057.
23 K. Okitsu, A. Yue, S. Tanabe et al., *Langmuir*, **2001**, *17*, 7717.
24 K. Okitsu, A. Yue, S. Tanabe et al., *Bull. Chem. Soc. Jpn.*, **2002**, *75*, 2289.
25 Y. P. Sun, H. W. Rollins, R. Guduru, *Chem. Mater.*, **1999**, *11*, 7.
26 N. A. Dhas, C. P. Raj, A. Gedanken, *Chem. Mater.*, **1998**, *10*, 3278.
27 F. Grieser, *Stud. Surf. Sci. Catal.*, **1997**, *103*, 57.
28 K. S. Suslick, M. M. Fang, T. Hyeon, *J. Am. Chem. Soc.*, **1996**, *118*, 11960.
29 K. V. P. M. Shafi, S. Wizel, T. Prozorov et al., *Thin Solid Films*, **1998**, *318*, 38.
30 K. V. P. M. Shafi, A. Gedanken, R. Prozorov, *Adv. Mater.*, **1998**, *10*, 590.
31 R. A. Salkar, P. Jeevanandam, G. Kataby et al., *J. Phys. Chem. B*, **2000**, *104*, 893.
32 K. Barbour, M. Ashokkumar, R. A. Caruso et al., *J. Phys. Chem. B*, **1999**, *103*, 9231.
33 R. A. Caruso, M. Ashokkumar, F. Grieser, *Colloid Surf. A*, **2000**, *169*, 219.
34 R. A. Caruso, M. Ashokkumar, F. Grieser, *Langmuir*, **2002**, *18*, 7831.
35 F. Grieser, *Stud. Surf. Sci. Catal.*, **1997**, *103*, 57.
36 K. S. Suslick, T. Hyeon, M. Fang et al., *Mater. Res. Soc. Symp. Proc.*, **1994**, *351*, 201.
37 K. V. P. M. Shafi, A. Gedanken, R. B. Goldfarb et al., *J. Appl. Phys.*, **1997**, *81*, 6901.
38 K. V. P. M. Shafi, A. Gedanken, R. Prozorov, *J. Mater. Chem.*, **1998**, *8*, 769.
39 K. E. Gonsalves, S. P. Rangarajan, A. Garcia Ruiz et al., *J. Mater. Sci. Lett.* **1996**, *15*, 1261.
40 R. Oshima, T. A. Yamamoto, Y. Mizukoshi et al., *Nanostruct. Mater.*, **1999**, *12*, 111.
41 S. W. Liu, L. X. Yin, Yu. Koltypin et al., *J. Magn. Magn. Mater.*, **2001**, *233*, 195.
42 K. V. P. M. Shafi, A. Gedanken, R. Prozorov et al., *J. Mater. Res.*, **2000**, *15*, 332.
43 S. Ramesh, Yu. Koltypin, R. Prozorov et al., *Chem. Mater.*, **1997**, *9*, 546.
44 W. Stober, A. Fink, E. Bohn, *J. Colloid. Interface Sci.*, **1968**, *26*, 62.
45 S. Ramesh, Y. Cohen, D. Aurbach et al., *Chem. Phys. Lett.*, **1998**, *287*, 461.
46 S. Ramesh, Y. Cohen, R. Prozorov et al., *J. Phys. Chem. B*, **1998**, *102*, 10234.
47 Z. Y. Zhong, Y. Mastai, Yu. Koltypin et al., *Chem. Mater.*, **1999**, *11*, 2350.
48 M. L. Breen, A. D. Dinsmore, R. H. Pink et al., *Langmuir*, **2001**, *17*, 903.
49 Y. Kitamoto, M. Abe, *Nanostruct. Mater.*, **1999**, *12*, 41.
50 A. Dokoutchaev, J. T. James, S. C. Koene et al., *Chem. Mater.*, **1999**, *11*, 2389.
51 V. G. Pol, D. N. Srivastava, O. Palchik et al., *Langmuir*, **2002**, *18*, 3352.
52 S. J. Doktycz, K. S. Suslick, *Science*, **1990**, *247*, 4946.
53 K. Prabhakaran, K. V. P. M. Shafi, A. Ulman et al., *Adv. Mater.*, **2001**, *13*, 1859.
54 K. Prabhakaran, K. V. P. M. Shafi, A. Ulman et al., *Surf. Sci.*, **2002**, *506*, L250.
55 K. Prabhakaran, K. V. P. M. Shafi, Y. Yamauchi et al., *Appl. Surf. Sci.*, **2002**, *190*, 161.
56 F. Papadimitrakopoulos, T. Phely-Bobin, P. Wisniecki, *Chem. Mater.*, **1999**, *11*, 522.
57 S. Wizel, R. Prozorov, Y. Cohen, D. Aurbach et al., *J. Mater. Res.*, **1998**, *13*, 211.
58 S. Wizel, S. Margel, A. Gedanken et al., *J. Mater. Res.*, **1999**, *14*, 3913.
59 A. Dutta, P. K. Mahato, N. N. Dass, *Eur. Polym. J.*, **1991**, *27*, 465.
60 S. Wizel, S. Margel, A. Gedanken, *Polym. Int.*, **2000**, *49*, 445.
61 R. V. Kumar, Y. Mastai, Y. Diamant et al., *J. Mater. Chem.*, **2001**, *11*, 1209.

62 R. V. Kumar, Yu. Koltypin, O. Palchik et al., *Appl. Polym. Sci.*, **2002**, *86*, 160.
63 K. S. Suslick, T. Hyeon, M. Fang et al., *Mater. Sci. Forum*, **1996**, *225*, 903.
64 K. S. Suslick, T. Hyeon, M. M. Fang et al., *Mater. Sci. Eng. A*, **1995**, *204*, 186.
65 K. Okitsu, Y. Mizukoshi, H. Bandow et al., *J. Phys. Chem. B*, **1997**, *101*, 5470.
66 N. A. Dhas, H. Cohen, A. Gedanken, *J. Phys. Chem. B*, **1997**, *101*, 6834.
67 Y. Koltypin, A. Fernandez, T. C. Rojas et al., *Chem. Mater.*, **1999**, *11*, 1331.
68 J. Walter, M. Nishioka, S. Hara, *Chem. Mater.*, **2001**, *13*, 1828.
69 A. Koshio, M. Yudasaka, M. Zhang et al., *Nano Lett.*, **2001**, *1*, 361.
70 S. Nikitenko, Yu. Koltypin, I. Felner et al., *Angew. Chem. Int. Ed. Engl.*, **2001**, *40*, 4447.
71 X. Cao, R. Prozorov, Y. Koltypin et al., *J. Mater. Res.*, **1997**, *12*, 402.
72 M. Sugimoto, *J. Magn. Magn. Mater.*, **1994**, *133*, 460.
73 K. Tanaka, K. Hirao, N. Soga, *J. Appl. Phys.*, **1991**, *69*, 7752.
74 M. Sugimoto, N. Hiratsuka, *J. Magn. Magn. Mater.*, **1983**, *31*, 1533.
75 N. A. Dhas, Y. Koltypin, A. Gedanken, *Chem. Mater.*, **1997**, *9*, 3159.
76 N. A. Dhas, A. Gedanken, *J. Phys. Chem. B*, **1997**, *101*, 9495.
77 R. Vijayakumar, Yu. Koltypin, X. N. Xu et al., *J. Appl. Phys.*, **2001**, *89*, 6324.
78 T. Prozorov, R. Prozorov, Yu. Koltypin et al., *J. Phys. Chem.*, **1998**, *102*, 10165.
79 K. V. P. M. Shafi, Y. Koltypin, A. Gedanken et al., *J. Phys. Chem. B*, **1997**, *101*, 6409.
80 K. V. P. M. Shafi, I. Felner et al., *J. Phys. Chem. B*, **1999**, *103*, 3358.
81 A. Patra, E. Sominska, S. Ramesh et al., *J. Phys. Chem. B*, **1999**, *103*, 3361.
82 A. Gedanken, R. Reisfeld, L. Sominski et al., *Appl. Phys. Lett.*, **2000**, *77*, 945.
83 A. Gedanken, R. Reisfeld, E. Sominski et al., *J. Phys. Chem. B*, **2000**, *104*, 7057.
84 V. G. Pol, R. Reisfeld, A. Gedanken, *Chem. Mater.*, **2002**, *14*, 3920.
85 V. G. Pol, O. Palchik, A. Gedanken et al., *J. Phys. Chem. B*, **2002**, *106*, 9737.
86 F. G. Sherif, F. A. Via, (Akzo America Inc.), *U.S. Pat.*, 4 764 357, 1988.
87 S. Avivi, Y. Mastai, G. Hodes et al., *J. Am. Chem. Soc.*, **1999**, *121*, 4196.
88 H. R. Craig, S. Y. Tyree, *Inorg. Chem.*, **1969**, *8*, 591.
89 S. Avivi, Y. Mastai, A. Gedanken, *Chem. Mater.*, **2000**, *12*, 1229.
90 S. Avivi, Y. Mastai, A. Gedanken, *J. Am. Chem. Soc.*, **2000**, *122*, 4331.
91 J. J. Zhu, Z. H. Lu, S. T. Aruna et al., *Chem. Mater.*, **2000**, *12*, 2557.
92 D. Aurbach, A. Nimberger, E. Levi et al., *Chem. Mater.*, **2002**, *14*, 4155.
93 C. N. R. Rao, *J. Mater. Chem.*, **1999**, *9*, 1.
94 A. Gedanken, N. Perkas, Yu. Koltypin et al., *Chem. Eur. J.*, **2001**, *7*, 4546.
95 Y. Wang, X. Tang, L. Yin et al., *Adv. Mater.*, **2000**, *12*, 1137.
96 H. X. Tang, S. W. Liu, Y. Wang et al., *Chem. Commun.*, **2000**, 2119.
97 Y. Q. Wang, L. X. Yin, O. Palchik et al., *Langmuir*, **2001**, *17*, 4131.
98 Y. Wang, L. Yin, A. Gedanken, *Ultrason. Sonochem.*, **2002**, *9*, 285.
99 C. T. Kresge, M. E. Leonowicz, W. J. Roth et al., *Nature*, **1992**, *359*, 710.
100 a) R. Ryoo, J. M. Kim, C. H. Ko et al., *J. Phys. Chem. B*, **1996**, *100*, 17718; b) R. Ryoo, S. Jun, *J. Phys. Chem. B*, **1997**, *101*, 317.
101 M. V. Landau, L. Vradman, M. Herskowitz et al., *J. Catal.*, **2001**, *201*, 22.
102 N. Perkas, Y. Wang, Yu. Koltypin et al., *Chem. Commun.*, **2001**, *11*, 988.
103 K. S. Suslick, in *Ultrasound: Its Chemical, Physical and Biological Effects*, ed. K. S. Suslick, VCH, Weinheim 1988.
104 W. Chen, W. P. Cai, Z. P. Zhang et al., *Chem. Lett.*, **2001**, *2*, 152.
105 W. Chen, W. P. Cai, G. Z. Wang et al., *Appl. Surf. Sci.*, **2001**, *174*, 51.

106 W. Chen, W. P. Cai, C. H. Liang et al., *Mater. Res. Bull.*, **2001**, *36*, 335.
107 W. Chen, W. P. Cai, L. Zhang et al., *J. Colloid Interface Sci.*, **2001**, *238*, 291.
108 W. Chen, W. P. Cai, Z. X. Chen et al., *Ultrason. Sonochem.*, **2001**, *8*, 335.
109 W. Chen, W. P. Cai, Y. Lei et al., *Mater. Lett.*, **2001**, *50*, 53.
110 J. C. Yu, L. Z. Zhang, J. G. Yu, *New J. Chem.*, **2002**, *26*, 416.
111 J. C. Yu, L. Z. Zhang, J. G. Yu, *Chem. Mater.*, **2002**, *14*, 4647.
112 D. N. Srivastava, S. Chappel, O. Palchik et al., *Langmuir*, **2002**, *18*, 4160.
113 D. N. Srivastava, N. Perkas, A. Gedanken et al., *J. Phys. Chem. B*, **2002**, *106*, 1878.
114 Y. Q. Wang, S. G. Chen, X. H. Tang et al., *J. Mater. Chem.*, **2001**, *11*, 521.
115 Rohit Kumar Rana, Y. Mastai, A. Gedanken, *Adv. Mater.*, **2002**, *14*, 1414.
116 S. S. Kim, W. Zhang, T. J. Pinnavaia, *Science*, **1998**, *282*, 1302.
117 P. Jeevanandam, Y. Koltypin, A. Gedanken, *Mater. Sci. Eng. B*, **2002**, *90*, 125.
118 X. Y. Liang, Z. Ma, Z. C. Bai et al., *Acta Phys. Chim. Sin.*, **2002**, *18*, 567.
119 I. Shames, E. Rozenberg, G. Gorodetsky et al., *Appl. Phys.*, **2002**, *91*, 7929.
120 P. Jeevanandam, Y. Koltypin, A. Gedanken, *Nano Lett.*, **2001**, *1*, 263.
121 X. Xia, L. L. Shen, Z. P. Guo et al., *J. Nanosci. Nanotechnol.*, **2002**, *2*, 45.
122 P. Jeevanandam, Y. Koltypin, A. Gedanken et al., *J. Mater. Chem.*, **2002**, *10*, 511.
123 Y. C. Zhu, H. L. Li, Y. Koltypin, A. Gedanken, *J. Mater. Chem.*, **2002**, *12*, 729.
124 J. C. Yu, J. G. Yu, W. K. Ho et al., *Chem. Commun.*, **2001**, 1942.
125 W. P. Huang, X. H. Tang, Y. Q. Wang et al., *Chem. Commun.*, **2000**, 415.
126 L. Guo, X. K. Wang, Z. M. Lin et al., *Chem. J. Chin. Univ.*, **2002**, *23*, 1592.
127 Y. C. Zhu, H. L. Koltypin, Y. R. Hacohen et al., *Chem. Commun.*, **2001**, *24*, 2616.
128 N. Sukpirom, M. M. Lerner, *Mater. Sci. Eng. A*, **2002**, *333*, 218.
129 K. V. P. M. Shafi, A. Ulman, X. Z. Yan et al., *Langmuir*, **2001**, *17*, 1726.
130 C. S. S. R. Kumar, H. Modrow, J. H. Hormes, *Part Part Syst. Char.*, **2002**, *19*, 336.
131 H. Wang, J. J. Zhu, J. M. Zhu et al., *Phys. Chem. Chem. Phys.*, **2002**, *4*, 3794.
132 L. X. Yin, Y. Q. Wang, G. S. Pang et al., *J. Colloid Interface Sci.*, **2002**, *246*, 78.
133 L. X. Yin, Y. Q. Wang, G. S. Pang et al., *J. Colloid Interface Sci.*, **2002**, *246*, 78.
134 E. V. Kumar, Y. Mastai, A. Gedanken, *Chem. Mater.*, **2002**, *12*, 3892.
135 Yu. Koltypin, S. I. Nikitenko, A. Gedanken, *J. Mater. Chem.*, **2002**, *12*, 1107.
136 R. V. Kumar, Y. Diamant, A. Gedanken, *Chem. Mater.*, **2000**, *12*, 2301.
137 Yu. Koltypin, X. Cao, R. Prozorov et al., *J. Mater. Chem.*, **1997**, *7*, 2453.
138 T. H. Hyeon, M. M. Fang, K. S. Suslick, *J. Am. Chem. Soc.*, **1996**, *118*, 5492.
139 S. I. Nikitenko, P. Moisy, I. A. Tcharushnikova et al., *Ultrason. Sonochem.*, **2000**, *7*, 177.
140 J. Lu, Y. Xie, X. C. Jiang, W. He et al., *J. Mater. Chem.*, **2001**, *11*, 3281.
141 H. L. Li, Y. C. Zhu, O. Palchik et al., *Inorg. Chem.*, **2002**, *41*, 637.
142 R. Katoh, Y. Tasaka, E. Sekreta et al., *Ultrason. Sonochem.*, **1999**, *6*, 185.
143 R. G. Compton, J. C. Eklund, F. Marken, *Electronanalysis*, **1997**, *9*, 509.
144 G. Hodes, Y. Mastai, *Electrodepostion of Nanocrystalline Semiconductor Materials*, in *Encyclopedia on Electrochemistry*, Vol. 6: Semiconductor Electrodes and Photoelectrochemistry, ed. S. Licht, Wiley-VCH, Weinheim, 2001, p. 173.
145 J. Reisse, H. Francois, J. Vandercammen et al., *Electrochim. Acta*, **1994**, *39*, 37.
146 A. Durant, J. L. Delplancke, R. Winand et al., *Tetrahedron Lett.*, **1995**, *36*, 4257.
147 J. Reisse, T. Caulier, C. Deckerkheer et al., *Ultrason. Sonochem.*, **1996**, *3*, S147.

148 Y. Mastai, R. Polsky, Yu. Koltypin et al., *J. Am. Chem. Soc.*, **1999**, *121*, 10047.
149 J. J. Zhu, S. W. Liu, O. Palchik et al., *Langmuir*, **2000**, 16, 6396.
150 S. W. Liu, W. P. Huang, S. G. Chen et al., *J. Non-Cryst. Solids*, **2001**, *283*, 231.
151 J. Zhu, S. T. Aruna, Y. Koltypin et al., *Chem. Mater.*, **2000**, *12*, 143.
152 J. J. Zhu, Q. F. Qiu, H. Wang et al., *Inorg. Chem. Commun.*, **2002**, *4*, 242.
153 Y. Socol, O. Abramson, A. Gedanken et al., *Langmuir*, **2002**, *18*, 4736.
154 Y. Mastai, M. Homyonfer, A. Gedanken et al., *Adv. Mater.*, **1999**, *11*, 1010.
155 D. M. P. Mingos, *Chem. Ind.*, **1994**, 596.
156 S. Caddick, *Tetrahedron*, **1995**, *51*, 10403.
157 D. M. P. Mingos, D. R. Baghurst, *Chem. Soc. Rev.*, **1991**, *20*, 1.
158 K. J. Rao, B. Vaidhyanathan, M. Ganguli et al., *Chem Mater.*, **1999**, *11*, 882.
159 R. Dagani, *Chem. Eng. News*, **1997**, *75*, 26.
160 V. Sridar, *Curr. Sci.* **1998**, *74*, 446.
161 D. M. P. Mingos, A. G. Whittaker, *J. Chem. Soc., Dalton Trans.*, **1995**, *2073*, 1995.
162 A. G. Whittaker, D. M. P. Mingos, *J. Chem. Soc., Dalton Trans.*, **1992**, *18*, 2751.
163 C. Gabriel, S. Gabriel, E. H. Grant et al., *Chem. Soc. Rev.*, **1988**, *27*, 213.
164 C. H. Chou, J. Phillips, *J. Mater. Res.*, **1992**, *7*, 2107.
165 Z. L. Jiang, Z. W. Feng, X. C. Shen, *Chin. Chem. Lett.*, **2001**, *12*, 551.
166 I. Pastoriza-Santos, L. M. Liz-Marzan, *Langmuir*, **2002**, *18*, 2888.
167 D. L. Boxall, G. A. Deluga, E. A. Kenik et al., *Chem. Mater.*, **2001**, *13*, 891.
168 a) F. Fievet, J. P. Lagier, M. Figlarz, *Mater. Res. Bull.*, **1989**, Dec. 29; b) G. Viau, F. Fievet-Vincent, F. Fievet, *J. Mater. Chem.*, **1996**, *6*, 1047; c) G. Viau, F. Fievet-Vincent, F. Fievet, *Solid State Ionics*, **1996**, *84*, 259.
169 F. Bonet, C. Guery, D. Guyromard et al., *Int. J. Inorg. Mater.*, **1999**, *1*, 47.
170 W. Y. Yu, W. X. Tu, H. F. Liu, *Langmuir*, **1999**, *15*, 6.
171 W. X. Tu, H. F. Liu, *J. Mater. Chem.*, **2000**, *10*, 2207.
172 W. X. Yu, H. Y. Liu, *Chem. Mater.*, **2000**, *12*, 564.
173 S. Komarneni, D. Li, B. Newalkar et al., *Langmuir*, **2002**, *18*, 5959.
174 A. M. Qin, Z. L. Jiang, Q. Y. Liu et al., *Chin. J. Anal. Chem.*, **2002**, *10*, 254.
175 W. X. Chen, J. Y. Lee, Z. L. Liu, *Chem. Commun.*, **2002**, 2588.
176 R. He, X. Qian, J. Yin, Z. Zhu, *J. Mater. Chem.*, **2002**, *12*, 3783.
177 S. Komarneni, R. Roy, L. I. Qh, *Mater. Res. Bull.*, **1992**, *27*, 1393.
178 D. C. Dong, P. J. Hong, S. S. Dai, *Mater. Res. Bull.*, **1995**, *30*, 531.
179 Q. Li, Y. Wei, *Mater. Res. Bull.*, **1998**, *33*, 779.
180 O. Palchik, I. Felner, G. Kataby et al., *J. Mater. Res.*, **2000**, *15*, 2176.
181 X. B. Han, L. Huang, Z. Hui, *J. Inorg. Mater*, **1999**, *14*, 669.
182 Z. B. Jia, Y. Wei, H. M. Wang, *J. Inorg. Mater*, **2000**, *15*, 926.
183 X. H. Liao, J. J. Zhu, W. Zhong et al., *Mater. Lett.*, **2001**, *50*, 341.
184 H. Katsuki, S. Komarneni , *J. Am. Ceram. Soc.*, **2001**, *84*, 2313.
185 S. S. Manoharan, S. J. Swati, Prasanna et al., *J. Am. Ceram. Soc.*, **2002**, *85*, 2469.
186 S. R. Dhage, Y. B. Khollam, H. S. Potdar et al., *Mater. Lett.*, **2002**, *57*, 457.
187 T. Caillot, D. Aymes, D. Stuerga et al., *J. Mater. Sci.*, **2002**, *37*, 5153.
188 Y. B. Khollam, S. R. Dhage, H. S. Potdar et al., *Mater. Lett.*, **2002**, *56*, 571.
189 S. Komarneni, M. C. D'Arrigo, C. Leonelli et al., *J. Am. Ceram. Soc.*, **1998**, *81*, 3041.
190 X. H. Liao, J. M. Zhu, J. J. Zhu et al., *Chem. Commun.*, **2001**, 937.
191 F. Bondioli, A. M. Ferrari, C. Leonelli et al., *J. Am. Ceram. Soc.*, **2001**, *84*, 2728.
192 K. Bellon, D. Chaumont, D. Stuerga, *J. Mater. Res.*, **1999**, *16*, 2619.

193 X. Jiang, F. Li, Yadong Li, *Inorg. Chem.*, **2002**, *41*, 3602.
194 J. J. Zhu, J. M. Zhu, X. H. Liao et al., *Mater. Lett.*, **2002**, *53*, 12.
195 D. S. Wu, C. Y. Han, S. Y. Wang et al., *Mater. Lett.*, **2002**, *53*, 155.
196 S. Komarneni, R. K. Rajha, H. Katsuki, *Mater. Chem. Phys.*, **1999**, *61*, 50.
197 T. Yamamoto, Y. Wada, H. Yin et al., *Chem. Lett.*, **2002**, 964.
198 O. Palchik, J. J. Zhu, A. Gedanken, *J. Mater. Chem.*, **2000**, *10*, 1251.
199 O. Palchik, S. Avivi, D. Pinkert et al., *Nanostruct. Mater.*, **1999**, *11*, 415.
200 H. Wang, J. Z. Xu, J. J. Zhu et al., *J. Cryst. Growth*, **2002**, *244*, 88.
201 M. M. Chowdry, D. M. P. Mingos, A. J. P. White et al., *Chem. Commun.*, **1996**, 899.
202 E. M. Gordon, *Organometallics*, **1993**, *12*, 5052.
203 M. Larhed, A. Hallberg, *J. Org. Chem.*, **1996**, *61*, 9582.
204 R. Laurent, A. Laporterie, J. Dubac et al., *J. Org. Chem.*, **1992**, *57*, 7099.
205 K. D. Raner, C. R. Strauss, F. Vyskoc et al., *J. Org. Chem.*, **1993**, *58*, 950.
206 S. A. Galema, *Chem. Soc. Rev.*, **1997**, *26*, 233.

# 7
# Solvothermal Synthesis of Non-Oxide Nanomaterials

*Y. T. Qian, Y. L. Gu, and J. Lu*

## 7.1
## Introduction

Inorganic non-oxide materials, such as III–V and II–VI group semiconductors, carbides, nitrides, phosphides, and borides, are traditionally prepared by solid state or gas-phase reactions. They have also been prepared via the pyrolysis of organometallic precursors but the products may be amorphous or poorly crystalline and a crystallization treatment at higher temperature is necessary. This treatment, however, may result in the crystalline size being beyond the nanometer scale. Exploration of milder techniques for preparing non-oxide nanomaterials with controlled shapes and sizes is very important for material science.

Hydrothermal synthesis is one of the important methods for producing fine powders of oxides. A hydrothermal system is usually maintained at a temperature beyond 100 °C and the autogenous pressure of water exceeds the ambient pressure, which is favorable for the crystallization of products. Recent research indicates that the hydrothermal method is also a practical means for preparing chalcogenide and phosphide nanomaterials, and hydrothermal treatment is an effective method for passivating porous silicons. Similar to hydrothermal synthesis, in a solvothermal process, a non-aqueous solvent, which is sealed in an autoclave and maintained in its superheated state, is the reaction medium, where the reactants and products are prevented effectively from oxidation and volatilization and the reaction and crystallization can be realized simultaneously. Furthermore, organic solvents may be favorable for the dispersion of non-oxide nanocrystallites and may stabilize some metastable phases.

In this chapter, we briefly review our recent progress in the solvothermal preparation of non-oxide nanomaterials. $\gamma$-ray irradiation and room temperature synthesis of chalcogenide nanocrystallites are also briefly described but first we present some progress in hydrothermal synthesis.

In different hydrothermal conditions, $TiO_2$ nanocrystallites (anatase and rutile) with nine kinds of morphologies have been prepared [1, 2]. $CdWO_4$ nanorods [3], which showed very strong photoluminescence at 486 nm ($\lambda_{ex} = 253$ nm) at room temperature, were prepared by a hydrothermal method at 130 °C. Lead tungstate

*The Chemistry of Nanomaterials: Synthesis, Properties and Applications, Volume 1.* Edited by C. N. R. Rao, A. Müller, A. K. Cheetham

ISBN: 3-527-30686-2

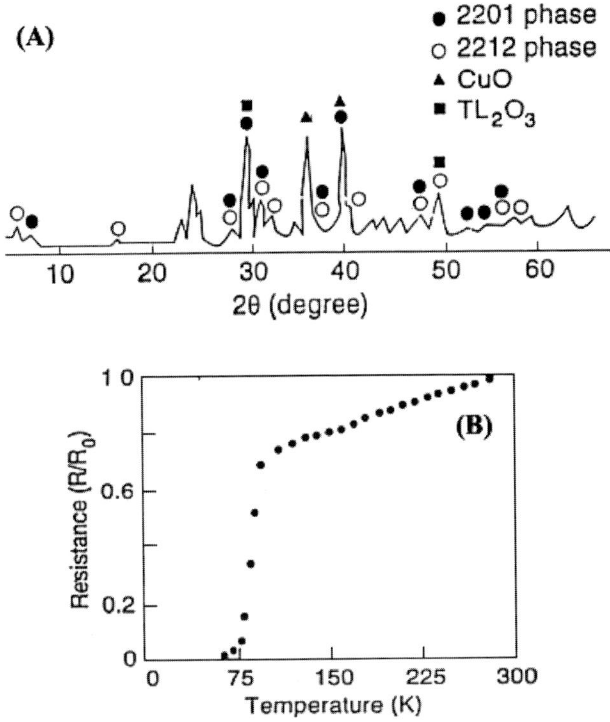


**Fig. 7.1.** The XRD pattern (A) and the *R–T* curve (B) of the Tl-based superconductor prepared by the hydrothermal method.

($PbWO_4$), an important inorganic scintillator, was also hydrothermally synthesized [4]. By using LiOH and $Nb_2O_5$ as the reactants, 40–100 nm flake-like $LiNbO_3$ nanocrystallites [5] were prepared.

To avoid the toxicity of thallium oxides in high temperature solid state reactions, a hydrothermal reaction containing $Tl(NO_3)_3$, BaO, CuO and $CaCO_3$ with $H_2O_2$ at 160 °C was designed and a Tl-based superconductor, $Tl_2O_3$–CaO–$BaCu_3O_4$, with $T_c$ onset near 95 K and zero resistance at 65 K has been prepared [6]. X-ray diffraction (XRD) studies showed that the products consisted of at least two superconducting phases (2201 and 2212). Figure 7.1 shows the XRD pattern and the *R–T* curve of the Tl-based superconductor. As a mixed valent magnetite, cubic perovskite structure $La_{0.5}Ba_{0.5}MnO_3$ single-crystal nanowires with [110] orientation were grown hydrothermally at 270 °C [7]. They show an enhanced magnetoresistance effect as compared to the bulk crystals.

Fe-passivated porous silicon (PS) with non-degrading photoluminescence (PL) was prepared by treating $P/P^+$-type, boron-doped single crystal (111) Si wafers at 140 °C in an aqueous solution with 40 wt% HF and 0.3 M $Fe(NO_3)_3$ (volume ratio, 7/6) [8, 9]. As shown in Figure 7.2, for freshly prepared samples, the PL peak intensity is 2–2.5 times stronger than that of ordinary PS. Furthermore, unlike

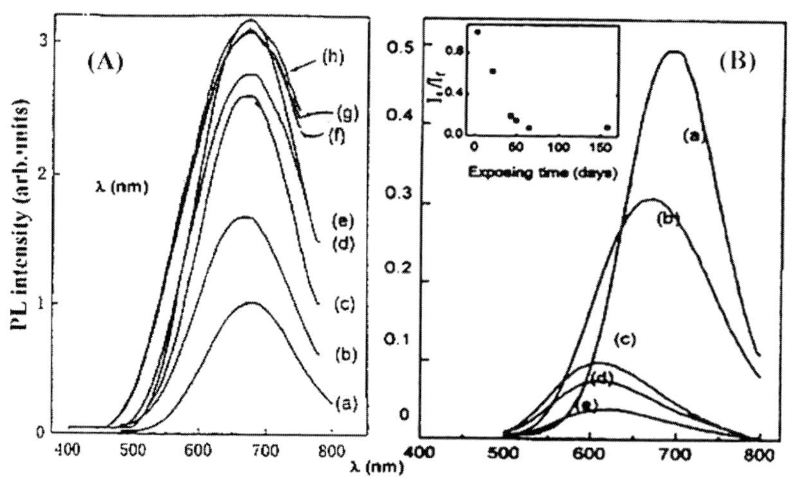


Fig. 7.2. Time evolution of the PS spectra of the Fe-passivated PS (A) and an ordinary PS (B) exposed to ambient air at room temperature for (A) (a) 2 h, (b) 45 days, (c) 75 days, (d) 110 days, (e) 140 days, (f) 170 days, (g) 235 days, (h) 500 days; (B) (a) 0 days, (b) 17 days, (c) 39 days, (d) 50 days, (e) 65 days.

ordinary PS, the PL intensity of the as-prepared PS increases during the first four months and then saturates when exposed to ambient air. No PL degradation was observed for eight months, and the peak position remained unchanged. This PL stability is attributed to the formation of stable Fe–Si bonds on the surface of the PS.

The hydrothermal method has been used to prepare monodispersed ZnS (6 nm) [10] and CdS nanocrystals (16 nm) [11]. By hydrothermal polymerization and simultaneous sulfidation processes, nanocomposites CdS/poly(vinyl acetate) nanorods [12] and nanospheres [13] were synthesized. In aqueous hydrazine solutions, nonstoichiometric metal telluride nanocrystallites such as $Cu_{2.86}Te_2$, $Cu_7Te_5$, $Cu_{2-x}Te$, and $Ag_7Te_4$ [14], and cubic $Co_9S_8$ were hydrothermally synthesized [15]. Other transition metal chalcogenides, such as single-molecular-layer $MoS_2$ [16] and $MoSe_2$ [17] were also prepared under hydrothermal conditions.

As shown in Figure 7.3, novel millimeter-sized tubular crystals of $Ag_2Se$ were grown via a hydrothermal reaction route from AgCl, Se and NaOH at 155 °C [18]. Similarly, CuS millimeter-scale tubular crystals were synthesized at 185 °C [19]. Reacting $Bi_2Te_3$ with $I_2$ via an iodine transport hydrothermal method at temperatures ranging from 190–200 °C resulted in BiTeI submicrometer hollow spheres with diameters of 200–300 nm [20]. The sphere wall is composed of BiTeI nanoparticles with average diameter 5 nm and thickness about 10 nm. The transmission electron microscope (TEM) and high-resolution transmission electron microscope (HRTEM) images of BiTeI spheres are shown in Figure 7.4. SbSI rodlike crystals [21] and nanorods [22] with diameters of 20–50 nm and lengths up to several micrometers were synthesized at about 200 °C.

**Fig. 7.3.** SEM images of the tubular $Ag_2Se$ hydrothermally prepared from AgCl, Se and NaOH at 155 °C.

A series of ternary sulfides were prepared by hydrothermal methods. $CdInS_2$ nanorods (20–25 nm × 400–450 nm) were prepared by reacting CdS nanorods with $InCl_3$ and thiourea [23]. Other hydrothermally prepared ternary sulfide nanocrystallites include $MInS_2$ (M = Cu [24], Ag [25]), $CuGaS_2$ [26], $AgGaS_2$ (6–12 nm), $\alpha$-$Ag_9GaS_6$ [27], $Ag_3CuS_2$ [28], $PbSnS_3$ [29], and $CuFeS_2$ [30].

Recently, InP nanocrystals were synthesized via the hydrothermal reaction of $InCl_3{\cdot}4H_2O$ and white phosphorus in aqueous ammonia with 0.01 mol $L^{-1}$ potassium stearate at 170 °C [31]. The XRD pattern can be indexed as the cubic phase. The secondary particles, 180 nm in size, consisting of fine InP nanocrystals, 15 nm in size, were found from the TEM images (Figure 7.5(a)). Alternatively, using 1,2-dimethoxyethane instead of water as the reaction medium, nanocrystalline InP 10 nm in size was synthesized through the solvothermal reaction of anhydrous $InCl_3$ and $Na_3P$ at 150 °C [32]. The TEM image is shown in Figure 7.5(b). Obviously, the solvothermal method resulted in better particle morphology. In fact, solvothermal

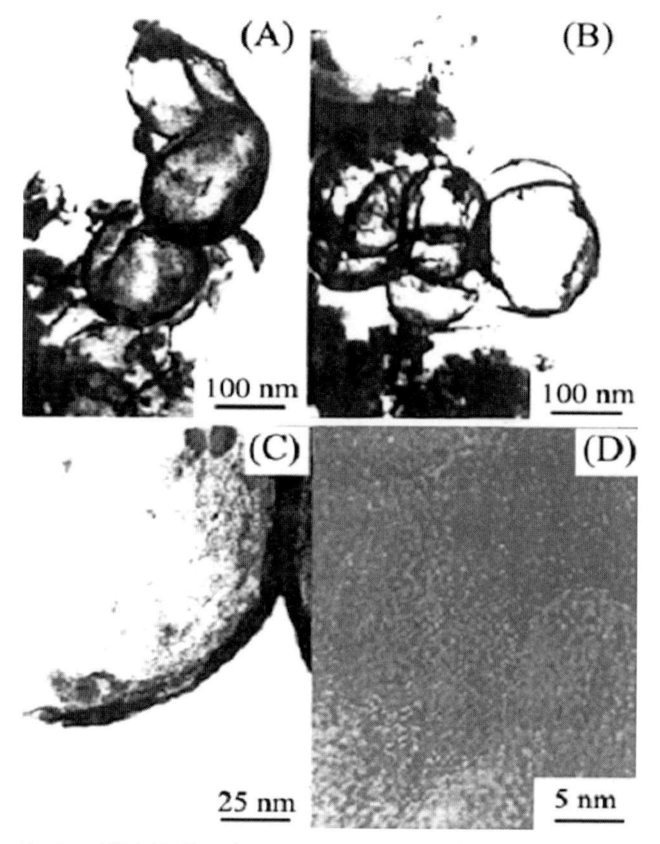


**Fig. 7.4.** TEM (A, B) and HRTEM (C, D) images of BiTeI spheres.

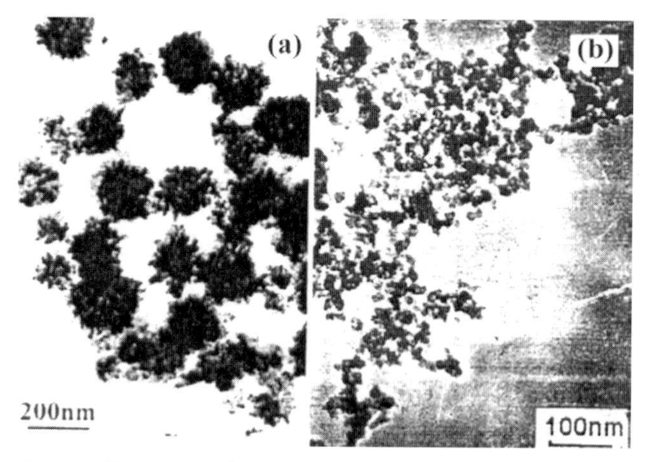


**Fig. 7.5.** TEM images of InP nanocrystals prepared by hydrothermal (a) and solvothermal (b) methods.

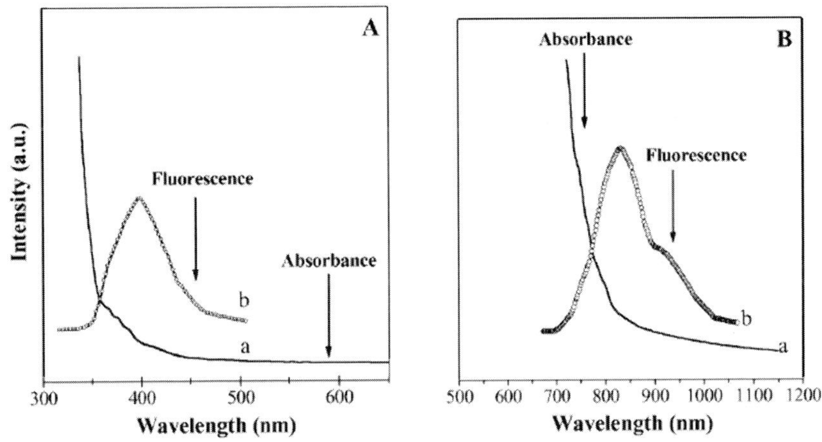


**Fig. 7.6.** Absorption and photoluminescence (PL) spectra of as-prepared products, GaP nanocrystals (A) and InP nanocrystals (B).

synthesis manifests many advantages over the hydrothermal method for preparing non-oxide materials. For example, it is effective in the preparation of nitride nanomaterials, which commonly cannot be prepared by the hydrothermal method. Recently, quantum-confined 5 nm GaP and 8 nm InP nanocrystals were hydrothermally synthesized at 120–160 °C [33]. As shown in Figure 7.6B an emission peak at 850 nm (1.55 eV) was observed in the photoluminescence spectrum of the as-prepared InP nanocrystals, which revealed the blue shift of the band gap with 0.2 eV, for GaP nanocrystals, an emission peak at 400 nm (3.1 eV) was also found (Figure 7.6A) which shows a pronounced quantum confinement effect.

## 7.2 Solvothermal Synthesis of III–V Nanomaterials

GaN is a direct band-gap semiconductor, which has potential applications in light-emitting devices in the visible and ultraviolet region. Wurtzite-type GaN was traditionally prepared via a gas-phase reaction in ammonia at 600–1000 °C [34]. It can also be produced via pyrolysis of single source precursors, such as $[H_2GaNH_2]_3$, $Ga(C_2H_5)_3NH_3$, which already have a Ga–N bond, followed by treatment at temperatures >500 °C [35]. On the other hand, nitrides of lanthanide or transition metals could be synthesized through a solid-state metathesis reaction (Reaction (1)) [36]:

$$MCl_n + Li_3N \xrightarrow{600-1100\,^\circ C} MN + 3LiCl \tag{1}$$

$$GaCl_3 + Li_3N \xrightarrow[\text{benzene}]{280\,^\circ C} GaN + 3LiCl \tag{2}$$

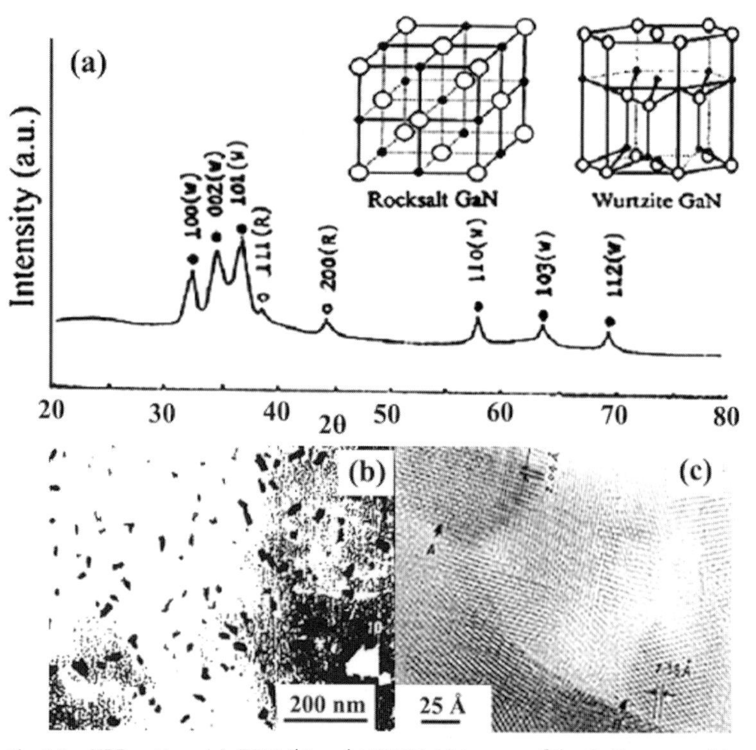


Fig. 7.7. XRD pattern (a), TEM (b) and HRTEM (c) images of the GaN nanoparticles.

However, this solid-state metathesis reaction failed for GaN [37] since $GaCl_3$ is hydroscopic at ambient temperature. When the metathesis reaction of $GaCl_3$ and $Li_3N$, Reaction (2), was carried out in benzene at 280 °C under pressure in an autoclave, 30 nm GaN nanocrystallites in ~80% yield (Figure 7.7(b)) were obtained [38]. As shown in Figure 7.7(a), the XRD pattern indicates that the product consists mainly of wurtzite-type GaN with a small fraction of rocksalt-type GaN [39]. This rocksalt-type phase with a lattice constant $a = 4.100$ Å, was observed directly with high-resolution electron microscopy (Figure 7.7(c)). It had been previously observed at ultrahigh pressure (at least 37 Gpa) via a pressure-induced phase transition from wurtzite-type GaN [40]; when the pressure was decreased to about 25 Gpa, it was transformed into the wurtzite-type GaN again. The size of the as-prepared GaN, 30 nm, is beyond its exciton Bohr diameter (11 nm), therefore, no quantum confinement effect was observed. Using $I_2$ as a heat sink and diluting reagents, nanocrystalline GaN was also synthesized from the reaction of $GaI_3$ and $NaN_3$ (Reaction (3)).

$$GaI_3 + 3NaN_3 + I_2 \xrightarrow{380\,^\circ C} GaN + 3NaI + 4N_2 + I_2 \qquad (3)$$

As can be seen in Figure 7.8, the XRD pattern shows that the product is of

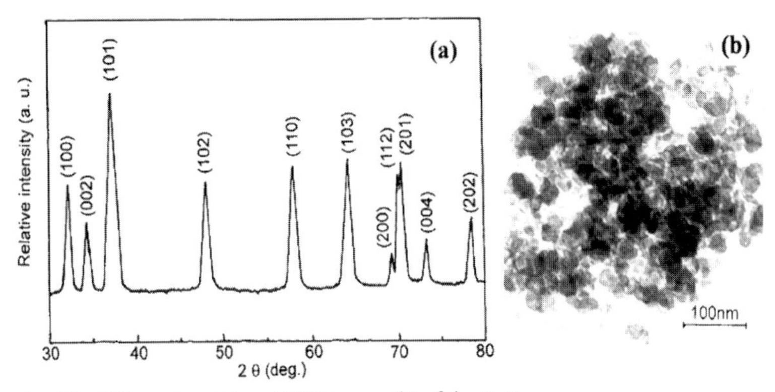


**Fig. 7.8.** XRD pattern (a) and TEM image (b) of the GaN powders prepared with molar ratios of $GaI_3$, $NaN_3$ and $I_2$ of 1/4.5/5.

wurtzite-type and the TEM image reveals spherical particles with an average size of 30 nm [41].

AlN nanocrystallites were synthesized via the benzene-solvothermal method [42], from a mixture of aluminum, $I_2$, $NaN_3$ after reacting at 350 °C for 10 h (Reaction (4)).

$$2Al + I_2 + 2NaN_3 \xrightarrow[350\ ^\circ C]{\text{benzene}} 2AlN + 2N_2 + 2NaI \tag{4}$$

The XRD pattern (Figure 7.9(a)) of the product can be indexed as the hexagonal AlN, with lattice constants $a = 3.1175$ and $c = 4.9813$ Å. TEM (Figure 7.9(b)) shows that the particle sizes range from 5 to 20 nm.

In xylene, 15 nm InAs nanocrystallites (Figure 7.10B) were prepared by zinc coreduction of $InCl_3{\cdot}4H_2O$ and $AsCl_3$ at 160 °C [43]. The XRD pattern indicates that the products were sphalerite-type cubic InAs (Figure 7.10A). As shown in Reactions (5)–(8), a trace amount of water resulting from the reactants may be helpful to the process.

$$3Zn + 6H_2O \rightarrow 3Zn(OH)_2 + 6H \tag{5}$$

$$InCl_3 + AsCl_3 + 6H \rightarrow InAs + 6HCl \tag{6}$$

$$3Zn(OH)_2 + 6HCl \rightarrow 3ZnCl_2 + 6H_2O \tag{7}$$

$$InCl_3 + AsCl_3 + 3Zn \xrightarrow[160\ ^\circ C]{\text{xylene}} InAs + 3ZnCl_2 \tag{8}$$

Compounds containing arsenic are usually very toxic, the order of toxicity being [44]: $R_3As$ (R = H, Me, Cl, etc.) > $As_2O_3$ (As (III)) > $(RAsO)_n$ > $As_2O_5$ (As(V)) > $R_nAsO(OH)_{3-n}$ > $R_4As^+$ > As(0). A safer route reacting $InCl_3{\cdot}4H_2O$ and $KBH_4$ in the presence of arsenic was proposed for the preparation of nanocrystalline InAs [45], see Reaction (9).

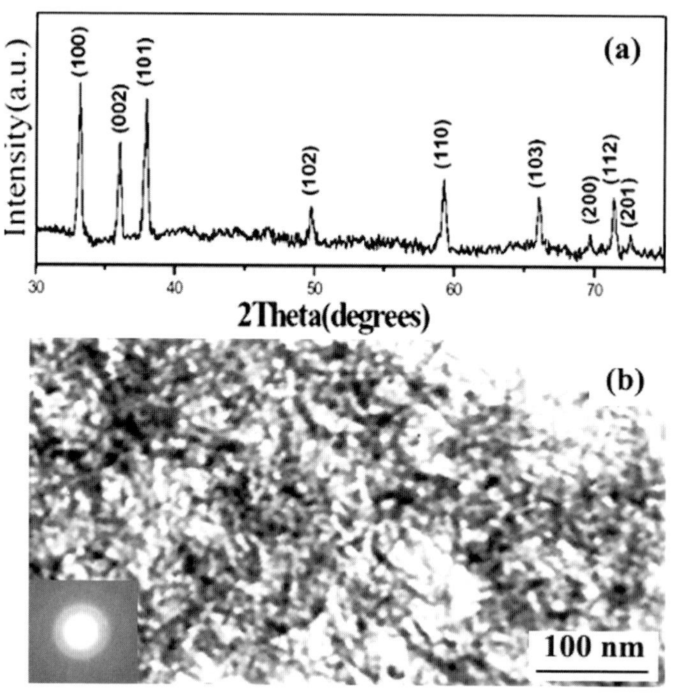


Fig. 7.9. XRD pattern (a) and TEM images (b) of the as-prepared nanocrystalline AlN.

$$2InCl_3 + 2As + 6KBH_4 \xrightarrow[160\ ^\circ C]{\text{ethylenediamine}} 2InAs + 6KCl + 6BH_3 + 3H_2 \qquad (9)$$

TEM shows that the products are polycrystalline fibers or near-single-crystal whiskers with widths of 15–100 nm and lengths of 150–1000 nm (Figure 7.10E). When the reaction temperature was reduced to 120 °C, which is below the melting point of indium (156.6 °C), spherical InAs nanocrystals of 20–50 nm were obtained (Figure 7.10D). However, there is trace indium co-existing in the products, indicating that the reaction proceeds through a metallic indium intermediate.

Similarly, as shown in Reactions (10) and (11), under high-intensity ultrasonic irradiation, InP nanocrystallites were synthesized in ethanol/benzene mixed solvent at room temperature [46].

$$2InCl_3 + 6KBH_4 \xrightarrow{\text{ultrasonic irradiation}} 2In + 6KCl + 6BH_3 + 3H_2 \qquad (10)$$

$$4In + P_4 \xrightarrow{\text{room temperature}} 4InP \qquad (11)$$

The XRD pattern (Figure 7.11(a)) indicates that the product is sphalerite-type cubic InP. The TEM images (Figure 7.11(b)) show that the products are 9 nm spherical

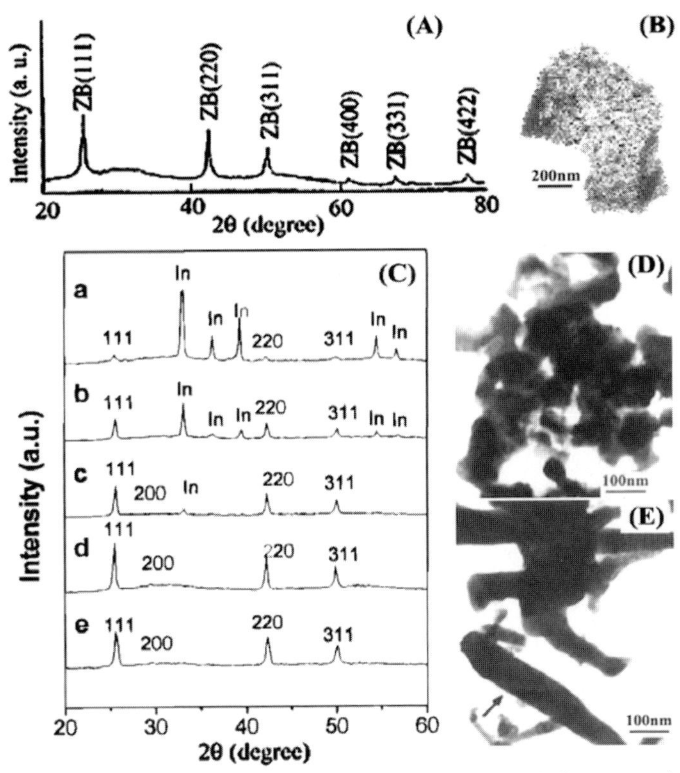


Fig. 7.10. XRD patterns and TEM images of the InAs samples prepared by the solvothermal zinc co-reduction route (A, B) and by the safe low temperature solvothermal route (C, D, E). (a) 100 °C, before 1 M HCl treatment; (b) 120 °C, before 1 M HCl treatment; (c) 160 °C, before 1 M HCl treatment; (d) 120 °C, after 1 M HCl treatment; (e) 160 °C, after 1 M HCl treatment.

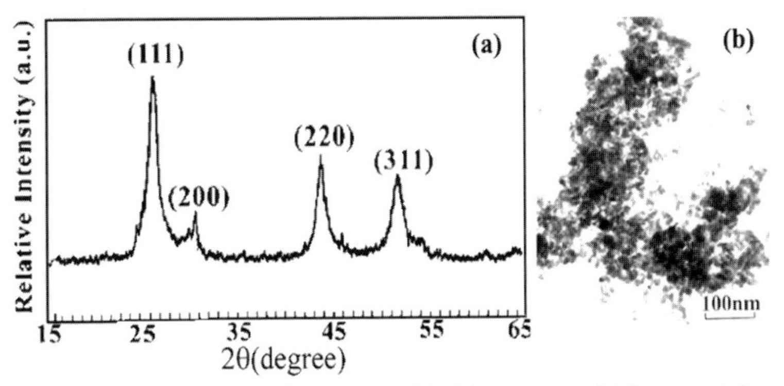


Fig. 7.11. XRD pattern (a) and TEM image (b) of the as-prepared InP nanocrystals.

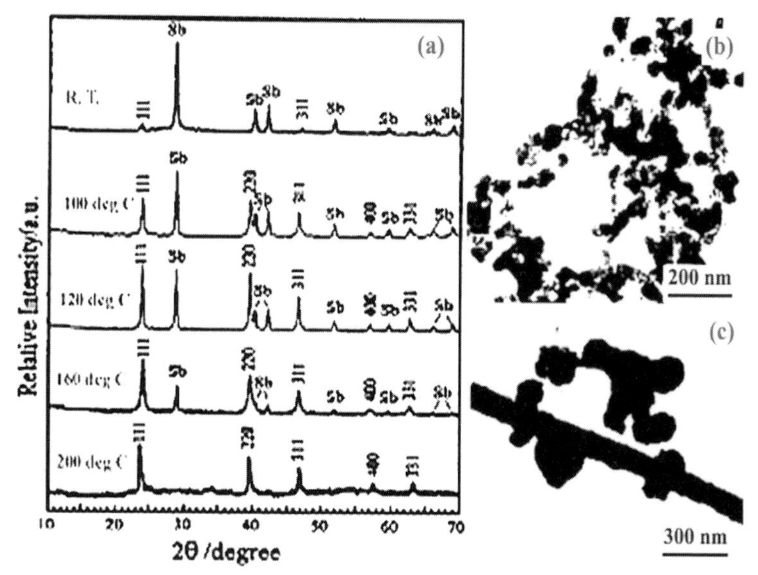


Fig. 7.12. XRD patterns (a) and TEM images (b) and (c) of the InSb nanocrystals prepared by $NaBH_4$ reduction of $InCl_3$ in the presence of Sb at 100 to 200 °C in ethylenediamine.

grains. Alternatively, from the same reaction, 15–20 nm InP nanocrystals were prepared at about 120 °C in ethylenediamine (en) [47].

A similar reaction was also used to prepare InSb nanomaterials by $NaBH_4$ reduction of $InCl_3$ in the presence of metallic antimony in en (Reaction (12)).

$$2InCl_3 + 2Sb + KBH_4 \xrightarrow[200\ °C]{\text{ethylenediamine}} 2InSb + 6KCl + 6BH_3 + H_2 \qquad (12)$$

As shown in Figure 7.12(a), the mean nanocrystalline dimensions of 40 nm are estimated from the half widths of the XRD peaks by the Scherr equation. Spherical grains of 40–60 nm and rod-like grains with a size of 120 × 1500 nm were found in the TEM images (Figure 7.12(b) and (c)). The reaction also proceeds through an indium intermediate, but a pure phase of nanocrystalline InSb was obtained at 200 °C when $InCl_3/Sb = 4/1$ [48].

Recently, a novel Ullmann-like reaction was designed to prepare one-dimensional InP and GaP nanocrystallites [49] (see Reaction (13)).

$$Ph_3P + In\ (\text{or } Ga) \rightarrow Ph\text{-}Ph + InP\ (\text{or } GaP) \qquad (13)$$

As a result of using organic phosphorus starting material, crystal growth was appropriately controlled. Nanowires of sphalerite-type GaP and InP (Figure 7.13) were prepared, suggesting a promising method for the solvothermal synthesis of one-dimensional III–V nanocrystallites.

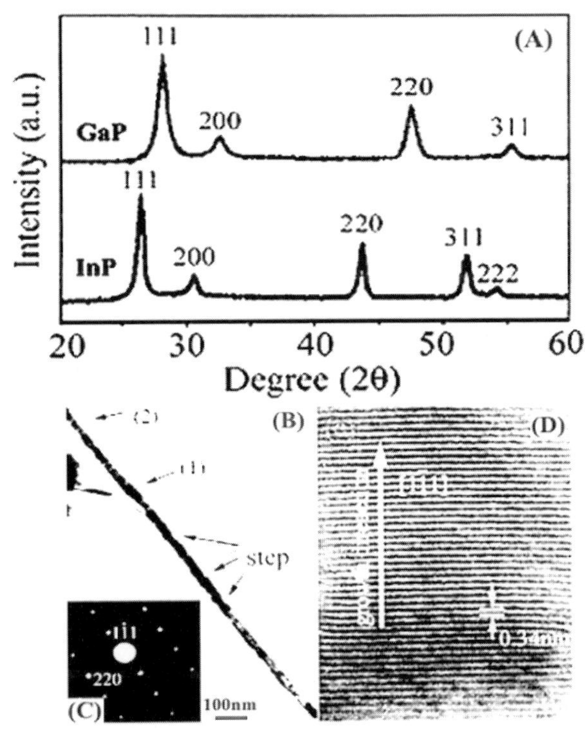


**Fig. 7.13.** XRD patterns of GaP and InP nanowires (A), TEM (B), ED (insert C) and HRTEM (D) images of InP nanowires prepared by an Ullmann-like reaction.

## 7.3
## Synthesis of Diamond, Carbon Nanotubes and Carbides

Diamonds were first synthesized through a high-temperature and high-pressure process [50]. Using diamond seeds, micrometer-sized diamonds have been grown in a hydrothermal process at 800 °C and 1.4 kbar [51]. Chemical vapor deposition (CVD) is a low-pressure technique for the preparation of polycrystalline diamonds films using $CH_4$ or $C_2H_2$ as a carbon source [52].

In traditional organic synthesis, the Wurtz reaction:

$$R^1X + R^2X + 2Na \rightarrow R^1\text{-}R^2 + NaX \qquad (14)$$

can be used to link two alkyl groups by forming a C–C single bond. Employing a Wurtz-type reaction coupling alkyl halides with excess metal sodium as both reductant and flux, fine carbon powders containing a trace amount of diamond (yield ~ 2%) (Figure 7.14(c)) were synthesized via a reduction–pyrolysis–catalysis route by reacting $CCl_4$ with metallic sodium at 700 °C [53]. This temperature is

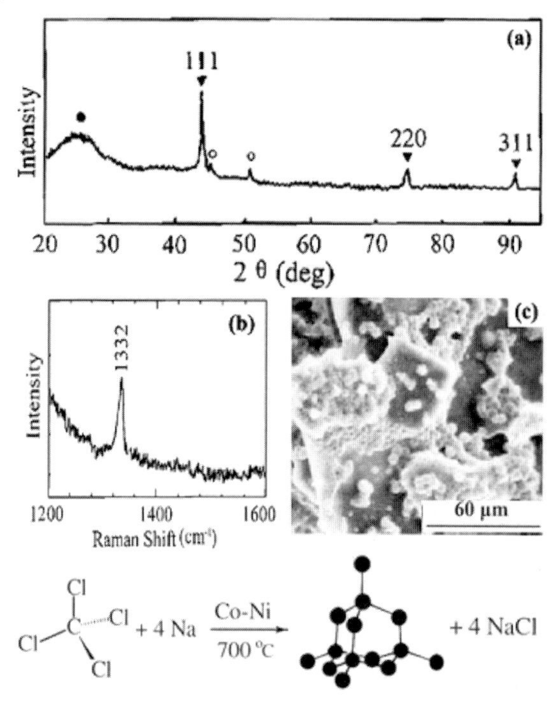


**Fig. 7.14.** XRD pattern (a), Raman spectrum (b) and SEM image (c) of the diamond sample synthesized by the reduction–pyrolysis–catalysis route.

much lower than that of traditional methods. The three strongest diffraction peaks of diamond were found in the XRD pattern (Figure 7.14(a)). The Raman spectrum (Figure 7.14(b)) shows a sharp peak at 1332 $cm^{-1}$, which is characteristic of diamond.

Carbon nanotubes are usually prepared by arc-discharge [54], laser evaporation of graphite [55], disproportionation of carbon monoxide [56], and pyrolysis of hydrocarbons [57]. In all these methods, no typical chemical reactions describing the synthetic process of the carbon nanotubes could be given. Since the walls of a carbon nanotube, both the cylindrical stories of a multiwall nanotube, and the planar sheet of a single wall nanotube, are built from a hexagonal lattice of $sp^2$ bonded carbon [58], a catalytic-assembly solvothermal route using the planar aromatic hexachlorobenzene ($C_6Cl_6$) was developed to prepare multiwalled carbon nanotubes [59], see Reaction (15):

$$C_6Cl_6 + 6K \xrightarrow[\text{benzene}]{350\,^{\circ}\text{C}} \text{Carbon nanotube} + 6KCl \tag{15}$$

In the reaction, the freshly formed free $C_6$ species can assemble into hexagonal carbon clusters, which can grow into nanotubes at the surface of the catalyst par-

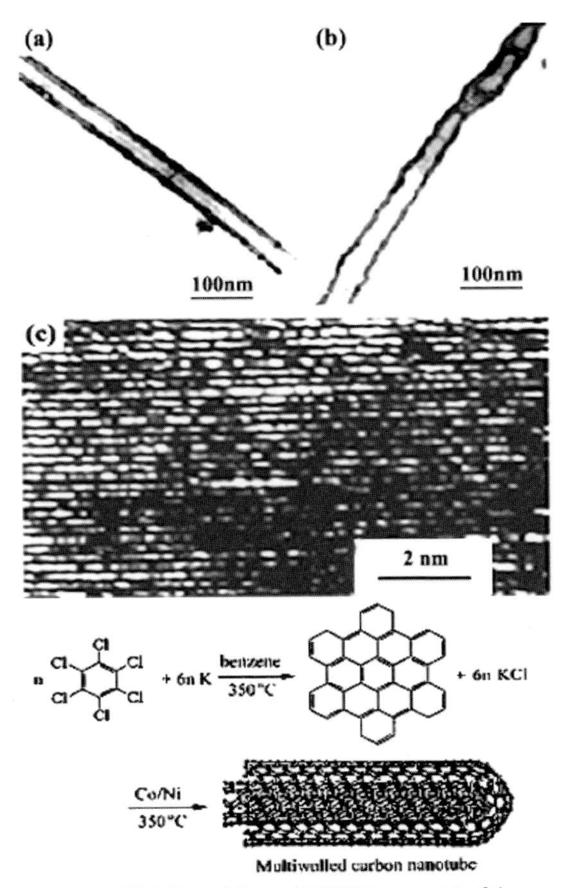


Fig. 7.15. TEM (a) and (b) and HRTEM image (c) of the multiwall carbon nanotubes prepared by a catalytic-assembly solvothermal route.

ticles (Co/Ni). Figure 7.15 shows the TEM and HRTEM images of the as-prepared multiwalled nanotubes. Carbon nanotubes with inner/outer diameter, 5–15 nm/10–35 nm (Figure 7.16(a)) were also prepared using commercial benzene as carbon source and solvent at 480 °C, with micro-size Fe/Ni alloy (46 wt% Ni) as the catalyst [60]. A typical HRTEM image (Figure 7.16(b)) reveals that the nanotube walls are comprised of about 40 graphite monolayers. The least spacing between the lattice stripes is 0.337 nm, which is characteristic of the typical (002) interlayer distance of hexagonal graphite. Tetrachloroethene ($C_2Cl_4$) was also used as carbon source for the solvothermal synthesis of carbon nanotubes at 200 °C [61], at the surface of Fe/Au catalyst particles, carbon nanotubes with an inner/outer diameter of 60 nm/80 nm were found, and a Y-junction carbon nanotube was also observed in the TEM images.

Cubic- or $\beta$-SiC nanorods are very attractive semiconductors with potential ap-

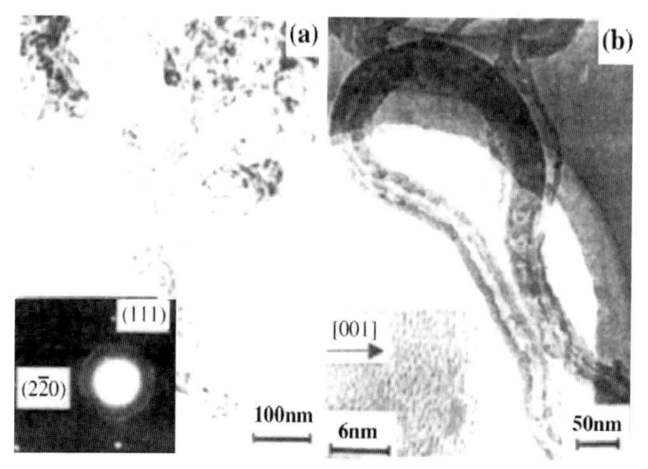


**Fig. 7.16.** (a) TEM image of many carbon nanotubes with catalyst in the ends, (insert) SAED pattern of catalyst in the tip of the nanotube, and (b) TEM image of a carbon nanotube of length up to 1.5 μm and (insert) HRTEM image of carbon nanotube.

plications in very-high-speed, high-temperature and high-power devices. The reduction of $SiCl_4$ with sodium has already been used to prepare crystalline Si at 375 °C under high pressure [62]. Ritter [63] has reported a sodium co-reduction of $SiCl_4$ and $CCl_4$ in a non-polar solvent at 130 °C to give an amorphous precursor containing Si and C, which required heating at 1450–1750 °C to form crystalline SiC. When this reaction with excess sodium metal as reductant and flux was carried out in an autoclave at 400 °C (Reaction (16)), $\beta$-SiC nanorods were grown [64]. The TEM images (Figure 7.17(a) and (b)) show that the product consists of nanorods with diameters from 10–40 nm and lengths up to several micrometers. When activated carbon and $SiCl_4$ were used as the starting materials (Reaction (17)), nanocrystalline $\beta$-SiC was also synthesized at 600 °C [65]. The TEM images indicate that the SiC powders consist only of spherical particles with an average diameter of 25 nm. Conversely, when using silicon powders and $CCl_4$ as the silicon and carbon sources (Reaction (18)), 3C–SiC nanowires (Figure 7.17(c)) of 15–20 nm diameter and length 5–10 μm were prepared through a reduction–carburization route at 700 °C [66]. In addition, an interesting tubular-like SiC nanowire (Figure 7.17(d)) was also found co-existing in the sample.

$$SiCl_4 + CCl_4 + 8Na \xrightarrow{400\,^\circ C} SiC + 8NaCl \quad (16)$$

$$C + SiCl_4 + 4Na \xrightarrow{600\,^\circ C} SiC + 4NaCl \quad (17)$$

$$Si + CCl_4 + 4Na \xrightarrow{700\,^\circ C} SiC + 4NaCl \quad (18)$$

TiC nanocrystallites were similarly synthesized at 450 °C (Reaction (19)).

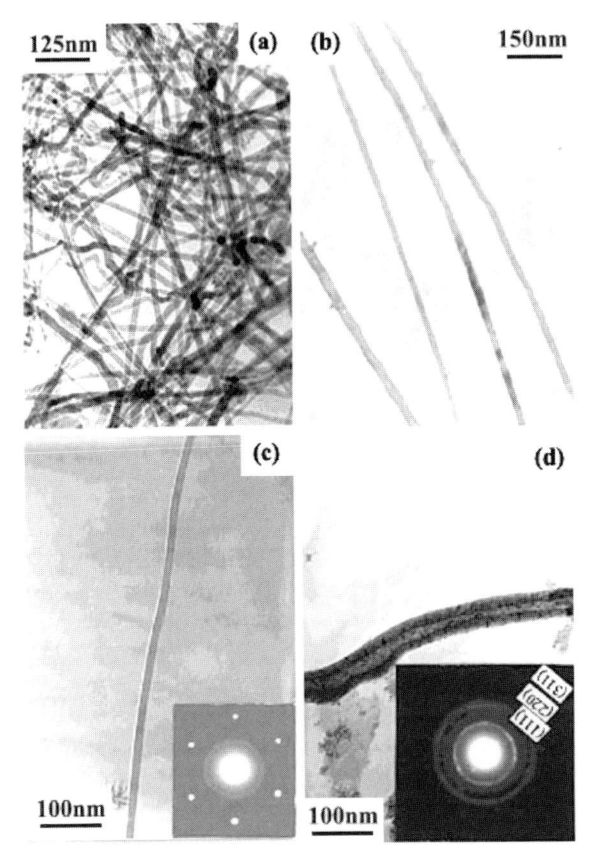


**Fig. 7.17.** (a) and (b) TEM images of SiC nanowires prepared by sodium co-reduction of $SiCl_4$ and $CCl_4$ at 400 °C, (c) TEM image of a straight single SiC nanowire and (insert) its ED pattern, and (d) TEM image, and (insert) its ED pattern, of a tubular-like or hollow-structure SiC nanowire prepared by sodium reduction of $CCl_4$ in the presence of silicon powders at 700 °C.

$$TiCl_4 + CCl_4 + 8Na \xrightarrow{450\,^{\circ}C} TiC + 8NaCl \quad (19)$$

The XRD pattern (Figure 7.18(a)) indicates that the obtained sample is cubic phase TiC, with cell constant $a = 4.34$ Å. TEM imaging (Figure 7.18(b)) reveals that it consists of nanoparticles of size 10–20 nm [67]. The co-reduction carburization method may be used to prepare other transition metal carbides. For example, as shown in Reaction (20), nanocrystalline ZrC, size 10–25 nm was synthesized on heating at 550 °C for 12 h [68].

$$ZrCl_4 + CCl_4 + 8Na \xrightarrow{550\,^{\circ}C} ZrC + 8NaCl \quad (20)$$

When carbon was used instead of $CCl_4$, 8–12 nm nanocrystalline TiC was also obtained at 650 °C [69].

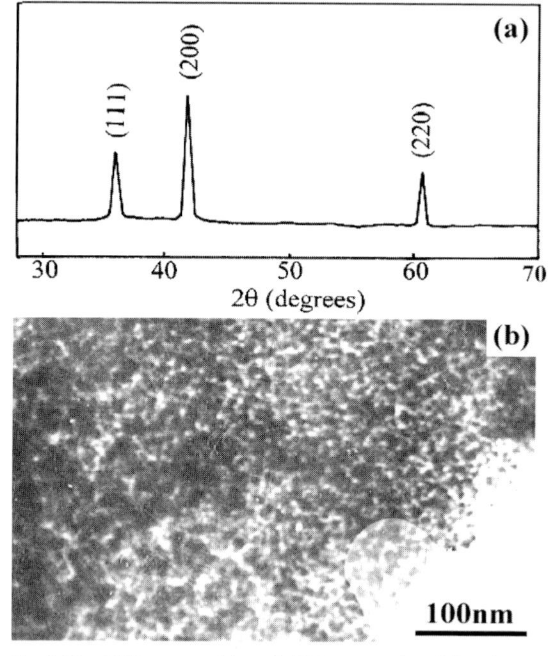


**Fig. 7.18.** XRD pattern (a) and TEM image (b) of the obtained TiC sample.

## 7.4 Synthesis of $Si_3N_4$, $P_3N_5$, Metal Nitrides and Phosphides

As shown in Reaction (21), in the liquid–solid reaction of $CrCl_3$ and $Li_3N$, nanocrystalline CrN with average particle size of about 25 nm was prepared via a benzene-thermal method in the temperature range 350–420 °C [70]. By using lithium nitride ($Li_3N$) instead of explosive $NaN_3$, ultrafine cubic ZrN powders of size about 180 nm were prepared in benzene [71].

$$CrCl_3 + Li_3N \xrightarrow[350\text{–}420\,^\circ C]{\text{benzene}} CrN + 3LiCl \tag{21}$$

Nanocrystalline TiN of size 50 nm [72] and nanocrystalline ZrN of size 10–20 nm [73] were prepared through the benzene-thermal reaction with $NaN_3$ (Reaction (22)).

$$MCl_4 + 4NaN_3 \xrightarrow[350\text{–}380\,^\circ C]{\text{benzene}} MN + 4NaCl + 11/2\ N_2 \quad M = Ti, Zr \tag{22}$$

Silicon nitride ($Si_3N_4$) is an important material for high-temperature engineering applications due to its chemical stability, high-temperature strength, and excellent creep resistance. A low-temperature preparation of crystalline $Si_3N_4$ has been developed that avoids the elevated temperatures above 1200 °C, which are necessary

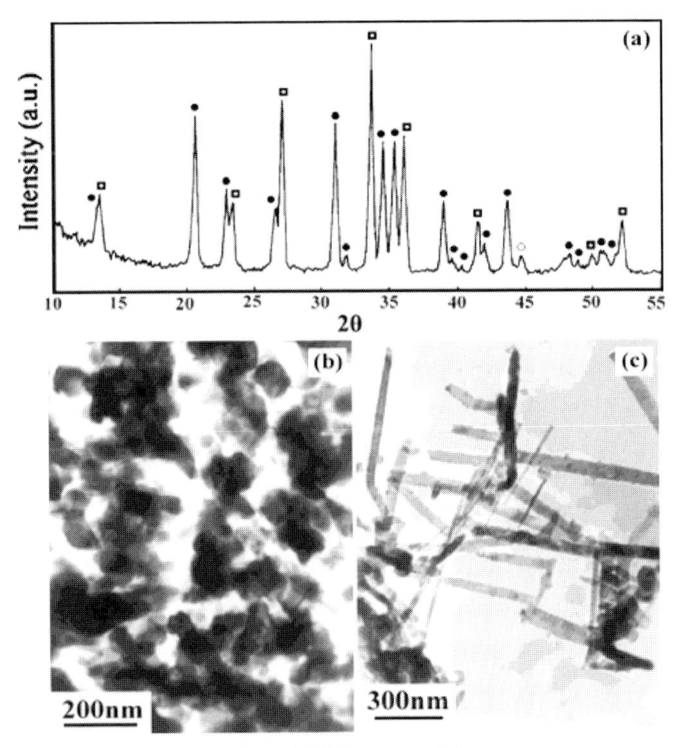


**Fig. 7.19.** XRD patterns (a) and TEM images of the as-prepared $Si_3N_4$ nanoparticles (b) and nanorods (c). ●, $\alpha$-$Si_3N_4$; ◘, $\beta$-$Si_3N_4$; ○, $Fe_3Si$.

in traditional methods. At 600–700 °C, with excess $SiCl_4$ as reactant and solvent, 40–60 nm nanoparticles (Figure 7.19(b)) and 60 × 750 nm nanorods (Figure 7.19(c)) of $Si_3N_4$ were obtained in the same sample (Reaction (23)).

$$3\ SiCl_4 + 12\ NaN_3 \xrightarrow{600-700\ ^\circ C} Si_3N_4 + 12\ NaCl + 16\ N_2 \quad (23)$$

The XRD patterns (Figure 7.19(a)) can be indexed as the hexagonal cell, with lattice parameters $a = 7.739$, $c = 5.622$ Å ($\alpha$-$Si_3N_4$). $\beta$-$Si_3N_4$ was the coexisting phase, with lattice constants $a = 7.578$, $c = 2.9075$ Å. The coexisting $Fe_3Si$ may result from the reaction between $SiCl_4$ with the surface of the stainless steel reactor [74].

Amorphous phosphorus nitride ($P_3N_5$) with flake-like morphology [75] was synthesized from the reaction of $PCl_5$ and $NaN_3$ in an autoclave at 190–300 °C (Reaction (24)).

$$3PCl_5 + 15NaN_3 \xrightarrow{190-300\ ^\circ C} P_3N_5 + 15NaCl + 20N_2 \quad (24)$$

When benzene was used as solvent and a hydrogen element source, microtubes, solid balls, hollow balls, and square frameworks (Figure 7.20(a)–(d)) of amorphous

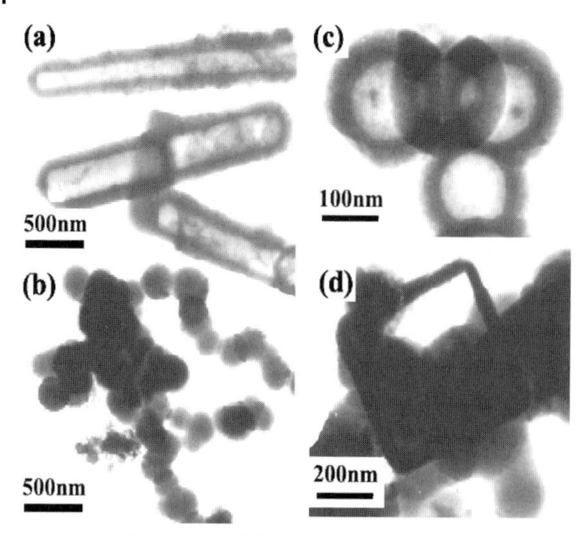


**Fig. 7.20.** TEM images of the amorphous $HPN_2$ microtubes (a), solid balls (b), hollow balls (c) and square frameworks (d) prepared at 190–250 °C for 0.5 to 10 days by reacting $PCl_5$ with $NaN_3$ in benzene.

phosphorus nitride imide ($HPN_2$) were prepared under mild conditions [76]. As an inorganic polymer, amorphous $HPN_2$ with these interesting morphologies may see potential uses in industry.

Metal phosphides are made conventionally by direct combination of the elements or by the reaction of highly toxic phosphines with metal or metal hydride via the metal organic chemical vapor deposition or high-temperature self-propagation synthesis route. Based on metathesis reactions, phosphides can be obtained under milder conditions through a solvothermal process. For example, reacting $SnCl_2$ with $Na_3P$ synthesized $Sn_4P_3$ nanorods at 120–140 °C (Reaction (25)) using en as the solvent and template [77].

$$12SnCl_2 + 8Na_3P + P \xrightarrow[120\text{–}140\,^\circ C]{en} 3Sn_4P_3 + 24NaCl \tag{25}$$

The XRD pattern (Figure 7.21(a)) is indexed as a hexagonal cell with $a = 3.98$ and $c = 35.34$ Å. The TEM image (Figure 7.21(b)) indicates that the diameter of the nanorods is about 20 nm and the length up to 400 nm.

As shown in Reactions (26)–(28), similar routes can be used to prepare iron group phosphides.

$$FeCl_3 + Na_3P \xrightarrow[180\text{–}190\,^\circ C]{benzene} FeP + 3NaCl \tag{26}$$

$$3CoCl_2 + 2Na_3P \xrightarrow[150\,^\circ C]{benzene} CoP + Co_2P + 6NaCl \tag{27}$$

$$6NiCl_2 + 4Na_3P \xrightarrow[150\,^\circ C]{benzene} 3Ni_2P + 12NaCl + P \tag{28}$$

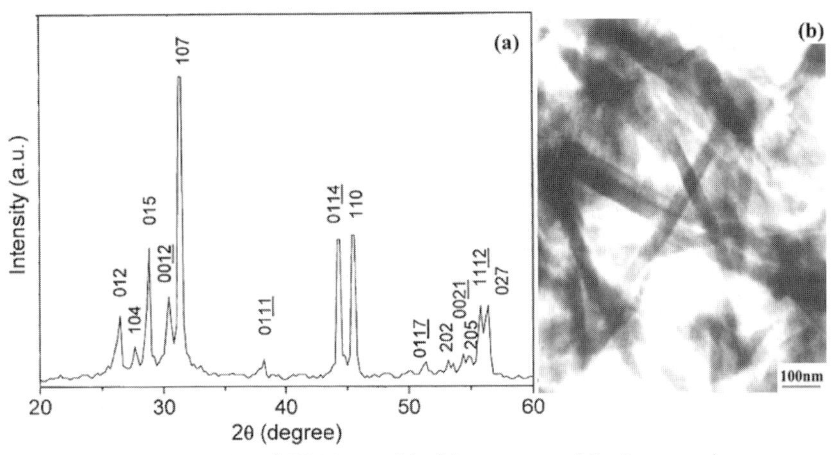


Fig. 7.21. XRD pattern (a) and TEM image (b) of the as-prepared $Sn_4P_3$ nanorods.

Figure 7.22 shows the FeP [78], $Co_2P$/CoP [79] and $Ni_2P$ [80] synthesized by solvothermal metathesis reactions.

Metal phosphide nanoparticles of $Co_2P$, $Ni_2P$ and $Cu_3P$ were also prepared in en at 80–140 °C, when the diamine solvent serves as an N-chelating ligand and reagent for scavenging chlorine from the metal salts [81]. Recently, titanium phosphides were synthesized via solvothermal co-reduction of $PCl_3$ and $TiCl_4$ at 300 and 350 °C using metallic sodium as reductant (Reaction (29)) [82].

$$PCl_3 + TiCl_4 + Na \xrightarrow[300,\,350\ ^\circ C]{PCl_3} TiP/TiP_{0.63} + NaCl \quad (29)$$

## 7.5 Synthesis of BN, $B_4C$, BP and Borides

Nanomaterials of boron-containing inorganic compounds and borides have received considerable attention due to their potential application in electronics, op-

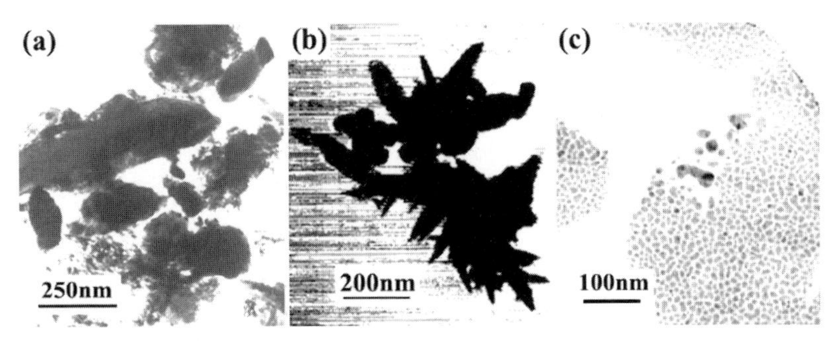


Fig. 7.22. TEM images of FeP (a), $Co_2P$/CoP (b) and $Ni_2P$ (c) prepared by solvothermal metathesis reactions.

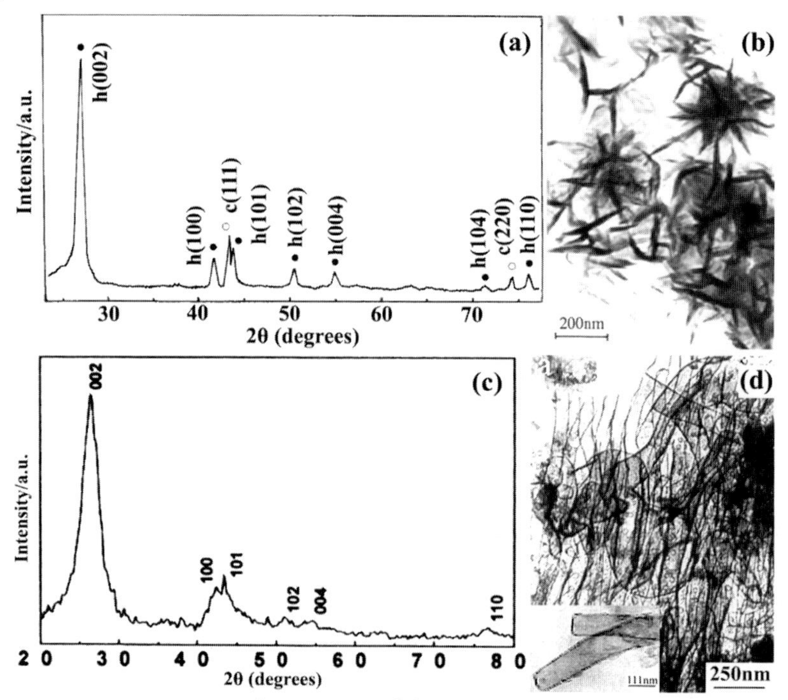


**Fig. 7.23.** XRD patterns and TEM images of the nanocrystalline BN (a) and (b) and BN nanotubes (c) and (d).

tics, catalysis, and magnetic storage. Unlike the traditional methods, which involve using high temperatures, toxic organometallic precursors, or complicated reactions and post-treatments, the solvothermal method may be a mild synthetic route to these materials.

Cubic BN is widely used in cutting tools and as grinding, and abrasive material. Nanocrystalline BN was prepared by the reaction of $KBH_4$ and $NH_4Cl$ at 650 °C [83], Reaction (30). The XRD pattern (Figure 7.23(a)) indicates that the product consists of hexagonal and cubic BN. The TEM image (Figure 7.23(b)) shows the BN powders consist of whisker-like particles with an average size of 250 nm × 10 nm. Cui and co-workers reported that much more of the cubic phase of BN can be obtained via the benzene-thermal reaction of $BBr_3$ and $Li_3N$ at about 450 °C [84, 85]. BN nanotubes were prepared via a precursor-pyrolysis route at 500 to 600 °C by reacting $KBH_4$, $NH_4BF_4$ and $NaN_3$ in the presence of catalysts (Reaction (31)). Figure 7.23(c) and (d) shows the XRD pattern and TEM image of the as-prepared tubostratic BN and BN nanotubes [86].

$$KBH_4 + NH_4Cl \xrightarrow{650\ ^\circ C} BN + KCl + 4H_2 \tag{30}$$

$$KBH_4 + NH_4BF_4 + NaN_3 \xrightarrow[\text{catalysts}]{500\text{–}600\ ^\circ C} \text{turbostratic BN and BN nanotubes} \tag{31}$$

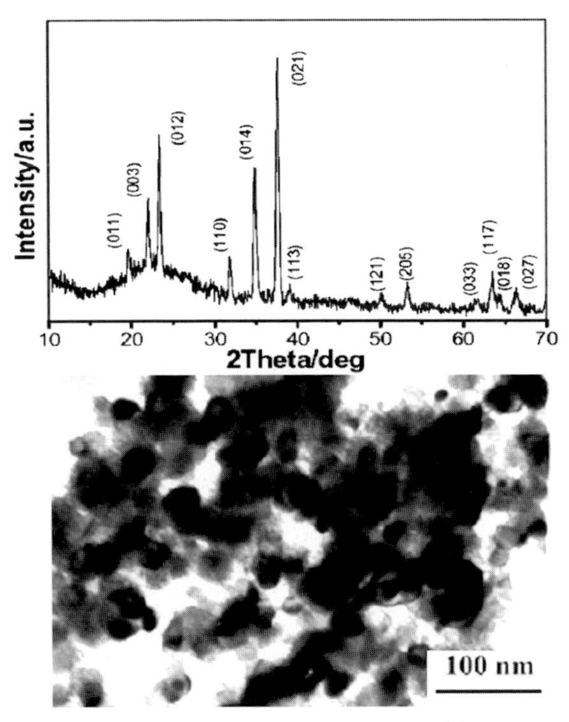


**Fig. 7.24.** XRD pattern (a) and TEM image (b) of the $B_4C$ sample prepared by reduction of $CCl_4$ in the presence of amorphous boron powder at 600 °C.

Boron carbide ($B_4C$) is one of the hardest known materials with excellent properties of low density, very high chemical and thermal stability, and high neutron absorption cross-section. Bulk $B_4C$ is conventionally synthesized by high temperature (up to 2400 °C) reactions, such as the carbothermal reduction of boric acid or boron oxide. Nanocrystalline $B_4C$ was solvothermally synthesized in $CCl_4$ at 600 °C (Reaction (32)).

$$4B(\text{amorphous}) + CCl_4 + 4Li \xrightarrow[600\ ^\circ C]{CCl_4} B_4C + 4LiCl \tag{32}$$

As shown in Figure 7.24, the XRD pattern can be indexed to the hexagonal $B_4C$ phase with lattice constants $a = 5.606$ and $c = 12.089$ Å. The TEM study shows that the $B_4C$ nanocrystallites are slightly agglomerated, with a particle size of 15–40 nm [87].

Boron phosphide (BP), a semiconductor with an indirect band gap of 2 eV, is one of the most promising high-temperature thermoelectric materials due to its outstanding chemical, mechanical, optical and thermal properties. Traditionally, BP is

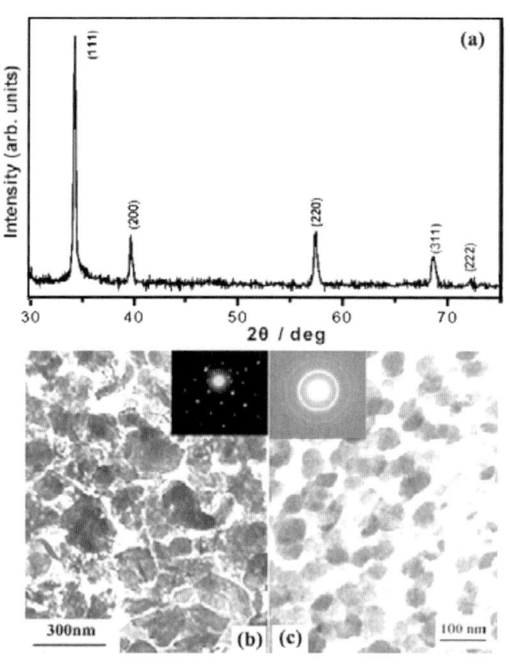


**Fig. 7.25.** XRD pattern and TEM images of the BP samples prepared by reduction of $PCl_3$ in the presence of amorphous boron powders at 350 °C (a) and (b) and by Na co-reduction of $PCl_3$ and $BBr_3$ at 300 °C in benzene (c).

prepared by various high-temperature reactions (up to 1000 °C). Through metal lithium reduction of $PCl_3$ in the presence of amorphous boron powders at 350 °C (Reaction (33)), ultrafine cubic BP powders (Figure 7.25(a) and (b)) were synthesized via a benzene-thermal method [88]. When $BBr_3$ was used instead of amorphous boron powders (Reaction (34)), nanocrystalline BP powders of about 30 nm in size (Figure 7.25(c)) were prepared under milder conditions [89].

$$\mathrm{B(amorphous)} + \mathrm{PCl_3} + 3\mathrm{Li} \xrightarrow[350\,^\circ\mathrm{C}]{\mathrm{PCl_3}} \mathrm{BP} + 3\mathrm{NaCl} \tag{33}$$

$$\mathrm{BBr_3} + \mathrm{PCl_3} + 6\mathrm{Li} \xrightarrow[300\,^\circ\mathrm{C}]{\mathrm{benzene}} \mathrm{BP} + 6\mathrm{NaCl} \tag{34}$$

In a similar way, nanocrystalline $TiB_2$ was prepared at 350 °C (Reaction (35)).

$$\mathrm{TiCl_4} + 2\mathrm{B(amorphous)} + 4\mathrm{Na} \xrightarrow{400\,^\circ\mathrm{C}} \mathrm{TiB_2} + 4\mathrm{NaCl} \tag{35}$$

As shown in Figure 7.26, the XRD pattern can be indexed to the hexagonal cell of $TiB_2$, with lattice constants $a = 3.027$ and $c = 3.228$ Å. The TEM image indicates that the particle size is in the range 15 to 40 nm [90].

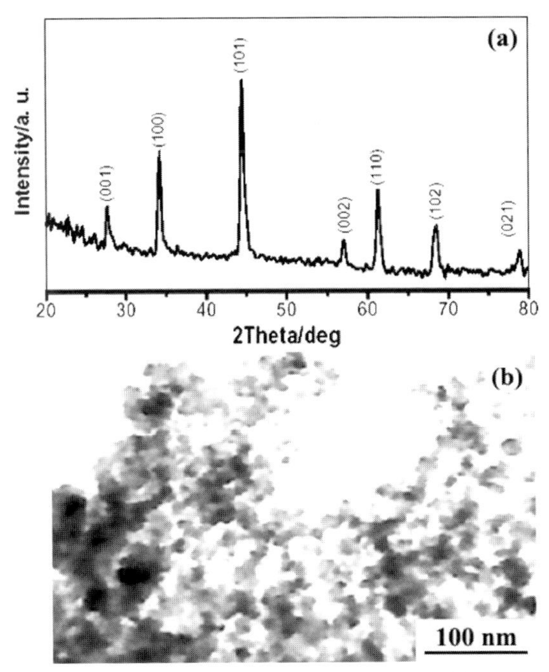


**Fig. 7.26.** XRD pattern (A) and TEM image (B) of the $TiB_2$ sample prepared by Na reduction of $TiCl_4$ in the presence of amorphous boron powders at 350 °C.

## 7.6
## Synthesis of One-Dimensional Metal Chalcogenide Nanocrystallites

Morphology control is very important for nanomaterials preparation due to the remarkable properties and potential applications of these materials ranging from microscopy probes to nanoelectronic devices. Therefore, current attention has focused on the development of convenient approaches to the preparation of nanomaterials with controlled shapes and sizes. The solvothermal method has been developed into a practical means of synthesis and controlling the morphology. By designing reactions, choosing solvents, controlling reaction temperatures and so forth, a series of metal chalcogenide nanomaterials, especially those with one-dimensional and special shapes, have been prepared.

Quantum-confined CdS nanowires (Figure 7.27) (4 nm × 150–250 nm) were grown from the decomposition of the precursor, cadmium bis(diethyldithiocarbamate), in en at 117 °C for 2 min [91]. Using this synthetic route, $CdIn_2S_4$ nanorods were prepared in an en/ethanol mixed solvent at 180 °C (Figure 7.28). The ethanol acts as the transportation reagent for $InCl_3$ while the en serves as the nucleophile, which causes scission of the thione groups in the cadmium bis(diethyldithiocarbamate) molecules and forms inorganic $Cd_2S_2$ cores. These newly formed cores as-

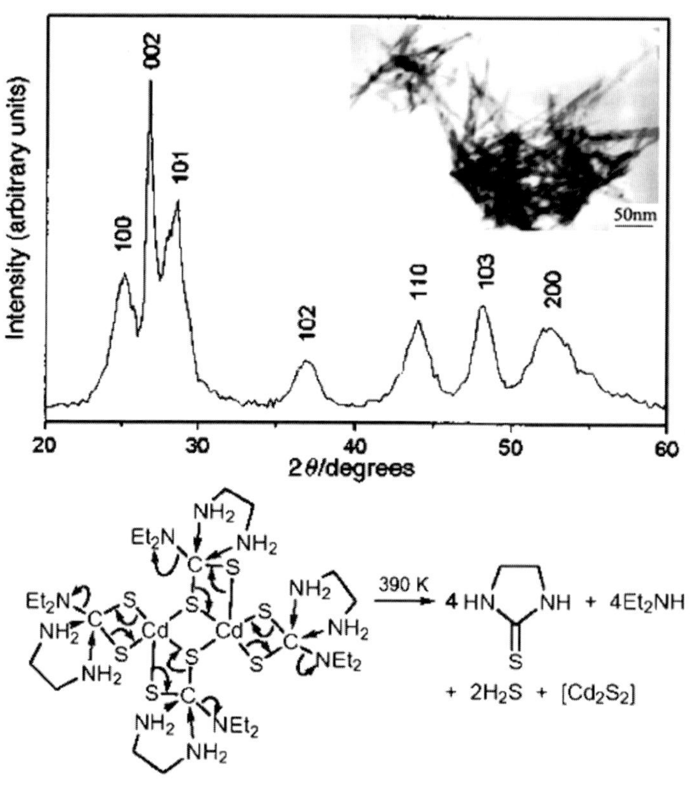


Fig. 7.27. XRD pattern, TEM image (insert) of the CdS nanowires prepared by precursor decomposition in organic solvent.

semble into one-dimensional $[CdS]_n$ clusters and act as intermediate templates for the subsequent growth of CdS nanorods [92].

Using thiosemicarbamide as the sulfur source, CdS nanowhiskers (60 nm × 12 μm) were grown via a solvothermal route [93]. Rod-, twinrod- and tetrapod-shaped CdS nanocrystals were obtained from spherical CdS nanocrystals via a solvothermal recrystallization technique [94]. Nanowires of CdS/CdSe core/sheath nanostructure were prepared by treating CdS nanowires with selenium in tributylphosphine at 100 °C [95]. Nanorods of CdE (E = S, Se or Te) with 10–40 nm diameters and several micrometers in length, were grown by choosing coordinating solvents such as en and 1,6-diaminohexane [96, 97]. When metal salts are used as starting materials, reducing reagents such as metallic sodium or hydrazine may be used as reductants. MSe (M = Zn or Cd) nanorods were prepared at 80–100 °C in en [98].

By choosing ligand solvents, nanocrystalline CdE can be synthesized by the reactions of $CdC_2O_4$ with E (E = S, Se or Te) in polyamines such as en, diethylenetriamine and triethylenetetramine. CdS nanorods (Figure 7.29(a)) and nano-

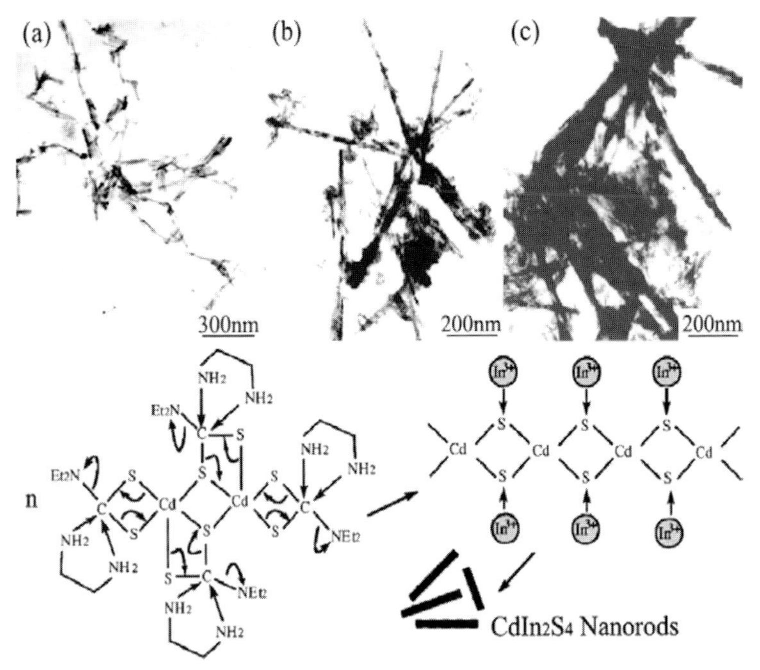


**Fig. 7.28.** TEM images of the $CdIn_2S_4$ nanorods prepared in mixed solvents with different ethanol/ethylenediamine ratios: 1/3 (a), 1/1 (b) and 3/1 (c).

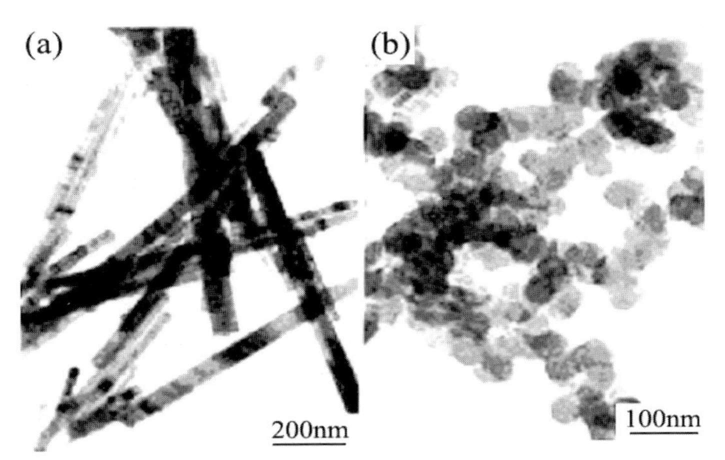


**Fig. 7.29.** TEM images of the CdS nanocrystals prepared in different solvents: (a) ethylenediamine, (b) pyridine.

particles (Figure 7.29(b)) were synthesized at 180 °C in en and pyridine, respectively [99].

The en molecules play an important role in controlling the nucleation and growth of the CdS nanorods. As a bidentate ligand, en molecules may react with $Cd^{2+}$ ions and form relatively stable complexes. Under appropriate solvothermal conditions, the complexes become unstable and decompose, which results in the formation of CdS nanorods [100]. A mono-dentate ligand, n-butylamine, was found to be a shape controller for nanorods of CdS and MSe (M = Zn, Cd or Pb) [101]. Similarly, the precursor of $ZnE(en)_{0.5}$ (E = S, Se) could also form in en which then is converted to ZnSe nanoparticles via pyrolysis in solvothermal conditions [102]. The coordinating ability of the solvent was found to play an important role in the nucleation and growth of nanocrystallites [103].

Polymer-shape-controlled solvothermal synthesis is efficient for fabricating one-dimensional nanomaterials. Using polyacrylamide as growth controller, CdS nanowires with homogeneous size distribution were grown at 170 °C for 10 days in en [104]. As shown in Figure 7.30, the SEM and TEM images show lengths up to 100 μm with diameters of about 40 nm. In the synthetic process, $Cd^{2+}$ ions were well distributed in the polymer matrix. The polymer may absorb solvent and form

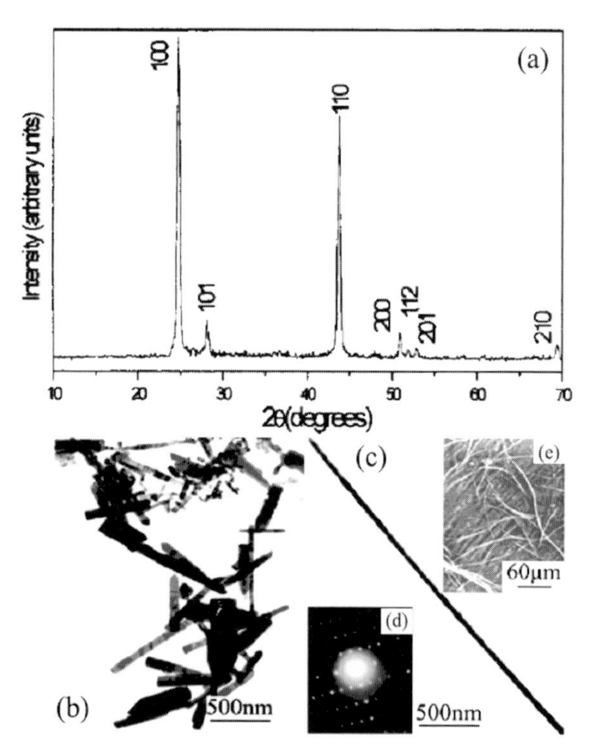


Fig. 7.30. XRD pattern (a), TEM image (b), (c), ED pattern (d) and SEM image (e) of the CdS nanowires prepared by a polyacrylamide-controlled solvothermal method.

a gel with small pores. It was supposed that the growth of CdS nanocrystallites was confined within the pores to form nanowires. Sphalerite-type CdSe nanowires were fabricated through a poly(vinyl alcohol) (PVA) assisted solvothermal reaction at 160–180 °C in en [105]. These as-prepared nanowires exhibit straight or zigzag shapes and most of them are grown with [111] orientation.

Rod-shaped PbS nanocrystals with diameters of 30–160 nm and lengths of up to several micrometers were also prepared by a biphasic solvothermal interface reaction at 140–160 °C [106]. Closed PbS nanowires with regular geometric morphologies (ellipse and parallelogram shape) were prepared with shape control by the polymer poly[*N*-(2-aminoethyl) acrylamide] in en and $H_2O$ (3:1, v/v) solvent at 110–150 °C [107]. VA group chalcogenides, nanorods or nanowires, were also solvothermally prepared, e.g., $Sb_2E_3$ [108, 109], $Bi_2E_3$ (E = S or Se) [110, 111, 112] and $Bi_3Se_4$ [113].

Interestingly, a novel α-NiS layer-rolled structure, intermediate between a two-dimensional layered structure and three-dimensional isotropic material, was prepared by adding $CS_2$ step-by-step to an aqueous ammonia solution containing $Ni(NH_3)_6{}^{2+}$ ions at about 60 °C [114]. By reacting $CS_2$ with $Ni(NH_3)_6{}^{2+}$, a new complex, $Ni(NH_3)_3(CS_3)$, may form. These complexes orientate and become linked in the least destabilizing mutual arrangement due to the NH–$H_2O$ hydrogen bonds. These linked complexes decompose on heating to produce the novel structure. The TEM images (Figure 7.31(a) and (b)) show strong contrast between the dark edges and the pale center, indicating the hollow nature of the particles. The contrast at the tip opening is evidence of an open rolled structure, like a broken tube. HRTEM (Figure 7.31(c) and (d)) shows different lattice features in the edge and center parts, indicating a multi-walled structure.

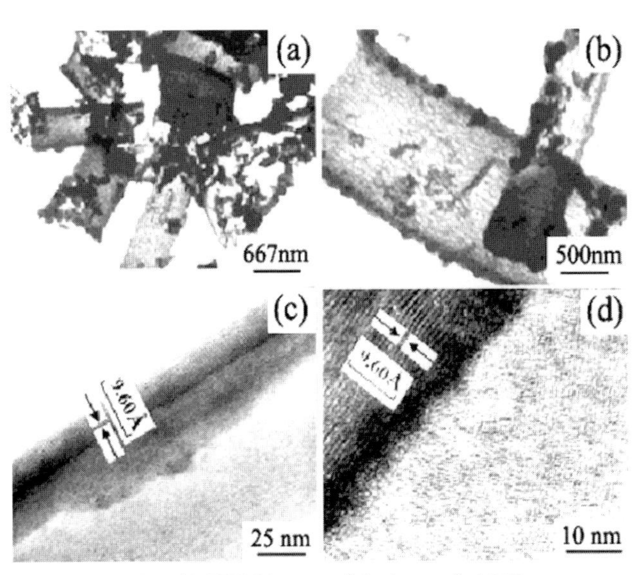


**Fig. 7.31.** TEM and HRTEM images of the layer-rolled NiS structures.

Reaction temperature and solvent conditions affect the phase transition and morphology of metal chalcogenides. With increasing temperature, $NiSe_2$ nanocrystals prepared through solvothermal-reduction at low temperatures transform from an initial filament to a final octahedron [115]. In different solvents, the well-crystallized nickel selenides obtained, such as $NiSe_2$, $Ni_{0.85}Se$ and $Ni_3Se_2$, showed different morphologies [116]. The synthesis of NiS, using en and hydrazine hydrate as solvent, resulted in a rod-like nanocrystalline product, whereas spherical nanoparticles were obtained in aqueous ammonia [117]. Two different phases of $\alpha$-(hexagonal, $P6_3/mmc$) and $\beta$-(rhombohedral, $R3m$) NiS nanocrystals were obtained by hydrazine reduction of $NiCl_2{\cdot}6H_2O$ in the presence of sulfur at 110 °C in ethanol and pyridine, respectively [118].

The metastable $\beta$- and $\gamma$-MnS crystallites were obtained at about 200 °C in tetrahydrofuran and benzene, whereas in water, ammonia liquor, en, the metastable phases converted to the stable phase of $\alpha$-MnS [119]. However, only the stable phase of $\alpha$-MnSe can be obtained by solvothermal reaction at 190 °C in en [120].

In the solvothermal synthesis of the I–III–$VI_2$ ternary compound semiconductors, organic amines are normally used as solvents for the reactants and as ligands to form metal complexes. In the reaction system $InCl_3{\cdot}4H_2O$, $CuCl_2{\cdot}2H_2O$ and elemental selenium at 180 °C, $CuInSe_2$ nanowhiskers with widths of 3–6 nm and lengths of 30–80 nm were prepared in en, whereas in diethylamine and pyridine, the product was only spherical nanoparticles [121]. Therefore, bidentate ligands may be more effective than monodentate ligands for directional growth of chalcogenide nanocrystals. Nanorods of $CuME_2$ ($M = In$ or Ga, $E = S$ or Se) were also prepared by elemental reactions in en at 200–280 °C [122, 123]. From a stoichiometric mixture of the single-molecule precursors $M(S_2CNEt_2)_3$ ($M = In$, Cu, Ag), nanorods of $MInS_2$ ($M = Cu$ or Ag) were prepared by removing the thione groups with en solvent at 195 °C [124]. Nanocrystalline $AgGaS_2$ and $AgInS_2$ with particle sizes ranging from 5 to 12 nm were prepared by reacting AgCl, sulfur, elemental gallium or indium in en at 180–230 °C [125].

Some other ternary chalcogenides have also been prepared by the solvothermal method, such as $Cu_{5.5}FeS_{6.5}$ nanotubes [126], $AgBiS_2$ nanowhiskers [127], $Cu_2SnS_3$ [128], $Cu_2SnSe_4$ [129], $CuSbS_2$ and $Ag_3SbS_3$ [130] nanoparticles.

## 7.7
## Room Temperature Synthesis of Nanomaterials

Room-temperature synthesis of nanomaterials can be realized by designing suitable reactions under some special solution conditions, or with the assistance of additional energy supplies such as $\gamma$-ray irradiation and ultrasound.

Traditionally, selenides have been synthesized at high temperature (>500 °C) by solid-state reaction [131] or self-propagating synthesis [132]. A high-energy ball milling method at room temperature has also been used [133], however, the product quality was difficult to control. In liquid ammonia, a low-temperature route to selenides was developed [134]. Organometallic precursors have been used to obtain

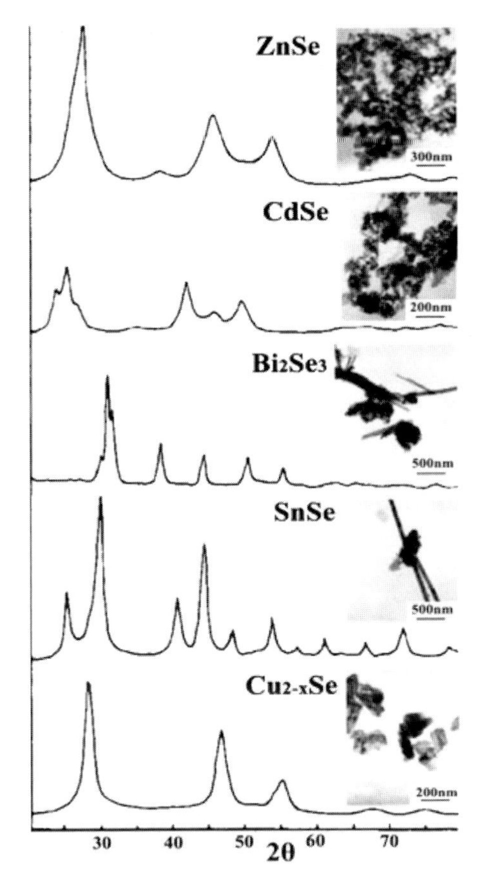


**Fig. 7.32.** XRD patterns and TEM images of ZnSe, CdSe, $Bi_2Se_3$, SnSe and $Cu_{2-x}Se$ nanomaterials prepared in ethylenediamine at room temperature.

selenides, but crystallization-treatment at 250–500 °C was necessary [135]. Therefore, it is necessary to explore a milder and simpler synthetic route for metal selenide nanocrystallites.

By using $KBH_4$, nanocrystalline MSe (M = Sn, Bi, Zn, Cd or Cu) (see Figure 7.32) was prepared from mixtures of $MCl_n$ ($SnCl_2$, $BiCl_3$, $ZnCl_2$, $CdCl_2$ or $CuCl_2$), and elemental selenium at ambient temperature (about 20 °C) and pressure in en [136]. It was supposed that the $KBH_4$ reduced elemental selenium to $Se^{2-}$ ions, which then reacted with metal ions $M^{n+}$ to form MSe. A similar process was used to prepare PbSe nanowires from $PbCl_2$, elemental selenium and $KBH_4$ sealed in en at 10 °C for 4 h [137].

Some novel solution routes have been designed for preparing cadmium chalcogenides with special morphologies. By an unusual in-situ source–template–interface method, in which $CdCl_2$ reacted with $H_2S$, generated by adding $CS_2$ to an

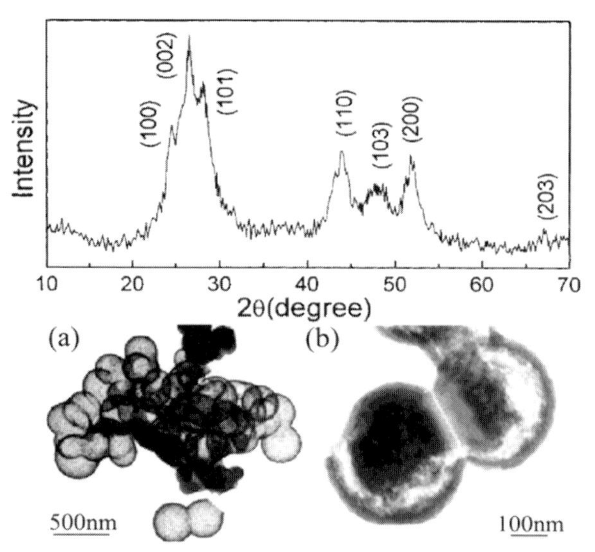


Fig. 7.33. XRD pattern and TEM images of the hollow spheres (a) and the peanut-like (b) structures of CdS.

aqueous en solution, forming CdS on the $CS_2$–water interface, hollow spheres [138] and a novel peanut-like structures of CdS [139] were prepared (Figure 7.33) under mild conditions.

Some other nanocrystallites with different morphologies were also prepared by room-temperature processes, such as a novel non-layered compound $Cu_{2-x}Se$ nanotubes (Figure 7.34), CuE and AgE (E = Se or Te) nanocrystallites (nanotubes,

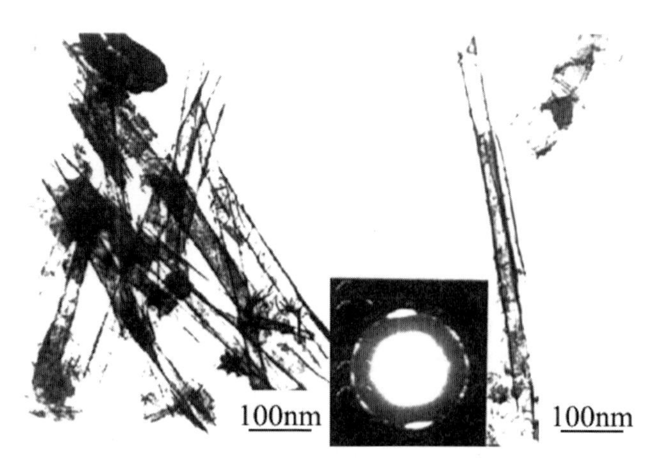


Fig. 7.34. TEM and TED images of $Cu_{2-x}Se$ nanotubes from hydrazine, $CuCl_2$ and selenium in ethylenediamine at room temperature.

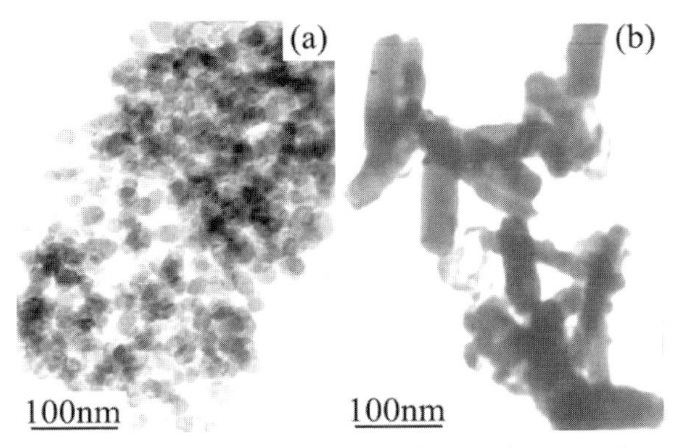


**Fig. 7.35.** TEM images of $Ag_2Se$ (a) and $Ag_2Te$ (b) prepared at room temperature in pyridine and in the mixed solvent of ethylenediamine–hydrazine hydrate, respectively.

nanorods and nanoparticles) [140], rod-like nanocrystalline $Ag_2Te$ (30 nm × 120 nm, Figure 7.35) [144], and spherical nanoparticles of $Ag_2Se$ (~25 nm) [142].

The $\gamma$-ray irradiation synthesis method, which can be carried out at ambient temperature and pressure in aqueous or non-aqueous solutions, has been developed to prepare nanomaterials of metals, alloys, elemental chalcogens, chalcogenide semiconductors and inorganic/polymer nanocomposites.

Nanocrystalline cobalt powders (14–42 nm) were prepared from aqueous ammonia solutions of $CoSO_4 \cdot 7H_2O$ or $CoCl_2 \cdot 6H_2O$ and surfactants (polyvinyl alcohol or sodium dodecyl sulfate) at ambient conditions by $\gamma$-ray irradiation (a dose rate, 67 Gy $min^{-1}$ in a $^{60}Co$ $\gamma$-ray field, $8 \times 10^6$ eV) [143]. In the process, $OH^{\cdot}$ and $H^{\cdot}$ radicals, $H_2O_2$, $H_2$, $H_3O^+$ and hydrated electrons ($e_{aq}^-$) may be generated from $H_2O$ upon $\gamma$-ray irradiation. The hydrogen radicals and hydrated electrons reduce $Co^{2+}$ and $Co^+$, forming metallic cobalt atoms, which then aggregate to give nanoparticles, which are stabilized by the surfactants, and precipitate from the aqueous solution. In a similar way, nanocrystalline Ag [144], Cd [145], Se [146] and ultrafine Cu–Pd alloy [147] were also prepared by a $\gamma$-ray irradiation method.

30 nm PbSe nanoparticles [148] and nanorods (60 nm × 700 nm) [149] were prepared by a $\gamma$-ray irradiation method in en–$H_2O$ solution at room temperature. In the process, selenium disproportionates by $OH^-$ to form $Se^{2-}$ and $SeO_3{}^{2-}$, followed by co-reduction of the $Se^{2-}$ and $Pb^{2+}$ by the hydrogen radicals (H) and/or hydrated electrons ($e_{aq}^-$) forming PbSe. Crystalline CdS, PbS, $Cu_2S$, and $Ag_2S$ nanoparticles were also prepared in aqueous en media under ambient conditions under $\gamma$-ray irradiation [150].

By $\gamma$-ray irradiation at room temperature, ZnS nanowires were grown, in which the crystal growth was in situ templated by an inverted hexagonal liquid crystal formed from oligo(ethylene oxide)oleyl ether amphiphiles, n-hexane, n-hexanol/isopropanol (2/1) and water [151]. Nanocrystalline $\beta$-ZnS (sphalerite) was prepared in aqueous isopropanol solution [152]. The TEM images (Figure 7.36) indicate that the product from mercaptoethanol is most well dispersed, and that from thiourea

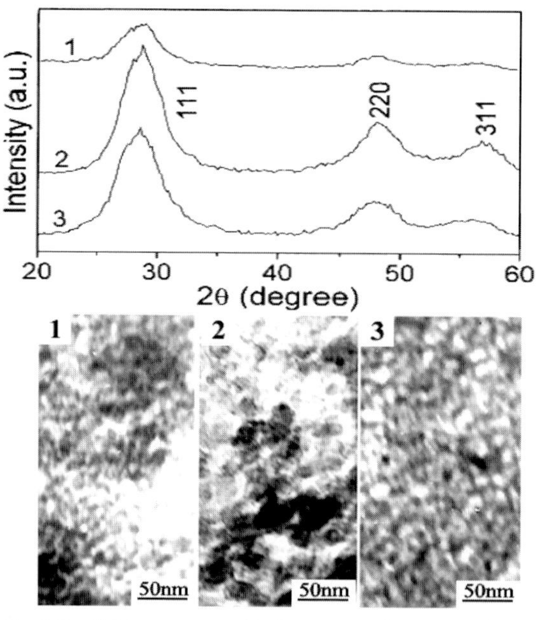


**Fig. 7.36.** XRD patterns and TEM images of nanocrystalline $\beta$-ZnS prepared by a $\gamma$-irradiation method with different sulfur sources: sodium thiosulfate (1), thiourea (2) and mercapto-ethanol (3).

aggregates most heavily. By using $CS_2$ as the sulfur source, nanocrystalline CdS of sphalerite and wurtzite structure [153] and nanocrystalline MS (M = Pb or Cu) [154] were prepared. Similarly, when $NaSeSO_3$ was used as the selenium source, 10 nm nonstoichiometric $Cu_{2-x}Se$ was obtained [155].

As various organic monomers, such as acrylamide, acrylonitrile, vinyl acetate, maleic acid and styrene, can polymerize at the same time as forming inorganic nanomaterials, various kinds of polymer–inorganic nanocomposites can be prepared at room temperature by a $\gamma$-irradiation synthesis method.

The $\gamma$-ray irradiation method offers a practical means for fabrication of inorganic/polymer core/sheath nanostructures under ambient conditions. From a heterogeneous system of vinyl acetate momomer, $CdSO_4 \cdot 8H_2O$ and $Na_2SeSO_3$ in aqueous isopropanol solution, novel nanocables of CdSe-core/poly(vinyl acetate)-sheath were prepared under $\gamma$-irradiation at room temperature [156]. As shown in Figure 7.37, the TEM images show that the as-prepared nanocable has an inner diameter as small as 6 nm, an outer diameter of about 80 nm and is some 5 μm in overall length. The image contrast of CdSe and amorphous polymer can be clearly observed. The electron diffraction pattern indicates the crystalline CdSe core. The HRTEM image reveals the lattice fringes of CdSe nanowire and the amorphous polymer sheath, in which plane the intervals of the CdSe area exactly agree with the (200) plane of cubic CdSe. In the synthetic process, the organic monomers may self-organize and build amphiphilic structures with tubule shapes having coelen-

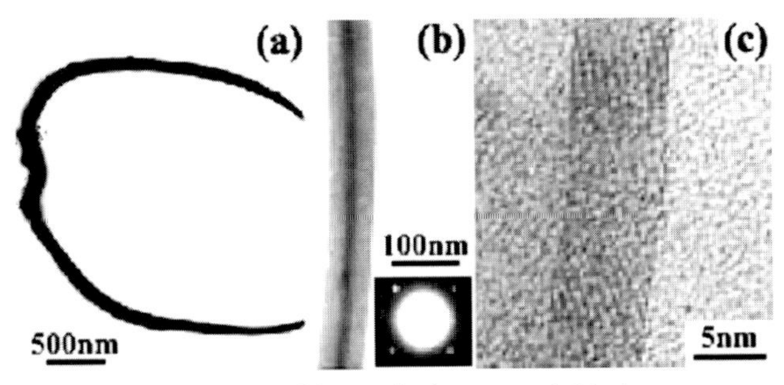


**Fig. 7.37.** Solution route fabricated CdSe-wire/poly(vinyl acetate)-sheath nanocable by $\gamma$-irradiation. (a) TEM image at lower magnification; (b) TEM image at higher magnification and inserted ED pattern of the same area; (c) HRTEM image of the nanocable.

terate cavities enveloped by hydrophilic or polar groups of vinyl acetate and coelenterate walls filled with the monomers. Meanwhile, CdSe is generated in the cavities and proceeds with confined growth into crystalline nanowires, the monomers simultaneously become solidified by $\gamma$-ray irradiation polymerization. Therefore, the pre-organized architecture of monomers eventually leads to the core/sheath nanostructures. Similar nanostructures of PbS/poly(vinyl acetate) were also obtained by the $\gamma$-ray irradiation method [157].

It is reasonable that, in the synthesis of polymer nanocomposites, the $\gamma$-ray irradiation method is convenient for growing nanofibers and nanowires of metal chalcogenides due to the shape-control of the macromolecules formed in situ. Figure 7.38 shows some of the resulting nanofiber-dispersed polymer composites,

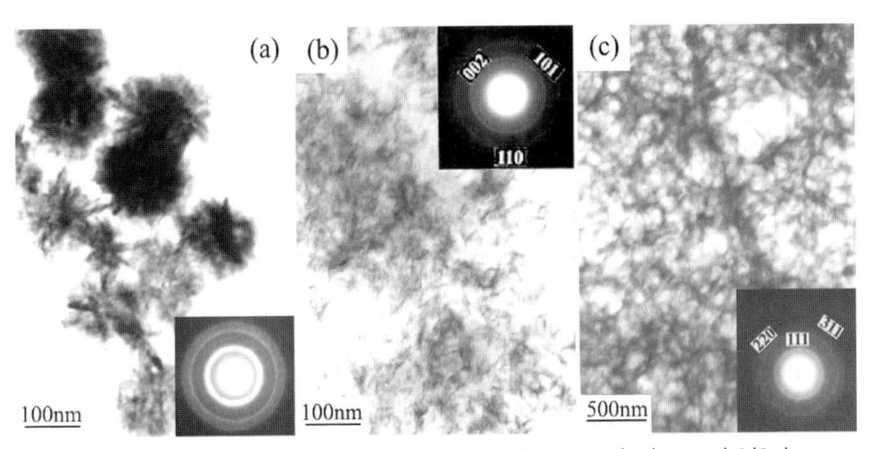


**Fig. 7.38.** TEM images and inserted ED patterns of nanofiber-dispersed polymer composites prepared by a $\gamma$-irradiation method. (a) Spherical assembled CdS nanofibers in poly(vinyl acetate); (b) dispersed CdS short nanofibers in poly(styrene-alt-maleic anhydride); (c) very long, entangled CdSe nanofibers in polyacrylamide.

which include spherical assembled CdS nanofibers in poly(vinyl acetate) [158], dispersed CdS short nanofibers in poly(styrene-alt-maleic anhydride) [159] and very long, entangled CdSe nanofibers in a polyacrylamide matrix [160]. Nanocomposites with nanoparticles dispersed in the polymer matrix, such as polyacrylamide/Ag [161], poly(vinyl acetate)/PbS [162], polyacrylamide/CdS [163] and polyacrylonitrile/CdS [164] were also prepared by $\gamma$-ray irradiation.

## References


1 Y. T. Qian, Q. W. Chen, Z. Y. Chen et al., *J. Mater. Chem.* **1993**, *3* (2), 203–205.

2 Q. W. Chen, Y. T. Qian, Z. Y. Chen et al., *Mater. Lett.* **1995**, *22* (1–2), 77–80.

3 H. W. Liao, Y. F. Wang, X. M. Liu et al., *Chem. Mater.* **2000**, *12* (10), 2819.

4 C. H. An, K. B. Tang, G. Z. Shen et al., *Mater. Lett.* **2002**, *57* (3), 565–568.

5 C. H. An, K. B. Tang, C. R. Wang et al., *Mater. Res. Bull.* **2002**, *37* (11), 1791–1796.

6 Q. W. Chen, Y. T. Qian, Z. Y. Chen et al., *Physica C*, **1994**, *224* (3–4), 228–230.

7 D. L. Zhu, H. Zhu, Y. H. Zhang, *Appl. Phys. Lett.* **2002**, *80*, 1634.

8 Q. W. Chen, G. Zhou, J. S. Zhu et al., *Phys. Lett.* **1996**, *A224* (1–2), 133–136.

9 Y. H. Zhang, X. J. Li, L. Zheng et al., *Phys. Rev. Lett.* **1998**, *81* (8), 1710–1713.

10 Y. T. Qian, Y. Su, Y. Xie et al., *Mater. Res. Bull.* **1995**, *30* (5), 601–605.

11 Y. F. Liu, J. H. Zhan, M. Ren et al., *Mater. Res. Bull.* **2001**, *36* (7–8), 1231–1236.

12 J. H. Zeng, J. Yang, Y. Zhu et al., *Chem. Commun.* **2001**, (15), 1332–1333.

13 J. H. Zeng, Y. Zhu, Y. F. Liu et al., *Mater. Sci. Eng.* **2002**, *B94* (2–3), 131–135.

14 J. Yang, S. H. Yu, Z. H. Han et al., *J. Solid State Chem.* **1999**, *146* (2), 387–389.

15 J. H. Zhan, X. G. Yang, Y. Xie et al., *J. Mater. Res.* **1999**, *14* (11), 4418–4420.

16 Y. Y. Peng, Z. Y. Meng, C. Zhong et al., *J. Solid State Chem.* **2001**, *159* (1), 170–173.

17 Y. Y. Peng, Z. Y. Meng, C. Zhong et al., *Chem. Lett.* **2001**, (8), 772–773.

18 J. Q. Hu, B. Deng, Q. Y. Lu et al., *Chem. Commun.* **2000**, (8), 715–716.

19 C. R. Wang, K. B. Tang, Q. Yang et al., *Chem. Lett.* **2001**, (6), 494–495.

20 C. R. Wang, K. B. Tang, Q. Yang et al., *J. Mater. Chem.* **2002**, *12* (8), 2426–2429.

21 Q. Yang, K. B. Tang, C. R. Wang et al., *J. Cryst. Growth* **2001**, *233* (4), 774–778.

22 C. R. Wang, K. B. Tang, Q. Yang et al., *Inorg. Chem. Commun.* **2001**, *4* (7), 339–341.

23 J. Q. Hu, B. Deng, W. X. Zhang et al., *Inorg. Chem.* **2001**, *40* (13), 3130–3133.

24 J. P. Xiao, Y. Xie, R. Tang et al., *J. Solid State Chem.* **2001**, *161* (2), 179–183.

25 J. Q. Hu, B. Deng, K. B. Tang et al., *J. Mater. Res.* **2001**, *16* (12), 3411–3415.

26 J. Q. Hu, B. Deng, C. R. Wang et al., *Solid State Commun.* **2002**, *121* (9–10), 493–496.

27 J. Q. Hu, B. Deng, K. B. Tang et al., *Solid State Sci.* **2001**, *3* (3), 275–278.

28 J. Q. Hu, B. Deng, W. X. Zhang et al., *Int. J. Inorg. Mater.* **2001**, *3* (7), 639–642.

29 C. R. Wang, K. B. Tang, Q. Yang et al., *J. Solid State Chem.* **2001**, *160* (1), 50–53.

30 J. Q. Hu, Q. Y. Lu, B. Deng et al., *Inorg. Chem. Commun.* **1999**, *2* (12), 569–571.

31 S. Wei, J. Lu, L. L. Zeng et al., *Chem. Lett.* **2002**, (10), 1034–1035.

32 Y. Xie, W. Z. Wang, Y. T. Qian et al., *Chin. Sci. Bull.* **1999**, *41* (23), 1964–1968.

33 S. M. Gao, J. Lu, N. Chen et al., *Chem. Commun.* **2002**, 3064.

34 A. Addamiano, *J. Electrochem. Soc.* **1961**, *108*, 1072.

35 J. W. Hwuang et al., *Chem. Mater.* **1995**, *7*, 517.

36 J. B. Wiley, R. B. Kaner, *Science* **1992**, *255*, 1093.

37 J. O. Fitzmaunce, A. Hector, *Polyhedron* **1994**, *13*, 235.

38 Y. Xie, Y. T. Qian, W. Z. Wang et al., *Science* **1996**, *272*, 1926.

39 Y. Xie, Y. T. Qian, S. Y. Zhang et al., *Appl. Phys. Lett.* **1996**, *69* (3), 334–336.

40 H. Xia, Q. Xia, A. L. Rouff, *Phys. Rev.* B **1993**, *47*, 12925.

41 J. Q. Hu, B. Deng, W. K. Zhang et al., *Chem. Phys. Lett.* **2002**, *351* (3–4), 229–234.

42 Z. F. Li, Y. L. Gu, H. Z. Gu et al., unpublished work.

43 Y. D. Li, X. F. Duan, Y. T. Qian et al., *J. Am. Chem. Soc.* **1997**, *119* (33), 7869–7870.

44 W. R. Penrose, *CRC Crit. Rev. Environ. Control* **1974**, 465.

45 Y. Xie, P. Yan, J. Lu et al., *Chem. Mater.* **1999**, *11* (9), 2619–2622.

46 B. Li, Y. Xie, J. X. Huang et al., *Ultrasonics Sonochem.* **2001**, *8* (4), 331–334.

47 P. Yan, Y. Xie, W. Z. Wang et al., *J. Mater. Chem.* **1999**, *9* (8), 1831–1833.

48 J. Lu, Y. Xie, X. X. Jiang et al., *Can. J. Chem.* **2001**, *79* (2), 127–130.

49 Q. Yang, Y. T. Qian, unpublished work.

50 F. P. Bundy, H. T. Hall, H. M. Strong et al., *Nature*, **1955**, *176*, 51.

51 X. Z. Zhao, R. Roy, K. A. Cherian et al., *Nature* **1997**, *385*, 513.

52 S. J. Harris, *Appl. Phys. Lett.* **1990**, *56*, 2298; S. S. Lee, D. W. Minsek, D. J. Vestyck et al., *Science*, **1994**, *263*, 1596; J. J. Wu, F. C. N. Hong, *Appl. Phys. Lett.* **1997**, *70*, 185.

53 Y. Li and Y. T. Qian et al., *Science.* **1998**, *281* (5374), 246–247.

54 S. Iijima, *Nature*, **1991**, *354*, 56.

55 T. Guo, P. Nikolaev, *Chem. Phys. Lett.* **1996**, *260*, 471.

56 S. Herreyre, P. Gadelle, *Carbon* **1995**, *33*, 234.

57 N. M. Rodriguez, *J. Mater. Res.* **1993**, *8*, 3233.

58 S. Iijima, T. Ichihashi, *Nature*, **1993**, *363*, 603.

59 Y. Jiang, Y. Wu, S. Zhang et al., *J. Am. Chem. Soc.* **2000**, *122*, 12383.

60 M. W. Shao, Q. Li, J. Wu et al., *Carbon* **2002**, *40* (15), 2961–2963.

61 X. J. Wang, J. Lu, Y. Xie et al., *J. Phys. Chem.* B **2002**, *106*, 933.

62 J. R. Heath, *Science*, **1992**, *258*, 1131.

63 J. J. Ritter, *Adv. Ceram.* **1987**, *21*, 21.

64 Q. Y. Lu, J. Q. Hu, K. B. Tang et al., *Appl. Phys. Lett.* **1999**, *75* (4), 507–509.

65 J. Q. Hu, Q. Y. Lu, K. B. Tang et al., *Chem. Mater.* **1999**, *11* (9), 2369–2371.

66 J. Q. Hu, Q. K. Lu, K. B. Tang et al., *J. Phys. Chem. B* **2000**, *104* (22), 5251–5254.

67 Q. Y. Lu, J. Q. Hu, K. B. Tang et al., *Chem. Phys. Lett.* **1999**, *314* (1–2), 37–39.

68 C. Li, X. G. Yang, Z. Y. Zhao et al., *Chem. Lett.* **2002**, (11), 1088–1089.

69 J. Q. Hu, Q. Y. Lu, K. B. Tang et al., *Chem. Lett.* **2000**, (5), 474–475.

70 X. F. Qian, X. M. Zhang, C. Wang et al., *Mater. Res. Bull.* **1999**, *34* (3), 433–436.

71 Y. L. Gu, F. Guo, Y. T. Qian et al., *Mater. Lett.* **2003**, *57* (11), 1679.

72 J. Q. Hu, Q. Y. Lu, K. B. Tang et al., *J. Am. Chem. Soc.* **2000**, *83* (2), 430–432.

73 J. Q. Hu, Q. Y. Lu, K. B. Tang et al., *Chem. Lett.* **2000**, (1), 74–75.

74 K. B. Tang, J. Q. Hu, Q. Y. Lu et al., *Adv. Mater.* **1999**, *11* (8), 653.

75 Z. Y. Meng, Y. Y. Peng, Z. P. Yang et al., *Chem. Lett.* **2000**, (11), 1252–1253.

76 Z. Y. Meng, Y. Y. Peng, Y. T. Qian, *Chem. Commun.* **2001**, (5), 469–470.

77 Y. Xie, H. L. Su, B. Li et al., *Mater. Res. Bull.* **2000**, *35* (5), 675–680.

78 Y. L. Gu, F. Guo, Y. T. Qian et al., *Mater. Res. Bull.* **2002**, *37* (6), 1101–1105.

79 X. F. Qian, Y. Xie, Y. T. Qian et al., *Mater. Sci. Eng.* **1997**, *B49* (2), 135–137.

80 X. F. Qian, X. M. Zhang, C. Wang et al., *Mater. Res. Bull.* **1998**, *33* (5), 669–672.
81 Y. Xie, H. L. Su, X. F. Qian et al., *J. Solid State Chem.* **2000**, *149*, 88–89.
82 Y. L. Gu, L. Y. Chen, Y. T. Qian et al., *J. Mater. Sci. Lett.*, in press.
83 J. Q. Hu, Q. Y. Lu, K. B. Tang et al., *J. Solid State Chem.* **1999**, *148* (2), 325–328.
84 X. P. Hao, D. L. Cui, G. X. Shi et al., *Chem. Mater.* **2001**, *13* (8), 2457–2459.
85 X. P. Hao, M. Y. Yu, D. L. Cui et al., *J. Cryst. Growth.* **2002**, *241* (1–2), 124–128.
86 L. Q. Xu, Y. T. Qian et al., *Chem. Mater.* **2003**, *15* (13), 2675.
87 Y. L. Gu, L. Y. Chen, Y. T. Qian et al., *J. Am. Ceram. Soc.*, in press.
88 Y. L. Gu, H. G. Zheng et al., *Chem. Lett.* **2002**, (7), 724–725.
89 Y. L. Gu, L. Y. Chen, Y. T. Qian et al., *Bull. Chem. Soc. Jpn.*, **2003**, *76* (7), 1469.
90 Y. L. Gu, Y. T. Qian, L. Y. Chen et al., *J. Alloys Compd.*, **2003**, *352* (1–2), 325.
91 P. Yan, Y. Xie, Y. T. Qian et al., *Chem. Commun.* **1999**, (14), 1293–1294.
92 J. Lu, Y. Xie, G. A. Du et al., *J. Mater. Chem.* **2002**, *12* (1), 103–106.
93 J. Wu, Y. Jiang, Q. Li et al., *J. Cryst. Growth* **2000**, *235* (1–4), 421–424.
94 M. Chen, Y. Xie, J. Lu et al., *J. Mater. Chem.* **2002**, *12* (3), 748–753.
95 Y. Xie, P. Yan, J. Lu et al., *Chem. Commun.* **1999**, (19), 1969–1970.
96 Y. D. Li, H. W. Liao, Y. Ding et al., *Inorg. Chem.* **1999**, *38* (7), 1382–1387.
97 Y. D. Li, H. W. Liao, Y. Fan et al., *Mater. Chem. Phys.* **1999**, *58* (1), 87–89.
98 W. Z. Wang, Y. Geng, P. Yan et al., *Inorg. Chem. Commun.* **1999**, *2* (3), 83–85.
99 S. H. Yu, Y. S. Wu, J. Yang et al., *Chem. Mater.* **1998**, *10* (9), 2309–2312.
100 J. Yang, J. H. Zeng, S. H. Yu et al., *Chem. Mater.* **2000**, *12* (11), 3259–3263.
101 J. Yang, C. Xue, S. H. Yu et al., *Angew. Chem. Int. Ed. Engl.* **2002**, *41* (24), 4697–4700.
102 J. H. Zhan, X. G. Yang, W. X. Zhang et al., *J. Mater. Res.* **2000**, *15* (3), 629–632.
103 Y. D. Li, Y. Ding, Y. T. Qian et al., *Inorg. Chem.* **1998**, *37* (12), 2844–2845.
104 J. H. Zhan, X. G. Yang, D. W. Wang et al., *Adv. Mater.* **2000**, *12* (18), 1348–1351.
105 Q. Yang, K. B. Tang, C. R. Wang et al., *J. Phys. Chem. B* **2002**, *106* (36), 9227–9230.
106 M. S. Mo, M. W. Shao, H. M. Hu et al., *J. Cryst. Growth* **2002**, *244* (3–4), 364–368.
107 D. B. Yu, D. B. Wang, Z. Y. Meng et al., *J. Mater. Chem.* **2002**, *12* (3), 403–405.
108 J. Yang, J. H. Zeng, S. H. Yu et al., *Chem. Mater.* **2000**, *12* (10), 2924–2929.
109 D. B. Wang, D. B. Yu, M. W. Shao et al., *Chem. Lett.* **2002**, (10), 1056–1057.
110 S. H. Yu, L. Shu, J. A. Yang et al., *J. Mater. Res.* **1999**, *14* (11), 4157–4162.
111 Q. Li, M. W. Shao, J. Wu et al., *Inorg. Chem. Commun.* **2002**, *5* (11), 933–936.
112 H. L. Su, Y. Xie, P. Gao et al., *Chem. Lett.* **2000**, (7), 790–791.
113 Y. F. Liu, J. H. Zeng, W. X. Zhang et al., *J. Mater. Res.* **2001**, *16* (12), 3361–3365.
114 X. C. Jiang, Y. Xie, J. Lu et al., *Adv. Mater.* **2001**, *13* (16), 1278–1281.
115 J. Yang, G. H. Cheng, J. H. Zeng et al., *Chem. Mater.* **2001**, *13* (3), 848–853.
116 Z. H. Han, S. H. Yu, Y. P. Li et al., *Chem. Mater.* **1999**, *11* (9), 2302.
117 N. Chen, W. Q. Zhang, W. C. Yu et al., *Mater. Lett.* **2002**, *55* (4), 230–233.
118 Z. Y. Meng, Y. Y. Peng, L. Q. Xu et al., *Mater. Lett.* **2002**, *53* (3), 165–167.
119 J. Lu, P. F. Qi, Y. Y. Peng et al., *Chem. Mater.* **2001**, *13* (6), 2169–2172.
120 T. Qin, J. Lu, S. Wei et al., *Inorg. Chem. Commun.* **2002**, *5* (5), 369–371.
121 B. Li, Y. Xie, J. X. Huang et al., *Adv. Mater.* **1999**, *11* (17), 1456–1459.
122 Q. Y. Lu, J. Q. Hu, K. B. Tang et al., *Inorg. Chem.* **2000**, *39* (7), 1606.
123 Y. Jiang, Y. Wu, X. Mo et al., *Inorg. Chem.* **2000**, *39* (14), 2964.
124 Y. Cui, J. Ren, G. Chen et al., *Chem. Lett.* **2001**, (3), 236–237.

125 J. Q. Hu, Q. Y. Lu, K. B. Tang et al., *Chem. Commun.* **1999**, (12), 1093–1094.

126 Y. Y. Peng, Z. Y. Meng, C. Zhong et al., *New J. Chem.* **2001**, *25* (11), 1359–1361.

127 B. Xie, S. W. Yuan, Y. Jiang et al., *Chem. Lett.* **2002**, (6), 612–613.

128 B. Li, Y. Xie, J. X. Huang et al., *J. Solid State Chem.* **2000**, *153* (1), 170–173.

129 B. Li, Y. Xie, J. X. Huang et al., *Solid State Ionics* **1999**, *126* (3–4), 359–362.

130 H. L. Su, Y. Xie, S. K. Wan et al., *Solid State Ionics* **1999**, *123* (1–4), 319–324.

131 R. Coustal, *J. Chim. Phys.* **1958**, *38*, 277.

132 I. P. Parkin, *Chem. Soc. Rev.* **1996**, *25*, 199.

133 T. Ohtani, M. Motoki, *Mater. Res. Bull.* **1995**, *30*, 1495.

134 G. Henshaw, I. P. Parkin, *Chem. Commun.* **1996**, 1095.

135 M. L. Steigerwald, A. P. Alivasatos, J. M. Gibson, *J. Am. Chem. Soc.* **1988**, *110*, 3046; C. B. Murray, D. J. Norris, M. G. Bawendi, *J. Am. Chem. Soc.* **1993**, *115*, 8706.

136 W. Z. Wang, Y. Geng, P. Yan et al., *J. Am. Chem. Soc.* **1999**, *121* (16), 4062–4063.

137 W. Z. Wang, Y. T. Qian et al., *Adv. Mater.* **1998**, *10* (17), 1479.

138 J. X. Huang, Y. Xie, B. Li et al., *Adv. Mater.* **2000**, *12* (11), 808–811.

139 Y. Xie, J. X. Huang, B. Li et al., *Adv. Mater.* **2000**, *12* (20), 1523–1526.

140 Y. Jiang, B. Xie, J. Wu et al., *J. Solid State Chem.* **2002**, *167* (1), 28–33.

141 Y. Jiang, Y. Wu, Z. P. Yang et al., *J. Cryst. Growth* **2001**, *224* (1–2), 1–4.

142 W. Z. Wang, Y. Geng, Y. T. Qian et al., *Mater. Res. Bull.* **1999**, *34* (6), 877–882.

143 Y. P. Liu, Y. J. Zhu, Y. H. Zhang et al., *J. Mater. Chem.* **1997**, *7* (5), 787–789.

144 Q. Yang, F. Wang, K. B. Tang et al., *Mater. Chem. Phys.* **2002**, *78* (2), 495–500.

145 Y. J. Zhu, Y. T. Qian, M. W. Zhang et al., *Mater. Trans. JIM* **1995**, *36* (1), 80–81.

146 Y. J. Zhu, Y. T. Qian, H. A. Hai et al., *Mater. Lett.* **1996**, *28* (1–3), 119–122.

147 H. R. Liu, Z. C. Zhang, Y. T. Qian et al., *Chin. J. Inorg. Chem.* **1999**, *15* (3), 388–392.

148 Y. Xie, Z. P. Qiao, M. Chen et al., *Chem. Lett.* **1999**, (9), 875–876.

149 M. Chen, Y. Xie, J. C. Lu et al., *J. Mater. Chem.* **2001**, *11* (2), 518–520.

150 M. Chen, Y. Xie, H. Y. Chen et al., *J. Colloid Interface Sci.* **2001**, *237* (1), 47–53.

151 X. Jiang, Y. Xie, J. Lu et al., *Chem. Mater.* 2001, *13* (4), 1213–1218.

152 Z. P. Qiao, Y. Xie, Y. T. Qian et al., *Mater. Phys. Chem.* **2000**, *62* (1), 88–90.

153 Z. P. Qiao, Y. Xie, X. J. Li et al., *J. Mater. Chem.* **1999**, *9* (3), 735–738.

154 Z. P. Qiao, Y. Xie, J. G. Xu et al., *J. Colloid Interface Sci.* **1999**, *214* (2), 459–461.

155 Z. P. Qiao, Y. Xie, J. G. Xu et al., *Can. J. Chem.-Revue Canadienne de Chimie* **2000**, *78* (9), 1143–1146.

156 Y. Xie, Z. P. Qiao, M. Chen et al., *Adv. Mater.* **1999**, *11* (18), 1512.

157 J. H. Zeng, Y. Zhu, J. Yang et al., *Chem. Lett.* **2001**, (10), 1000–1001.

158 Y. Xie, Z. P. Qiao et al., *Nanostructured Mater.* **1999**, *11* (8), 1165–1169.

159 M. Chen, Y. Xie, Z. P. Qiao et al., *J. Mater. Chem.* **2000**, *10* (2), 329–332.

160 Z. P. Qiao, Y. Xie, J. X. Huang et al., *Radiat. Phys. Chem.* **2000**, *58* (3), 287–292.

161 Y. J. Zhu, Y. T. Qian, X. J. Li et al., *Chem. Commun.* **1997**, (12), 1081–1082.

162 Z. P. Qiao, Y. Xie, M. Chen et al., *Chem. Phys. Lett.* **2000**, *321* (5–6), 504–507.

163 Z. P. Qiao, Y. Xie, G. Li et al., *J. Mater. Sci.* **2000**, *35* (2), 285–287.

164 Z. P. Qiao, Y. Xie, J. G. Xu et al., *Mater. Res. Bull.* **2000**, *35* (8), 1355–1360.

# 8
# Nanotubes and Nanowires

*A. Govindaraj and C. N. R. Rao*

**Abstract**

Carbon nanotubes (CNTs) were discovered as an electron microscopic marvel in 1991. Since then, there has been intense activity related to the synthesis, structure, properties and applications of CNTs. The discovery of CNTs has triggered much research on other one-dimensional nano-objects such as inorganic nanotubes and nanowires. This chapter covers the highlights of the entire variety of nanotubes and nanowires.

## 8.1
## Introduction

Diamond and graphite are the two well-known forms of crystalline carbon. Diamond has four-coordinate $sp^3$ carbon atoms forming an extended three-dimensional network, whose motif is the chair conformation of cyclohexane. Graphite has three-coordinate $sp^2$ carbons forming planar sheets, whose motif is the flat six-membered benzene ring. The new carbon allotrope, the fullerenes, are closed-cage carbon molecules with three-coordinate carbon atoms tiling the spherical or nearly-spherical surfaces, the best known example being $C_{60}$, with a truncated icosahedral structure formed by twelve pentagonal rings and twenty hexagonal rings. Fullerenes were discovered by Kroto et al. [1] in 1985 while investigating the nature of carbon present in interstellar space. The coordination at every carbon atom in fullerenes is not planar, but slightly pyramidalized, with some $sp^3$ character present in the essentially $sp^2$ carbons. The key feature is the presence of five-membered rings which provide the curvature necessary to form a closed-cage structure. In 1990, Krätschmer et al. [2] found that the soot produced by arcing graphite electrodes contained $C_{60}$ and other fullerenes. It was the ability to generate fullerenes in gram quantities in the laboratory, using a relatively simple apparatus, that gave rise to intense research activity on these molecules and caused a renaissance in the study of carbon. Iijima [3] observed in 1991 that nanotubules of graphite were deposited on the negative electrode during the direct current arcing

*The Chemistry of Nanomaterials: Synthesis, Properties and Applications, Volume 1.* Edited by C. N. R. Rao, A. Müller, A. K. Cheetham

ISBN: 3-527-30686-2

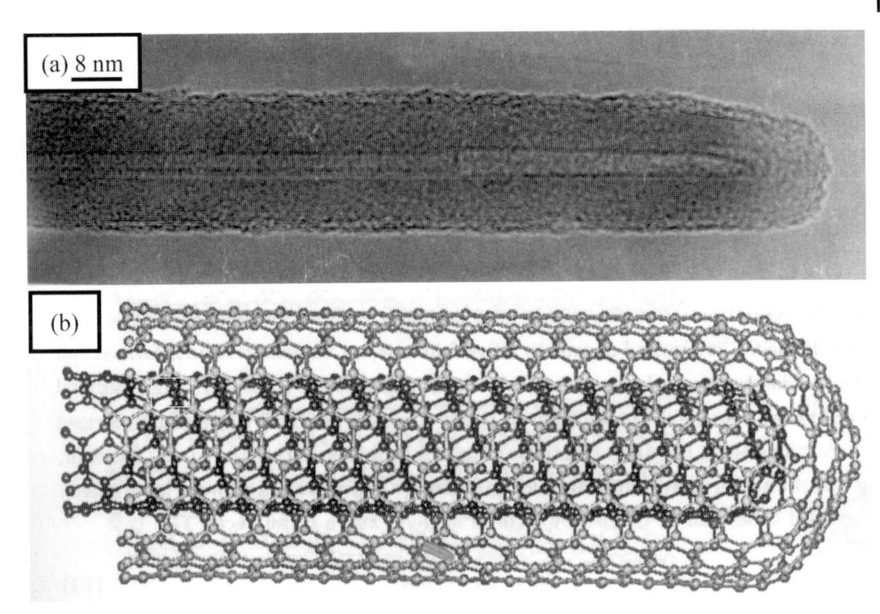


**Fig. 8.1.** (a) A TEM image of a multi-walled carbon nanotube; (b) Minimum energy structure of a double-walled carbon nanotube. Reproduced from ref. [20a], with permission.

of graphite for the preparation of fullerenes. These nanotubes are concentric graphitic cylinders closed at either end due to the presence of five-membered rings. Nanotubes can be multi-walled with a central tubule of nanometric diameter surrounded by graphitic layers separated by ~3.4 Å. Unlike the multi-walled nanotubes (MWNTs), in single-walled nanotubes (SWNTs), there is only the tubule and no graphitic layers. A transmission electron microscope (TEM) image of a MWNT is shown in Figure 8.1(a). In this nanotube, graphite layers surround the central tubule. Figure 8.1(b) shows the structure of a nanotube formed by two concentric graphitic cylinders, obtained by force field calculations. A single-walled nanotube can be visualized by cutting $C_{60}$ along the center and spacing apart the hemispherical corannulene end-caps by a cylinder of graphite of the same diameter. Carbon nanotubes are the only form of carbon with extended bonding and yet with no dangling bonds. Since carbon nanotubes are derived from fullerenes, they are referred to as tubular fullerenes or bucky tubes.

Ever since the discovery of the carbon nanotubes, several ways of preparing them have been explored. Besides MWNTs, SWNTs have been prepared [4, 5], the various methods including electrochemical synthesis [6] and pyrolysis of precursor molecules [7]. The structure of carbon nanotubes has been extensively investigated by high-resolution electron microscopy [8–10]. The nanotubes, as prepared by arc vaporization of graphite, are closed at either end, but can be opened by various oxidants [11, 12]. There has been considerable success in filling the nanotubes with various materials [13]. Apart from opening and filling, carbon nanotubes have been doped with boron and nitrogen, giving rise to p-type and n-type materials respectively. By employing carbon nanotubes as removable templates, oxidic, carbidic and other nanostructures have been prepared. One of the recent developments is the synthesis of aligned nanotube bundles for specific applications. A variety of properties and phenomena as well as several applications of carbon nanotubes, some potential and some likely, have been reported. It is no wonder, therefore, that these nanomaterials have elicited such great interest. There have been several review articles, special issues of journals and conference proceedings [14–20] dealing with carbon nanotubes in the literature, together with a book which appeared in 1996 [17]. Some of the reviews present possible technological applications with focus on the electronic properties [19, 20].

Since the discovery of the carbon nanotubes, there has been considerable work on other layered materials such as $MoS_2$, $WS_2$ and BN to explore the formation of nanotubes of these materials. Indeed several of them have been synthesized and characterized [21–23]. Similarly, nanowires of various inorganic materials have also been made [21]. In this chapter, we shall present the various important aspects of carbon nanotubes including their preparation, structure, mechanism of formation, chemical substitution, properties and applications. The methodologies developed for synthesizing nanowires and nanotubes of various inorganic materials as well as their salient features will also be discussed [21–24].

## 8.2 Carbon Nanotubes

### 8.2.1 Synthesis

#### 8.2.1.1 Multi-Walled Nanotubes

Carbon nanotubes are readily prepared by striking an arc between graphite electrodes in 2/3 atm ($\sim$500 torr) of helium, considerably higher than the pressure of helium used in the production of fullerene soot. A current of 60–100 A across a potential drop of about 25 V gives high yields of carbon nanotubes. The arcing process can be optimized such that the major portion of the carbon anode is deposited on the cathode in the form of carbon nanotubes and graphitic nanoparticles [25]. Carbon nanotubes have been produced by using plasma arc-jets [26] and in large quantities, by optimizing the quenching process in an arc between a graphite anode and a cooled copper electrode [27]. Scanning tunneling microscope (STM) studies show that the deposition of carbon vapor on cooled substrates of highly oriented pyrolytic graphite gives rise to tube-like structures [28]. Carbon nanotubes are also produced by carrying out electrolysis in molten halide salts with carbon electrodes in an argon atmosphere [29a]. In addition, MWNTs with well-ordered graphitic structures have been synthesized under hydrothermal conditions using a polyethylene and water mixture in the presence of nickel catalyst at around 800 °C under 60–100 MPa pressure [29b]. Besides the conventional arc-

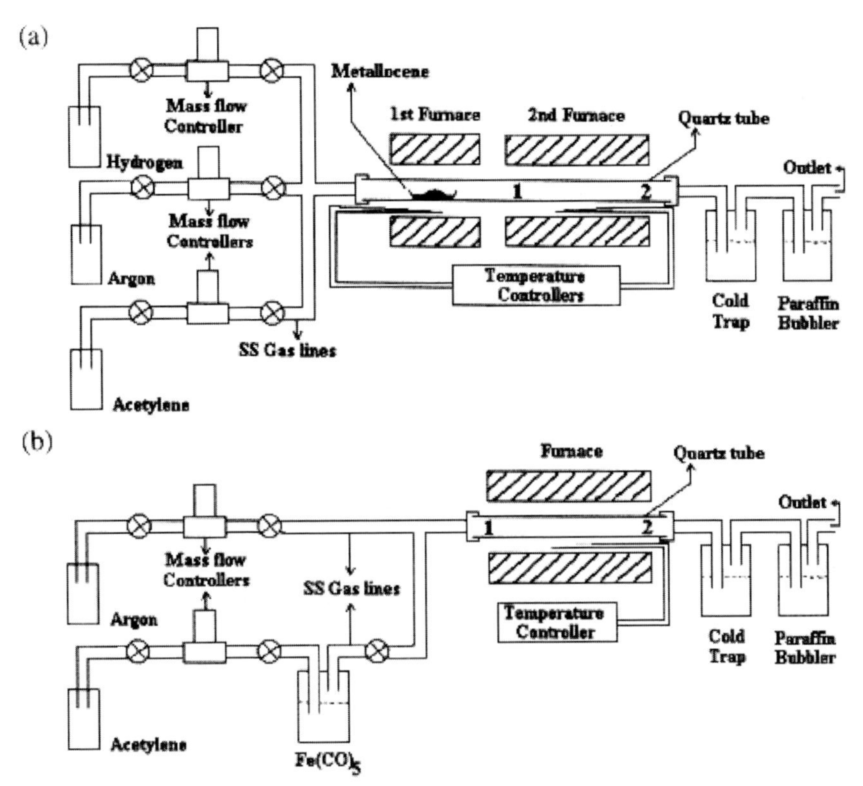


**Fig. 8.2.** Pyrolysis apparatus employed for the synthesis of carbon nanotubes by pyrolysis of mixtures of (a) metallocene + $C_2H_2$ and (b) $Fe(CO)_5 + C_2H_2$. The numbers 1 and 2 indicated in the figure represent the inlet and outlet, respectively. Reproduced from ref. [44, 45], with permission.

evaporation technique, carbon nanotubes are produced by the decomposition of hydrocarbons such as $C_2H_2$ under inert conditions at around 700 °C, over Fe/graphite [30], Co/graphite [31] or Fe/silica [32] catalysts. The presence of transition metal particles is essential for the formation of nanotubes by the pyrolysis process, and the diameter of the nanotube is determined by the size of the metal particles [33]. Sen et al. [34] prepared carbon nanotubes and metal-filled onion-like structures by the pyrolysis of ferrocene, cobaltocene and nickelocene under reductive conditions, wherein the precursor acts as a source of the metal as well as carbon (Figure 8.2). They also showed that the pyrolysis of benzene or acetylene in the presence of ferrocene or $Fe(CO)_5$ gives high yields of nanotubes (Figure 8.2), the wall thickness depending on the proportion of the carbon source and the metal precursor [35]. Figure 8.3(a) shows the TEM image of MWNTs obtained by the pyrolysis of mixture of $C_2H_2$ and ferrocene. Under similar conditions, nickelocene with benzene also gives MWNTs (Figure 8.3(b)).

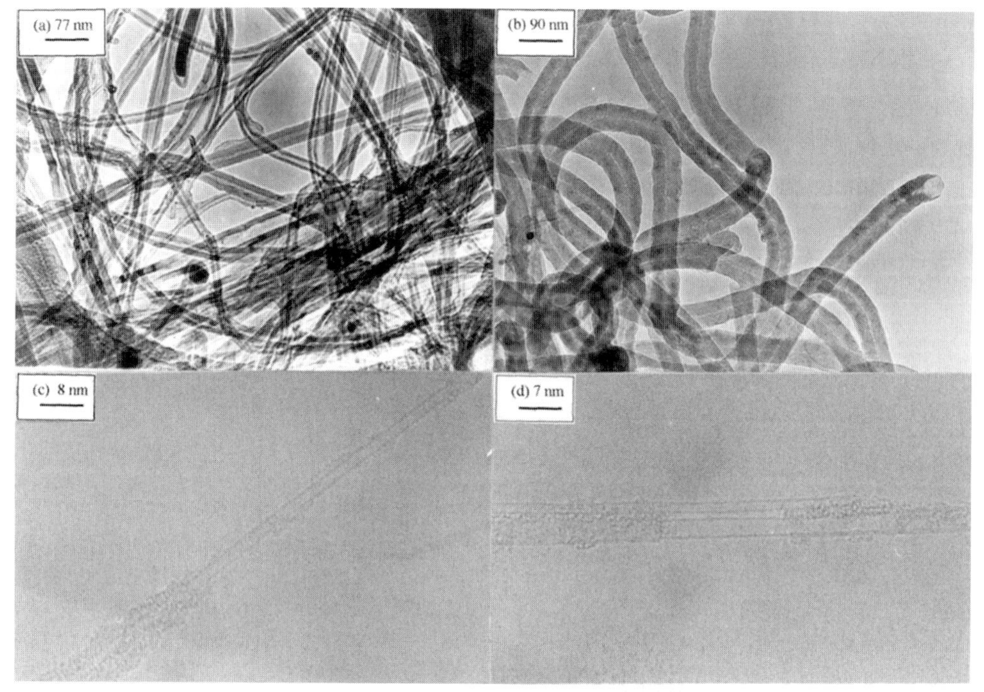


**Fig. 8.3.** (a) TEM image of MWNTs obtained by pyrolysis of a mixture of $C_2H_2$ (25 sccm) and ferrocene at 1100 °C at 1000 sccm Ar flow. (b) TEM image of MWNTs obtained by pyrolysis of a mixture of nickelocene and benzene at 900 °C in 85% Ar and 15% $H_2$ mixture at a flow rate of 1000 sccm. (c) HREM image of SWNTs obtained by the pyrolysis of nickelocene and $C_2H_2$ at 1100 °C in a flow of Ar (1000 sccm) with $C_2H_2$ flow rate of 50 sccm. (d) HREM image of SWNTs obtained by the pyrolysis of ferrocene and $CH_4$ at 1100 °C in a flow of Ar (990 sccm) with $CH_4$ flow rate of 10 sccm. Reproduced from ref. [34, 45], with permission.

#### 8.2.1.2 **Aligned Carbon Nanotube Bundles**

Carbon nanotubes are potential candidates for use as field emitters [36]. Of particular relevance to this application is the synthesis of aligned nanotube bundles. Aligned nanotube bundles have been obtained by chemical vapor deposition (CVD) over transition metal catalysts embedded in the pores of mesoporous silica or the channels of alumina membranes [37, 38]. Terrones et al. [39, 40] prepared aligned nanotubes over silica substrates, laser-patterned with cobalt. Ren et al. [41] employed plasma-enhanced chemical vapor deposition on nickel-coated glass, using acetylene and ammonia mixtures, for this purpose. The mechanism of the growth of nanotubes by this method and the exact role of the metal particles are not clear, although a nucleation process involving the metal particles is considered to be important. Fan et al. [42] have obtained aligned nanotubes by employing CVD on porous silicon and plain silicon substrates patterned with Fe films. The role of the transition metal particles assumes importance in view of the report of Pan et al. [43] that aligned nanotubes can be obtained by the pyrolysis of acetylene over iron/

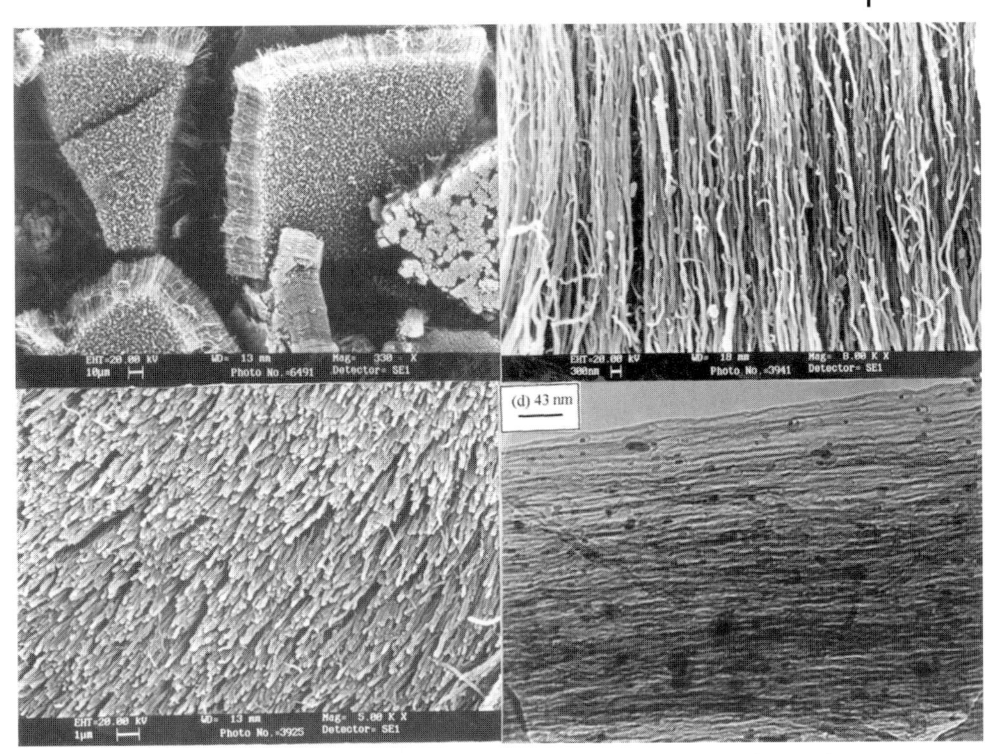


**Fig. 8.4.** SEM image (a) showing the bundles of aligned nanotubes obtained by pyrolysing ferrocene with butane (50 sccm) at 1100 °C in an Ar flow of 950 sccm; (b) and (c) show views of the aligned nanotubes perpendicular to and along the axis of the nanotubes respectively; (d) TEM image of part of an aligned nanotube bundle obtained from the pyrolysis of acetylene (85 sccm) and ferrocene mixture at 110 °C in an Ar flow of 1000 sccm. Reproduced from ref. [45], with permission.

silica catalyst surfaces. In the light of the earlier work on the synthesis of carbon nanotubes by the pyrolysis of mixtures of organometallic precursors and hydrocarbons, [34, 35] one would expect that the transition metal nanoparticles produced in situ in the pyrolysis, may not only nucleate the formation of carbon nanotubes but also align them. This aspect has been examined by carrying out the pyrolysis of metallocenes along with additional hydrocarbon sources, in a suitably designed apparatus (Figure 8.2) [44, 45]. Scanning electron microscope (SEM) images of aligned nanotubes obtained by the pyrolysis of ferrocene are shown in Figure 8.4(a)–(c). The image in Figure 8.4(a) shows larger bundles of aligned nanotubes. The image in Figure 8.4(b) shows the side-view and the image, whereas Figure 8.4(c) shows the top-view of the aligned nanotubes, wherein the nanotube tips are seen. A TEM image of a part of an aligned nanotube bundle obtained from the pyrolysis of acetylene–ferrocene mixture is shown in Figure 8.4(d). The average length of the nanotubes is generally around 60 μm with methane and acetylene. Andrews et al. [46a] have carried out the pyrolysis of ferrocene–xylene mixtures to obtain aligned carbon nanotubes. Pyrolysis of Fe(II)phthalocyanine also yields aligned nanotubes [46b]. Hexagonally ordered arrays of nanotubes are produced by using alumina templates with ordered pores [47a]. By employing catalytic chemical vapor deposition (CCVD), Mukhopadhyay et al. [47b] have obtained quasi-aligned carbon nanotubes using metal impregnated zeolite templates. The advantage of the precursor method is that the aligned bundles are produced in one step, at a relatively low cost, without prior preparation of substrates. The precursor route to carbon nanotubes has been discussed recently by Rao and Govindaraj [48a].

TEM observations of aligned nanotubes produced by ferrocene + hydrocarbon pyrolysis show the presence of iron nanorods encapsulated inside the carbon nanotubes, the proportion of the nanorods depending on the proportion of ferrocene. Typical TEM images of such nanorods are shown in Figure 8.5(a), (b) and (c). The inset in Figure 8.5(b) shows the selected area electron diffraction (SAED) pattern of the nanorods showing spots due to the (010) and (011) planes of $\alpha$-Fe. The high-resolution electron microscope (HREM) image of the iron nanorod shows well-resolved (011) planes of $\alpha$-Fe in single-crystalline form. X-ray diffraction studies also show the presence of $\alpha$-Fe with a small portion of $Fe_3C$ as the minor phase. In addition to the nanorods, there are iron nanoparticles (20–40 nm diameter) encapsulated inside the graphite layers. Both iron nanorods and nanoparticles are well protected against oxidation by the graphitic layers. The iron nanorods also exhibit a complex behavior with respect to magnetization reversal, showing Barkhausen jumps [48b]. Iron-filled carbon nanotubes could be useful as probes in magnetic force microscopy.

#### 8.2.1.3 **Single-Walled Carbon Nanotubes**

The nanotubes generally obtained by the arc method or hydrocarbon pyrolysis are multi-walled, having several graphitic sheets or layers (Figure 8.1). Single-walled nanotubes (SWNTs) were first prepared by metal-catalyzed dc arcing of graphite rods [4, 5] in a He atmosphere. The graphite anode was filled with metal powders (Fe, Co or Ni) and the cathode was made of pure graphite. SWNTs generally occur in the web-like material deposited behind the cathode. Various metal catalysts have been used to make SWNTs by this route. Dai et al. [49] prepared SWNTs by the disproportionation of CO at 1200 °C over Mo particles of a few nanometers diameter dispersed in a fumed alumina matrix. Saito et al. [50] compared SWNTs produced by using different catalysts and found that a Co or a Fe/Ni bimetallic catalyst gives rise to tubes forming a highway-junction pattern. SWNTs are also prepared by using various oxides $Y_2O_3$, $La_2O_3$, $CeO_2$ as catalysts [51]. The arc discharge technique, though cheap and easy to implement, gives low yields of SWNTs. Journet et al. [52] obtained ~80% yield of SWNTs in the arc, by using a mixture of 1 at.% Y and 4.2 at.% Ni as catalyst. Arc evaporation of graphite rods filled with Ni and $Y_2O_3$ in a He atmosphere (660 torr) gives rise to web-like deposits on the chamber walls near the cathode, consisting of SWNT bundles [45]. HREM images show bundles consisting of 10–50 SWNTs forming highway junctions (Figure 8.6). The average diameter of the SWNTs was around 1.4 nm and the length extended upto 10 μm. SWNTs have been produced in more than 70% yield by the conden-

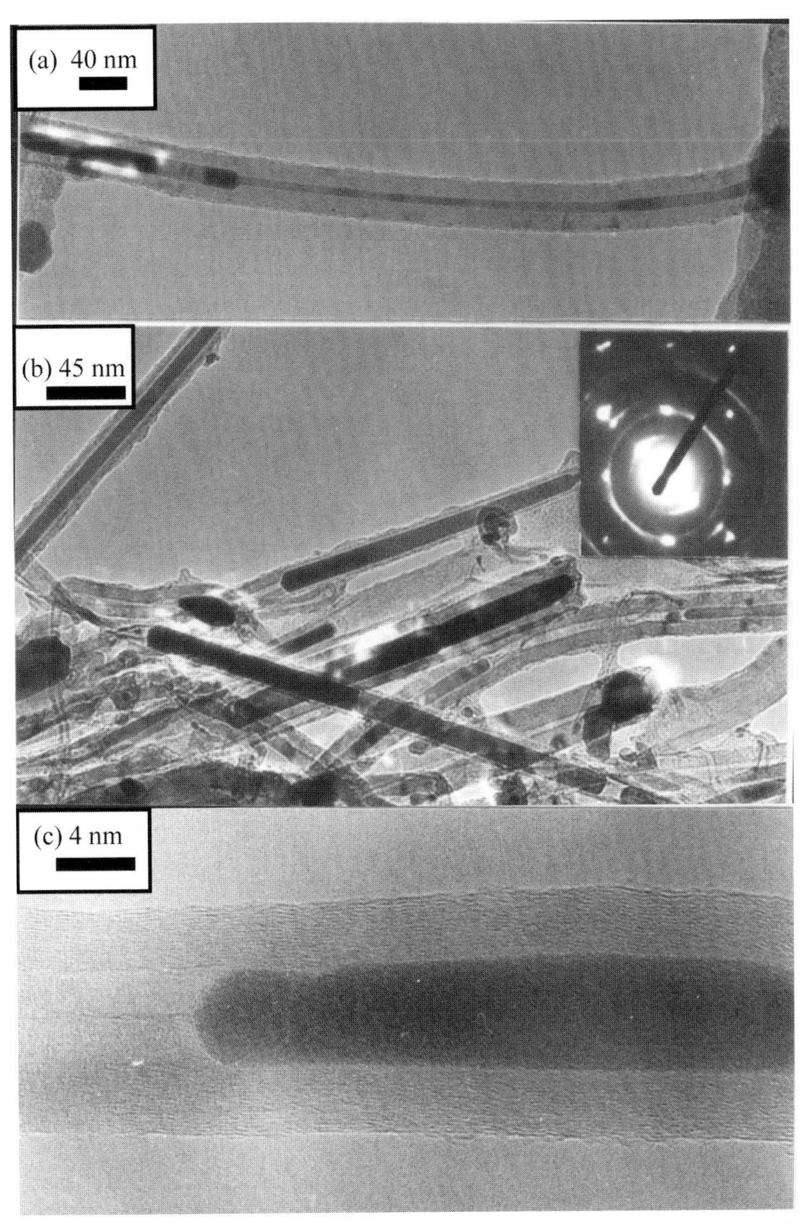


**Fig. 8.5.** (a), (b) TEM images of the iron nanorods encapsulated inside the carbon nanotubes from aligned nanotube bundles. (c) HREM image of a single-crystal iron nanorod encapsulated inside a carbon nanotube. The inset in (b) represents the selected area electron diffraction (SAED) pattern of an iron nanorod. Reproduced from ref. [45], with permission.

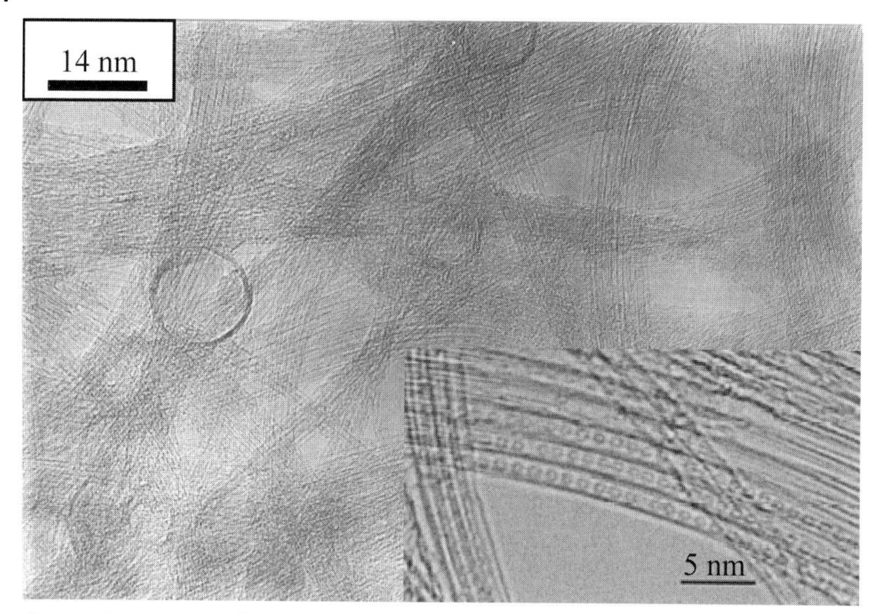


**Fig. 8.6.** HREM image of SWNTs obtained by arcing graphite electrodes filled with Ni and $Y_2O_3$ under a He atmosphere (660 Torr). Inset: The HREM image of encapsulated fullerenes inside the SWNTs; scale bar is 5 nm. Reproduced from ref. [45], with permission.

sation of a laser-vaporized carbon–nickel–cobalt mixture at 1200 °C [53]. These SWNTs were nearly uniform in diameter and self-assemble into ropes which consist of 100 to 500 tubes in a 2D triangular lattice.

Under controlled conditions of pyrolysis, dilute hydrocarbon–organometallic mixtures yield SWNTs [45, 54]. Pyrolysis of metallocene–acetylene mixtures at 1100 °C yields SWNTs [54, 55], shown in the TEM image in Figure 8.3(c). The diameter of the SWNT in Figure 8.3(c) is 1.4 nm. Figure 8.3(d) shows the SWNTs obtained similarly by the pyrolysis of a ferrocene–$CH_4$ mixture at 1100 °C. It may be recalled that the pyrolysis of nickelocene in admixture with benzene under similar conditions primarily yields MWNTs. The bottom portion of the SWNT in Figure 8.3(c) shows an amorphous carbon coating around the tube, common with such preparations. This can be avoided by reducing the proportion of the hydrocarbon $C_2H_2$ and mixing hydrogen in the Ar stream. Pyrolysis of acetylene in mixture with $Fe(CO)_5$ at 1100 °C gives good yields of SWNTs. Pyrolysis of ferrocene–thiophene mixtures also yield SWNTs, but the yield appears to be somewhat low. Pyrolysis of benzene and thiophene along with ferrocene gives a high yield of SWNTs [56].

Laplaze et al. [57] have demonstrated that concentrated solar energy can be employed to vaporize graphite to synthesize SWNTs. Nikolaev et al. [58] have obtained SWNTs using a gas-phase catalytic method involving the pyrolysis of $Fe(CO)_5$ and CO. The decomposition of CO on a silica-supported Co–Mo catalyst also yields

SWNTs [59]. Colomer et al. [60] obtained SWNTs in high yield by the decomposition of methane over transition metal supported MgO substrates. Flahaut et al. [61] have synthesized SWNTs by passing a $H_2$–$CH_4$ mixture over transition metal containing oxide spinels, obtained by the combustion route. The quality of SWNTs has been characterized on the basis of adsorption measurements. Zeolites containing one-dimensional channels have also been investigated for the synthesis of monosized SWNTs [62]. As-prepared SWNTs contain several contaminants such as amorphous carbon and nanometer-size catalyst particles coated with carbon. The amorphous carbon can be burnt away by heating the nanotubes in air at around 300 °C. Bandow et al. [63] use microfiltration to clean SWNTs of the other contaminants. Size exclusion chromatography of the surfactant-stabilized raw SWNTs is also employed to purify the nanotubes from amorphous carbon, metal particles etc [64]. In addition to purification, size separation has also been achieved. Since SWNTs occur as large bundles with lengths of the order of microns, it is desirable to break them from the bundles, for purposes of further manipulation. Liu et al. [65] have employed chemical processing based on ultrasound treatment, wherein the SWNTs in an acidic medium were subjected to sonication so that the bundles break up into open-ended small fragments of 100–300 nm length. The smaller fragments were functionalized.

### 8.2.2
### Structure and Characterization

Transmission electron microscope observations show that the nanotubes prepared by the arcing process generally consist of multi-layered, concentric cylinders of single graphitic (graphene) sheets. The diameter of the inner tubes is of the order of a few nanometers. The outermost tubes could be as large as 10–30 nm as shown in Figure 8.1(a). During the curling of a graphene sheet into a cylinder, helicity is introduced. Electron diffraction studies establish the presence of helicity, suggesting that the growth of nanotubes occurs as in the spiral growth of crystals. The separation between concentric cylinders in MWNTs is about 3.45 Å, which is close to the separation between the (002) planes of graphite. These are the lowest energy surfaces of graphite with no dangling bonds, so that the nanotubes are in fact the expected structures. In the electron microscope images, one typically observes nanotubes along their lengths, with the electron beam falling perpendicular to the axis of the nanotube. In high-resolution images, it is possible to see spots due to the lattice planes running along the length of the nanotubes. Iijima [66] has published such an image for the (110) planes separated by 2.1 Å. Ring-like patterns are found due to individual tubes comprising cylindrical graphitic sheets which are independently oriented (with no registry between the sheets) with helical symmetry for the arrangement of the hexagons. Graphitic cylinders would have dangling bonds at the tips, but the carbon nanotubes are capped by dome-shaped hemispherical fullerene-type units. The capping units consist of pentagons to provide the curvature necessary for closure. Ajayan et al. [8a] studied the distribution of pentagons at the caps of carbon nanotubes, finding that the caps need not be perfectly conical or hemispherical, but can form skewed structures. The simplest

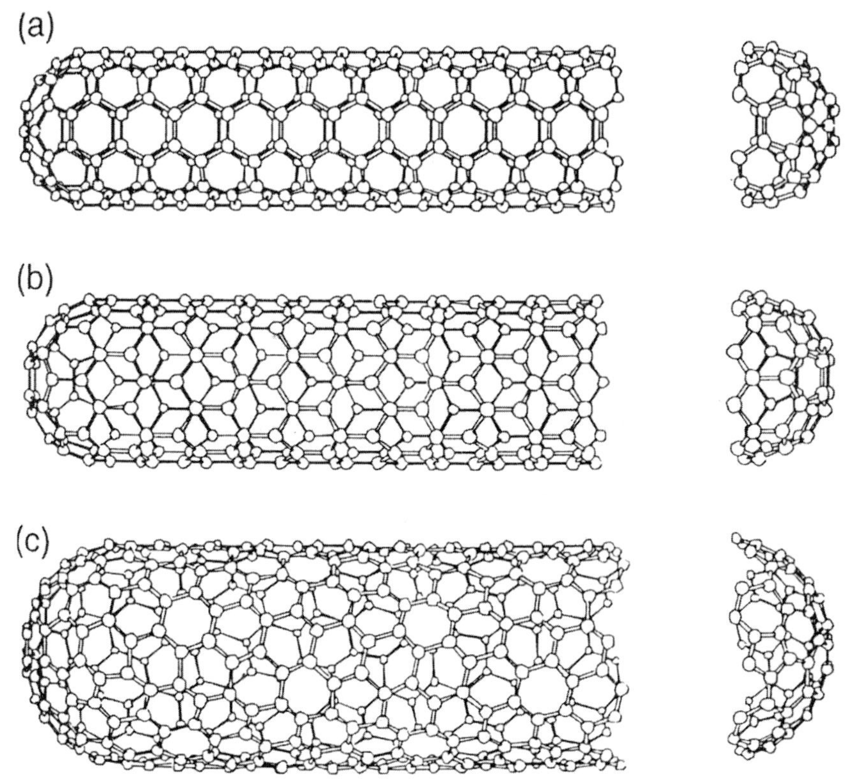


**Fig. 8.7.** Models of (a) armchair, (b) zigzag, and (c) chiral nanotubes. Reproduced from ref. [17], with permission.

possible single-walled carbon nanotube can be visualized by cutting the $C_{60}$ structure across the middle and adding a cylinder of graphite of the same diameter. If $C_{60}$ is bisected normal to a five-fold axis, an armchair tube is formed, and if it is bisected normal to a three-fold axis, a zigzag tube is formed. Armchair and zigzag tubes are non-chiral. In addition to these, a variety of chiral tubes can be formed with the screw axis along the axis of the tube. In Figure 8.7 we show the models of the three types of nanotubes formed by bisecting the $C_{60}$ molecule and adding a cylinder of graphite. Nanotubes can be defined by a chiral angle $\theta$ and a chiral vector $\boldsymbol{C}_{\mathrm{h}}$, given by Eq. (1), where $\boldsymbol{a}_1$ and $\boldsymbol{a}_2$ are unit vectors in a 2D graphene lattice and $n$ and $m$ are integers.

$$\boldsymbol{C}_{\mathrm{h}} = n\boldsymbol{a}_1 + m\boldsymbol{a}_2 \tag{1}$$

The vector $\boldsymbol{C}_{\mathrm{h}}$ connects two crystallographically equivalent sites on a 2D graphene sheet and the chiral angle is the angle it makes with respect to the zigzag direction, Figure 8.8. A tube is formed by rolling up the graphene sheet in such a way that the two points connected by the chiral vector coincide. A number of possible chiral

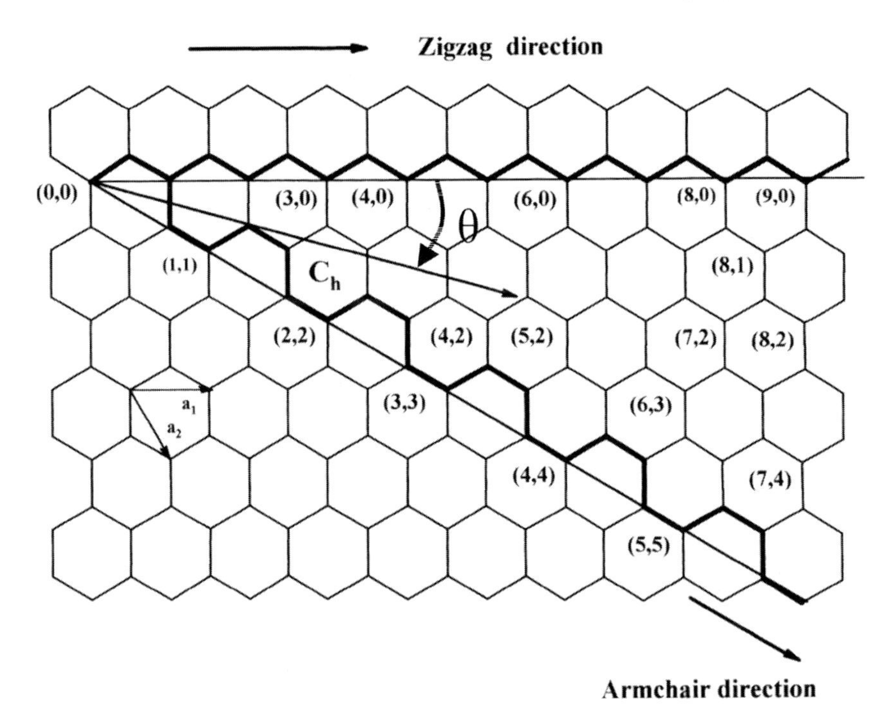


**Fig. 8.8.** A 2D graphene sheet showing chiral vector $\mathbf{C}_h$ and chiral angle.

vectors can be specified by Eq. (1) in terms of pairs of integers $(n, m)$. Many such pairs are shown in Figure 8.8 and each pair $(n, m)$ defines a different way of rolling up the graphene sheet to form a carbon nanotube of certain chirality. The limiting cases are $n \neq 0$, $m = 0$ (zigzag tube), and $n = m \neq 0$, (armchair tube). For a carbon nanotube defined by the index $(n, m)$, the diameter, $d$, and the chiral angle, $\theta$, are given by following Eq. (2) and (3) where $a = 1.42(3)^{1/2}$ and $0 \leq \theta \leq 30°$.

$$d = a(m^2 + mn + n^2)^{1/2}/\pi \tag{2}$$

$$\theta = \arctan(-(3)^{1/2}m)/2n + m) \tag{3}$$

Armchair SWNTs are metals; those with $n - m = 3k$, where $k$ is a nonzero integer, are semiconductors with a tiny band gap; and all others are semiconductors with a band gap that inversely depends on the nanotube diameter [8b, 17, 18]. The MWNTs consist of capped concentric cylinders separated by 3.45 Å (which is slightly larger than the interlayer spacing in graphite) because the number of carbon atoms increases as we go from an inner cylinder to an outer cylinder and it is not possible to maintain perfect ABAB.... stacking as in graphite. Thus, an interlayer spacing close to that in turbostratic graphite is observed in MWNTs. In addition to pentagons and hexagons, carbon nanotubes can also have heptagons. Pentagons impart a positive curvature whereas heptagons give rise to a negative curvature to the otherwise flat graphene sheet made of hexagons. Thus, nanotubes with pentagons and heptagons will have unusual curvatures and shapes. Bent nanotubes arising from the presence of pentagons and heptagons on opposite sides of the tube have been observed [66]. Based on force field calculations, Tersoff and Ruoff [67] suggest that the nanotubes will form cylindrical bundles in a crystal and large tubes will be hexagonal to maximize the van der Waals contact between the tubes. Simulation studies indicate radial p-orbital character and large pyramidalization angles at the sites of local deformation [68].

An interesting observation with SWNTs is the presence of rings in the electron microscopic images [69]. The rings are formed during the ultrasound treatment in an acidic medium, followed by settling of nanotube dispersions on the substrate. Similar ring morphologies are observed in atomic force microscope (AFM) and SEM studies of catalytically produced multi-walled nanotubes [70a]. Huang et al. [70b] have observed "crop circles" of aligned nanotubes in a direction normal to the substrate surface, on pyrolysing Fe(II)phthalocyanine. Yet another important nanostructure discovered recently is the presence of encapsulated fullerenes inside SWNTs, as observed by high resolution TEM [71, 72]. Fullerenes were observed inside the nanotubes after annealing laser-synthesized SWNTs. We present a typical TEM image showing the presence of $C_{60}$ in SWNTs in Figure 8.6 (inset). Analysis of the size distribution of the encapsulated fullerenes inside arc-produced SWNTs shows that there are indeed significant quantities of fullerenes in the size range $C_{36}$–$C_{120}$ in the nanotube capillaries [73].

X-ray diffraction (XRD) measurements have been employed to characterize carbon nanotubes [74, 75]. The XRD patterns of nanotubes show only the ($hk0$) and (001) reflections but no general ($hkl$) reflections. This is the case in turbostratically modified graphites [76]. Warren [77] has suggested special methods for the analysis of the ($hk0$) reflections and such studies support the electron microscopy data in showing that structural correlations exist along the direction perpendicular to the carbon nanotube axis as well as within each individual tube, but not in any combination of these. The correlation lengths obtained from the analysis of the XRD patterns are in the same regime as that from microscopy.

Ebbesen et al. [78] find that the nanotube bundles are self-similar in the sense that large cylindrical bundles comprise smaller ones and the smaller ones are made up of nanotubes and so on. STM has been used to probe the electronic structure of carbon nanotubes deposited on various substrates [81–83]. STM has also been used to probe $sp^3$ defect structures, closure of the tips and pentagon-induced changes in the electronic structure in carbon nanotubes [84]. Venema et al. [85] have obtained atomically resolved STM images of SWNTs, wherein the chirality of the nanotubes is unambiguously determined, which in turn influences the electronic property of the nanotubes. Raman spectroscopy has provided important insights into the structure of nanotubes. Jishi et al. [86] calculated the Raman-active phonon modes using a zone-folding method for a 2D graphene sheet, and demonstrated that there are 15 allowed Raman modes for each diameter of the tube. The frequency of the allowed mode depends on the tube diameter and the chiral angle, the number of modes being independent of the diameter. Hiura et al. [87] surprisingly found that the linewidth of phonon peaks in the Raman spectrum was narrow, of the order of 20 $cm^{-1}$. The Raman phonon frequency of nanotubes is softer than that of HOPG, probably due to the curvature of the nanotubes. Softening of phonon modes can be related to the larger *c*-axis lattice parameter in the nanotube as compared to graphite. Holden et al. [88] have examined the spectra of single-walled carbon nanotubes produced by using Co catalysts and have compared them with the predictions of Jishi et al. [86]. Studies on SWNTs by Rao et al. [89] reveal many of the characteristic normal modes of an armchair (n,n) carbon nanotubes and also show a diameter-selective resonance behavior. The resonance results from the one-dimensional quantum confinement of electrons in the nanotubes. Kasuya et al. [90] have provided the first evidence for a diameter-dependent dispersion arising from the cylindrical symmetry of the nanotubes. They carried out Raman scattering studies on SWNTs with mean diameters of 1.1, 1.3 and 2 nm and found size-dependent multiple splitting of the optical phonon peak corresponding to the $E_{2g}$ mode of graphite. In Figure 8.9, we show typical Raman spectra of laser-synthesized SWNTs. Assignment of bands due to nanotubes of different diameters is indicated in the figure. Polarized Raman

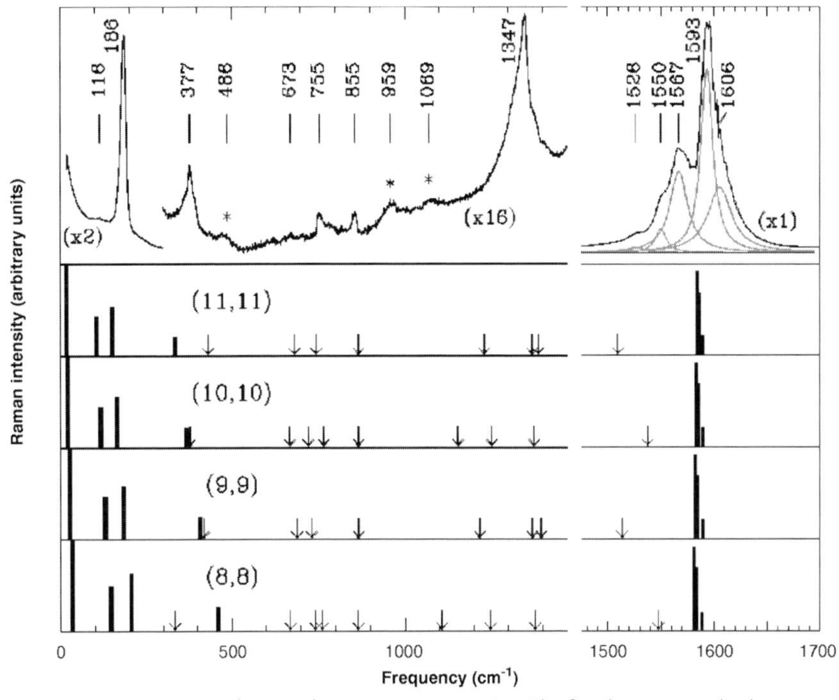


**Fig. 8.9.** Raman spectra showing the diameter-dependent scattering in SWNTs. An asterisk in the spectrum refers to a feature assigned tentatively to second-order Raman scattering. The four bottom panels show calculated Raman spectra for armchair (*n*,*n*) nanotubes ($n = 8$–11). Reproduced from ref. [89], with permission.

studies on aligned multi-walled carbon nanotubes show a strong dependence of the graphite-like G-band and disorder induced D-band on the polarization geometry [91].

Pressure-induced phase transformations under static and dynamic loading between the many allotropes of carbon like diamond, graphite, $C_{60}$ and $C_{70}$ and their polymeric and amorphous forms are of academic and practical importance. Pressure-effects on SWNT bundles have been probed by Raman spectroscopy up to a maximum pressure of 25.9 GPa (1 GPa $= 10^9$ N m$^{-2}$) in a diamond anvil cell [92, 93]. The spectra arising from the radial and tangential modes at 0.1 GPa are similar to those reported earlier at atmospheric pressure [89]. The two dominant radial bands in the spectrum of the sample recorded at 0.1 GPa were at 172 and 182 cm$^{-1}$. For an isolated SWNT, the calculated frequencies of the radial mode $\omega_R[\mathrm{cm}^{-1}]$ for a tube of diameter $d$[nm] fit to $\omega_R = 223.75/d$, irrespective of the nature of the tube [94]. This gives $\omega_R = 164$ cm$^{-1}$ for the (10,10) tube and 183 cm$^{-1}$ for the (9,9) tube. The inclusion of van der Waals interaction between the (9,9) tubes shifts the radial mode frequency from 171.8 cm$^{-1}$ (for an isolated tube) to 186.2 cm$^{-1}$. This blue shift of 14.4 cm$^{-1}$ is due to intertube interaction, and is independent of the tube diameter [95]. Accordingly, the empirical relation for the diameter dependence of the radial mode frequency in a SWNT bundle is given by, $\omega_R = 14.4 + 209.9/d$, which retains the $1/d$ dependence of $\omega_R$ and reproduces $\omega_R = 186.2$ cm$^{-1}$ for the (9,9) tube [96]. The tangential modes are assigned in terms of the irreducible representations of $D_{nh}$ ($D_{nd}$) for even $n$ (odd $n$), with 1531 cm$^{-1}$ as $E_{1g}$, 1553 and 1568 cm$^{-1}$ as $E_{2g}$, 1594 cm$^{-1}$ with unresolved doublet $A_{1g} + E_{1g}$ and 1606 cm$^{-1}$ with $E_{2g}$ symmetry [97, 98]. The intensities of the radial modes fall rapidly with increasing pressure, and were not discernible beyond 2.6 GPa, but the features are reversible. The intensities of the tangential modes also decrease with pressure. The modes at $\omega_T = 1568$ cm$^{-1}$ and 1594 cm$^{-1}$ show softening between ~10–16 GPa, beyond 16 GPa, the band position increases with pressure. Remarkably, when the pressure is reduced from the highest pressure of 25.9 GPa, the peak positions follow the same trend as in the increasing pressure run. Studies of SWNTs under high pressure confirm the potential of these materials as the strongest ever carbon nanofibers and also their remarkable resilience [96, 99].

### 8.2.3
### Mechanism of Formation

Several growth models are proposed for the carbon nanotubes prepared by the pyrolysis of hydrocarbons on metal surfaces. Baker and Harris [100] suggested a four-step mechanism. In the first step, the hydrocarbon decomposes on the metal surface to release hydrogen and carbon, which dissolves in the particle. The second step involves the diffusion of the carbon through the metal particle and its precipitation on the rear face to form the body of the filament. The supply of carbon onto the front face is faster than the diffusion through the bulk, causing an accumulation of carbon on the front face, which must be removed to prevent the physical blocking of the active surface. This is achieved by surface diffusion and the carbon forms a skin around the main filament body, in step three. In the fourth step, overcoating and deactivation of the catalyst and termination of tube growth takes place. Oberlin et al. [101] proposed a mechanism in which bulk diffusion is insignificant and carbon is entirely transported around the particle by surface diffusion. Dai et al. [49] proposed a mechanism wherein carbon forms a hemispherical graphene cap on the catalyst particle and the nanotubes grow from such a yarmulke. The diameter of the nanotube is controlled by the size of the catalytic particle, nanometer size particles yielding SWNTs. A crucial feature of this model is that it avoids dangling bonds at all stages of growth. SWNTs produced by arc vaporization may also be formed by the yarmulke mechanism.

A number of models have been proposed for the growth of MWNTs in the arc. Endo and Kroto [102], based on the observation of $C_2$ ejection from $C_{60}$ in mass spectrometry, suggest that tube formation processes are a consequence of the formation of fullerenes. Smalley [103], however, pointed out that only the growth of outer layers of multi-walled tubes would be permitted by such a mechanism. Iijima et al. [104] presented electron microscopy evidence for the open-ended growth of carbon nanotubes and suggested that the termination of incomplete layers of carbon seen on the tube surface may arise because of the extension and thickening of the nanotubes by the growth of graphite islands on the surfaces of existing tubes. The nucleation of pentagons and heptagons on the open tube ends results in a change in the direction of the growing tube and some novel morphologies, including one where the tube turns around 180° during the growth, have been observed. The growth is self-similar and fractal-like with the inner tubules telescoping out of the larger ones, with logarithmic scaling of the size.

Isotope scrambling experiments of Ebbesen et al. [105] show that under the conditions of fullerene formation, the plasma has vaporized atoms of carbon. Based on tube morphologies, a mechanism similar to that of Saito et al. [106], wherein the carbonaceous material reaching the cathode anneals into polyhedral particles, was suggested. Given the right conditions, the tip might open and continue to grow. Such a growth could occur from the outside inwards. Ebbesen et al. [105] suggest the possibility of tubes forming directly from the closing of a large graphene sheet. Such a suggestion gains credence from the simulations of Robertson et al. [107] who have examined the curling and closure of small graphitic ribbons. Amelinckx et al. [108a] introduce the concept of a spatial velocity hodograph to describe the extrusion of a carbon tubule from a catalytic particle. The model is consistent with the observed tubule shapes and explains how spontaneous plastic deformation of the tubule can occur. Amelinckx et al. [108b] propose a model in which the graphene sheets can form both concentric cylinders and scroll-type structures. They also propose that the nanotubes nucleate from a large fullerene type dome. Maiti et al. [109] propose a model wherein nanometer-sized protrusions on the metal particle surface lead to the nucleation of SWNTs.

TEM examination of the carbonaceous products obtained from the pyrolysis of hydrocarbons and organometallic precursors indicates that the size of the catalyst particle plays an important role (see Figure 8.10).

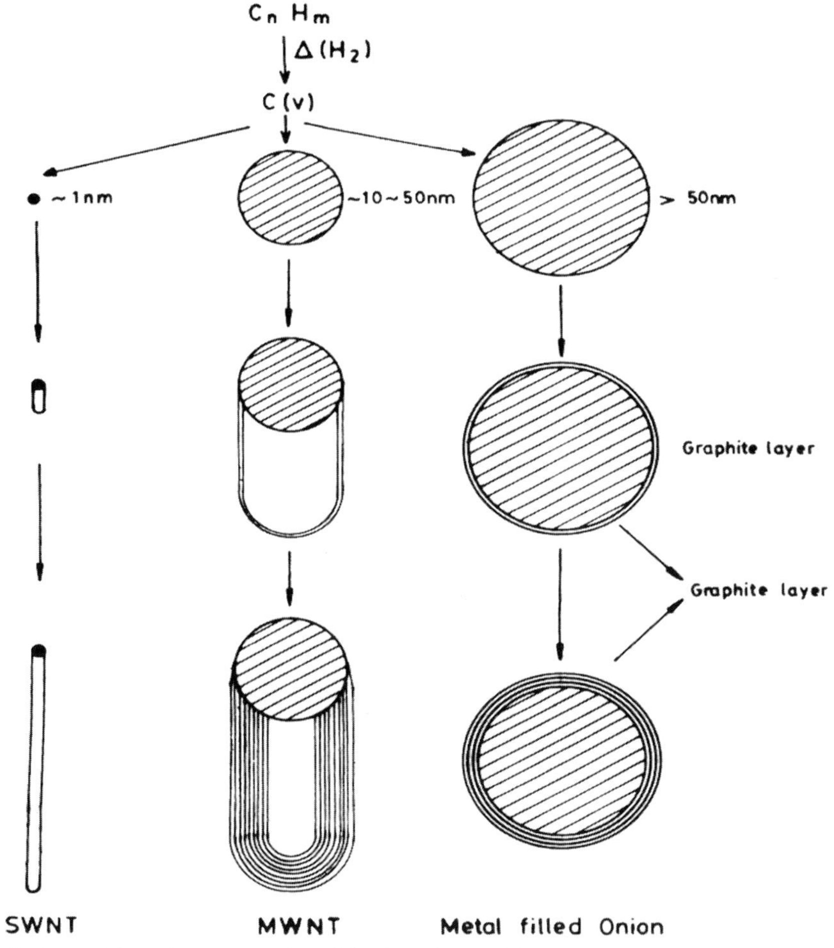


**Fig. 8.10.** Schematic representation of the dependence of the carbon nanostructure obtained by hydrocarbon pyrolysis on the size of the metal nanoparticles. Reproduced from ref. [20a], with permission.

## 8.2.4
## Chemically Modified Carbon Nanotubes

### 8.2.4.1 Doping with Boron and Nitrogen

Since the discovery of carbon nanotubes, there has been immense interest in substituting carbon with other elements. Accordingly, boron–carbon (B–C), boron–carbon–nitrogen (B–C–N) and carbon–nitrogen (C–N) nanotubes have been prepared and characterized. Boron-substitution in the carbon nanotubes gives rise to p-type doping and nitrogen-doped carbon nanotubes correspond to n-type doping. Novel electron transport properties are expected of such doped nanotubes [110].

Boron-doped carbon nanotubes have been synthesized by carrying out the pyrolysis of mixtures of acetylene and diborane and characterized by employing microscopic and spectroscopic techniques [111]. The average composition of these nanotubes is $C_{35}B$. B–C–N nanotubes have been prepared by striking an arc between a graphite anode filled with B–N and a pure graphite cathode in a helium atmosphere [112]. B–C–N nanotubes have also been obtained by laser ablation of a composite target containing B–N, carbon, Ni and Co at 1000 °C under flowing nitrogen [113]. Terrones et al. [114] pyrolyzed the addition compound, $CH_3CN{:}BCl_3$, over Co powder at 1000 °C to obtain B–C–N nanotubes. B–C–N as well as C–N nanotubes were prepared by Sen et al. [115] by the pyrolysis of appropriate precursors. Pyrolysis of aza-aromatics such as pyridine over Co catalysts gives C–N nanotubes ($C_{33}N$ on average). Pyrolysis of the 1:1 addition compound of $BH_3$ with $(CH_3)_3N$ produces B–C–N nanotubes. Typical TEM images of a few nanotubes are shown in Figure 8.11, exhibiting bamboo-shaped, nested cone-shaped cross sections as well as unusual morphology, including coiled nanotubes. The composition of the B–C–N nanotubes varies with the preparation. Furthermore, considerable variability exists in the composition in any given batch of B–C, B–C–N or C–N nanotubes obtained by the pyrolysis of precursors.

Aligned carbon nanotubes are considered ideal for field emission properties. The availability of aligned bundles of doped carbon nanotubes may provide further improvement in field-emission characteristics. Nath et al. [116a] have obtained aligned carbon–nitrogen nanotube bundles by the pyrolysis of pyridine over sol–gel derived iron/silica or cobalt/silica substrates. Employing anodic alumina, Sung et al. [116b] synthesized C–N nanotubes by electron cyclotron resonance CVD, using $C_2H_2$ and $N_2$. Suenaga et al. [117] carried out CVD of Ni-phthalocyanine to obtain aligned C–N nanotubes.

Goldberg et al. [118] have employed a method wherein SWNTs were thermally treated with boron trioxide in a nitrogen atmosphere to obtain boron or boron and nitrogen doped SWNTs. EELS analysis showed the boron content to be ~10 at.% in B–C nanotubes. An interesting aspect of the B–C–N nanostructures is that phase separation occurs in which the $BC_3$ islands segregate in the graphene sheets. Tunnelling conductance measurements of doped nanotubes demonstrate acceptor-like states, near the Fermi level, arising out of the $BC_3$ islands [119]. Efsarjani et al. [120] propose that a nanotube with donor atoms at one side and acceptor atoms on the other can function as a nano-diode. An experimental situation near to this effect is the observation of rectification in a SWNT [121]. The presence of an impurity in one of the segments of a SWNT influences its nonlinear transport behavior.

#### 8.2.4.2 **Opening, Filling and Functionalizing Nanotubes**

Multi-walled nanotubes are generally closed at either end, the closure being made possible by the presence of five-membered rings. MWNTs can be uncapped by oxidation with carbon dioxide or oxygen at elevated temperatures [74, 122, 123]. High yields of uncapped MWNTs are, however, obtained by boiling them in concentrated $HNO_3$. Filling the opened nanotubes with metals has been accomplished. The well-known method [124] involves the treatment of the nanotubes with boiling

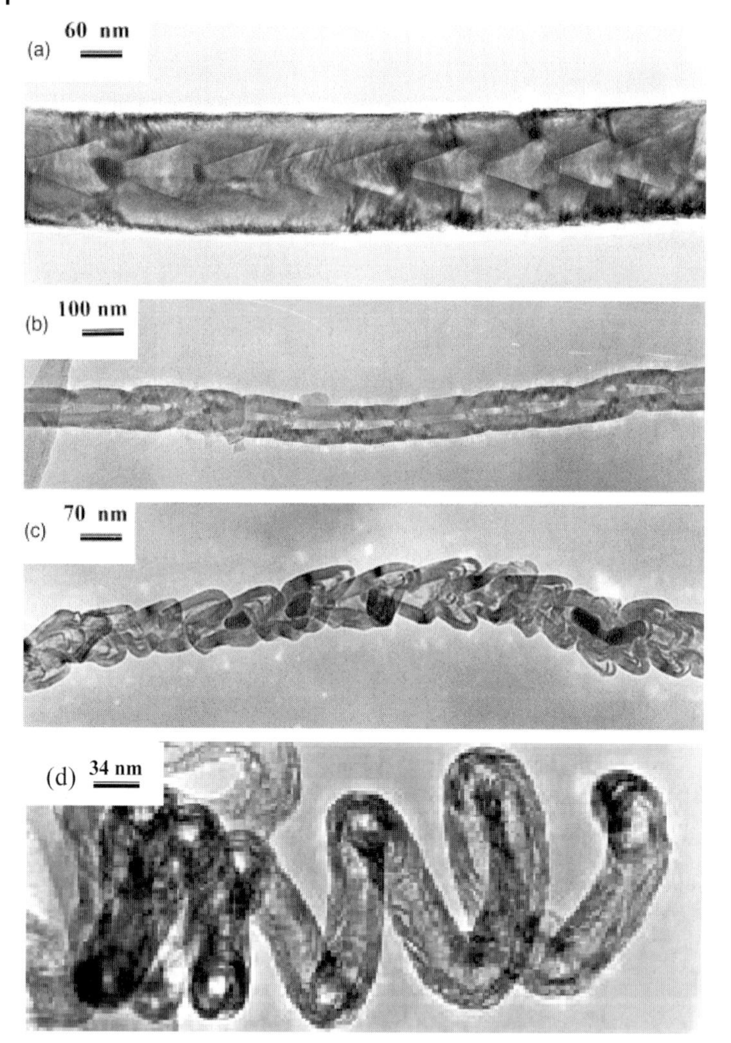


**Fig. 8.11.** TEM images of carbon nanotubes obtained by the pyrolysis of pyridine (flow rate = 30 $cm^3$ $min^{-1}$) over $Fe/SiO_2$ substrates at 900 °C for 1.5 h under Ar (120 $cm^3$ $min^{-1}$) flow. The nanotubes show (a) bamboo shape, (b) nested cone, and (c) other unusual morphologies. (d) The TEM image of a coiled nanotube obtained by pyridine pyrolysis over Co. Reproduced from ref. [116a], with permission.

$HNO_3$ in the presence of metal salts such as $Ni(NO_3)_2$. The nanotubes are opened by $HNO_3$ and filled by the metal salt. On drying and calcination, the metal salt transforms to the metal oxide and reduction of the encapsulated oxide in hydrogen at around 400 °C gives rise to the metal inside the nanotubes. MWNTs have been opened by using a variety of oxidants [125–127] and the opened nanotubes have been filled with Ag, Au, Pd, or Pt by different chemical means, rather than by reduction with hydrogen at high temperatures [127]. By employing in situ techniques in a TEM, Bower et al. [128] observed alkali metal intercalation into the SWNTs. Sealed tube reactions of SWNTs and metal salts are also known to yield metal intercalated SWNTs [129]. In an effort to realize the conversion of $sp^2$ carbon of nanotubes into $sp^3$, Hsu et al. [130] treated potassium intercalated MWNTs with $CCl_4$ hydrothermally, and obtained crystallization of KCl inside the nanotubes and within the tube walls. Possible ways of closing the nanotubes, opened by oxidants, have been examined [127]. Besides opening, filling and closing nanotubes, highly functionalized MWNTs have been prepared by treatment with acids [127, 131]. SWNTs are readily opened by mild treatment with acids and filled with metals [45, 132]. Acid-treated nanotube surfaces can be decorated by nanoparticles of metals such as Au, Ag or Pt [45, 133].

Chen et al. [134] have derivatized SWNT fragments with halogen and amine moieties in order to dissolve them in organic solvents. Doping has been carried out to modify their electronic properties in the solution phase. Fluorination of SWNTs has been carried out by Mickelson et al. [135]. Fluorinated nanotubes can be solvated in alcohol media and precipitated back by reaction with hydrazine. STM studies of fluorinated SWNTs reveal an interesting banded structure followed by atomically resolved regions, indicating sidewall functionalization [136]. Starting from fluorinated SWNTs, Boul et al. [137] have carried out alkylation by reaction with alkylmagnesium bromides or alkylithium. Individual SWNTs have been deposited controllably on chemically functionalized nanolithographic templates [138].

## 8.2.5 Electronic Structure, Properties and Devices

### 8.2.5.1 Electronic Structure and Properties

As with the fullerenes, the curvature of the graphitic sheets in the nanotubes would be expected to influence the electronic structure. The electronic properties of perfect MWNTs are rather similar to those of perfect SWNTs, because the coupling between the cylinders is weak in MWNTs. Calculations show that nanotubes may be as good conductors as copper, although a combination of the degree of helicity and the number of six-membered rings per turn around the tube can tune the electronic properties in the metal–semiconductor range [139–141]. Low-temperature STM and scanning tunneling spectroscopy (STS) studies of SWNTs reveal the atomically resolved images of the graphene cylinders and their size-specific transport properties [142, 143], in agreement with theoretical predictions. Collins et al. [144] used STM to explore local electrical characteristics of SWNTs. Well-defined positions where the current changes abruptly from a graphite-like response to a highly nonlinear response, were found, including near-perfect rectification. STM studies conducted in our laboratory in ultra-high vacuum (UHV) show a variable conductivity and gap along the length of the nanotubes. Because of the nearly one-dimensional electronic structure, electron transport in metallic SWNTs and MWNTs occurs ballistically over long nanotube lengths, enabling them to carry high currents with essentially no heating [145, 146]. Phonons also propagate easily along the nanotube. The measured room temperature thermal conductivity for an individual MWNT (>3000 W $m^{-1}$ $K^{-1}$) is greater than that of natural diamond and the basal plane of graphite (both 2000 W $m^{-1}$ $K^{-1}$) [147]. Superconductivity has also been observed, but only at low temperatures, with transition temperatures of ~0.55 K for 1.4 nm diameter SWNTs [148] and ~5 K for 0.5 nm diameter SWNTs grown in zeolites [149].

#### 8.2.5.2 Electronic and Electrochemical Devices

**Electrochemical Devices** Because of the high electrochemically accessible surface area of porous nanotube arrays, combined with their high electronic conductivity and useful mechanical properties, these materials are attractive as electrodes for devices that use electrochemical double-layer charge injection. Examples include "supercapacitors," which have giant capacitances in comparison with those of ordinary dielectric-based capacitors, and electromechanical actuators that may eventually be used in robots. Like ordinary capacitors, carbon nanotube supercapacitors [150–152] and electromechanical actuators [153] typically comprise two electrodes separated by an electronically insulating material, which is ionically conducting in electrochemical devices. Because this separation is about a nanometer for nanotubes, as compared with the micrometer or larger separations in ordinary dielectric capacitors, very large capacitances result from the high nanotube surface area accessible to the electrolyte. These capacitances (typically between ~15 and ~200 F $g^{-1}$, depending on the surface area of the nanotube array) result in large amounts of charge injection when only a few volts are applied [150–153]. This charge injection can be used for energy storage in nanotube supercapacitors and to provide electrode expansions and contractions that can do mechanical work in electromechanical actuators. The capacitances (180 and 102 F $g^{-1}$ for SWNT and MWNT electrodes, respectively) and power densities (20 kW $kg^{-1}$ at energy densities of ~7 W h $kg^{-1}$ for SWNT electrodes) [150, 151] are attractive, especially because performance can likely be improved by replacing SWNT bundles and MWNTs with unbundled SWNTs. An extraordinarily short discharge time of 7 ms was reported [152] for 10 MWNT capacitors connected in series, which operated at up to 10 V. Nanotube electromechanical actuators function at a few volts, compared with the ~100 V used for piezoelectric stacks and the ≥1000 V used for electrostrictive actuators. Nanotube actuators have been operated at temperatures up to 350 °C. Operation above 1000 °C should be possible, on the basis of SWNT thermal stability and industrial carbon electrode electrochemical application above this temperature [20b]. From observed nanotube actuator strains that can exceed 1%, order-of-magnitude advantages over commercial actuators in work per cycle and stress generation capabilities are predicted if the mechanical properties of nanotube sheets can be increased to close to the inherent mechanical properties of the individual nanotubes [20b]. The maximum observed isometric actuator stress of SWNT actuators is presently 26 MPa [20b]. This is >10 times the stress initially reported for these actuators and ~100 times that of the stress generation capability of natural muscle, and it approaches the stress generation capability of high-modulus commercial ferroelectrics (~40 MPa). However, the ability to generate stress is still >100 times lower than that predicted for nanotube fibers with the modulus of the individual SWNTs. The success of actuator technology based on carbon nanotubes will depend on improvements in the mechanical properties of nanotube sheets and fibers with a high surface area by increasing nanotube alignment and the binding between nanotubes. The use of nanotubes as electrodes in lithium batteries is a possibility because of the high reversible component of storage capacity at high discharge rates. The maximum reported reversible capacity is 1000 mA h $g^{-1}$ for SWNTs that are mechanically milled in order to enable the filling of nanotube cores, as compared to 372 mA h $g^{-1}$ for graphite [154] and 708 mA h $g^{-1}$ for ball-milled graphite [155].

**Nanometer-Sized Electronic Devices** The possible use of carbon nanotubes in nanoelectronics has aroused considerable interest. Dramatic recent advances have fueled speculation that nanotubes (SWNTs) will be useful for downsizing circuit dimensions. Because of their unique electronic properties, SWNTs can be interfaced with other materials to form novel heterostructures [156]. The simplest device one can imagine with carbon nanotubes is that involving a bend or a kink, arising from the presence of a diametrically opposite pentagon–heptagon pair. The resultant junction connects two nanotubes of different chirality and hence of different electronic structure, leading to the realization of an intramolecular device. Such a device in SWNTs is found to behave like a diode rectifier [157]. Silicon nanowire–carbon nanotube heterojunctions do indeed exhibit a rectification behavior [158].

The current-induced electromigration causes conventional metal wire interconnects to fail when the wire diameter becomes too small. The covalently bonded structure of carbon nanotubes militates against similar breakdown of nanotube wires and, because of ballistic transport, the intrinsic resistance of the nanotube should essentially vanish. Experimental results show that metallic SWNTs can carry up to $10^9$ A $cm^{-2}$, whereas the maximum current densities for normal metals are $10^5$ A $cm^{-2}$ [145, 159]. Unfortunately, the ballistic current carrying capability is less useful for presently envisioned applications because of necessarily large contact resistances. An electronic circuit involving electrical leads to and from a SWNT will have a resistance of at least $h/4e^2$ or 6.5 kΩ, where $h$ is Planck's constant and $e$ is the charge of an electron [160]. Contacting all layers in a MWNT could reduce this contact resistance, but it cannot be totally eliminated. In nanotube field effect transistors (NT-FETs), gating has been achieved by applying a voltage to a submerged gate beneath a SWNT (Figure 8.12(a) and (b)), which was contacted at opposite nanotube ends by metal source and drain leads [161]. A typical nanoelectronic device of NT-FET consists of a semiconducting nanotube, which is on top of an insulating aluminum oxide layer, connected at both ends to a gold electrode. The nanotube is switched by applying a potential to the aluminum gate under the nanotube and aluminum oxide. The transistors were fabricated by lithographically applying electrodes to the nanotubes that were either randomly distributed on a silicon substrate or positioned on the substrate with an atomic

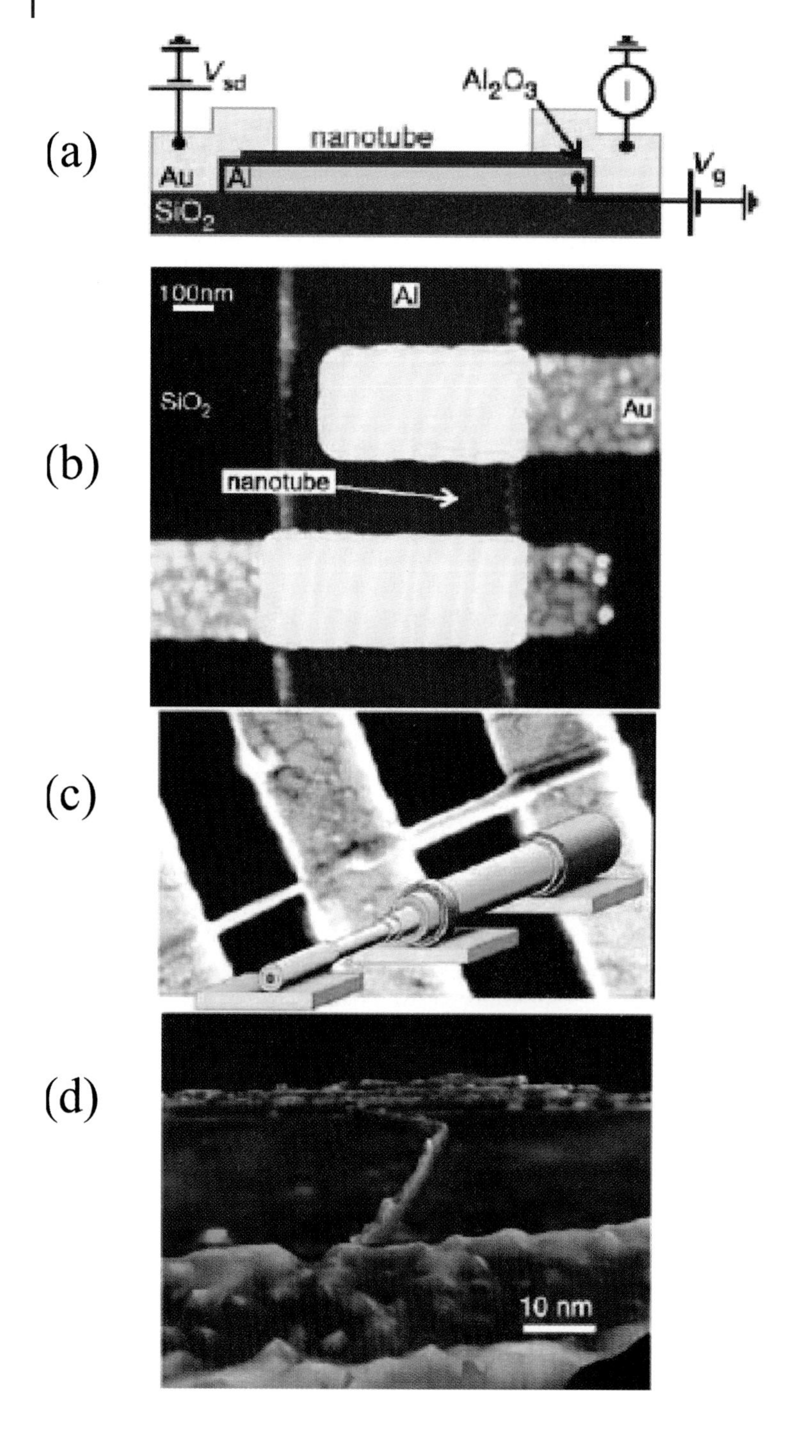
(a)
$V_{sd}$
$Al_2O_3$
I
nanotube
Au
Al
$V_g$
$SiO_2$
(b)
100nm
Al
$SiO_2$
Au
nanotube
(c)
(d)
10 nm

force microscope [162, 163]. A transistor assembled in this way may or may not work, depending on whether the chosen nanotube is semiconducting or metallic, over which the operator generally has no control. It is possible to selectively peel outer layers from a MWNT (Figure 8.12(c)) until a nanotube cylinder with the desired electronic properties is obtained [164], but this process is not yet very reliable and is probably unsuitable for mass production. Research toward nanoscopic NT-FETs aims to replace the source-drain channel structure with a nanotube. A more radical approach is to construct entire electronic circuits from interconnected nanotubes. Because the electronic properties depend on helicity, it should be possible to produce a diode, for example, by grafting a metallic nanotube to a semiconducting nanotube. Such a device has been demonstrated. The bihelical nanotube was not, however, rationally produced; rather, it was fortuitously recognized, in a normal nanotube sample, by its kinked structure (Figure 8.12(d)), which was caused by the helicity change [157]. The development of rational synthesis routes to multiply branched and interconnected low-defect nanotubes with targeted helicity would be a revolutionary advance for nanoelectronics.

With crossed SWNTs, three- and four-terminal electronic devices have been made [165], as well as a nonvolatile memory that functions like an electromechanical relay [166]. For such applications it is important to be able to connect the nanotubes of different diameters and chirality [167]. Complex three-point nanotube junctions have been proposed as the building blocks of nanoelectronics and in this regard Y- and T-junctions have been considered as prototypes [168, 169]. The Y- and T-junctions appear to defy the conventional models in favor of an equal number of five- and seven-membered rings to create nanotube junctions. Instead, the Y- and T-junctions can be created with an equal number of five- and eight-membered rings [169]. However, junctions consisting of crossed nanotubes have been fabricated to study their transport characteristics [165, 170]. Y-junction nanotubes have been produced by using Y-shaped nanochannel alumina as a template [171]. By carrying out a simple pyrolysis of a mixture of nickelocene with thiophene, Y-junction carbon nanotubes have been synthesized recently in good quantities [172]. A TEM image of such Y-junction nanotubes is shown in Figure 8.13(a) and (c). A TEM image revealing the presence of several Y-junction carbon nanotubes is shown in Figure 8.13(b). STM and STS studies of Y-junction carbon nanotubes show interesting diode-like device characteristics at the junctions. The $I-V$ plot at the junction is asymmetric with respect to bias polarity, unlike that along the arm. Such asymmetry is characteristic of a junction diode and this in

← **Fig. 8.12.** Nanoelectronic devices: (a) Schematic diagram [163] for a carbon NT-FET. $V_{sd}$, source-drain voltage; $V_g$, gate voltage. Reproduced from ref. [163], with permission. (b) Scanning tunneling microscope (STM) picture of a SWNT field-effect transistor made using the design of (a); the aluminum strip is overcoated with aluminum oxide. (c) Image and overlaying schematic representation for the effect of electrical pulses in removing successive layers of a MWNT, so that layers having desired transport properties for devices can be revealed. Reproduced from ref. [164], with permission. (d) STM image of a nanotube having regions of different helicity on opposite sides of a kink, which functions as a diode; one side of the kink is metallic, and the opposite side is semiconducting. The indicated scale bar is approximate. Reproduced from ref. [157], with permission.

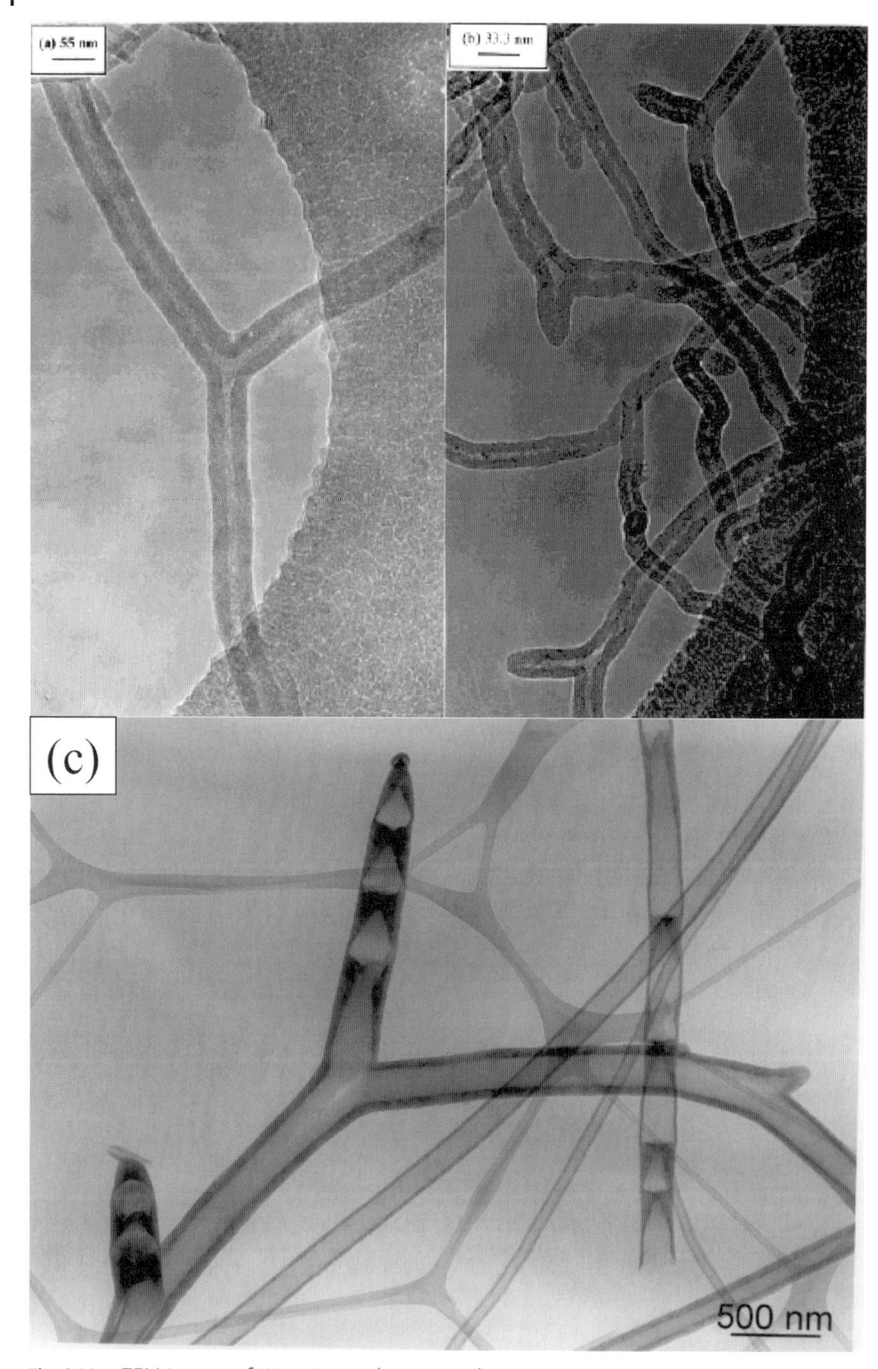


**Fig. 8.13.** TEM images of Y-junction carbon nanotubes obtained by the pyrolysis of nickelocene and thiophene at 1000 °C. Reproduced from ref. [172], with permission.

turn indicates the existence of intramolecular junctions in the carbon nanotubes. The spectrum of interesting findings discussed above opens up the possibility of assembling carbon nanotubes, possessing such novel device-like properties [157, 161, 162, 172] into multi-functional circuits, and ultimately towards the realization of a carbon-nanotube-based computer chip. Rueckes et al. [166] have recently described the concept of carbon-nanotube-based nonvolatile random access memory for molecular computing. The viability of the concept has also been demonstrated. Integrated nanotube devices involving two nanotube transistors have been reported [163, 166], providing visions of large-scale integration. Patterned growth of SWNTs on a 4 in (10 cm) silicon wafer [173] may prove an important step toward integrated nanotube electronics. IBM expects that nanotube electronics will be realized in about a decade [174].

**Sensors and Probes** Possible chemical sensor applications of nonmetallic nanotubes are interesting, because nanotube electronic transport and thermopower (voltages between junctions caused by interjunction temperature differences) are very sensitive to substances that affect the amount of injected charge [175, 176]. The main advantages are the minute size of the nanotube sensing element and the correspondingly small amount of material required for a response. Major challenges remain, however, in making devices that differentiate between absorbed species in complex mixtures and provide rapid forward and reverse responses. Nishijima et al. [177] have developed such a novel microprocess incorporated in a SEM, to attach individual nanotubes to scanning probe microscope tips, which are later used as probes to image biological and industrial specimens. Carbon nanotube scanning probe tips for atomic probe microscopes are commercially available. The cylindrical shape and small tube diameter enable imaging in narrow, deep crevices and improve resolution in comparison to conventional nanoprobes, especially for high sample feature heights [178, 179]. Covalently modifying the nanotube tips, such as by adding biologically responsive ligands, enables the mapping of chemical and biological functions [180]. Nanoscopic tweezers have been made by Kim and Lieber that are driven by the electrostatic interaction between two nanotubes on a probe tip [181]. They attached carbon nanotubes to electrodes fabricated on pulled glass micropipettes. Voltages are applied to the electrodes to achieve closing and opening of the free ends of the nanotubes, to facilitate the grabbing and manipulation of submicron clusters and nanowires. They may be used as nanoprobes for assembly. These uses may not have the business impact of other applications, but they increase the value of measurement systems for characterization and manipulation on the nanometer scale. Arie et al. [182] use $Ni_3C$-filled MWNTs as probes in a magnetic force microscope (MFM) and to image magnetic recording media. The resolution, however, needs improvement through the optimization of ferromagnetic particle size, trace height etc. Karl and Tomanek [183] propose a molecular pump based on carbon nanotubes for the transport of atoms. Using SWNTs attached to an AFM cantilever as the probe and atomically flat titanium surfaces on an $\alpha$-$Al_2O_3$ substrate, Cooper et al. [184] demonstrate an areal data storage density of the order of terabits per square inch. This method

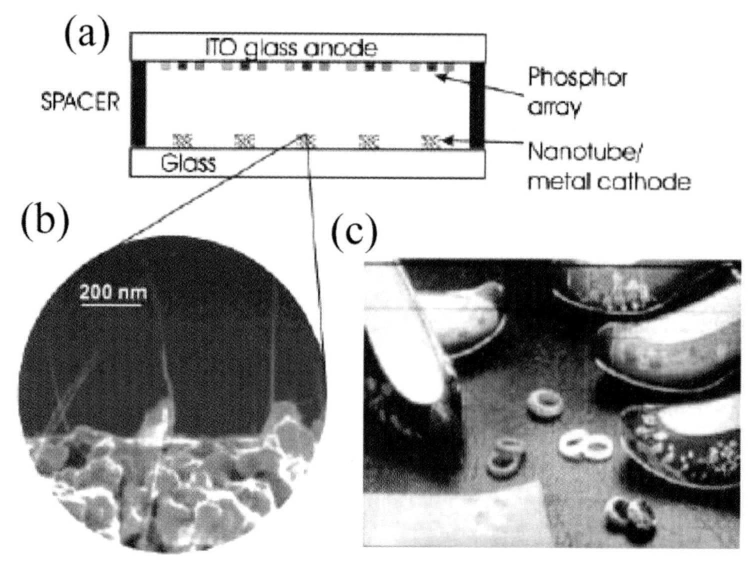


**Fig. 8.14.** (a) Schematic illustration of a flat panel display based on carbon nanotubes. ITO, indium tin oxide. (b) SEM image of an electron emitter for a display, showing well-separated SWNT bundles protruding from the supporting metal base. (c) Photograph of a 5 in (13 cm) nanotube field emission display made by Samsung. Reproduced from ref. [187], with permission.

employs SWNT based lithography which offers sub-10 nm nanofabrication capabilities. In contrast to conventional electronic devices operating on the basis of charge transport, spin-electronic devices operate upon the concept of spin transport.

**Field Emission** Industrial and academic research activity on electronic devices has focused principally on using SWNTs and MWNTs as field emission electron sources [185, 186] for flat panel displays [187], lamps [188], gas discharge tubes providing surge protection [189], and X-rays [190]. A potential applied between a carbon nanotube-coated surface and an anode produces high local fields, as a result of the small radius of the nanofiber tip and the length of the nanofiber. These local fields cause electrons to tunnel from the nanotube tip into the vacuum. Electric fields direct the field-emitted electrons toward the anode, where a phosphor produces light for the flat panel display application (Figure 8.14). However, the complete picture is not nearly so simple. Unlike for ordinary bulk metals, nanotube tip electron emission arises from discrete energy states, rather than continuous electronic bands [191]. Also, the emission behavior depends critically on the nanotube tip structure: Enhanced emission results from opening SWNT [186] or MWNT tips [188]. Nanotube field-emitting surfaces are relatively easy to manufacture by screen-printing nanotube pastes and do not deteriorate in moderate vacuum ($10^{-8}$ torr). These are advantages over tungsten and molybdenum tip arrays, which require a vacuum of $10^{-10}$ torr and are more difficult to fabricate [192]. Nanotubes provide stable emission, long lifetimes, and low emission threshold potentials [185, 188]. Current densities as high as 4 A $cm^{-2}$ have been obtained, compared with the 10 mA $cm^{-2}$ needed for flat panel field emission displays and the >0.5 A $cm^{-2}$ required for microwave power amplifier tubes [193]. Flat panel displays are one of the more lucrative nanotube applications being developed by industry. However, they are also technically the most complex, requiring concurrent advances in electronic addressing circuitry, the development of low-voltage phosphors, methods for maintaining the required vacuum, spacers withstanding the high electric fields, and the elimination of faulty pixels. The advantages of nanotubes over liquid crystal displays are a low power consumption, high brightness, a wide viewing angle, a fast response rate, and a wide operating temperature range. Samsung has produced several generations of prototypes (Figure 8.14), including a 9 in (23 cm) red–blue–green color display that can reproduce moving images [187].

Aligned carbon nanotubes are considered to be ideal for the purpose because of the high packing density and hence their use as high brightness field emitters. MWNT-based field emission lighting devices have been built and their luminescence characteristics studied [194]. One such field emission lighting device is shown in Figure 8.15. Choi et al. [195a] have assembled a sealed 4.5 $in^2$ field-emission display device using vertically aligned SWNTs along with organic binders. The display in three primary colors has an emission current of 1.5 mA at 3 V $\mu m^{-1}$, with a brightness of 1800 Cd $m^{-2}$. Lee et al. [195b] have just shown that aligned nanotube bundles exhibit a high emission current density of around 2.9 mA $cm^{-2}$ at 3.7 V $\mu m^{-1}$. Lovall et al. [196] have investigated the emission properties of SWNT ropes by employing field ion microscopy. The field-emitted electron energy distribution (FEED) of SWNT field emitters shows a large density of states near the Fermi energy. Emission characteristics of CVD produced MWNTs as well as SWNTs have been examined by Groning et al. [197], who obtained as emission site density of 10000 emitters $cm^{-2}$ at fields around 4 V $\mu m^{-1}$. A work function of 5 eV was obtained with MWNTs, and a smaller value with the SWNTs.

The use of dense, quasi-aligned carbon nanotubes produced by the pyrolysis of ferrocene on a pointed tungsten tip exhibit high emission current densities with good performance characteristics [198]. In Figure 8.16(a) we show a typical $I$–$V$ plot for the carbon nanotube covered tungsten tip for currents ranging from 0.1 nA to 1 mA. The applied voltage was 4.3 kV for a total current of 1 μA and 16.5 kV for 1000 μA. The Fowler–Nordheim (F–N) plot shown in Figure 8.16(b) has two distinct regions. The behavior is metal-like in the low-field region, while it saturates at higher fields as the voltage is increased. We have obtained a field emission current density of 1.5 A $cm^{-2}$ at a field of 290 V $mm^{-1}$, a value considerably higher than that found with planar cathodes. Accordingly, the field enhancement factor calculated from the slope of the F–N plot in the low-field region is also large. The field emission micrographs reveal the lobe structure symmetries typical of carbon nanotube bundles. The emission current is remarkably stable over an operating period of more than 3 h for various current values in the 10–500 mA range. The relative fluctuations decrease with increasing current level, and the emitter can be

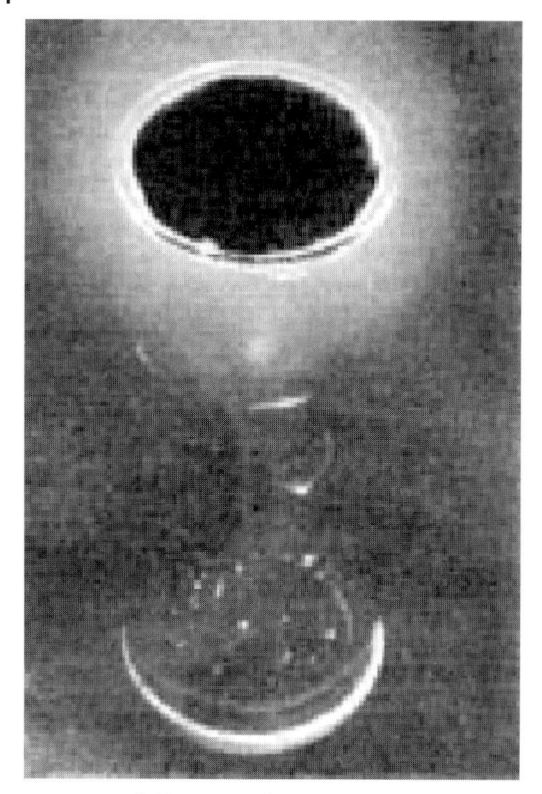

**Fig. 8.15.** A field-emission fluorescent display based on carbon nanotubes. The anode current is 200 μA at a voltage of 10 kV. Reproduced from ref. [194], with permission.

operated continuously at the high current levels for at least 3 h without any degradation in the current.

**Other Properties of Carbon Nanotubes** Optical limiting properties of the carbon nanotubes are considered important for applications involving high power lasers. Optical limiting behavior of visible nanosecond laser pulses in the SWNT suspensions occurs mainly due to nonlinear scattering [199]. Nanotube composites with conducting polymers seem to possess novel electrical and optical properties. Yoshino et al. [200] observed increased conductivity at relatively low nanotube concentrations in the composites and an enhancement of photoconductivity, implying possible use in optoelectronic devices. SWNT-coated molybdenum electrodes appear to be alternatives for gas discharge tubes in advanced telecom networks, due to their lower dc breakdown voltage and higher reliability [189]. Electromechanical actuators based on SWNT-sheets appear to generate stresses higher than natural muscles and higher strains than the ferroelectric counterparts [153]. This behavior of nanotubes may be useful in the direct conversion of electrical energy to mechanical energy, relevant to applications in robotics. Wood and Wagner [201]

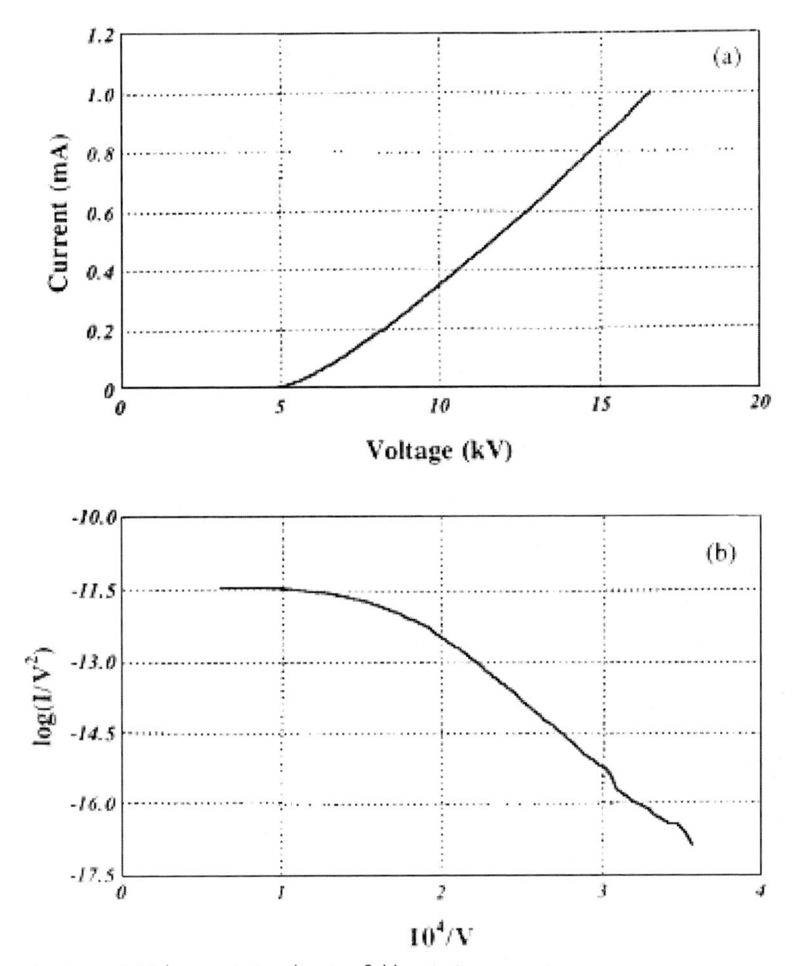


**Fig. 8.16.** *I–V* characteristics showing field emission currents in the range 0.1 nA to 1 mA. (b) Fowler–Nordheim plot corresponding to the data in (a). Reproduced from ref. [198], with permission.

observe a significant shift of the peaks in the Raman spectrum of SWNTs upon immersion in liquids. This allows the use of nanotubes as molecular sensors. Kong et al. [176a] have measured the sensitivity of the electron transport properties of SWNTs to gaseous molecules such as $NO_2$ or $NH_3$. The nanotubes exhibit faster response and a higher sensitivity than the available solid state sensors at room temperature; there is also good sensor reversibility. Ghosh et al. [176b] have reported that the flow of a liquid on single-walled carbon nanotube bundles induces a voltage/current in the sample along the direction of the flow. They found that the voltage so produced fits a logarithmic velocity dependence over nearly six decades of velocity. The magnitude of the voltage/current depends sensitively on the ionic conductivity and the polar nature of the liquid. Their measurements suggest that the dominant mechanism responsible for this highly nonlinear response should

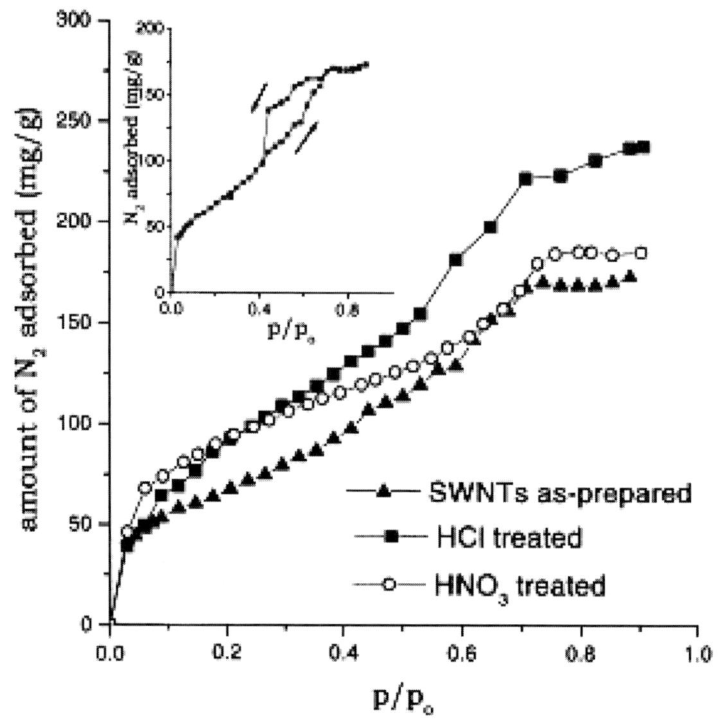


**Fig. 8.17.** $N_2$ adsorption isotherms of SWNTs at 77 K; as-prepared (▲), HCl treated (■), and $HNO_3$ treated (○). Inset: The hysteresis in the adsorption–desorption isotherms for SWNTs. Reproduced from ref. [202], with permission.

involve a direct forcing of the free charge carriers in the nanotubes by the fluctuating Coulombic field of the liquid flowing past it. They propose an explanation based on pulsating asymmetric ratchets. Their work highlights the device potential for nanotubes as sensitive flow sensors and for energy conversion.

SWNTs have nano-sized channels which can facilitate adsorption of liquids or gases. Eswaramoorthy et al. [202] have studied the adsorption properties of SWNTs with respect to methane, benzene and nitrogen. The studies indicate that SWNTs are good microporous materials with a total surface area above 400 $m^2$ $g^{-1}$. In Figure 8.17, we show typical adsorption isotherms for SWNTs. The unique hexagonal packing of the SWNTs in the bundles offers ideal channels, thus allowing the realization of one-dimensional (1D) adsorbates.

**Hydrogen Storage** Carbon nanotubes are considered to be good hosts for hydrogen storage (for example, for fuel cells that power electric vehicles or laptop computers), although there is some controversy about the magnitude of the hydrogen uptake [20a, 203]. However, experimental reports of high storage capacities are so controversial that it is impossible to assess the applications potential [203–207]. Reversible adsorption of molecular hydrogen in carbon nanotubes was first reported by Dillon et al. [208]. These workers measured the hydrogen adsorption capacity of the as-prepared SWNT bundles (0.1–0.2 wt.%) containing unidentified carbonaceous materials as well as large fractions of cobalt catalyst particles (20 wt.%). Composition (H/C) versus pressure isotherms at 80 K (−197 °C) of as-prepared SWNTs, sonicated SWNTs and a high surface area carbon (Saran) are reported by Ye et al. [207]. These workers find the hydrogen storage capacity in arc-derived SWNTs to be 8.25 wt.% at 80 K and ~4 MPa. A hydrogen storage capacity of 4.2 wt.% for SWNTs was reported by Liu et al. [209] at 27 °C and 10.1 MPa. The SWNTs used in this study had a large mean diameter of 1.85 nm. Moreover, 78.3% of the adsorbed hydrogen (3.3 wt.%) could be released under ambient pressure at room temperature, while the release of the residual hydrogen (0.9%) required heating of the sample. A comparative study of high-pressure hydrogen adsorption experiments along with electrochemical hydrogen storage has been carried out by Gundiah et al. on various carbon nanotube samples [210]. The carbon samples that they used for hydrogen storage studies are as follows: SWNTs synthesized by the arc-discharge method (as-synthesized), I; SWNTs synthesized by the arc-discharge method (treated with conc. $HNO_3$), II; MWNTs synthesized by the pyrolysis of acetylene (as-synthesized), III; MWNTs synthesized by the pyrolysis of acetylene (treated with conc. $HNO_3$), IV; MWNTs synthesized by the arc-discharge method, V; aligned MWNT bundles synthesized by the pyrolysis of ferrocene (as-synthesized), VI; aligned MWNT bundles synthesized by the pyrolysis of ferrocene (treated with acid), VII; aligned MWNT bundles synthesized by the pyrolysis of ferrocene and acetylene (as-synthesized), VIII; and aligned MWNT bundles synthesized by the pyrolysis of ferrocene and acetylene (treated with acid), IX. Figure 8.18(a), shows the plots of hydrogen adsorption versus time for the various carbon nanostructured samples studied by them. By eliminating most of the common errors encountered in these experiments, they achieved a maximum storage capacity of 3.75 wt.% (143 bar, 27 °C) in the case of densely aligned nanotubes, prepared by the pyrolysis of ferrocene–hydrocarbon mixtures. SWNTs and MWNTs (arc-generated) showed a high-pressure hydrogen storage capacity, which is much less than 3 wt.%. In Figure 8.18(b) we show the plots of electrochemical charging capacity of various types of carbon nanotubes. Electrodes made out of aligned MWNTs, clearly demonstrate higher electrochemical charging capacities up to 1100 mA h $g^{-1}$ which correspond to a hydrogen storage capacity of 3.75 wt.%. SWNTs and MWNTs (arc-generated), however, show capacity in the range 2–3 wt.%.

## 8.3 Inorganic Nanotubes

### 8.3.1 Preliminaries

Several layered inorganic compounds possess structures comparable to the structure of graphite, the metal dichalcogenides (sulfides, selenides, and tellurides),

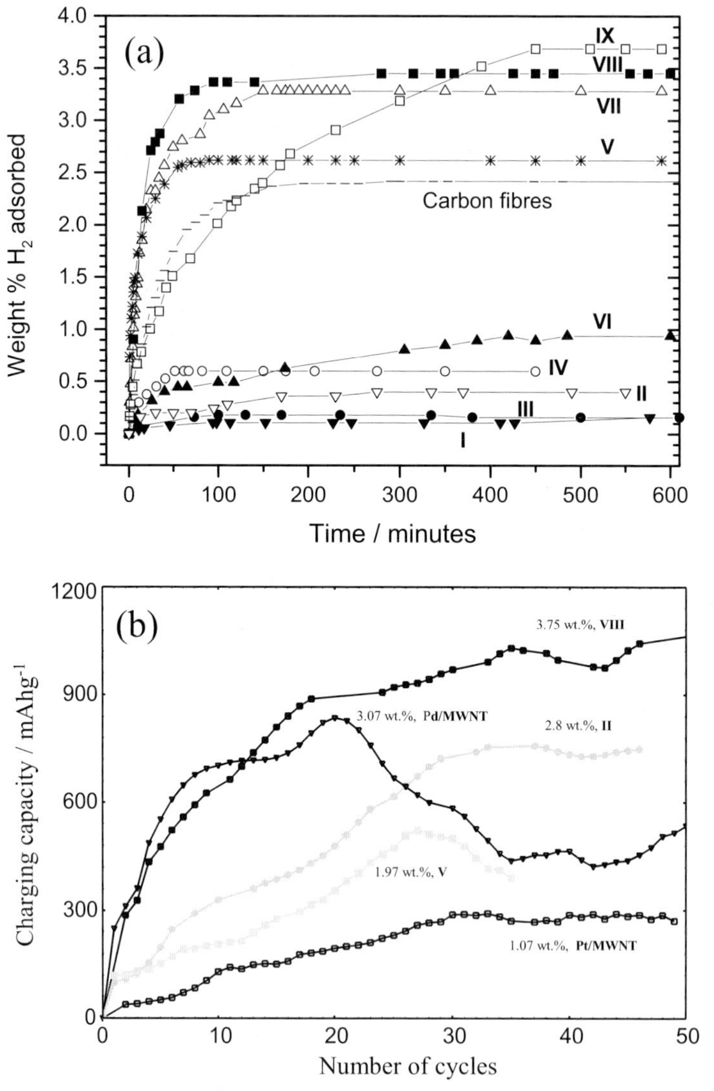


**Fig. 8.18.** (a) Amount of hydrogen adsorbed, in wt%, as a function of time for the various carbon nanostructures (I–IX). The broken curve represents the blank data obtained in the absence of a carbon sample. (b) Plot of the charging capacity against the number of cycles for different carbon nanostructures. Also shown are the corresponding weight percentages of $H_2$ stored. Reproduced from ref. [210], with permission.

halides (chlorides, bromides, and iodides), oxides, and numerous ternary (quaternary) compounds being important examples. The metal dichalcogenides, $MX_2$ (M = Mo, W, Nb, Hf; X = S, Se) contain a metal layer sandwiched between two chalcogen layers with the metal in a trigonal pyramidal or octahedral coordination

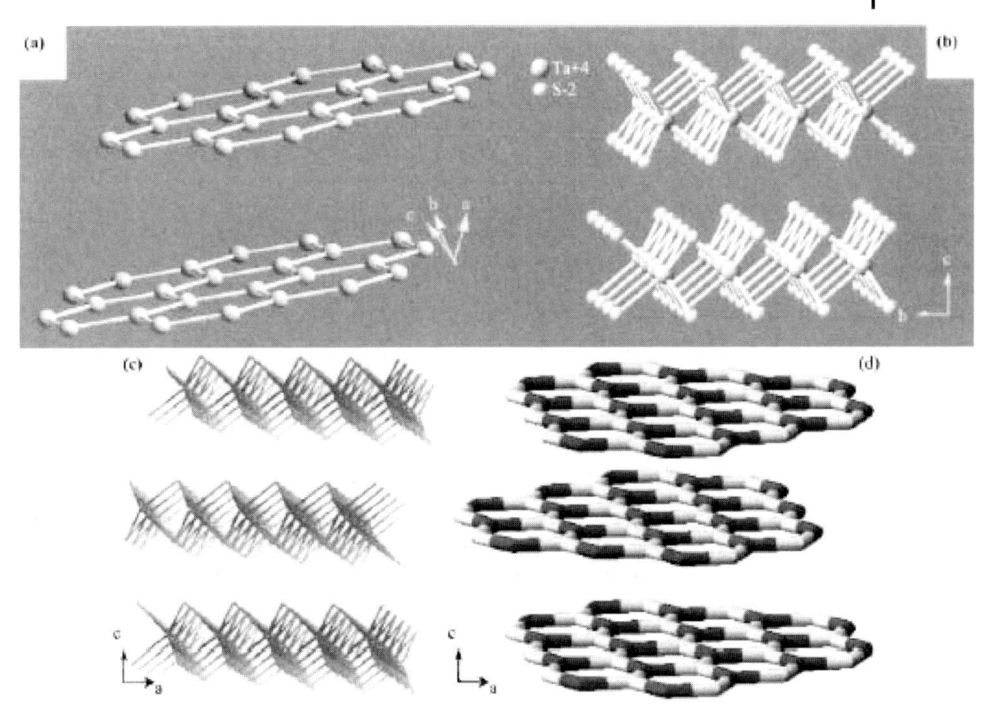


**Fig. 8.19.** Comparison of the structures of (a) graphite and inorganic layered compounds such as (b) $NbS_2/TaS_2$; (c) $MoS_2$; (d) BN. In the layered dichalcogenides, the metal is in trigonal prismatic ($TaS_2$) or octahedral coordination ($MoS_2$). Reproduced from ref. [22], with permission.

mode [211]. The $MX_2$ layers are stacked along the *c*-direction in ABAB fashion. The $MX_2$ layers are analogous to the single graphene sheets in the graphite structure (Figure 8.19). However, in contrast to graphite, each molecular sheet consists of multiple layers of different atoms chemically bonded together. When viewed parallel to the *c*-axis, the layers show the presence of dangling bonds due to the absence of an X or M atom at the edges. Such unsaturated bonds at the edges of the layers also occur in graphite. The dichalcogenide layers are unstable towards bending and have a high propensity to roll into curved structures. If the dimensions of the dichalcogenide layers are small, then they form hollow, closed clusters designated as inorganic fullerene-like (IF) structures. Folding in the layered transition metal chalcogenides (LTMCs) was recognized as early as 1979 by Chianelli et al., well before the discovery of the carbon nanotubes [212]. They reported rag-like and tubular structures of $MoS_2$ and studied their usefulness in catalysis. The observed folded sheets appear as crystalline needles in low magnification transmission electron microscope (TEM) images, and were described as layers that fold onto themselves. These structures indeed represent those of nanotubes. Tenne et al. [213] first demonstrated that Mo and W dichalcogenides are capable of forming

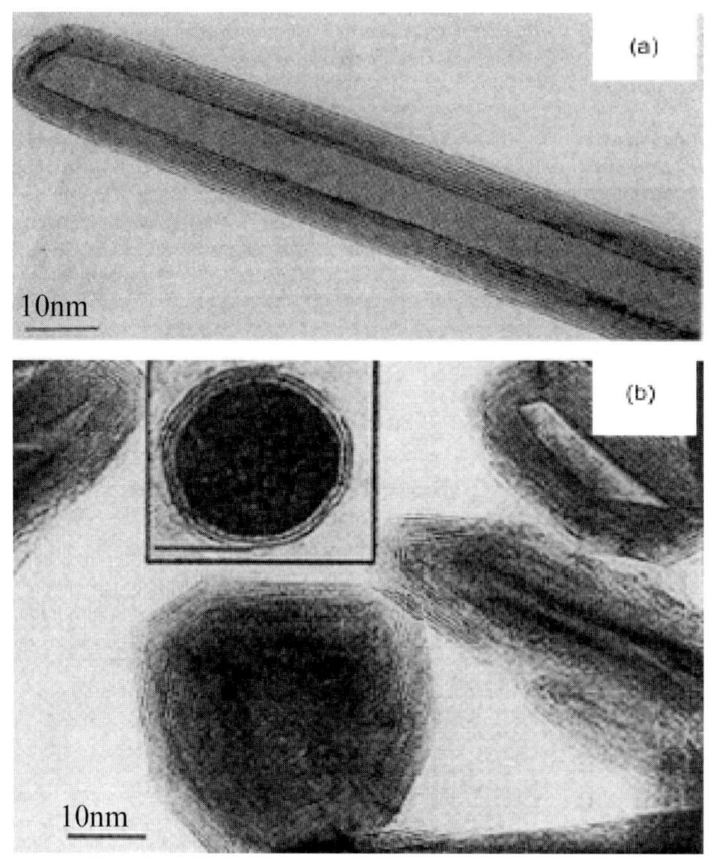


Fig. 8.20. TEM images of (a) a multi-walled nanotube of $WS_2$ and (b) hollow particles (inorganic fullerenes) of $WS_2$. Reproduced from ref. [213], with permission.

nanotubes (Figure 8.20(a)). Closed fullerene-type structures (inorganic fullerenes) also formed along with the nanotubes (Figure 8.20(b)). The dichalcogenide structures contain concentrically nested fullerene cylinders, with a less regular structure than in the carbon nanotubes. Accordingly, $MX_2$ nanotubes have varying wall thickness and contain some amorphous material on the exterior of the tubes. Nearly defect-free $MX_2$ nanotubes are rigid as a consequence of their structure and do not permit plastic deformation. The folding of a $MS_2$ layer in the process of forming a nanotube is shown in the schematic in Figure 8.21. There has been some speculation on the cause of folding and curvature in the LTMCs. Stoichiometric LTMC chains and layers such as those of $TiS_2$ possess an inherent ability to bend and fold, as observed in intercalation reactions. The existence of alternate coordination and therefore of stoichiometry in the LTMCs may also cause folding. Lastly, a change in the stoichiometry within the material would give rise to closed rings. Considerable progress has been made in the synthesis of the nanotubes of

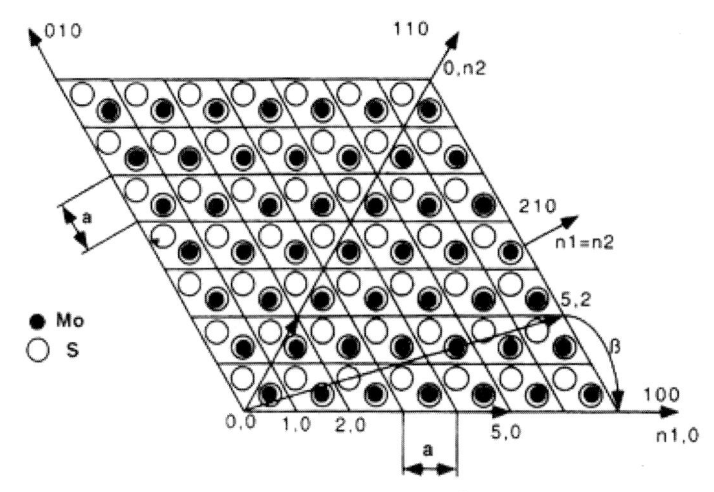


**Fig. 8.21.** Schematic illustration of the bending of a $MoS_2$ layer. Reproduced from ref. [229], with permission.

Mo and W dichalcogenides in the last few years. Detailed studies on the synthesis, structural characterization and applications of these inorganic nanotubes can be found in a series of review articles devoted to this topic [20a, 22–24, 214].

Transition metal chalcogenides possess a wide range of interesting physical properties. They are widely used in catalysis and as lubricants. They have both semiconducting and superconducting properties. With the synthesis and characterization of the fullerenes and nanotubes of $MoS_2$ and $WS_2$, a wide field of research has opened up enabling the successful synthesis of nanotubes of other metal chalcogenides. It may be recalled that the dichalcogenides of many of the Group 4 and 5 metals have layered structures suitable for forming nanotubes.

Curved structures are not only limited to carbon and the dichalcogenides of Mo and W. Perhaps the most well-known example of a tube-like structure with diameters in the nanometer range is formed by the asbestos mineral (chrysotil) whose fibrous characteristics are determined by the tubular structure of the fused tetrahedral and octahedral layers. The synthesis of mesoporous silica with well-defined pores in the 2–20 nm range was reported by Beck and Kresge [215]. The synthetic strategy involved the self-assembly of liquid crystalline templates. The pore size in zeolitic and other inorganic porous solids is varied by a suitable choice of the template. However, in contrast to the synthesis of porous compounds, the synthesis of nanotubes is somewhat more difficult.

Nanotubes of oxides of several transition metals, as well as of other metals, have been synthesized by employing different methodologies [24, 216–220]. Silica nanotubes were first produced as a spin-off product during the synthesis of spherical silica particles by the hydrolysis of tetraethylorthosilicate (TEOS) in a mixture of water, ammonia, ethanol and D,L-tartaric acid [216]. Since self-assembly reactions are not straightforward with respect to the desired product, particularly its morphology, templated reactions have been employed using carbon nanotubes to obtain nanotube structures of metal oxides [217, 218]. Oxides such as $V_2O_5$ have good catalytic activity in the bulk phase. Redox catalytic activity is also retained in the nanotubular structure. There have been efforts to prepare $V_2O_5$ nanotubes by chemical methods as well [219].

Boron nitride (BN) crystallizes in a graphite-like structure and can be simply viewed as replacing a C–C pair in the graphene sheet with the iso-electronic B–N pair. It can, therefore, be considered as an ideal precursor for the formation of BN nanotubes. Replacement of the C–C pairs partly or entirely by the B–N pairs in the hexagonal network of graphite leads to the formation of a wide array of two-dimensional phases that can form hollow cage structures and nanotubes. The possibility of replacing C–C pairs by B–N pairs in the hollow cage structure of $C_{60}$ was predicted [221] and verified experimentally [222]. BN-doped carbon nanotubes have been prepared [112, 115]. Pure BN nanotubes have been generated by employing several procedures, yielding nanotubes with varying wall thickness and morphology [223–225]. It is therefore quite possible that nanotube structures of other layered materials can be prepared as well. For example, many metal halides (e.g., $NiCl_2$), oxides (e.g., ZnO) and nitrides (e.g., GaN) crystallize in layered structures and nanotubes of such materials have indeed been characterized [226]. Nanotubes of elemental materials such as Te [227a] and Ni [227b] have also been prepared. There is considerable interest at present in preparing exotic nanotubes and studying their properties. In this section, we discuss the synthesis and characterization of nanotubes of chalcogenides of Mo, W and other metals, metal oxides, BN and other materials and present the current status of the subject. We briefly examine some of the important properties of the inorganic nanotubes and indicate possible future directions.

### 8.3.2
### General Synthetic Strategies

Several strategies have been employed for the synthesis of carbon nanotubes [20a]. In addition to arc evaporation and pyrolysis methods, carbon nanotubes have been prepared by laser ablation of graphite, electrochemical and templating (using porous alumina membrane) techniques [228]. The above methods broadly fall into two categories. Methods such as the arc evaporation of graphite employ processes which are far from equilibrium. The chemical routes are generally closer to equilibrium conditions. Nanotubes of metal chalcogenides and boron nitride are also prepared by employing techniques similar to those of carbon nanotubes, although there is an inherent difference in that the nanotubes of inorganic materials such as $MoS_2$ or BN would require reactions involving the component elements or compounds containing the elements. Decomposition of precursor compounds containing the elements is another possible route.

Nanotubes of dichalcogenides such as $MoS_2$, $MoSe_2$ and $WS_2$ are also obtained by employing processes far from equilibrium, such as arc discharge and laser ablation [229]. By far the most successful routes employ appropriate chemical reactions. Thus, $MoS_2$ and $WS_2$ nanotubes are conveniently prepared by starting with the stable oxides, $MoO_3$ and $WO_3$ [213]. The oxides are first heated at high temperatures in a reducing atmosphere and then reacted with $H_2S$. Reaction with $H_2Se$ is used to obtain the selenides [230]. Recognizing that the trisulfides $MoS_3$ and $WS_3$ are likely to be the intermediates in the formation of the disulfide nanotubes, the trisulfides have been directly decomposed to obtain the disulfide nanotubes [231]. Diselenide nanotubes have been obtained from the metal triselenides [232]. The trisulfide route is indeed found to provide a general route for the synthesis of the nanotubes of many metal disulfides such as $NbS_2$ [233] and $HfS_2$ [234]. In the case of Mo and W dichalcogenides, it is possible to use the decomposition of the precursor ammonium salt, such as $(NH_4)_2MX_4$ ($X = S$, Se; $M = Mo$, W) as a means of preparing the nanotubes [231]. Other methods employed for the synthesis of dichalcogenide nanotubes include hydrothermal methods where the organic amine is taken as one of the components in the reaction mixture.

The hydrothermal route has been used for synthesizing nanotubes and related structures of a variety of other inorganic materials as well. Thus, nanotubes of several metal oxides (e.g., $SiO_2$ [235], $V_2O_5$ [219], ZnO [236]) have been produced hydrothermally. Nanotubes of oxides such as $V_2O_5$ are also conveniently prepared from a suitable metal oxide precursor in the presence of an organic amine or a surfactant [237]. Surfactant-assisted synthesis of CdSe and CdS nanotubes has been reported. Here the metal oxide reacts with the sulfidizing/selenidizing agent in the presence of a surfactant such as Triton X [238, 246].

Sol–gel chemistry is widely used in the synthesis of metal oxide nanotubes, a good example being that of silica [216] and $TiO_2$ [239]. Oxide gels in the presence of surfactants or suitable templates form nanotubes. For example, by coating carbon nanotubes (CNTs) with oxide gels and then burning off the carbon, one obtains nanotubes and nanowires of a variety of metal oxides including $ZrO_2$, $SiO_2$ and $MoO_3$ [218, 240]. Sol–gel synthesis of oxide nanotubes is also possible in the pores of alumina membranes. It should be noted that $MoS_2$ nanotubes are also prepared by the decomposition of a precursor in the pores of an alumina membrane [241].

Boron nitride nanotubes have been obtained by striking an electric arc between $HfB_2$ electrodes in a $N_2$ atmosphere [242]. BCN and BC nanotubes are obtained by arcing between B/C electrodes in an appropriate atmosphere. A greater effort has gone into the synthesis of BN nanotubes starting with different precursor molecules containing B and N. Decomposition of borazine in the presence of transition metal nanoparticles and the decomposition of the 1:2 melamine–boric acid addition compound yield BN nanotubes [223, 224]. Reaction of boric acid or $B_2O_3$ with $N_2$ or $NH_3$ at high temperature in the presence of activated carbon, carbon nanotubes or catalytic metal particles has been employed to synthesize BN nanotubes [225]. Goldberger et al. [226b] have synthesized single-crystal GaN nanotubes with inner diameters of 30–200 nm and wall thicknesses of 5–50 nm by employing 'epitaxial casting' approach. They used hexagonal ZnO nanowires as templates for the epitaxial overgrowth of thin GaN layers in a chemical vapor deposition system. In a typical experiment they used trimethylgallium and ammonia as precursors with argon or nitrogen as carrier gas and maintained the deposition temperature at

600–700 °C. The ZnO nanowire templates were subsequently removed by thermal reduction and evaporation, resulting in ordered arrays of GaN nanotubes on the substrates.

A wide range of examples of organic templates (organogels) and extensive coverage of the methods for the formation of morphologically interesting inorganic materials were given in a review by Shinkai's group [214]. The organogels present a wider range of morphologies, such as fibrous, tubular, ribbon-like, lamellar, hollow spherical, than have been transcribed so far. These organogels are comprised of an organic liquid and low concentrations ($< 0.5$ wt.%) of relatively low molecular weight molecules (namely gelators) [243], the morphology of which may be spherical or fibrous. These organogels are usually prepared by heating a mixture of a gelator and solvent until the solid dissolves, upon cooling the solution (sol) thickens to form a gel. The gelation of a number of compounds such as tetraethylorthosilicate (TEOS) along with a cholesterol-based gelator, under acidic pH conditions, followed by polycondensation, exhibits a network of fibers with diameter ranging from 50–200 nm [244]. Subsequent drying and calcination steps resulted in silica tubes without the presence of the original organic template.

### 8.3.3 Structures

Some of the important inorganic nanotubes synthesized and characterized in the last few years are the following [22]:

*Chalcogenides*: $MoS_2$, $WS_2$, $MoSe_2$, $WSe_2$, $NbS_2$, $NbSe_2$, $HfS_2$, $ZrS_2$
*Oxides*: $TiO_2$, $ZrO_2$, $VO_x$, $SiO_2$, ZnO, $Ga_2O_3$, $BaTiO_3$, $PbTiO_3$
*Nitrides*: BN, GaN
*Halides*: $NiCl_2$
*Metals*: Ni, Cu, Te, Co, Fe

In Figure 8.22 we show the TEM images of $MoS_2$ nanotubes prepared by the direct thermal decomposition of the ammonium thiomolybdate in $H_2$ atmosphere (flow) [231]. In addition to providing a direct method for the preparation of dichalcogenide nanotubes [231], the trichalcogenide or the ammonium chalcometallate route also enables the easy synthesis of nanotubes of the other layered dichalcogenides [232–234]. The structure of $MoS_2$ consists of disulfide layers stacked along the *c*-direction [211]. This implies that the S–S interaction between the $MoS_2$ slabs is weaker than the intralayer interactions. The S–S interlayer distances are therefore susceptible to distortions during the folding of the layers. This is exemplified by the slight expansion of the *c*-axis (2%) in the $MoS_2$ nanotubes [229]. High-resolution (HREM) images of the disulfide nanotubes show stacking of the (002) planes parallel to the tube axis. The distance between the layer fringes corresponds to the *d*(002) spacing. In Figure 8.23(a) we show the SEM image of nanotubes of $HfS_2$ which are quite lengthy, some being more than a micron long. Interestingly, a large proportion of these nanostructures are nanotubes. In Figure 8.23(b) and (c),

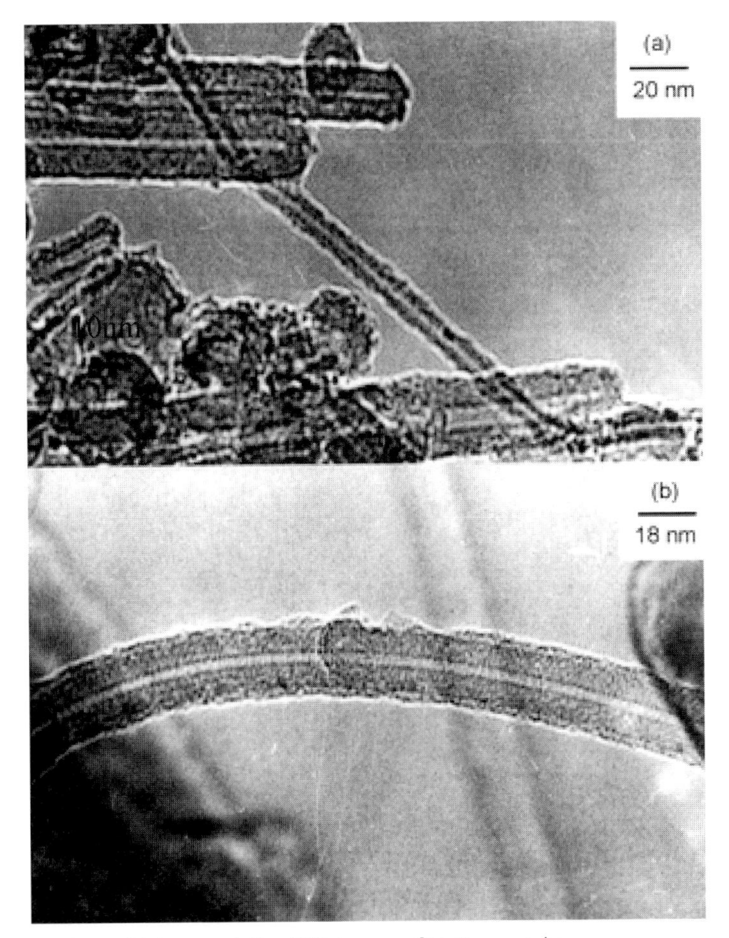


**Fig. 8.22.** (a) Low-resolution TEM images of $MoS_2$ nanotubes grown by the decomposition of ammonium thiomolybdate; (b) HREM image of the $MoS_2$ nanotube. Reproduced from ref. [231], with permission.

we show TEM images of these nanotubes. These $HfS_2$ nanotubes are obtained in good yield by the decomposition of $HfS_3$ [234]. The HREM image in Figure 8.23(d) shows a considerable number of defects and edge dislocations present along the length of the tube wall. The analysis of the electron diffraction (ED) pattern together with the HREM indicates that the growth axis of the nanotube is perpendicular to the *c*-direction.

Exhaustive studies have been carried out on the synthesis of BN nanotubes and nanowires by various CVD techniques [225]. The methods examined include heating boric acid with activated carbon, multi-walled carbon nanotubes, catalytic iron particles or a mixture of activated carbon and iron particles, in the presence of ammonia. With activated carbon, BN nanowires are obtained as the primary prod-

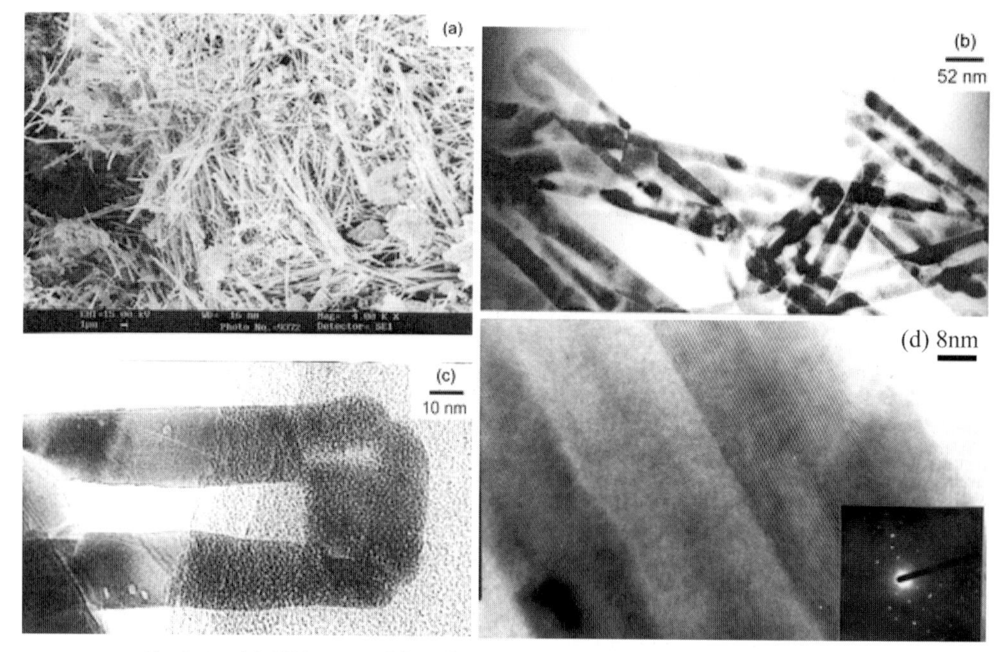


**Fig. 8.23.** (a) SEM image of the $HfS_2$ nanostructures; (b) and (c) low-resolution TEM images showing hollow nanotubes. The tube in (c) has a flat tip; (d) HREM image of the $HfS_2$ nanotubes, showing a layer separation of ~0.6 nm in the walls. Inset shows a typical ED pattern. Reproduced from ref. [234], with permission.

uct. However, with multi-walled carbon tubes, high yields of pure BN nanotubes are obtained as the major product. BN nanotubes with different structures were obtained on heating boric acid and iron particles in the presence of $NH_3$. Aligned BN nanotubes are obtained when aligned multi-walled nanotubes are used as the templates (Figure 8.24).

Needle-shaped $TiO_2$ (anatase) nanotubes could be precipitated from a gel containing a mixture of $SiO_2$ and $TiO_2$ [239]. A mixture of titanium isopropoxide and tetraethylorthosilicate (TEOS) was hydrolyzed and gelled in an incubator, and the gel further heated to 600 °C, resulting in the precipitation of fine $TiO_2$ (anatase) crystals. This was further treated with NaOH at 100 °C for 20 h to yield the $TiO_2$ nanotubular phase. An amorphous $SiO_2$-related phase present in the product was removed by chemical treatment. The nanotubes formed by this method had a diameter of ~8 nm and lengths upto 100 nm (Figure 8.25(a)). Much smaller $TiO_2$ nanotubes have recently been synthesized by a surprisingly simple procedure. $TiO_2$ with anatase or rutile structure was treated with NaOH and subsequently with HCl [239, 245]. The resulting $TiO_2$ nanotubes are 50–200 nm long and their diameter is about 10 nm (Figure 8.25(b)). The HREM image (Figure 8.25(c)) of

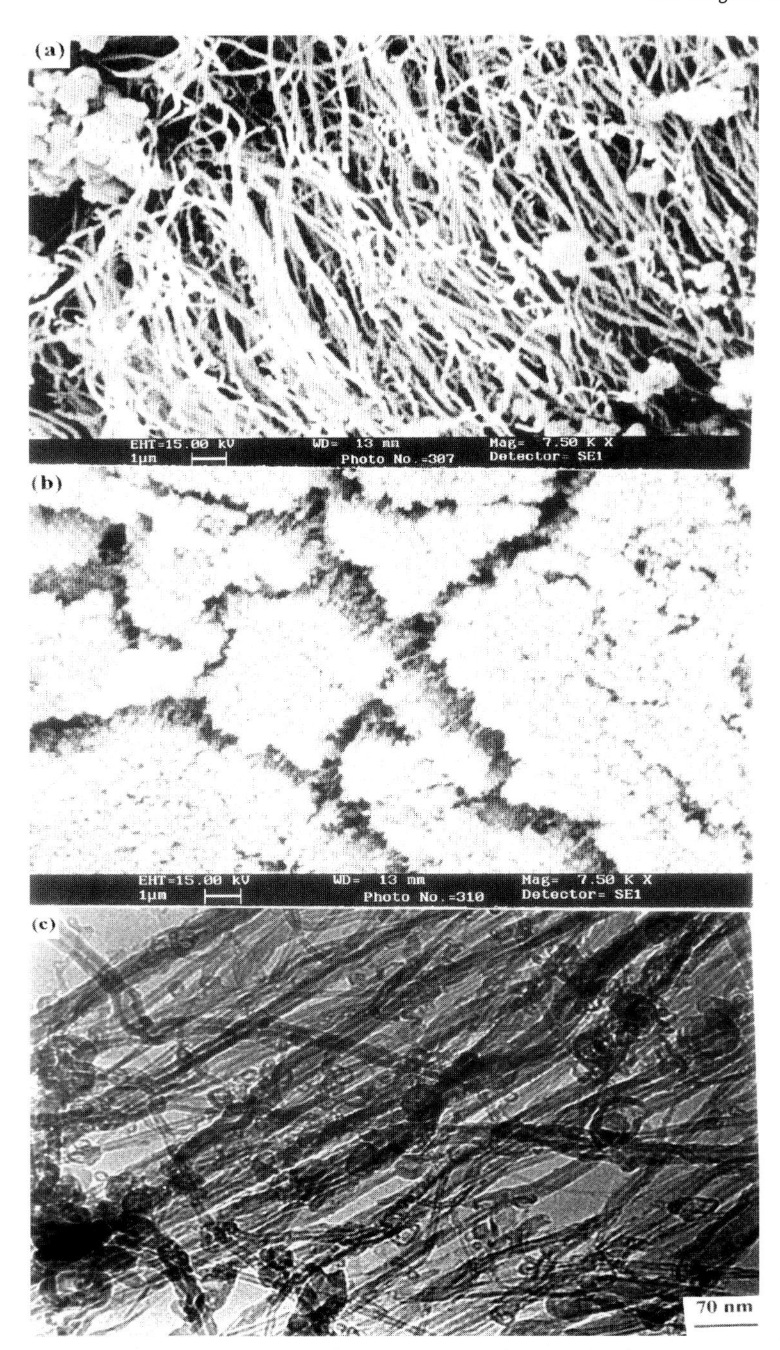


**Fig. 8.24.** SEM and TEM images of aligned BN nanotubes: (a) and (b) give side and top view SEM images, respectively; (c) TEM image of pure BN nanotube. Reproduced from ref. [225], with permission.

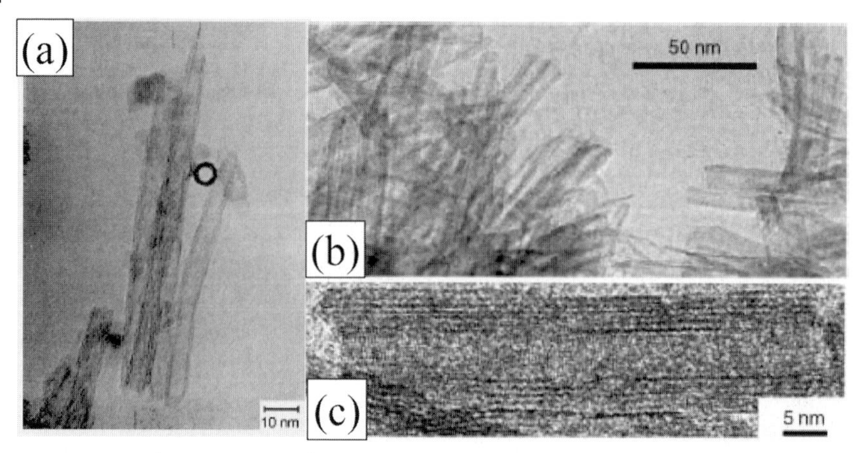


**Fig. 8.25.** (a), (b) TEM images of $TiO_2$ nanotubes. (c) HRTEM image of a well developed, ~50 nm long nanotube with a diameter of ~10 nm. Lattice fringes can be seen. Reproduced from ref. [24, 239], with permission.

such a $TiO_2$ nanotube shows that the tubes have an inner core and walls. The presence of lattice fringes indicates the crystalline structure of $TiO_2$ nanotubes. Parallel fringes in the walls correspond to a distance of about 7 nm, which can also be detected as a broad reflection by X-ray and electron diffraction.

Nanotubes of II–VI semiconductor compounds such as CdS and CdSe have been obtained by a soft chemical route involving surfactant-assisted synthesis [238, 246]. For CdS nanotubes, the metal oxide was reacted with the sulfidizing reagent in the presence of a surfactant such as Triton 100X in a basic medium. To obtain nanotubes of CdSe, a similar procedure was followed, except that NaHSe was used in place of thioacetamide as selenidizing reagent in the presence of a surfactant such as Triton 100X (Figure 8.26). Both the CdSe and CdS nanotubes seem to be polycrystalline, formed by aggregates of nanoparticles [247]. The nanotubes of CdSe, though extended in one direction show quantum confinement and the absorption band is blue-shifted to 550 nm from 650 nm in the bulk sample.

Nesper and co-workers [219, 248] synthesized nanotubules of alkylammonium intercalated $VO_x$ by hydrothermal means. The vanadium alkoxide precursor was hydrolyzed in the presence of hexadecylamine and the hydrolysis product (lamellar structured composite of the surfactant and the vanadium oxide) yielded $VO_x$ nanotubes along with the intercalated amine under hydrothermal conditions (Figure 8.27(a) and (b)). The interesting feature of this vanadium oxide nanotube is the presence of vanadium in the mixed valent state, thereby rendering it redox-active. The template could not be removed by calcination as the structural stability was lost above 250 °C. Nevertheless, it was possible to partially extract the surfactant under mildly acidic conditions. These workers have later shown that the alkylamine intercalated in the intertubular space could be exchanged with other alkylamines of varying chain lengths as well as $\alpha,\omega$-diamines [248]. The distance

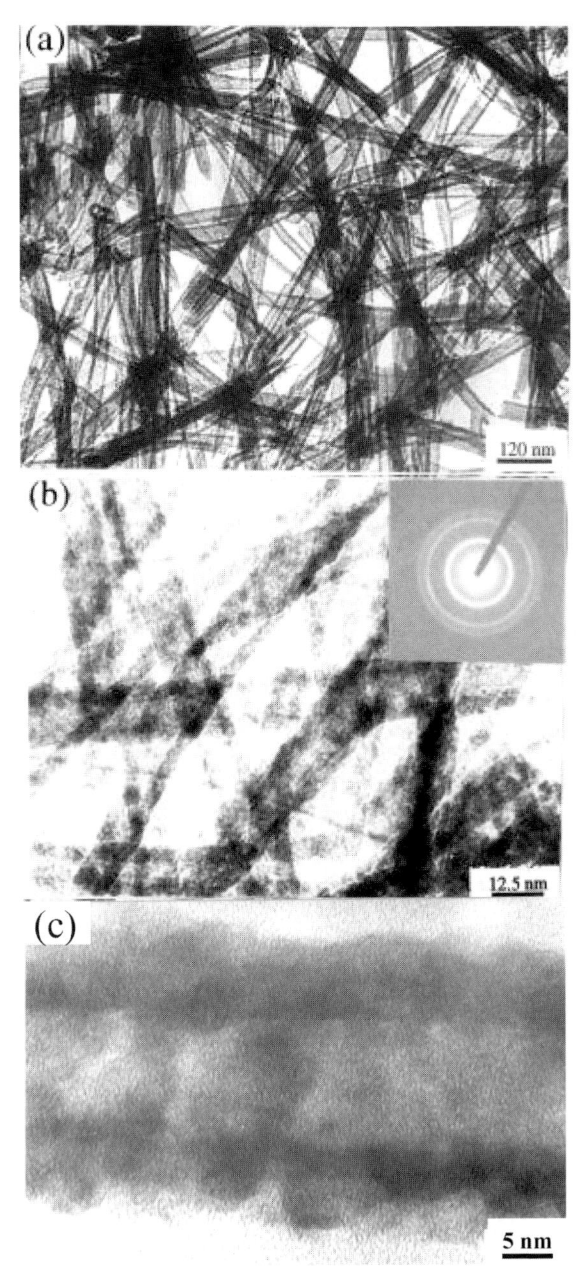


**Fig. 8.26.** (a) and (b) low-resolution TEM images of CdSe nanotubes. Inset shows a typical ED pattern; (c) HREM image of the CdSe nanotube showing walls containing several nanocrystallites. Reproduced from ref. [238, 246], with permission.

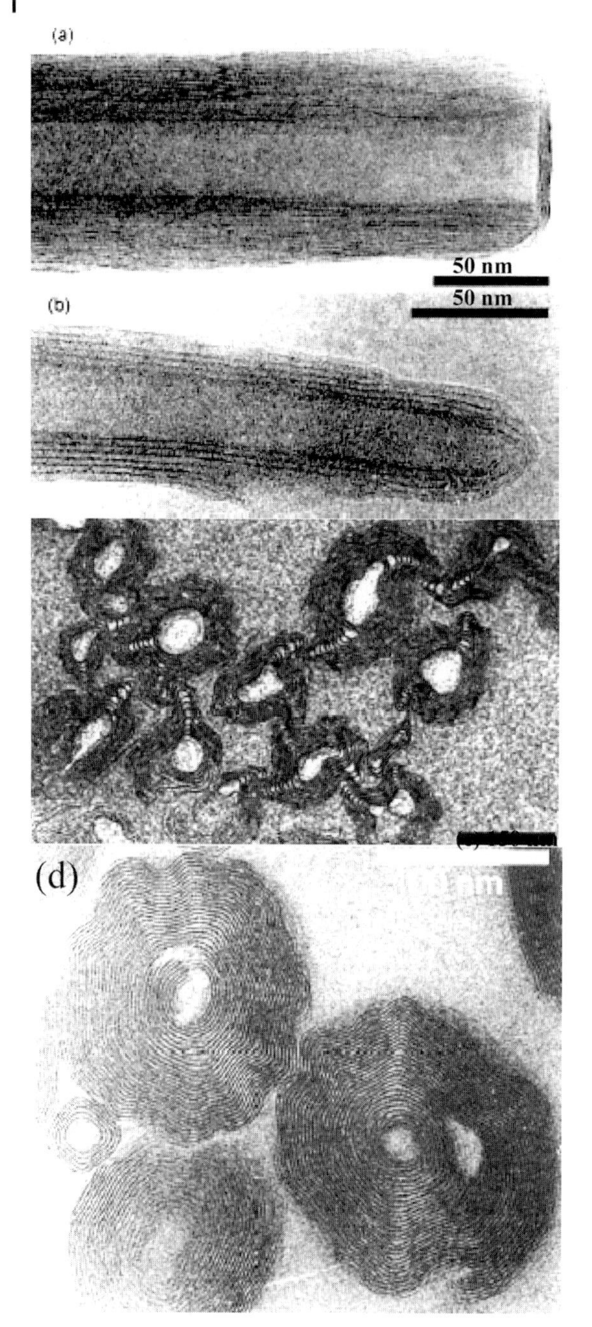


**Fig. 8.27.** TEM images of $VO_x$ nanotubes with intercalated amine having varying chain lengths; (a) $C_4VO_x$–NT; (b) $C_{16}$–$VO_x$–NT. The length of the bar is 50 nm. (c) Cross-sectional TEM images of monoamine-intercalated $VO_x$ nanotubes showing serpentine-like scrolls. (d) Cross-sectional TEM images of diamine-intercalated $VO_x$ nanotubes showing a larger thickness of the tube walls and a smaller inner core. Reproduced from ref. [248], with permission.

between the layers in the $VO_x$ nanotubes can be controlled by the length of the $-CH_2-$ chain in the amine template.

Most of the $VO_x$ nanotubes obtained by the hydrothermal method are open-ended. Very few closed tubes had flat or pointed conical tips. Cross-sectional TEM images of the nanotubular phases show that instead of concentric cylinders, (i.e. layers that fold and close within themselves), the tubes are made up of single or double layer scrolls providing a serpentine-like morphology [248, 249]. The scrolls are seen in the images as circles that do not close (Figure 8.27(c)). Non-symmetric fringe patterns in the tube walls exemplify that most of the nanotubes are not rotationally symmetric and carry depressions and holes in the walls. Diamine-intercalated $VO_x$ nanotubes are multilayer scrolls with narrow cores and thick walls, composed of packs of several vanadium oxide layers (Figure 8.27(d)). Many of the nanotubes formed by layered materials show various types of defects. They also exhibit unusual tip structures. The tips are not always spherical in these nanotubes. These aspects are discussed at length by Rao and Nath [22].

### 8.3.4 Useful Properties of Inorganic Nanotubes

The properties and applications of the inorganic nanotubes have not been investigated as extensively as would be desirable. The electronic structures of $MoS_2$ and $WS_2$ have been examined briefly and the semiconducting nature of the nanotubes confirmed [250, 251a]. It is necessary to investigate the optical, electrical and other properties of the various chalcogenide nanotubes. This is especially true of nanotubes of $NbS_2$ and such materials which are predicted to be metallic [251b]. $NbSe_2$ nanotubes have been found to be metallic at ordinary temperatures, becoming superconducting at lower temperatures [252]. Electronic and optical properties of the BN nanotubes have not yet been investigated in detail. Theoretical calculations suggest BN tubes to be insulating with a wide band gap of 5.5 eV [221].

Like carbon nanotubes, it would be worthwhile to look into the $H_2$ storage ability of some of the inorganic nanotubes [253]. The chalcogenide nanotubes with an $\sim$6 Å van der Waals gap between the layers, are potential candidates for storage capacity. It has been shown recently that BN nanotubes can store a reasonable quantity of $H_2$ [254]. Multi-walled BN nanotubes have been shown to possess a capacity of 1.8–2.6 wt% of $H_2$ uptake under $\sim$10 MPa at room temperature. This value, though smaller than that reported for CNTs, nevertheless suggests the possible use of BN nanotubes as a hydrogen storage system. $MoS_2$ nanotubes could be electrochemically charged and discharged with a capacity of 260 mA h $g^{-1}$ at 20 °C, corresponding to a formula of $H_{1.24}MoS_2$ [255]. The high storage capacity is believed to be due to the enhanced electrochemical–catalytic activity of the highly nanoporous structure. This may find wide applications in high energy batteries.

Mechanical properties of BN nanotubes would be worthy of exploration. Unlike carbon nanotubes, BN nanotubes are predicted to have stable insulating properties independent of their structure and morphology. Thus, BN tubes can be used as nano-insulating devices for encapsulating conducting materials like metallic wires. Filled BN nanotubes are expected to be useful in nanoscale electronic devices and for the preparation of nano-structured ceramics.

Electrochemical studies have been performed with the alkylammonium intercalated $VO_x$ nanotubes [256] as well as Mn intercalated $VO_x$ nanotubes [257]. Cyclic voltammetry studies of alkylammonium–$VO_x$ nanotubes showed a single reduction peak, which broadened, on replacing the amine with Na, with an additional peak. Li ion reactivity has also been tested with Mn–$VO_x$ nanotubes by reacting with n-butyllithium, and it was found that $\sim$2 lithiums per V ion are consumed. Electrochemical Li intercalation of Mn–$VO_x$ nanotubes shows that 0.5 Li ions per V atom were intercalated above 2 V [257]. This observation may be relevant to battery applications.

CNTs have been used as AFM tips and there appears to be every likelihood that extremely narrow structures can be probed [179]. $WS_2$ could be mounted on the ultrasharp Si tip following a similar methodology. These tips were tested in an AFM microscope by imaging a replica of high aspect ratio, and it was observed that these $WS_2$ nanotube tips provide a considerable improvement in the image quality compared to the conventional ultrasharp Si tips [258].

The most likely application of the chalcogenide nanotubes is as solid lubricants. Mo and W chalcogenides are widely used as solid lubricants. It has been observed that the hollow nanoparticles of $WS_2$ show better tribological properties and act as a better lubricant compared to the bulk phase in every respect (friction, wear and life-time of the lubricant) [259]. Tribological properties of 2H–$MoS_2$ and $WS_2$ powder can be attributed to the weak van der Waals forces between the layers, which allow easy shear of the films with respect to each other. The mechanism in the $WS_2$ nanostructures is somewhat different and the better tribological properties may arise from the rolling friction allowed by the round shape of the nanostructures.

Recently, open-tipped $MoS_2$ nanotubes were prepared by the decomposition of ball-milled ammonium thiomolybdate powder under a $H_2$–thiophene atmosphere, and used as catalyst for the methanation of CO with $H_2$ [260]. The conversion of CO to $CH_4$ was achieved at a much lower temperature compared to polycrystalline $MoS_2$ particles, and there was no deterioration even after 50 h of consecutive catalysis cycles. This observation is of importance in the context of energy conversion of global $CO_2$.

Inorganic nanotubes have emerged as a group of interesting materials. Although this area of research started with the layered metal chalcogenides, recent results suggest that other inorganic materials can also be prepared in the form of nanotubes, as typified by the metal oxides. It is likely that many new types of inorganic nanotubes will be made in the near future. These would include metal nanotubes as well as nanotubes of inorganic compounds such as $MgB_2$, $GeO_2$ and GaSe. Theoretical calculations indeed predict a stable nanotubular structure for GaSe [261]. Various layered materials could be explored for this purpose.

## 8.4 Nanowires

### 8.4.1 Preliminaries

One-dimensional (1D) nanostructures such as nanowires, nanorods and nanobelts, provide good models to investigate the dependence of electronic transport, optical, mechanical and other properties on size confinement and dimensionality. Nanowires are likely to play a crucial role as interconnects and active components in nanoscale devices. An important aspect of nanowires relates to the assembly of individual atoms into such unique 1D nanostructures in a controlled fashion. Excellent chemical methods have been developed for generating zero-dimensional nanostructures (nanocrystals or quantum dots) with controlled sizes and from a wide range of materials (see earlier chapters of this book). The synthesis of nanowires with controlled composition, size, purity and crystallinity, requires a proper understanding of the nucleation and growth processes at the nanometer regime. 1D nanostructures have been fabricated recently using nanolithographic techniques [262], such as electron-beam or focused-ion-beam writing [263], proximal-probe patterning [264] and X-ray lithography [265]. These methods are generally not very cost-effective and rapid for the purpose of making large quantities of the 1D nanostructures based on a broad range of materials. Chemical methods tend to be superior and provide an alternative strategy for generating 1D nanostructures [266]. There are a few reviews dealing with the synthesis, characterization, self-assembly and applications of nanowires [267–269]. Several ways of growing semiconductor nanowires, such as laser ablation, chemical vapor deposition (CVD) and template-assisted growth have been explored. Laser ablation and template-assisted approaches provide large quantities of nanowires, but do not provide control over the composition, size or orientation direction of the nanowires. Chemical methods, include solution and vapor based methods and precursor methods, as well as solvothermal, hydrothermal and carbothermal methods.

### 8.4.2 Synthetic Strategies

One of the aspects of 1D structures relates to crystallization [270], wherein the evolution of a solid from a vapor, a liquid, or a solid phase involves nucleation and growth. As the concentration of the building units (atoms, ions, or molecules) of a solid becomes sufficiently high, they aggregate into small nuclei or clusters through homogeneous nucleation. These clusters serve as seeds for further growth to form larger clusters. Several chemical strategies have been developed for 1D nanowires with different levels of control over the growth parameters [271–277]. These include: (i) the use of the anisotropic crystallographic structure of the solid to facilitate 1D nanowire growth; (ii) the introduction of a solid–liquid interface to reduce the symmetry of a seed, (iii) the use of templates (with 1D morphologies) to direct the formation of nanowires, (iv) the use of supersaturation control to modify the growth habit of a seed; (v) the use of capping agents to kinetically control the growth rates of various facets of a seed and (vi) self-assembly of 0D nanostructures. They can be usefully categorized into: (1) nanowire growth in the gas phase; (2) solution-based approaches to nanowires.

#### 8.4.2.1 Vapor Phase Growth of Nanowires

Vapor phase growth is commonly used to produce nanowires. Starting with the simple evaporation technique in an appropriate atmosphere to produce elemental or oxide nanowires, vapor–liquid–solid, vapor–solid and other processes are made use of.

**Vapor–Liquid–Solid Growth** The growth of a nanowire via a gas phase reaction involving a vapor–liquid–solid (VLS) process has been extensively studied. Wagner, during his studies on the growth of large single-crystalline whiskers, proposed in the 1960s, a mechanism for the growth via a gas phase reaction involving the so-called vapor–liquid–solid (VLS) process [278]. According to this mechanism, the anisotropic crystal growth is promoted by the presence of a liquid alloy–solid interface. His mechanism was widely accepted and applied to understanding the growth of the nanowires of Si, Ge and others. The growth of Ge nanowire using Au clusters as solvent at high temperature can be explained based on the Ge–Au binary phase diagram as shown in Figure 8.28(b). The Ge and Au will form a liquid alloy when the temperature is higher than the eutectic point (363 °C) as shown in Figure 8.28(a-I). The liquid surface has a large accommodation coefficient and is therefore a preferred deposition site for incoming Ge vapor. After the liquid alloy becomes supersaturated with Ge, Ge nanowire growth occurs by precipitation at the solid–liquid interface (Figure 8.28(a-II, a-III)). A real time observation of Ge nanowire growth conducted in an *in situ* high temperature transition electron microscope shows a sequence of TEM images which directly mirrors the proposed VLS mechanism [279]. This VLS method has been exploited in the past several decades to produce 1–100 μm diameter 1D structures (whiskers). By controlling the nucleation and growth, it is possible to produce semiconductor nanowhiskers (e.g., InAs, GaAs) using organometallic vapor phase epitaxy. There are reports on the VLS growth of elemental semiconductors (e.g., Si and Ge), III–V semiconductors (e.g., GaAs, InP, InAs), II–VI semiconductors (e.g., ZnS, CdS, CdSe), oxides (e.g., ZnO, $SiO_2$) [271–273, 277, 279–292]. Lieber and coworkers have developed and optimized a laser ablation based VLS process to produce semiconductor nanowires with many different compositions [273, 277]. TEM studies showed the product obtained after the VLS growth is primarily wire-like structures with remarkably uniform diameters of the order of 10 nm with lengths >1 μm. By knowing the equilibrium phase diagram one can predict the catalyst materials and growth conditions for the VLS approach. By following the VLS approach, Lee and coworkers [293a] have synthesized highly pure, ultra-long and uniform-sized semiconductor nanowires in bulk quantities by employing laser

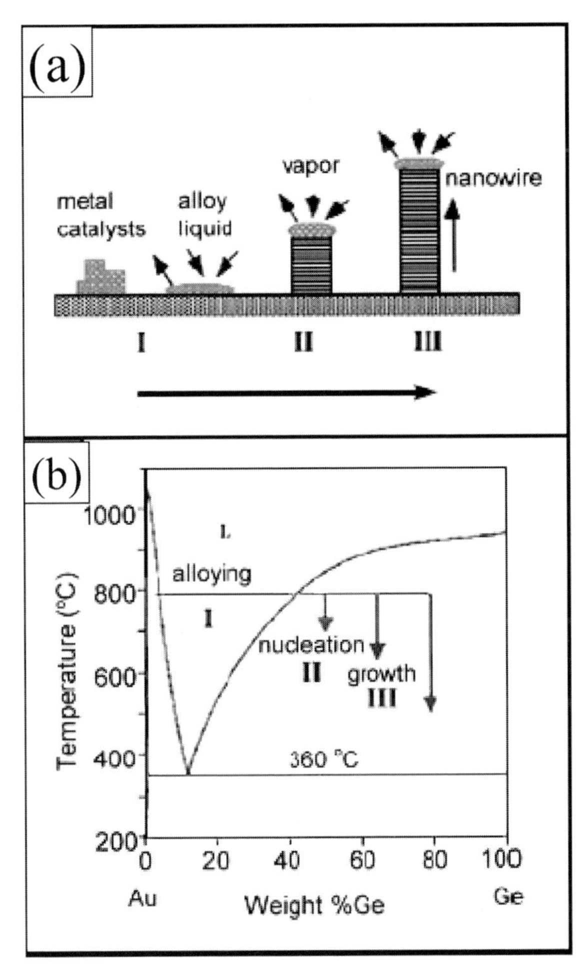


**Fig. 8.28.** (a) Schematic illustration of vapor–liquid–solid nanowire growth mechanism including three stages (I) alloying, (II) nucleation and (III) axial growth. The three stages are projected onto the conventional Au–Ge phase diagram; (b) shows the compositional and phase evolution during the nanowire growth process. Reproduced from ref. [269], with permission.

ablation and thermal evaporation of semiconductor powders mixed with metal or oxide catalysts. The solid target used for laser ablation is made of pure Si powder mixed with metals (Fe, Ni, or Co) and the temperature around the solid target was maintained in the range 1200 to 1400 °C. The temperature of the area around the substrate on which the nanowire grew was between 900 and 1100 °C. The transmission electron microscopy (TEM) investigations showed that Si nanowires obtained by this method were extremely long and highly curved with a typical diameter in the range 20–80 nm (Figure 8.29(a) and (b)). Each wire consisted of an outer layer of Si oxide and a crystalline Si core. A high density of defects, such as stacking faults and micro-twins, has been observed in the crystalline Si core. As

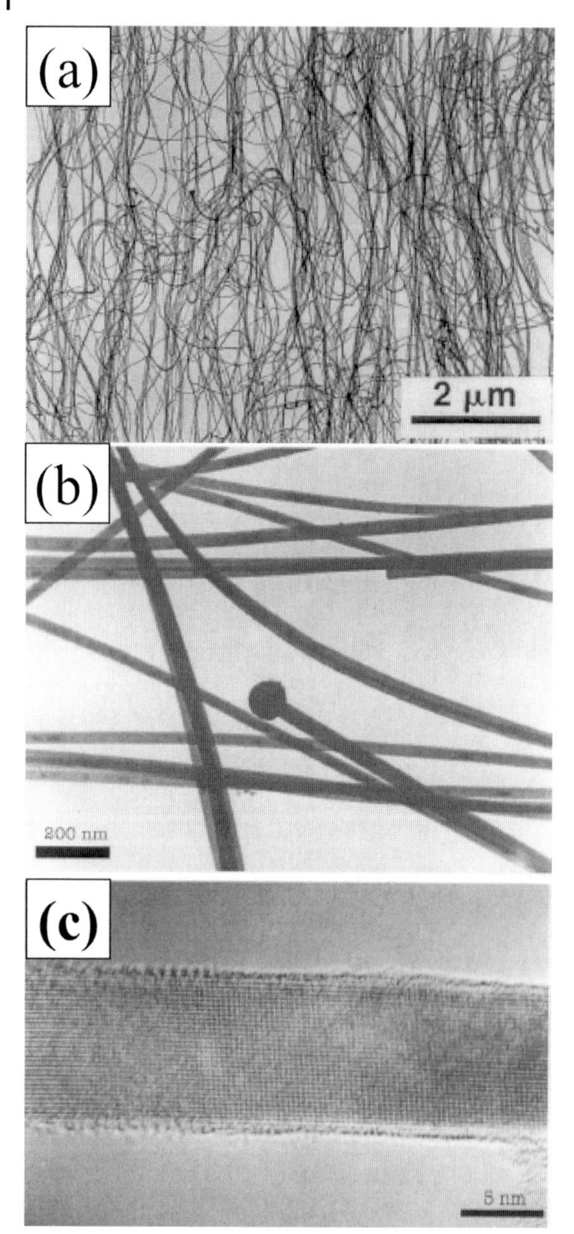


**Fig. 8.29.** (a) TEM micrograph of Si nanowires; (b) the morphology; and (c) a typical HREM image of the Si wires formed in the high temperature zone. Reproduced from ref. [293a], with permission.

identified by electron diffraction and high-resolution transmission electron microscopy (HREM) in Figure 8.29(c), the axis of the nanowires was generally along the ⟨112⟩ direction and the {111} surfaces of Si crystalline cores were parallel to the axis of the nanowire.

**Oxide-Assisted Growth** In contrast to the well-established VLS mechanism, Lee et al. have recently proposed a new nanowire growth route called oxide-assisted nanowire growth. They report the synthesis and optical characterization of GaAs nanowires obtained by oxide-assisted laser ablation of a mixture of GaAs and $Ga_2O_3$ [294a]. The GaAs nanowires have lengths up to tens of micrometers and diameter in the range 10–120 nm, with an average of 60 nm. The nanowires have a thin oxide layer covering and a crystalline GaAs core with a [111] growth direction. The oxide-assisted nanowire growth mechanism was further applied to the production of Si nanowires [293, 295]. The growth of Si nanowires was greatly enhanced when $SiO_2$-containing Si powder targets were used. Wang et al. have synthesized Si nanowires with uniform size by laser ablation of highly pure Si powder targets mixed with $SiO_2$ [295a]. A large quantity of Si nanowires was obtained by mixing 30%–70% of $SiO_2$ into the Si powder target. $SiO_2$ played a crucial role in enhancing the formation and growth of the Si nanowires. In this proposed oxide-assisted growth mechanism, the vapor phase of $Si_xO$ $(x > 1)$ generated by thermal evaporation or laser ablation is the key factor.

$$Si_xO(s) \rightarrow Si_{x-1}(s) + SiO(s) \quad (x > 1)$$

$$2SiO(s) \rightarrow Si(s) + SiO_2(s)$$

The Si nanowires synthesized by using different $SiO_2$ contents in the targets are similar in structure except that the outer Si-oxide surfaces of the nanowires synthesized from targets with high $SiO_2$ content are rough. The diameter of the nanowires measured from the TEM image ranges from 9 to 12 nm. Most nanowires are quasi-aligned. Obviously, the intensity of the cubic Si (111) diffraction ring exhibits a strong texture feature of the Si crystals in the nanowires. This indicates that the crystals in the nanowires should have a similar orientation, i.e., the Si nanowires should have a similar growth direction.

**Vapor–Solid Growth** Besides the VLS mechanism, the classical vapor–solid (VS) method for whiskers growth also merits attention for the growth of nanometer 1D materials [278]. In this process, the vapor is first generated by evaporation, chemical reduction or gaseous reaction. The vapor is subsequently transported and condensed onto a substrate. The VS method has been used to prepare oxide, metal whiskers with micrometer diameters. Hence it is possible to synthesize the 1D nanostructures using the VS process if one can control its nucleation and subsequent growth process. Using the VS method, synthesis of nanowires for oxides of Zn, Sn, In, Cd, Mg and Ga has been attempted. Seo et al. [293c] synthesized single crystalline gallium phosphide nanowires with mean diameter 40 nm and length up to 300 μm via sublimation of ball-milled gallium phosphide powder. Lee and coworkers [293d] have synthesized large quantities of Si nanowires (6–28 nm diameter, ~1 mm in length) by the simple sublimation of SiO powder. The thermal sublimation of SiO powders produced SiO vapor, which underwent a disproportionation reaction, was transported and deposited at ~930 °C to form nanowires containing a crystalline Si core and an amorphous $SiO_2$ sheath. The axis of the Si nanowires is approximately along the [211] direction. This method has the advantage over the laser-assisted catalytic growth method because it can produce high-purity Si nanowires without any metal contamination. Recently, Ma et al. [294b] have prepared small diameter (1–7 nm) Si nanowires using the above oxide-assisted procedure. They obtained stable, faceted Si nanowire surfaces terminated with hydrogen after removing the $SiO_2$ sheath by dipping these nanowires in hydrofluoric acid. Scanning tunneling microscopy (STM) of these Si nanowires showed atomically resolved images with two types of nanowire surfaces which they interpreted as hydrogen-terminated Si (111)-($1 \times 1$) and Si (001)-($1 \times 1$) surfaces corresponding to $SiH_3$ on Si (111) and $SiH_2$ on Si (001), respectively. Interestingly these hydrogen terminated Si nanowire surfaces are more oxidation-resistant than similarly treated Si wafer surfaces. The scanning tunneling spectroscopy (STS) measurements showed that the electronic energy gaps were found to increase with decreasing Si nanowire diameter from 1.4 eV for 7 nm to 3.5 eV for 1.3 nm. Wang et al. reported the synthesis of other nanostructures such as oxide nanobelts by simply evaporating the commercial metal oxide powders at high temperatures [296–298].

**Carbothermal Reactions** It is noteworthy that a variety of oxides, nitrides and elemental nanowires can be synthesized by carbothermal reactions. For example, carbon (activated carbon or carbon nanotubes) in mixture with an oxide produces oxide or suboxide vapor species, which react with other reactants ($O_2$, $N_2$ or $NH_3$) to produce the desired nanowires. Thus GaN nanowires are produced by heating a mixture of $Ga_2O_3$ and carbon in $N_2$ or $NH_3$. Silicon nanowires can also be made by heating $SiO_2$ with carbon in a suitable atmosphere.

Yang et al. reported the synthesis of MgO, $Al_2O_3$, ZnO, $SnO_2$ nanowires via a carbothermal reduction process [299, 300]. Gundiah et al. [301] employed an indirect vapor–phase method via a carbothermal reduction processes, for the synthesis of silicon carbide, silicon oxynitride and silicon nitride nanowires. The simplest method to obtain $\beta$-SiC nanowires involves heating silica gel with activated carbon at 1360 °C in $H_2$ or $NH_3$ (Figure 8.30). The same reaction, if carried out in the presence of catalytic iron particles, at 1200 °C gives $\alpha$-$Si_3N_4$ nanowires and $Si_2N_2O$ nanowires at 1100 °C. Another method to obtain $Si_3N_4$ nanowires is to heat MWNTs with silica gel at 1360 °C in an atmosphere of $NH_3$. In the presence of catalytic Fe particles, this method yields $Si_3N_4$ nanowires in pure form. The formation of carbide follows two steps (steps I and II shown below), initially carbon reduces the $SiO_2$ to the volatile suboxide of silicon and then the formation of carbide follows.

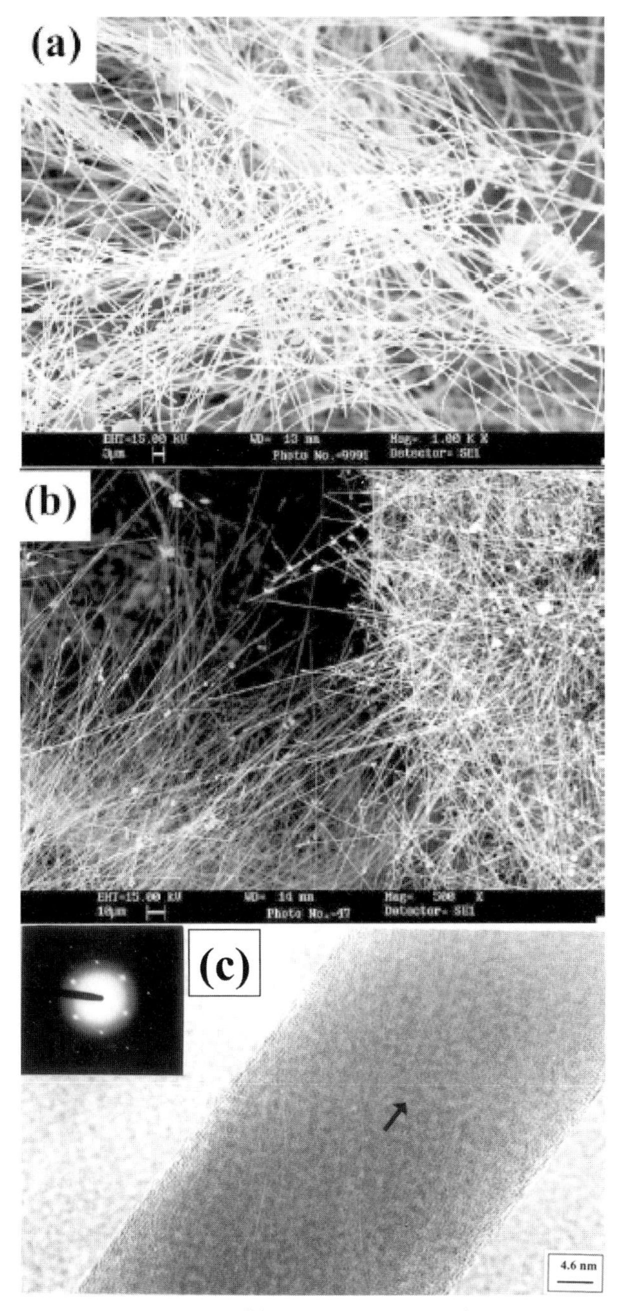


**Fig. 8.30.** SEM images of the $\beta$-SiC nanowires obtained by, (a) heating the gel containing the activated carbon and silica at 1360 °C in $NH_3$ for 4 h and (b) heating the gel prepared by the reaction of ethylene glycol with citric acid in the presence of TEOS at 1360 °C in $NH_3$ for 4 h. (c) HREM image of a $\beta$-SiC nanowire prepared by heating the gel containing the activated carbon and silica at 1360 °C in $NH_3$ for 7 h. Reproduced from ref. [301], with permission.

$$SiO_2 + C \rightarrow SiO + CO \qquad (I)$$

$$SiO + 2C \rightarrow SiC + CO \qquad (II)$$

Hence, by using the correct carbon source with the silica gel under carbothermal conditions it is possible to obtain nitride or carbide nanowires. With the carbon nanotubes the reaction follows one step (Step III shown below) to produce $Si_3N_4$ nanowires.

$$3SiO_2 + 6C + 4NH_3 \rightarrow Si_3N_4 + 6H_2 + 6CO \qquad (III)$$

The role of the catalytic iron particles in the above reactions is likely to be in facilitating the removal of oxygen from the silica. The iron oxide formed in such a reaction would readily be reduced back to metal particles in the reducing atmosphere. Similarly, Gundiah et al. carried out the conversion of $Ga_2O_3$ powder into nanosheets and nanobelts in addition to nanowires under similar carbothermal conditions [302a]. They were able to prepare different nanostructures of $\beta$-$Ga_2O_3$ by the reaction of gallium oxide with activated carbon and carbon nanotubes (Figure 8.31(a) and (b)). The flow rate of the Ar gas determines the morphology of the final nanostructures: thin nanowires being favored by a high flow rate of argon whereas at very low flow rates of argon nanobelts of $Ga_2O_3$ were obtained in high yield. The Reaction of $Ga_2O_3$ powder with activated carbon mainly gives rise to nanosheets and nanorods. The procedures employed in this study are attractive since they give high yields of nanowires and nanobelts. The HREM image in Figure 8.31(c) shows that these $Ga_2O_3$ nanowires are single crystalline with the growth direction perpendicular to the (102) planes. Deepak et al. [302b] prepared gallium nitride nanowires by employing several procedures involving the use of carbon nanotube templates or catalytic Fe (Ni) metal particles. These GaN nanowires are single crystalline, with the wurtzite structure, and have high aspect ratios with lengths in the micron range.

#### 8.4.2.2 **Other Processes in the Gas Phase**

Chen et al. reported the synthesis of another class of semiconductor nanowires of metal silicide systems [303a,b]. In their preparation process, submonolayer amounts of Er deposited onto Si(001) react with the substrate to form epitaxial nanowires of crystalline $ErSi_2$. The $ErSi_2$ nanowires so deposited are $<1$ nm high, a few nanometers wide, close to a micron long, crystallographically aligned to Si $\langle 110 \rangle$ directions. Lauhon and co-workers have grown core–shell and core–multishell nanowire heterostructures using a chemical vapor deposition (CVD) method that provides increased control over the structure's composition [304]. Using their technique, the nanowires are grown by gradually building up thin, uniform shells around a nanometre-sized cluster of gold atoms (Figure 8.32). The nanowires had boron-doped silicon shells surrounding intrinsic silicon, as well as silicon wrapped around a silicon oxide core. These nanowires are only 50 nm in diameter, containing a germanium core surrounded by a silicon shell.

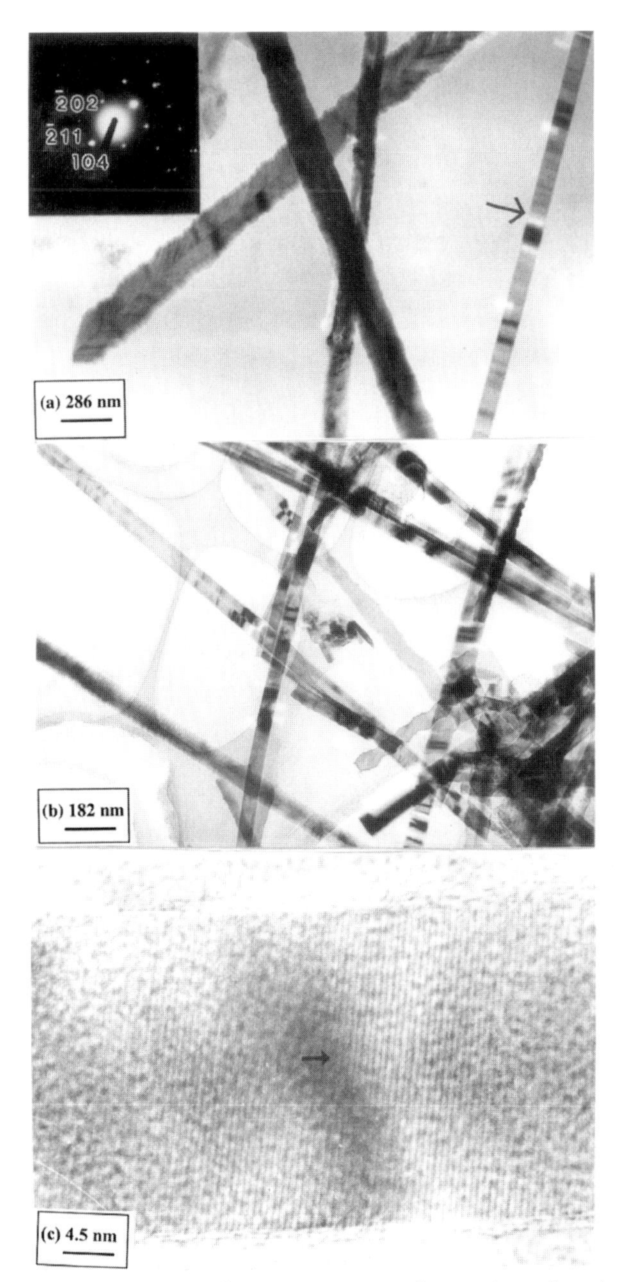


**Fig. 8.31.** Low magnification TEM images of $Ga_2O_3$ nanowires obtained by MWNTs as the source of carbon. (a) At a flow rate of Ar maintained at 40 sccm and the product collected at the inner tube. The arrow shows a nanobelt. Inset is the SAED pattern of the sample. (b) Flow rate of Ar maintained at 80 sccm and the product collected at the outlet. (c) HREM of a nanowire obtained on the inner tube by using the same procedure on maintaining the flow of the Ar gas at 60 sccm. The arrow indicates the growth direction that makes an angle of 6° with the normal to the ($\bar{1}02$) planes. Reproduced from ref. [302a], with permission.

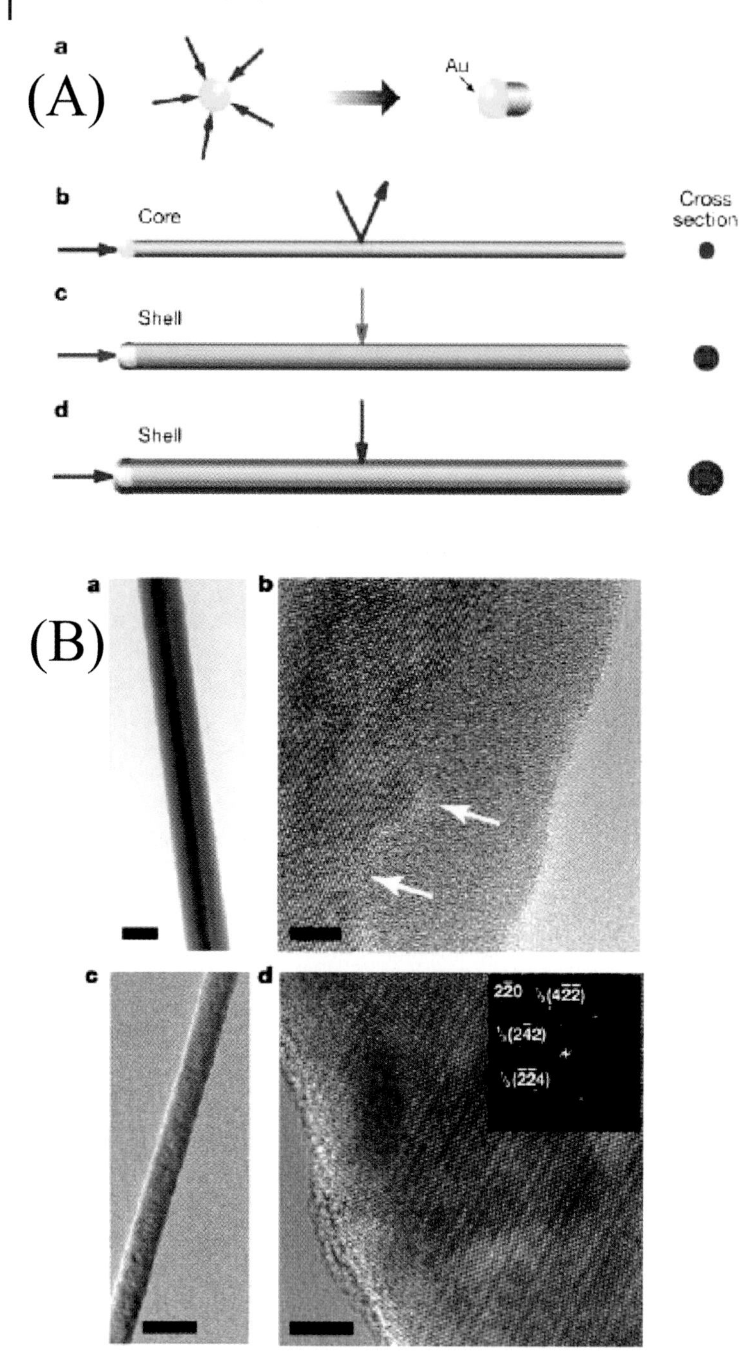
(A)
a
Au
b
Core
Cross section
c
Shell
d
Shell
(B)
a
b
c
d

#### 8.4.2.3 **Solution-Based Growth of Nanowires**

This synthetic strategy for nanowires makes use of anisotropic growth dictated by the crystallographic structure of a solid material; or confined and directed by templates; or kinetically controlled by supersaturation; or by the use of appropriate capping agents.

**Highly Anisotropic Crystal Structures by Non-Template Methods** Many solid materials such as polysulphur nitride, $(SN)_x$, grow into 1D nanostructures and this habit is determined by the highly anisotropic bonding in the crystallographic structure [305a,b]. Other materials such as selenium [306, 307b], tellurium [308] and molybdenum chalcogenides [309a,b], are also easily obtained as nanowires due to the anisotropic bonding, which makes the crystallization occur along the *c*-axis, favoring the stronger covalent bonds over the relatively weak van der Waals forces between the chains. Molybdenum chalcogenides, with the general formula $M_2Mo_6X_6$ (M = Li, Na; X = Se, Te) contain hexagonally close packed linear chains of formula $Mo_6X_6$. When dissolved in a highly polar solvent such as dimethylsulfoxide or *N*-methylformamide, they mainly exist as chains of ~2 nm diameter. Some chains may aggregate into bundles or fibres with cross sections of ~1 μm diameter and lengths up to ~1 μm. Yang and co-workers [309a,b] studied the self-organization of these molecular wires ($Li_2Mo_6Se_6$) into mesoscopic bundles in the presence of organic surfactants of opposite charges. By changing the length of the surfactant molecule, the spacing between these inorganic nanowires could be varied in the range 2–4 nm. Xia et al. [307] synthesized a spherical colloidal suspension/dispersion of amorphous (a-) selenium with diameters of ~300 nm by refluxing selenious acid and hydrazine at elevated temperatures. After cooling the suspension to room temperature a small amount of selenium dissolved in the solution precipitates out as nanocrystallites of triagonal Se (*t*-Se). During aging of this dispersion in the dark, the a-Se dissolves slowly in the solution and subsequently crystallizes out slowly on a *t*-Se seed. The intrinsic anisotropic nature of *t*-Se building blocks, that is extended, helical chains of Se atoms (Figure 8.33(a)) in the

← **Fig. 8.32.** (A) Synthesis of core–shell nanowires by chemical vapor deposition. (a) Gaseous reactants (red) catalytically decompose on the surface of a gold nanocluster leading to nucleation and directed nanowire growth. (b) One-dimensional growth is maintained as reactant decomposition on the gold catalyst is strongly preferred. (c) Synthetic conditions are altered to induce homogeneous reactant decomposition on the nanowire surface, leading to a thin, uniform shell (blue). (d) Multiple shells are grown by repeated modulation of reactants. (B) Si–Si homoepitaxial core–shell nanowires. (a), (b) Diffraction contrast and high-resolution TEM images, respectively, of an unannealed intrinsic silicon core and p-type silicon shell nanowire grown at 450 °C. Crystal facets in the high-resolution TEM image designated by arrows indicate initially epitaxial shell growth at low temperature. Scale bars are 50 nm and 5 nm, respectively. (c), (d) TEM images (analogous to (a) and (b)) of an –Si/p–Si core–shell nanowire annealed at 600 °C for 30 min after core–shell growth at 50 °C. Inset, two-dimensional Fourier transforms of the image depicting the [111] zone axis of the single crystal nanowire. The 1/3{422} reflections, although forbidden in bulk silicon, arise as a result of the finite thickness of the nanowire. Scale bar is 50 nm. Reproduced from ref. [304], with permission.

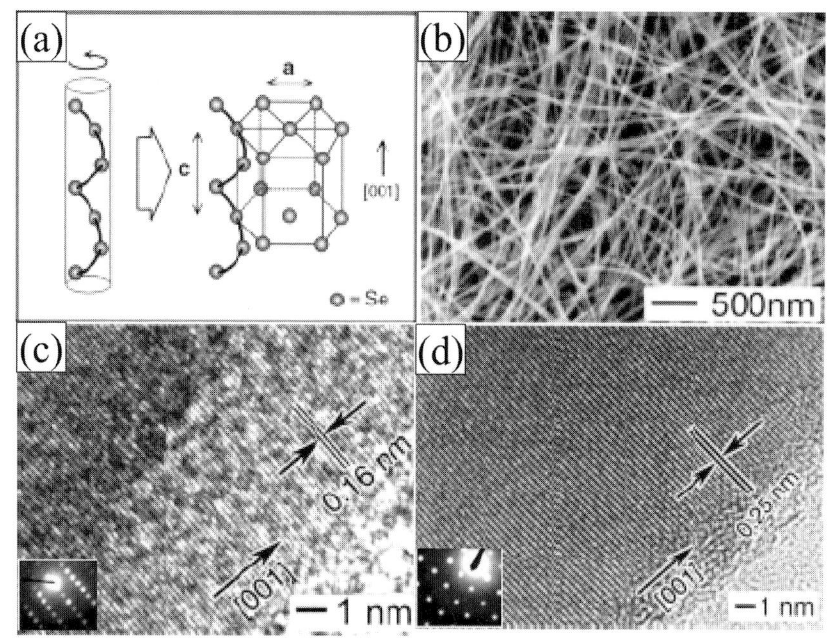


**Fig. 8.33.** (a) An illustration of the crystal structure of *t*-Se composed of hexagonally packed, helical chains of Se atoms parallel to each other along the *c*-axis. (b) Scanning electron microscopy (SEM) image of the T–Se nanowires of 32 nm mean diameter. (c) A high-resolution TEM image recorded from the edge of an individual Se nanowire (inset shows the ED pattern obtained from the middle portion of the nanowire, confirming the growth direction was along the ⟨100⟩ axis) and (d) high-resolution TEM image of an $\alpha$-$Ag_2Se$ nanowire (inset shows the ED pattern corresponding to tetragonal crystal structure). Reproduced from ref. [307, 316b,c], with permission.

trigonal phase assembles (crystallizes) ultimately into nanowires of *t*-Se (Figure 8.33(b)) [307]. Each nanowire is single crystalline, as shown by the high-resolution TEM image (in Figure 8.33(c)) recorded at the edge of a nanowire, showing a well resolved interference fringe spacing, 0.16 nm, that agrees well with the interplanar distance between {003} lattice planes. The electron diffraction pattern (ED) obtained from the middle portion of individual nanowires confirms that the growth direction (inset in Figure 8.33(c)) is along the ⟨001⟩ axis. The plausible explanation for the formation of the Se or Te nanowire is that the helical chains (shown as in the schematic Figure 8.33(a)) can be readily packed into a hexagonal lattice through van der Waals interactions. These non-template directed methods are readily extended to a range of other solid materials whose crystallographic structures are characterized by chain-like building blocks [305c–e]. Some other typical examples including SbSI, a ferroelectric and optoelectronic material [309c,d]; $K_2[Pt(CN)_4]$, a narrow band gap semiconductor [309e], crystallize in the form of whiskers.

**Template-Based Synthesis** Template directed synthesis represents a convenient and versatile method for generating 1D nanostructures. In this technique, the template simply serves as a scaffold against which other kinds of materials with similar morphologies are synthesized. In other words, the in situ generated material is shaped into a nanostructure with its morphology complementary to that of the template. These templates could be nanoscale channels within mesoporous materials or porous alumina and polycarbonate membranes. They can be filled using (i) a solution route or (ii) a sol–gel technique or (iii) an electrochemical route to generate 1D nanowires. The produced nanowires can be released from the templates by selectively removing the host matrix [274]. Unlike the polymer membranes fabricated by track itching, porous AAO membranes containing hexagonally packed 2D array of cylindrical pores with a uniform size are prepared using anodization of aluminum foils in an acidic medium (Figure 8.34(a)). Many materials have been fabricated into nanowires using porous anodic alumina membranes (AAM) in a templating process, including various inorganic materials such as Au, Ag, Pt, $TiO_2$, $MnO_2$, ZnO, $SnO_2$, electronically conducting polymers: polypyrrole, poly(3-methylthiophene), and polyaniline, and carbon nanotubules [310a]. Figure 8.34(b) shows the highly ordered $In_2O_3$ nanowires uniformly assembled into the hexagonally ordered nanochannels of the AAM by oxidizing the In nanowire arrays electrodeposited in the nanochannels of the AAM [310b]. Figure 8.34(c) shows a TEM image of $In_2O_3$ nanowires after removing the AAM from the $In_2O_3$/AAM samples by dissolving the AAM in NaOH solution followed by washing several times with distilled water. Besides alumina and polymer membranes, with their high surface areas and unifirm pore sizes, mesoporous silica materials (MCM-41 or SBA-15) have been successfully used as templates for the synthesis of polymer and inorganic nanowires [311–315]. Ag nanowires of uniform diameters of 5–6 nm and large aspect ratios between 100 and 1000 are synthesized by $AgNO_3$ solution impregnation in SBA-15 or MCM-41 template followed by thermal decomposition. Similarly Ge nanowires have been successfully synthesized within the mesochannels of MCM-41 [315a].

Mesophase structures self-assembled from surfactants (Figure 8.35) provide another class of useful and versatile templates for generating 1D nanostructures in relatively large quantities. It is well known that at critical micellar concentration (CMC) surfactant molecules spontaneously organize into rod-shaped micelles [315c]. These anisotropic structures can be used immediately as soft templates to promote the formation of nanorods when coupled with appropriate chemical or electrochemical reaction. The surfactant needs to be selectively removed to collect the nanorods/nanowires as a relatively pure sample. Based on this principle, nanowires of CuS, CuSe, CdS, CdSe, ZnS and ZnSe have been grown selectively by using surfactants such as Na-AOT or Triton X of known concentrations [238, 246].

The nanowires themselves can be used as templates to generate the nanowires of other materials. The template may be coated onto the nanowire (physically) forming coaxial nanocables [316a], or it may react with the nanowires forming a new material [317a,b]. In the physical (solution or sol–gel coating) approach, the surfaces of the nanowires could be directly coated with conformal sheaths made of a different material to form coaxial nanocables. Subsequent dissolution of the original nanowires could lead to nanotubes of the coated material. The sol–gel

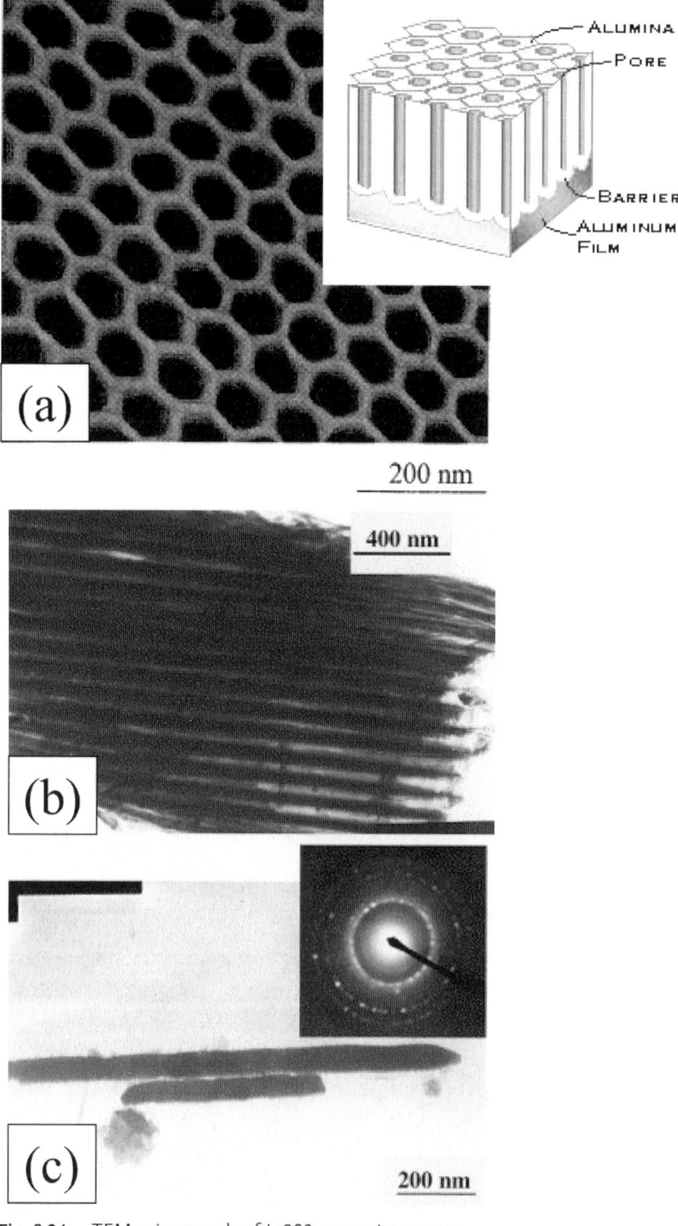


**Fig. 8.34.** TEM micrograph of In203 nanowire arrays embedded in AAM with channel diameters of 60 nm: (a) AFM image of AAO membrane (b) In203/AAM system. (c) In203 nanowires. Reproduced from ref. [310b], with permission.

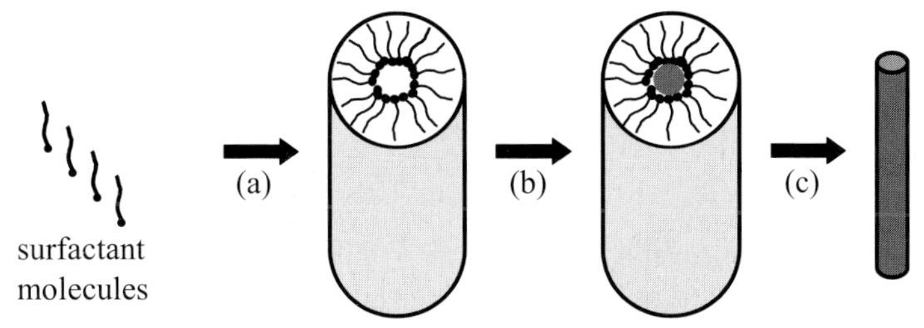


**Fig. 8.35.** Schematic illustrations showing the formation of nanowires by templating against mesostructures which are self-assembled from surfactant molecules; (a) formation of a cylindrical micelle, (b) formation of the desired material in the aqueous phase encapsulated by the cylindrical micelle, (c) removal of the surfactant molecule with an appropriate solvent (or by calcinations) to obtain an individual nanowire.

coating method is a generic route to synthesize coaxial nanocables that may contain electrically conductive metal cores and insulating sheaths [316a]. The thickness of the $SiO_2$ sheath could be controlled in the range 2–100 nm by varying the concentration of the precursor and the deposition time. Figure 8.36(a) shows the TEM image of a typical sample of coaxial nanocables, Ag in $SiO_2$, obtained by coating Ag nanowires with silica derived from a sol–gel precursor. Selective removal of the silver core by dissolving in ammonia gives a silica tube as shown in Figure 8.36(b). Single crystalline $Ag_2Se$ (tetragonal) nanowires of diameters less than 40 nm have been successfully synthesized through a novel topotactic reaction wherein *t*-Se single crystalline nanowire templates react with $AgNO_3$ solutions at room temperature [317a,b]. The high-resolution TEM image (Figure 8.33(d)) obtained from the edge of an individual nanowire (compared with Figure 8.33(c)) indeed shows the complete conversion of Se nanowire into single crystalline tetragonal $Ag_2Se$ nanowire. The fringe spacing of 0.25 nm corresponds to the interplanar distance of [200], implying the growth direction of this nanowire was $\langle 100 \rangle$. Beyond 40 nm diameter the orthorhombic structure becomes more stable. In some other cases this technique, however, intrinsically yields products of a polycrystalline nature.

Template-directed synthesis of metal nanorods covered by carbon and other materials has been reported in the literature [45, 273, 318]. By employing the arc vaporization method, Demoncy et al. [319], have shown the role of sulfur along with the transition metals in the formation of metal-filled MWNTs. Electrolytic formation of carbon-sheathed Sn–Pb nanowires with diameters in the 40–90 nm range has been reported [320]. Sloan et al. [321] find that SWNTs can be filled up to 50% by silver, by employing the KCl–$UCl_4$ and AgCl–AgBr eutectic systems, to produce nanowires. Govindaraj et al. [129], have demonstrated that a variety of metal nanowires of 1.0–1.4 nm diameter can be readily prepared by filling SWNTs, opened by acid treatment. Nanowires of Au, Pt, Pd and Ag have been synthesized by employing sealed-tube reactions as well as solution methods. In addition, incorporation of thin layers of metals in the intertubular space of the SWNT bundles

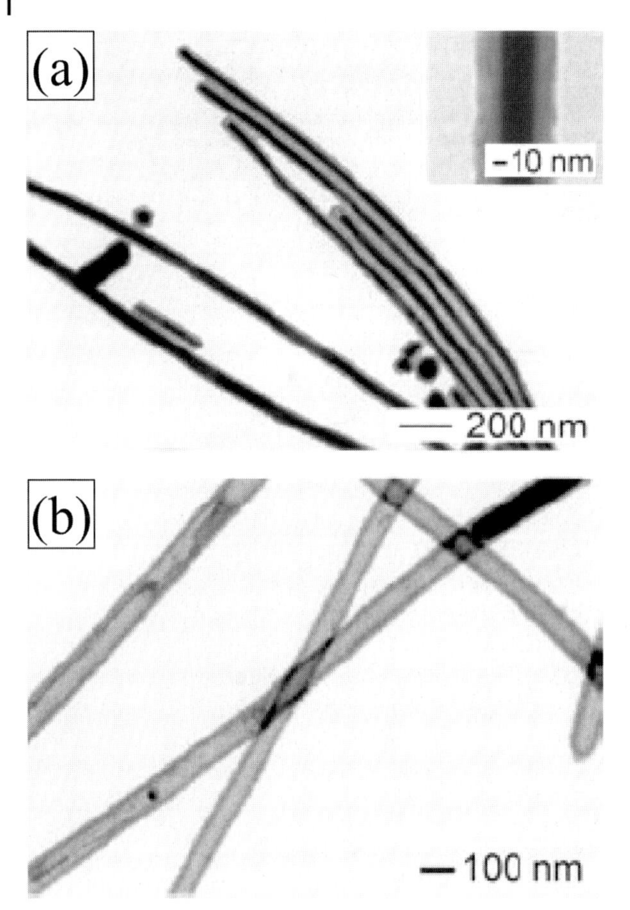


**Fig. 8.36.** (a) TEM image of $Ag/SiO_2$ coaxial nanocables prepared by coating Ag nanowires with amorphous silica using the sol–gel method. (b) TEM image of silica nanotubes prepared by selectively dissolving Ag cores of $Ag/SiO_2$ in ammonia solution (pH 11). Reproduced from ref. [316a], with permission.

has been observed. In Figure 8.37(a), we show a TEM image, which reveals the presence of a large number of gold nanowires obtained from the sealed tube reaction. The image clearly shows extensive filling of SWNT bundles. The length of the gold nanowires is in the range 15–70 nm with diameters in the range 1.0–1.4 nm. In some of the nanowires, however, the metal is single-crystalline, as revealed by the HREM image of Figure 8.37(b). This image reveals the resolved lattice of gold with a spacing of ~0.23 nm, corresponding to the (111) planes. Polycrystalline Au aggregates are transformed into the single crystalline form by annealing. The Au nanowires are decomposed to nanoparticles by electron-beam damage at a temperature of 300–350 °C. Electronic absorption spectra of dispersions of gold nanowires in ethanol show transverse and longitudinal plasmon absorption bands

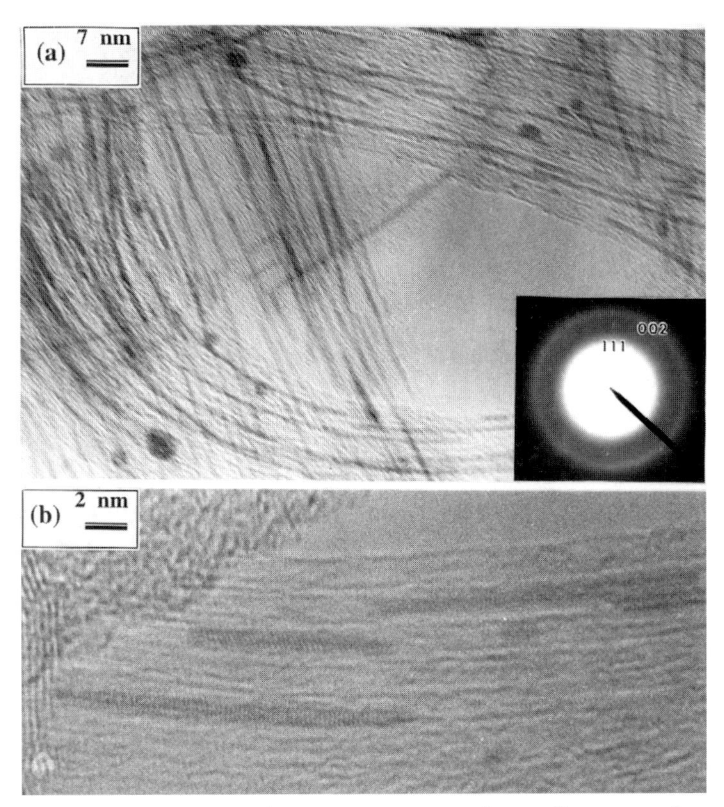


**Fig. 8.37.** (a) TEM image of Au nanowires inside SWNTs obtained by the sealed tube reaction. Inset: A SAED pattern of the nanowires (b) HRTEM image showing the single crystalline nature of some of the Au nanowires. Lattice planes with $d_{111} = 2.3$ Å can be seen. Reproduced from ref. [129], with permission.

[129], with the latter shifting to long wavelengths with the increasing aspect ratio, as predicted by Link et al. [322] Nanowires of bismuth with a diameter of 1 nm have been obtained by filling SWNTs during arc vaporization as well as by a solution method [323]. Interestingly, by employing the anodic aluminum oxide (AAO) templates and organometallic chemical vapor deposition (MOCVD), nickel nanowires of 4 nm diameter have been produced inside the carbon nanotubes [324]. However, the diameter of the AAO templates dictates the diameter of the nanowires, making it difficult to tune the nanowire diameter down to 1 nm. SWNTs with a narrow size distribution seem to have an advantage as ideal templates [323, 325].

**Solution–Liquid–Solid Process** Buhro and coworkers [276], have developed a low temperature solution–liquid–solid (SLS) method for the synthesis of highly crystalline nanowires of III–V semiconductors [276, 327]. In a typical procedure, a metal (e.g., In, Sn, Bi) with a low melting point was used as a catalyst, and the de-

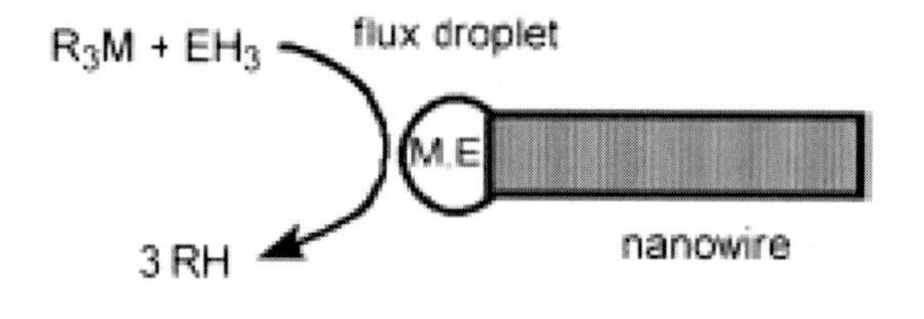


**Fig. 8.38.** (a) Schematic illustration showing the growth of nanowire through the solution–liquid–solid mechanism which is similar to the vapor–liquid–solid process. Reproduced from ref. [280, 333], with permission.

sired material was generated through the decomposition of organometallic precursors. They have grown InP, InAs and GaAs nanowhiskers by low temperature ($\leq$203 °C) solution phase reactions. The schematic illustration in Figure 8.38 clearly shows the growth of nanowires or whiskers through a solution–liquid–solid (SLS) method. The product was essentially single crystalline whiskers or filaments with dimensions of 10–150 nm and length up to several micrometers. A similar low temperature synthesis route is employed for the catalyzed growth of indium nitride fibres from azido-indium precursors [326]. Highly crystalline InP fibres of diameter 10–100 nm and length 50–1000 nm are grown by the methanolysis of $\{tert\text{-}Bu_2In[\mu\text{-}P(SiMe_3)_2]\}_2$ in aromatic solvent at 111–203 °C [326, 327a]. The key component of the synthesis was the decomposition of the organometallic precursor which proceeds through a sequence of isolated and fully characterized intermediates to yield a complex $[tert\text{-}Bu_2In(\mu\text{-}PH_2)]_3$; this complex subsequently underwent alkane elimination to generate the building blocks: $(InP)_n$ fragments. The fragments dissolved in a dispersion of droplets formed by molten indium, and recrystallized as InP fibers.

Korgel et al. [328a,b], by using the supercritical fluid–liquid–solid (SFLS) approach, have grown bulk quantities of defect-free silicon (Si) and germanium (Ge) nanowires with nearly uniform diameters ranging from 40–50 Å (Si), 50–300 Å (Ge) and with lengths of several micrometers. They used solvent-dispersed, size-monodisperse alkane thiol-capped gold (Au) nanocrystals to direct the Si nanowire growth with narrow wire diameter distributions [328a]. Sterically stabilized Au nanocrystals were dispersed in supercritical hexane with a silicon precursor, diphenysilane at a temperature of 500 °C and pressure of 270 bar. At this temperature, the diphinylsilane decomposes to Si atoms which dissolve into the sterically stabilized Au nanocrystals until reaching supersaturation, at which point they are expelled from the particle as a thin nanometer-scale wire. This supercritical fluid medium provides the high temperatures necessary to promote Si crystallization.

In addition to these solution routes to elemental and III–V semiconductor nanowires, it has recently been reported that, by exploring the selective capping capabilities of mixed surfactants, it is now possible to extend the well-established

II–IV semiconductor nanocrystal synthesis to the synthesis of semiconductor nanorods [329, 330a], a unique version of nanowires with relatively shorter aspect ratio. Xia and co-workers used a polyol method to generate the Ag nanowires by reducing silver nitrate with ethylene glycol in the presence of polyvinyl pyrrolidone (PVP) [330b–d]. The key to the formation of a 1D nanostructure is the use of PVP as a polymeric capping reagent and the introduction of a seeding step. Silver nitrate reduces in the presence of the seed (Pt, or Ag nanoparticles) to form silver nanoparticles with bimodal size distribution produced via homogeneous and heterogeneous nucleation processes. Ag nanorods grow at the expense of the small Ag nanoparticles, as directed by the capping reagent (PVP). In the presence of PVP, most silver particles can be confined and directed to grow into nanowires of uniform diameter. These nanowires have FCC structure with mean diameter ~40 nm.

**Solvothermal Synthesis** Solvothermal methodology has been extensively examined as a possible solution route to semiconductor nanowires and nanorods. In this process, a solvent is mixed with certain metal precursors and possibly crystal growth regulating or templating agents such as amines. This solution mixture was then placed in an autoclave kept at relatively high temperature and pressure to carry out the crystal growth and assembly process. The methodology seems to be quite versatile and has been demonstrated to be able to produce many different crystalline semiconductor nanorods and nanowires [331, 332]. Xia et al. [307] demonstrated a solution-phase approach by refluxing selenious acid and excess hydrazine for the synthesis of uniform nanowires of selenium with lateral dimensions controllable in the range 10–30 nm and with lengths of up to hundreds of micrometers.

#### 8.4.2.4 Growth Control

A significant challenge for the chemical synthesis is how to rationally control the nanostructure assembly so that the size, dimensionality, interfaces and ultimately the 2-D and 3-D superstructures can be tailormade to a desired functionality. Many physical and thermodynamic properties are diameter dependent. Yang et al. have used monodispersed Au clusters with sizes 15.3, 20, 25, 29, 52 nm and obtained uniform nanowires with sizes from 23 to 57 nm respectively.

Controlling the growth orientation is important for the applications of nanowires. By applying the conventional epitaxial crystal growth technique to the VLS process, a new vapor–liquid–solid epitaxy (VLSE) technique, has been developed for the controlled synthesis of nanowire arrays. Nanowires generally have preferred growth directions. For example, Si nanowires prefer to grow along the $\langle 111 \rangle$ direction. Hence if (111) Si wafer is used as substrate, Si nanowires will grow epitaxially and vertically on the substrate [333, 334]. Similarly ZnO nanowires prefer to grow along the $\langle 001 \rangle$ direction [335].

It is clear from the VLS nanowire growth mechanism that the positions of the nanowires can be controlled by the initial positions of the Au clusters or Au thin films. By creating desired patterns of Au using the lithographic technique it is possible to grow ZnO nanowires of the same designed pattern since they grow vertically only from the region that is coated with Au and form the designed patterns of ZnO nanowire arrays [333, 334]. Similarly, a network of nanowires with the precise placement of individual nanowires on the substrate with the desired configuration is achieved by the surface patterning strategy [333, 334].

Integration of nanowire building blocks into complex functional networks in a predictable and controlled way is a major scientific challenge. In the very first technique the direct one-step growth process was used [333, 334]. The other technique is to put the nanowire building blocks together into the functional structure to develop a suitable hierarchical assembly. The atomic force microscope has been used to push or deposit the nanotubes into the desired configuration [336]. By using a simple dubbed microfluidic assisted nanowire integration process wherein nanowire solution/suspension was filled in the microchannels formed between poly(dimethylsiloxane) (PDMS) micromold and a flat Si substrate, followed by evaporation of the solvent, nanowire surface patterning and alignment was achieved [337, 338]. Duan has shown by dispersing Si and InP nanowires between the fabricated metal electrode arrays and then by applying a bias (inducing electric field), that it is possible to get alignment of Si and InP nanowires [339]. Kim has used a Langmuir–Blodgett technique to get aligned, high-density nanowire assemblies [340].

### 8.4.3 Properties of Nanowires

Compared with bulk materials, low dimensional nanoscale materials, with their large surface area and possible quantum confinement effect, exhibit distinct electric, optical, chemical and thermal properties. Korgel et al. have shown that the absorption edge of the Si nanowires was strongly blue shifted from the bulk indirect band gap of 1.1 eV [328, 341]. They also observed sharp discrete absorbance features and relatively strong "band edge" photoluminescence. They attribute this observation to a quantum confinement effect as well as to lattice orientation of the nanowires. Lieber et al. studied the fundamental PL properties of individual isolated indium phosphide nanowires [342]. These polarization sensitive measurements reveal a striking anisotropy in the PL intensity recorded parallel and perpendicular to the long axis of the nanowire. This intrinsic anisotropy was used to create polarization-sensitive nanoscale photodetectors, which may be useful in optical switches, high-resolution detectors and integrated photonic circuits.

Huang et al. have demonstrated the room temperature ultra-violet lasing of ZnO nanowires [335]. The observed lasing action in these nanowire arrays without any fabricated mirror indicate these single-crystalline, facetted nanowire arrays can indeed function as natural resonance cavities. This nanowire lasing has been further confirmed with the optical characterization of single ZnO nanowire by near-field scanning microscopy (NSOM) [343]. Semiconductor nanowires have recently been used as building blocks for assembling a range of nanodevices including FETs, p–n diodes, bipolar junction transistors, and complementary inverters [344–348]. The nanotube and nanowires with sharp tips are promising materials for applications as cold cathode field emission devices. Field emission characteristics of both $\beta$-SiC and Si nanowires have been investigated using current voltage measurements and the Fowler–Northeim equation. These Si and SiC exhibit robost field emission with turn-on fields of 15 and 20 V $\mu m^{-1}$, respectively [286, 349], with current density of 0.01 mA $cm^{-2}$, which are comparable with those of other field emitters like diamond and carbon nanotubes.

## References


1 H. W. Kroto, J. R. Heath, S. C. O'Brien et al., *Nature* **1985**, *318*, 162–163.
2 W. Kratschmer, L. D. Lamb, K. Fostiropoulos et al., *Nature* **1990**, *347*, 354–358.
3 S. Iijima, *Nature* **1991**, *354*, 56–58.
4 S. Iijima, T. Ichihashi, *Nature* **1993**, *363*, 603–605.
5 D. S. Bethune, C. H. Kiang, M. S. de Vries et al., *Nature* **1993**, *363*, 605–607.
6 W. K. Hsu, J. P. Hare, M. Terrones et al., *Nature* **1995**, *377*, 687.
7 M. Endo, K. Takeuchi, S. Igarashi et al., *J. Phys. Chem. Solids* **1993**, *54*, 1841–1848.
8 (a) P. M. Ajayan, T. Ichihashi, S. Iijima, *Chem. Phys. Lett.* **1993**, *202*, 384–388; (b) S. G. Louie, *Top. Appl. Phys.* **2001**, *80*, 113–145.
9 S. Iijima, P. M. Ajayan, T. Ichihashi, *Phys. Rev. Lett.* **1992**, *69*, 3100–3103.
10 V. P. Dravid, X. Lin, Y. Wang et al., *Science* **1993**, *259*, 1601–1604.
11 P. M. Ajayan, T. W. Ebbesen, T. Ichihashi et al., *Nature* **1993**, *362*, 522–525.
12 S. C. Tsang, P. J. F. Harris, M. L. H. Green, *Nature* **1993**, *362*, 520–522.
13 S. C. Tsang, Y. K. Chen, P. J. F. Harris et al., *Nature* **1994**, *372*, 159–162.
14 C. N. R. Rao, R. Seshadri, A. Govindaraj et al., *Mater. Sci. Eng., R.* **1995**, *15*, 209–262; P. M. Ajayan, *Chem. Rev.* **1999**, *99*, 1787–1799.
15 Special issue on Carbon Nanotubes, *Appl. Phys. A.* **1998**, *67*, 1–119; *Appl. Phys. A.* **1999**, *69*, 245–312; Carbon Nanotubes, *J. Mater. Res.* **1998**, *13*, 2355–2423; articles in *J. Phys. Chem.* **2000**, 104 and *Physics World*, June **2000**.
16 *AIP Conf. Proc.* **1999**, *442*.
17 M. S. Dresselhaus, G. Dresselhaus, P. C. Eklund, in Science of Fullerenes and Carbon Nanotubes, Academic Press, San Diego 1996, 756–917.
18 Carbon Nanotubes: (special issue), *Acc. Chem. Res.* **2002**, *35*, 998–1113.
19 M. Terrones, W. K. Hsu, H. W. Kroto et al., *Top. Curr. Chem.* **1999**, *199*, 190–234.
20 (a) C. N. R. Rao, B. C. Satishkumar, A. Govindaraj et al., *Chem. Phys. Chem.* **2001**, *2*, 78–105; (b) R. H. Baughman, A. A. Zakhidov, W. A. de Heer, *Science* **2002**, *297*, 787–792.
21 P. Yang, Y. Wu, R. Fan, *Int. J. Nanoscience* **2002**, 1–39.
22 C. N. R. Rao, M. Nath, *Dalton Trans.* **2003**, 1–24.
23 R. Tenne, *Chem. Eur. J.* **2002**, *8* (23), 5296–5304.
24 G. R. Patzke, F. Krumeich, R. Nesper, *Angew. Chem. Int. Ed. Engl.* **2002**, *41*, 2446–2461.
25 T. W. Ebbesen, P. M. Ajayan, *Nature* **1992**, *358*, 220–222.
26 N. Hatta, K. Murata, *Chem. Phys. Lett.* **1994**, *217*, 398–402.
27 D. T. Colbert, J. Zhang, S. M. McClure et al., *Science* **1994**, *266*, 1218–1222.
28 M. Ge, K. Sattler, *Science* **1993**, *260*, 515–518.
29 (a) W. K. Hsu, M. Terrones, J. P. Hare et al., *Chem. Phys. Lett.* **1994**, *262*, 161–166; (b) Y. Gogotsi, J. A. Libera, M. Yoshimura, *J. Mater. Res.* **2000**, *15*, 2591–2594.

30 M. Jose-Yacaman, M. Miki-Yoshida, L. Rendon et al., *Appl. Phys. Lett.* **1993**, *62*, 202–204.
31 V. Ivanov, J. B. Nagy, Ph. Lambin et al., *Chem. Phys. Lett.* **1994**, *223*, 329–335.
32 K. Hernadi, A. Fonseca, J. B. Nagy et al., *Synth. Met.* **1996**, *77*, 31–34.
33 N. M. Rodriguez, *J. Mater. Res.* **1993**, *8*, 3233–3250.
34 R. Sen, A. Govindaraj, C. N. R. Rao, *Chem. Phys. Lett.* **1997**, *267*, 276–280.
35 R. Sen, A. Govindaraj, C. N. R. Rao, *Chem. Mater.* **1997**, *9*, 2078–2081.
36 W. A. de Heer, J. M. Bonard, K. Fauth et al., *Adv. Mater.* **1997**, *9*, 87–89.
37 W. Z. Li, S. S. Xie, L. X. Qian et al., *Science* **1996**, *274*, 1701–1703.
38 G. Che, B. B. Laxmi, C. R. Martin et al., *Chem. Mater.* **1998**, *10*, 260–267.
39 M. Terrones, N. Grobert, J. Olivares et al., *Nature (London)* **1997**, *388*, 52–55.
40 M. Terrones, N. Grobert, J. P. Zhang et al., *Chem. Phys. Lett.* **1998**, *285*, 299–305.
41 Z. F. Ren, Z. P. Huang, J. W. Xu et al., *Science* **1998**, *282*, 1105–1107.
42 S. Fan, M. C. Chapline, N. R. Franklin et al., *Science* **1999**, *283*, 512–514.
43 Z. W. Pan, S. S. Xie, B. H. Chang et al., *Chem. Phys. Lett.* **1999**, *299*, 97–102.
44 (a) B. C. Satishkumar, A. Govindaraj, C. N. R. Rao, *Chem. Phys. Lett.* **1999**, *307*, 158–162; (b) C. N. R. Rao, R. Sen, B. C. Satishkumar et al., *Chem. Commun.* **1998**, 1525–1526.
45 C. N. R. Rao, A. Govindaraj, R. Sen et al., *Mater. Res. Innov.* **1998**, *2*, 128–141.
46 (a) R. Andrews, D. Jacques, A. M. Rao et al., *Chem. Phys. Lett.* **1999**, *303*, 467–474; (b) S. Huang, A. W. H. Mau, T. W. Turney et al., *J. Phys. Chem. B* **2000**, *104*, 2193–2196.
47 (a) J. Li, C. Papadopoulos, J. M. Xu, M. Moskovits, *Appl. Phys. Lett.* **1999**, *75*, 367–369; (b) K. Mukhopadhyay, A. Koshio, T. Sugai et al., *Chem. Phys. Lett.* **1999**, *303*, 117–124.
48 (a) C. N. R. Rao, A. Govindaraj, *Acc. Chem. Res.* **2002**, *35*, 998–1007; (b) B. C. Satishkumar, A. Govindaraj, P. V. Vanitha et al., *Chem. Phys. Lett.* **2002**, *362*, 301–306.
49 H. Dai, A. Z. Rinzler, P. Nikolaev et al., *Chem. Phys. Lett.* **1996**, *260*, 471–475.
50 Y. Saito, M. Okuda, T. Koyama, *Surf. Rev. Lett.* **1996**, *3*, 863–864.
51 Y. Saito, K. Kawabata, M. Okuda, *J. Phys. Chem.* **1995**, *99*, 16076–16079.
52 C. Journet, W. K. Maser, P. Bernier et al., *Nature* **1997**, *388*, 756–758.
53 A. Thess, R. Lee, P. Nikolaev et al., *Science* **1996**, *273*, 483–487.
54 B. C. Satishkumar, A. Govindaraj, R. Sen et al., *Chem. Phys. Lett.* **1998**, *293*, 47–52.
55 S. Seraphin, D. Zhou, *Appl. Phys. Lett.* **1994**, *64*, 2087–2089.
56 H. M. Cheng, F. Li, G. Su, H. Y. Pan et al., *Appl. Phys. Lett.* **1998**, *72*, 3282–3284.
57 D. Laplaze, P. Bernier, W. F. Maser et al., *Carbon* **1998**, *36*, 685–688.
58 P. Nikolaev, M. Bronikowski, R. K. Bradley et al., *Chem. Phys. Lett.* **1999**, *313*, 91–97.
59 B. Kitiyanan, W. E. Alvarez, J. H. Harwell et al., *Chem. Phys. Lett.* **2000**, *317*, 497–503.
60 J. F. Colomer, C. Stefan, S. Lefrant et al., *Chem. Phys. Lett.* **2000**, *317*, 83–89.
61 E. Flahaut, A. Govindaraj, A. Peigney et al., *Chem. Phys. Lett.* **1999**, *300*, 236–242.
62 H. D. Sun, Z. K. Tang, J. Chen et al., *Appl. Phys. A* **1999**, *69*, 381–384.
63 S. Bandow, A. M. Rao, K. A. Williams et al., *Phys. Chem. B* **1997**, *101*, 8839–8842.
64 G. S. Duesberg, J. Muster, V. Krstic et al., *Appl. Phys. A* **1998**, *67*, 117–119.
65 J. Liu, A. G. Rinzler, H. Dai et al., *Science* **1998**, *280*, 1253–1256.
66 S. Iijima, *MRS Bull.* **1994**, *19*, 43–47.
67 J. Tersoff, R. S. Ruoff, *Phys. Rev. Lett.* **1994**, *73*, 676–679.
68 D. Srivastava, D. W. Brenner, J. D. Schall et al., *J. Phys. Chem. B* **1999**, *103*, 4330–4337.
69 R. Martel, H. R. Shea, Ph. Avouris, *J. Phys. Chem. B* **1999**, *103*, 7551–7556.

70 (a) M. Ahlskog, E. Seynaeve, R. J. M. Vullers et al., *Chem. Phys. Lett.* **1999**, *300*, 202–206; (b) S. Huang, L. Dai, A. W. H. Mau, *J. Mater. Chem.* **1999**, *9*, 1221–1222.
71 B. W. Smith, M. Monthioux, D. E. Luzzi, *Nature (London)* **1998**, *396*, 323–324.
72 B. Burteaux, A. Claye, B. W. Smith et al., *Chem. Phys. Lett.* **1999**, *310*, 21–24.
73 J. Sloan, R. E. Dunin-Borkowski, J. L. Hutchison et al., *Chem. Phys. Lett.* **2000**, *316*, 191–198.
74 R. Seshadri, A. Govindaraj, H. N. Aiyer et al., *Curr. Sci. (India)* **1994**, *66*, 839–847.
75 Y. Murakami, T. Shibata, K. Okuyama et al., *J. Phys. Chem. Solids* **1993**, *54*, 1861–1870.
76 W. Ruland, in *Chemistry and Physics of Carbon*, vol 4, ed. P. L. Walker, Marcel Dekker, New York 1968.
77 B. E. Warren, *Phys. Rev.* **1941**, *59*, 693–698.
78 T. W. Ebbesen, H. Hiura, J. Fujita et al., *Chem. Phys. Lett.* **1993**, *209*, 83–90.
79 Y. Ando, *Jpn. J. Appl. Phys. Lett.* **1993**, *32*, L1342–1345.
80 X. K. Wang, X. M. Lin, V. P. Dravid et al., *Appl. Phys. Lett.* **1993**, *62*, 1881–1883.
81 Z. Zhang, C. M. Lieber, *Appl. Phys. Lett.* **1993**, *62*, 2792–2794.
82 C. H. Olk, J. P. Haremans, *J. Mater. Res.* **1994**, *9*, 259–262.
83 R. Seshadri, H. N. Aiyer, A. Govindaraj et al., *Solid State Commun.* **1994**, *91*, 195–199.
84 D. L. Carroll, Ph. Redlich, P. M. Ajayan et al., *Phys. Rev. Lett.* **1997**, *78*, 2811–2814.
85 L. C. Venema, V. Meunier, Ph. Lambin et al., *Phys. Rev. B* **2000**, *61*, 2991–2996.
86 R. A. Jishi, L. Venkataraman, M. S. Dresselhaus et al., *Chem. Phys. Lett.* **1993**, *209*, 77–82.
87 H. Hiura, T. W. Ebbesen, K. Tanigaki et al., *Chem. Phys. Lett.* **1993**, *202*, 509–512.
88 J. M. Holden, P. Zhou, X. X. Bi et al., *Chem. Phys. Lett.* **1994**, *220*, 186–191.
89 A. M. Rao, E. Richter, S. Bandow et al., *Science* **1997**, *275*, 187–191.
90 A. Kasuya, Y. Sasaki, Y. Saito et al., *Phys. Rev. Lett.* **1997**, *78*, 4434–4437.
91 A. M. Rao, A. Jorio, M. A. Pimenta et al., *Phys. Rev. Lett.* **2000**, *84*, 1820–1823.
92 P. V. Teredesai, A. K. Sood, D. V. S. Muthu et al., *Chem. Phys. Lett.* **2000**, *319*, 296–302.
93 A. K. Sood, P. V. Teredesai, D. V. S. Muthu et al., *Phys. Status Solidi B* **1999**, *215*, 393–401.
94 S. Bandow, S. Asaka, Y. Saito et al., *Phys. Rev. Lett.* **1998**, *80*, 3779–3782.
95 U. D. Venkataraman, A. M. Rao, E. Richter et al., *Phys. Rev. B* **1999**, *59*, 10928–10934.
96 J. R. Wood, M. D. Frogley, E. R. Meurs et al., *J. Phys. Chem. B* **1999**, *103*, 10388–10392.
97 H. D. Sun, Z. K. Tang, J. Chen et al., *Solid State Commun.* **1999**, *109*, 365–369.
98 S. A. Chesnokov, V. A. Nalimova, A. G. Rinzler et al., *Phys. Rev. Lett.* **1999**, *82*, 343–346.
99 M. J. Peters, L. E. McNeil, J. P. Lu et al., *Phys. Rev. B* **2000**, *61*, 5939–5944.
100 R. T. K. Baker and P. S. Harris, in *Chemistry and Physics of Carbon*, eds. P. L. Walker, P. A. Thrower, Marcel Dekker, New York 1978, vol. 14.
101 A. Oberlin, M. Endo, T. Koyama, *J. Cryst. Growth.* **1976**, *32*, 335–349.
102 M. Endo, H. W. Kroto, *J. Phys. Chem.* **1992**, *96*, 6941–6944.
103 R. E. Smalley, *Mater. Sci. Eng., B* **1993**, *19*, 1–7.
104 S. Iijima, P. M. Ajayan, T. Ichihashi, *Phys. Rev. Lett.* **1992**, *69*, 3100–3103.
105 T. W. Ebbesen, J. Tabuchi, K. Tanigaki, *Chem. Phys. Lett.* **1992**, *191*, 336–338.
106 Y. Saito, T. Yoshikawa, M. Inagaki et al., *Chem. Phys. Lett.* **1994**, *204*, 277–282.
107 D. H. Robertson, D. W. Brenner, C. T. White, *J. Phys. Chem.* **1992**, *96*, 6133–6135.
108 (a) S. Amelinckx, X. B. Zhang, D. Bernaerts et al., *Science* **1994**, *265*, 635–639; (b) S. Amelinckx, D.

Bernaerts, X. B. Zhang et al., *Science* **1995**, *267*, 1334–1337.

109 A. Maiti, C. T. Brabec, J. Bernholc, *Phys. Rev. B* **1997**, *55*, R6097–6100.

110 Y. Miyamoto, A. Rubio, M. L. Cohen et al., *Phys. Rev. B* **1994**, *50*, 4976–4979.

111 B. C. Satishkumar, A. Govindaraj, K. R. Harikumar et al., *Chem. Phys. Lett.* **1999**, *300*, 473–477.

112 O. Stephan, P. M. Ajayan, C. Colliex et al., *Science* **1994**, *266*, 1683–1685.

113 Y. Zhang, H. Gu, K. Suenaga et al., *Chem. Phys. Lett.* **1997**, *279*, 264–269.

114 M. Terrones, A. M. Benito, C. Mantega-Diego et al., *Chem. Phys. Lett.* **1996**, *257*, 576–582.

115 (a) R. Sen, B. C. Satishkumar, A. Govindaraj et al., *Chem. Phys. Lett.* **1998**, *287*, 671–676; (b) R. Sen, B. C. Satishkumar, A. Govindaraj et al., *J. Mater. Chem.* **1997**, *7*, 2335–2337.

116 (a) M. Nath, B. C. Satishkumar, A. Govindaraj et al., *Chem. Phys. Lett.* **2000**, *322*, 333–340; (b) S. L. Sung, S. H. Tsai, C. H. Tseng et al., *Appl. Phys. Lett.* **1999**, *74*, 197–199.

117 (a) K. Suenaga, M. Yusadaka, C. Colliex et al., *Chem. Phys. Lett.* **2000**, *316*, 365–372; (b) K. Suenaga, M. P. Johansson, N. Hellgren et al., *Chem. Phys. Lett.* **1999**, *300*, 695–700.

118 D. Goldberg, Y. Bando, W. Han et al., *Chem. Phys. Lett.* **1999**, *308*, 337–342.

119 D. L. Carroll, Ph. Redlich, X. Blasé et al., *Phys. Rev. Lett.* **1998**, *81*, 2332–2335.

120 K. Efsarjani, A. A. Farajin, Y. Hashi et al., *Appl. Phys. Lett.* **1999**, *74*, 79–81.

121 R. D. Antonov, A. T. Johnson, *Phys. Rev. Lett.* **1999**, *83*, 3274–3276.

122 P. M. Ajayan, T. W. Ebbesen, T. Ichihashi et al., *Nature* **1993**, *362*, 522–524.

123 S. C. Tsang, P. J. F. Harris, M. L. H. Green, *Nature* **1993**, *362*, 520–522.

124 R. M. Lago, S. C. Tsang, K. L. Lu et al., *J. Chem. Soc., Chem. Commun.* **1995**, 1355–1356.

125 K. C. Hwang, *J. Chem. Soc., Chem. Commun.* **1995**, 173–174.

126 H. Hiura, T. W. Ebbesen, K. Tanigaki, *Adv. Mater.* **1995**, *7*, 275–277.

127 B. C. Satishkumar, A. Govindaraj, G. N. Subbanna et al., *J. Phys. B, Atm. Mol. Opt. Phys.* **1996**, *29*, 4925–4936.

128 C. Bower, S. Suzuki, K. Tanigaki et al., *Appl. Phys. A* **1998**, *67*, 47–52.

129 A. Govindaraj, B. C. Satishkumar, M. Nath et al., *Chem. Mater.* **2000**, *12*, 202–204.

130 W. K. Hsu, W. Z. Li, Y. Q. Zhu et al., *Chem. Phys. Lett.* **2000**, *317*, 77–82.

131 C. N. R. Rao, A. Govindaraj, B. C. Satishkumar, *Chem. Commun.* **1996**, 1525–1526.

132 J. Sloan, J. Hammer, M. Zwiefka-Sibley et al., *Chem. Commun.* **1998**, 347–348.

133 B. C. Satishkumar, E. M. Vogl, A. Govindaraj et al., *J. Phys. D, Appl. Phys.* **1996**, *29*, 3173–3176.

134 J. Chen, M. A. Hamon, H. Hu et al., *Science* **1998**, *282*, 95–98.

135 E. T. Mickelson, I. W. Chiang, J. L. Zimmerman et al., *J. Phys. Chem. B* **1999**, *103*, 4318–4322.

136 K. F. Kelly, I. W. Chiang, E. T. Mickelson et al., *Chem. Phys. Lett.* **1999**, *313*, 445–450.

137 P. J. Boul, J. Liu, E. T. Mickelson et al., *Chem. Phys. Lett.* **1999**, *310*, 367–372.

138 J. Liu, M. J. Casavant, M. Cox et al., *Chem. Phys. Lett.* **1999**, *303*, 125–129.

139 J. W. Mintmire, B. I. Dunlap, C. T. White, *Phys. Rev. Lett.* **1992**, *68*, 631–634.

140 N. Hamada, S. Sawada, A. Yoshiyama, *Phys. Rev. Lett.* **1992**, *68*, 1579–1581.

141 R. Saito, M. Fujita, G. Dresselhaus et al., *Appl. Phys. Lett.* **1992**, *60*, 2204–2206.

142 J. W. G. Wildoer, L. C. Venema, A. G. Rinzler et al., *Nature* **1998**, *391*, 59–62.

143 T. W. Odom, J. L. Huang, P. Kim et al., *Nature* **1998**, *391*, 62–64.

144 P. G. Collins, A. Zettl, H. Bando et al., *Science* **1997**, *278*, 100–103.

145 W. Liang, M. Bockrath, D. Bozovic et al., *Nature* **2001**, *411*, 665–669.

146 S. Frank, P. Poncharal, Z. L. Wang et al., *Science* **1998**, *280*, 1744–1746.
147 P. Kim, L. Shi, A. Majumdar et al., *Phys. Rev. Lett.* **2001**, *87*, 215502-1–215502-4.
148 M. Kociak, A. Yu. Kasumov, S. Guéron et al., *Phys. Rev. Lett.* **2001**, *86*, 2416–2419.
149 Z. K. Tang, L. Zhang, N. Wang et al., *Science* **2001**, *292*, 2462–2465.
150 K. H. An, W. S. Kim, Y. S. Park et al., *Adv. Funct. Mater.* **2001**, *11*, 387–392.
151 C. Niu, E. K. Sickel, R. Hoch et al., *Appl. Phys. Lett.* **1997**, *70*, 1480–1482.
152 C. Niu, J. Kupperschmidt, R. Hock, in *Proceedings of the 39th Power Sources Conference* (Maple Hill, NJ, 2000), pp. 314–317.
153 R. H. Baughman, C. Cui, A. A. Zakidov et al., *Science* **1999**, *284*, 1340–1344.
154 B. Gao, A. Kleinhammes, X. P. Tang et al., *Chem. Phys. Lett.* **1999**, *307*, 153–157.
155 F. Salver-Disma, C. Lenain, B. Beaudoin et al., *Solid State Ionics* **1997**, *98*, 145–158.
156 Y. Zhang, T. Ichihashi, E. Landree et al., *Science* **1999**, *285*, 1719–1722.
157 Z. Yao, H. W. Ch. Postma, L. Balents et al., *Nature* **1999**, *402*, 273–276.
158 J. Hu, M. Ouyang, P. Yang, C. M. Lieber, *Nature* **1999**, *399*, 48–51.
159 Z. Yao, C. L. Kane, C. Dekker, *Phys. Rev. Lett.* **2000**, *84*, 2941–2944.
160 Z. Yao, C. Dekker, Ph. Avouris, *Top. Appl. Phys.* **2001**, *80*, 147–171.
161 S. J. Tans, A. R. M. Verschueren, C. Dekker, *Nature* **1998**, *393*, 49–52.
162 R. Martel, T. Schmidt, H. R. Shea et al., *Appl. Phys. Lett.* **1998**, *73*, 2447–2449.
163 A. Bachtold, P. Hadley, T. Nakanishi et al., *Science* **2001**, *294*, 1317–1320.
164 P. G. Collins, M. S. Arnold, Ph. Avouris, *Science* **2001**, *292*, 706–709.
165 M. S. Fuhrer, J. Nygard, L. Shih et al., *Science* **2000**, *288*, 494–497.
166 T. Rueckes, K. Kim, E. Joseluich et al., *Science* **2000**, *289*, 94–97.
167 (a) L. Chico, V. H. Crespi, L. X. Benedict et al., *Phys. Rev. Lett.* **1996**, *76*, 971–974; (b) L. Kouwenhoven, *Science* **1997**, *275*, 1896–1897; (c) P. L. McEuen, *Nature (London)* **1998**, *393*, 15–17.
168 M. Menon, D. Srivastava, *Phys. Rev. Lett.* **1997**, *79*, 4453–4455.
169 M. Menon, D. Srivastava, *J. Mater. Res.* **1998**, *13*, 2357–2360.
170 J.-C. Charlier, *Acc. Chem. Res.* (Special issue) **2002**, *35*, 1063–1069.
171 J. Li, C. Papadopoulos, J. Xu, *Nature (London)* **1999**, *402*, 254–255.
172 B. C. Satishkumar, P. J. Thomas, A. Govindaraj et al., *Appl. Phys. Lett.* **2000**, *77*, 2530–2532.
173 N. R. Franklin, Y. Li, R. J. Chen et al., *Appl. Phys. Lett.* **2001**, *79*, 4571–4573.
174 "At IBM, a tinier transistor outperforms its silicon cousins," *New York Times*, 20 May 2002, p. C4.
175 P. G. Collins, K. Bradley, M. Ishigami et al., *Science* **2000**, *287*, 1801–1804.
176 (a) J. Kong, N. R. Franklin, C. Zhou et al., *Science* **2000**, *287*, 622–625; (b) S. Ghosh, A. K. Sood, N. Kumar, *Science* **2003**, *299*, 1042–1044.
177 H. Nishijima, S. Akita, Y. Nakayama, *Jpn. J. Appl. Phys.* **1999**, *38*, 7247–7252.
178 J. H. Hafner, C. L. Cheung, C. M. Leiber, *Nature* **1999**, *398*, 761–762.
179 H. Dai, J. H. Hafner, A. G. Rinzler et al., *Nature* **1996**, *384*, 147–150.
180 S. S. Wong, E. Joselevich, A. T. Woolley et al., *Nature* **1998**, *394*, 52–55.
181 P. Kim, C. L. Lieber, *Science* **1999**, *286*, 2148–2150.
182 T. Arie, H. Nishijima, S. Akita et al., *J. Vac. Sci. Technol. B* **2000**, *18*, 104–106.
183 P. Karl, D. Tomanek, *Phys. Rev. Lett.* **1999**, *82*, 5373–5376.
184 E. B. Cooper, S. R. Manalis, H. Fang et al., *Appl. Phys. Lett.* **1999**, *75*, 3566–3568.
185 W. A. de Heer, A. Chatelain, D. Ugarte, *Science* **1995**, *270*, 1179–1180.
186 A. G. Rinzler, J. H. Hafner, P. Nikolaev et al., *Science* **1995**, *269*, 1550–1553.
187 N. S. Lee, D. S. Chung, I. T. Han et

al., *Diamond Relat. Mater.* **2001**, *10*, 265–270.

**188** Y. Saito, S. Uemura, *Carbon* **2000**, *38*, 169–182.

**189** R. Rosen, W. W. Simendinger, C. Debbault et al., *Appl. Phys. Lett.* **2000**, *76*, 1668–1670.

**190** H. Sugie, M. Tanemura, V. Filip et al., *Appl. Phys. Lett.* **2001**, *78*, 2578–2580.

**191** J.-M. Bonard, T. Stöckli, F. Maier et al., *Phys. Rev. Lett.* **1998**, *81*, 1441–1444.

**192** J.-L. Kwo, C. C. Tsou, M. Yokoyama et al., *J. Vac. Sci. Technol. B* **2001**, *19*, 23–26.

**193** W. Zhu, C. Bower, O. Zhou et al., *Appl. Phys. Lett.* **1999**, *75*, 873–875.

**194** Y. Saito, K. Hamaguchi, S. Uemura et al., *Appl. Phys. A* **1998**, *67*, 95–100.

**195** (a) W. B. Choi, D. S. Chung, J. H. Kang et al., *Appl. Phys. Lett.* **1999**, *75*, 3129–3130; (b) C. J. Lee, J. Park, S. Y. Kang et al., *Chem. Phys. Lett.* **2000**, *326*, 175–180.

**196** D. Lovall, M. Buss, E. Graugnard et al., *Phys. Rev. B* **2000**, *61*, 5683–5691.

**197** O. Groning, O. M. Kuttel, Ch. Emmenegger et al., *J. Vac. Sci. Technol. B* **2000**, *18*, 665–678.

**198** R. B. Sharma, V. N. Tondare, D. S. Joag et al., *Chem. Phys. Lett.* **2001**, *344*, 283–286.

**199** S. R. Mishra, H. S. Rawat, S. C. Mehendale et al., *Chem. Phys. Lett.* **2000**, *317*, 510–514.

**200** K. Yoshino, H. Kajii, H. Araki, *Fullerene Sci. Technol.* **1999**, *7*, 695–711.

**201** J. R. Wood, H. D. Wagner, *Appl. Phys. Lett.* **2000**, *76*, 2883–2885.

**202** M. Eswaramoorthy, R. Sen, C. N. R. Rao, *Chem. Phys. Lett.* **1999**, *304*, 207–210.

**203** M. S. Dresselhaus, K. A. Williams, P. C. Eklund, *MRS Bull.* **2000**, *24*, 45–50.

**204** G. G. Tibbetts, G. P. Meisner, C. H. Olk, *Carbon* **2001**, *39*, 2291–2301.

**205** M. Hirscher, M. Becher, M. Haluska et al., *J. Alloys Compd.* **2002**, *330–332*, 654–658.

**206** C. Zandonella, *Nature* **2001**, *410*, 734–735.

**207** Y. Ye, C. C. Ahn, C. Witham et al., *Appl. Phys. Lett.* **1999**, *74*, 2307–2309.

**208** A. C. Dillon, K. M. Jones, T. A. Bekkedahl et al., *Nature* **1997**, *386*, 377–379.

**209** C. Liu, Y. Y. Fan, M. Liu et al., *Science* **1999**, *286*, 1127–1129.

**210** G. Gundiah, A. Govindaraj, N. Rajalakshmi et al., *J. Mater. Chem.* **2003**, *13*, 209–213.

**211** (a) P. Ratnasamy, L. Rodrigues, A. J. Leonard, *J. Phys. Chem.* **1973**, *77*, 2242–2245; (b) J. A. Wilson, A. D. Yoffe, *Adv. Phys.* **1969**, *269*, 193–335.

**212** R. R. Chianelli, E. Prestridge, T. Pecorano et al., *Science* **1979**, *203*, 1105–1107.

**213** (a) R. Tenne, L. Margulis, M. Genut et al., *Nature* **1992**, *360*, 444–446; (b) L. Margulis, G. Salitra, R. Tenne et al., *Nature* **1993**, *365*, 113–114; (c) Y. Feldman, E. Wasserman, D. J. Srolovitch et al., *Science* **1995**, *267*, 222–225.

**214** K. J. C. van Bommel, A. Friggeri, S. Shinkai, *Angew. Chem. Int. Ed. Engl.* **2003**, *42*, 980–999.

**215** (a) C. T. Kresge, M. E. Leonowicz, W. J. Roth et al., *Nature* **1992**, *259*, 710–712; (b) J. S. Beck, J. C. Vartulli, W. J. Roth et al., *J. Am. Chem. Soc.* **1992**, *114*, 10834–10843.

**216** (a) W. Stöber, A. Fink, E. Bohn, *J. Colloid Interface Sci.* **1968**, *26*, 62–69; (b) M. Nakamura, Y. Matsui, *J. Am. Chem. Soc.* **1995**, *117*, 2651–2652.

**217** P. M. Ajayan, O. Stephane, Ph. Redlich et al., *Nature* **1995**, *375*, 564–567.

**218** (a) B. C. Satishkumar, A. Govindaraj, E. M. Vogl et al., *J. Mater. Res.* **1997**, *12*, 604–606; (b) B. C. Satishkumar, A. Govindaraj, M. Nath et al., *J. Mater. Chem.* **2000**, *10*, 2115–2219.

**219** M. E. Spahr, P. Bitterli, R. Nesper et al., *Angew. Chem. Int. Ed. Engl.* **1998**, *37*, 1263–1265 (*Angew. Chem.* **1998**, *110*, 1339).

**220** L. Pu, X. Bao, J. Zou et al., *Angew. Chem. Int. Ed. Engl.* **2001**, *40*, 1490–1493.

**221** (a) M. L. Cohen, *Solid State Commun.* **1994**, *92*, 45–52; (b) A. Rubio, J. L.

Corkill, M. L. Cohen, *Phys. Rev. B* **1994**, *49*, 5081–5084; (c) Y. Miyamoto, A. Rubio, S. G. Louie et al., *Phys. Rev. B* **1994**, *50*, 18360–18366.

**222** (a) K. Kobayashi, N. Kurita, *Phys. Rev. Lett.* **1993**, *70*, 3542–3544; (b) Z. W. Sieh, K. Cherrey, N. G. Chopra et al., *Phys. Rev. B* **1994**, *51*, 11229–11232.

**223** P. Gleize, M. C. Schouler, P. Gadelle et al., *J. Mater. Sci.* **1994**, *29*, 1575–1580.

**224** (a) O. R. Lourie, C. R. Jones, B. M. Bertlett et al., *Chem. Mater.* **2000**, *12*, 1808–1810; (b) R. Ma, Y. Bando, T. Sato, *Chem. Phys. Lett.* **2001**, *337*, 61–64.

**225** F. L. Deepak, C. P. Vinod, K. Mukhopadhyay et al., *Chem. Phys. Lett.* **2002**, *353*, 345–352.

**226** (a) Y. R. Hacohen, E. Grunbaum, R. Tenne et al., *Nature* **1998**, *395*, 336–337; (a) J. Goldberger, R. He, Y. Zhang et al., *Nature* **2003**, *422*, 599–602.

**227** (a) J. Bao, C. Tie, Z. Xu et al., *Adv. Mater.* **2001**, *13*, 1631–1633; (b) B. Mayers, Y. Xia, *Adv. Mater.* **2002**, *14*, 279–282.

**228** J. C. Hulteen, C. R. Martin, *J. Mater. Chem.* **1997**, *7*, 1075–1087.

**229** R. Tenne, M. Homyonfer, Y. Feldman, *Chem. Mater.* **1998**, *10*, 3225–3238 and references therein.

**230** (a) M. Hershfinkel, L. A. Gheber, V. Volterra et al., *J. Am. Chem. Soc.* **1994**, *116*, 1914–1917; (b) T. Tsirlina, Y. Feldman, M. Homyonfer et al., *Fullerene Sci. Technol.* **1998**, *6*, 157–165.

**231** M. Nath, A. Govindaraj, C. N. R. Rao, *Adv. Mater.* **2001**, *13*, 283–286.

**232** M. Nath, C. N. R. Rao, *Chem. Commun.* **2001**, 2236–2237.

**233** M. Nath, C. N. R. Rao, *J. Am. Chem. Soc.* **2001**, *123*, 4841–4842.

**234** M. Nath, C. N. R. Rao, *Angew. Chem. Int. Ed. Engl.* **2002**, *41*, 3451–3454.

**235** A. P. Lin, C. Y. Mou, S. D. Liu, *Adv. Mater.* **2000**, *12*, 103–106.

**236** J. Zhang, L. Sun, C. Liao et al., *Chem. Commun.* **2002**, 262–263.

**237** M. Niederberger, H.-J. Muhr, F. Krumeich et al., *Chem. Mater.* **2000**, *12*, 1995–2000.

**238** C. N. R. Rao, A. Govindaraj, F. L. Deepak et al., *Appl. Phys. Lett.* **2001**, *78*, 1853–1855.

**239** T. Kasuga, M. Hiramatsu, A. Hason et al., *Langmuir* **1998**, *14*, 3160–3163.

**240** C. N. R. Rao, B. C. Satishkumar, A. Govindaraj, *Chem. Commun.* **1997**, 1581–1582.

**241** C. M. Zelenski, P. K. Dorhout, *J. Am. Chem. Soc.* **1998**, *120*, 734–742.

**242** A. Loiseau, F. Williame, N. Demonecy et al., *Phys. Rev. Lett.* **1996**, *76*, 4737–4740.

**243** P. Terech, R. G. Weiss, *Chem. Rev.* **1997**, *97*, 3133–3159.

**244** Y. Ono, K. Nakashima, M. Sano et al., *Chem. Commun.* **1998**, 1477–1478.

**245** T. Kasuga, M. Hiramutsu, A. Hoson et al., *Adv. Mater.* **1999**, *11*, 1307–1311.

**246** A. Govindaraj, F. L. Deepak, N. A. Gunari et al., *Israel J. Chem.* **2001**, *41*, 23–30.

**247** P. V. Teredesai, F. L. Deepak, A. Govindaraj et al., *J. Nanosci. Nanotechnol.* **2002**, *2*, 495–498.

**248** F. Krumeich, H.-J. Muhr, M. Niederberger et al., *J. Am. Chem. Soc.* **1999**, *121*, 8324–8331.

**249** H.-J. Muhr, F. Krumeich, U. P. Schönholzer et al., *Adv. Mater.* **2000**, *12*, 231–234.

**250** G. Seifert, H. Terrones, M. Terrones et al., *Phys. Rev. Lett.* **2000**, *85*, 146–149.

**251** (a) G. Seifert, H. Terrones, M. Terrones et al., *Solid State Commun.* **2000**, *114*, 245–248; (b) G. Seifert, H. Terrones, M. Terrones et al., *Solid State Commun.* **2000**, *115*, 635–638.

**252** M. Nath, S. Kar, A. K. Raychaudhuri et al., *Chem. Phys. Lett.* **2003**, *368*, 690–695.

**253** (a) H. M. Cheng, Q. H. Yang, C. Liu, *Carbon* **2001**, *39*, 1447–1454; (b) P. Hou, Q. Yang, S. Bai et al., *J. Phys. Chem. B* **2002**, *106*, 963–966 and references therein.

**254** R. Ma, Y. Bando, H. Zhu et al., *J. Am. Chem. Soc.* **2002**, *124*, 7672–7673.

**255** J. Chen, N. Kuriyama, H. Yuan et al., *J. Am. Chem. Soc.* **2001**, *123*, 11813–11814.

**256** M. E. Spahr, P. S. Bitterli, R.

Nesper et al., *J. Electrochem. Soc.* **1999**, *146*, 2780–2783.
257 A. Dobley, K. Ngala, S. Yang et al., *Chem. Mater.* **2001**, *13*, 4382–4386.
258 A. Rothschild, S. R. Cohen, R. Tenne, *Appl. Phys. Lett.* **1999**, *75*, 4025–4027.
259 L. Rapoport, Y. Bilik, Y. Feldman et al., *Nature* **1997**, *387*, 791–793.
260 J. Chen, S.-L. Li, Q. Xu et al., *Chem. Commun.* **2002**, 1722–1723.
261 M. Côté, M. L. Cohen, D. J. Chadi, *Phys. Rev. B* **1998**, *58*, R4277–R4280.
262 F. Cerrina, C. Marrian, *MRS Bull.* **1996**, *21* (December), 56.
263 (a) J. M. Gibson, *Phys. Today* **1997**, October, 56–61; (b) S. Matsui, Y. Ochiai, *Nanotechnology* **1996**, *7*, 247–258.
264 (a) S. H. Hong, J. Zhu, C. A. Mirkin, *Science* **1999**, *286*, 523–525; (b) J. A. Dagata, *Science* **1995**, *270*, 1625 (see reference therein).
265 (a) M. D. Levenson, P. J. Silverman, R. George et al., *Solid State Technol.* **1995**, *38*, 81–82, 84, 86, 88, 90, 92, 94, 96, 98; (b) P. N. Dunn, *Solid State Technol.* **1994**, *37*, 49–50, 52, 58, 61–62.
266 Y. Xia, J. A. Rogers, K. E. Paul et al., *Chem. Rev.* **1999**, *99*, 1823–1848.
267 D. Routkevitch, A. A. Tager, J. Haruyama et al., *IEEE Trans. Electron Devices* **1996**, *43*, 1646–1657.
268 P. Yang, Y. Wu, R. Fan, *Int. J. Nanosci.* **2002**, *1*, 1–39.
269 Y. Xia, P. Yang, Y. Sun et al., *Adv. Mater.* **2003**, *15*, 353–389.
270 E. I. Givargizov, Highly Anisotropic Crystals, eds. M. Senechal, S. College, Reidel, Dordrecht, The Netherlands 1987.
271 Y. Wu and P. Yang, *Chem. Mater.* **2000**, *12*, 605–607.
272 M. H. Huang, Y. Wu, H. Feick et al., *Adv. Mater.* **2001**, *13*, 113–116.
273 A. M. Morales, C. M. Lieber, *Science* **1998**, *279*, 208–211.
274 (a) C. R. Martin, *Science* **1994**, *266*, 1961–1966; (b) D. Almawlawi, C. Z. Liu, M. Moskovits, *J. Mater. Res.* **1994**, *9*, 1014–1018.
275 W. Han, S. Fan, W. Li et al., *Science* **1997**, *277*, 1287–1289.
276 T. J. Trentler, K. M. Hickman, S. C. Geol et al., *Science* **1995**, *270*, 1791–1794.
277 X. F. Duan, C. M. Lieber, *Adv. Mater.* **2000**, *12*, 298–302.
278 R. S. Wagner, in Whisker Technology, ed. A. P. Levitt, Wiley-Interscience, New York 1970, 47–119.
279 Y. Wu, P. Yang, *J. Am. Chem. Soc.* **2001**, *123*, 3165–3166.
280 M. S. Gudiksen, C. M. Lieber, *J. Am. Chem. Soc.* **2000**, *122*, 8801–8802.
281 C. C. Chen, C. C. Yeh, C. H. Chen et al., *J. Am. Chem. Soc.* **2001**, *123*, 2791–2798.
282 X. F. Duan, C. M. Lieber, *J. Am. Chem. Soc.* **2000**, *122*, 188–189.
283 W.-S. Shi, H.-Y. Peng, Y.-F. Zheng et al., *Adv. Mater.* **2000**, *12*, 1343–1345.
284 C. C. Tang, S. Fan, M. L. de la Chapelle et al., *Adv. Mater.* **2000**, *12*, 1346–1348.
285 G. Gu, M. Burghard, G. T. Kim, G. S. Dusberg et al., *J. Appl. Phys.* **2001**, *90*, 5747–5751.
286 Z. W. Pan, H. L. Lai, F. C. K. Au et al., *Adv. Mater.* **2000**, *12*, 1186–1190.
287 Y. Wu, B. Messer, P. Yang, *Adv. Mater.* **2001**, *13*, 1487–1489.
288 M. Yazawa, M. Koguchi, A. Muto et al., *Adv. Mater.* **1993**, *5*, 577–580.
289 J. Liu, X. Zhang, Y. J. Zhang et al., *J. Mater. Res.* **2001**, *16*, 3133–3138.
290 K. Hiruma, M. Yazawa, T. Katsuyama et al., *J. Appl. Phys.* **1995**, *77*, 447–462.
291 T. Shimada, K. Hiruma, M. Shirai et al., *Superlattices Microstruct.* **1998**, *24*, 453–458.
292 X. F. Duan, J. F. Wang, C. M. Lieber, *Appl. Phys. Lett.* **2000**, *76*, 1116–1118.
293 (a) S. T. Lee, N. Wang, C. S. Lee, *Mater. Sci. Eng. A* **2000**, *286*, 16–23; (b) N. Wang, Y. H. Tang, Y. F. Zhang et al., *Chem. Phy. Lett.* **1999**, *299*, 237–242; (c) H. W. Seo, S. Y. Bae, J. Park et al., *Chem. Commun.* **2002**, 2564–2565; (d) Y. F. Zhang, Y. H. Tang, C. Lam et al., *J. Cryst. Growth* **2000**, *212*, 115–118.
294 (a) S. T. Lee, N. Wang, Y. F. Zhang et al., *MRS Bull.* **1999**, *24*, 36–42; (b) D. D. D. Ma, C. S. Lee, F. C. K. Au et al., *Science* **2003**, *299*, 1874–1877.

**295** (a) N. Wang, Y. F. Zhang, Y. H. Tang et al., *Appl. Phys. Lett.* **1998**, *73*, 3902–3904; (b) W.-S. Shi, H.-Y. Peng, N. Wang et al., *J. Am. Chem. Soc.* **2001**, *123*, 11095–11096.

**296** Z. L. Wang, R. P. Gao, Z. W. Pan et al., *Adv. Eng. Mater.* **2001**, *3*, 657–661.

**297** Z. W. Pan, Z. R. Dai, Z. L. Wang, *Science* **2001**, *291*, 1947–1949.

**298** Z. R. Dai, Z. W. Pan, Z. L. Wang, *Solid State Commun.* **2001**, *118*, 351–354.

**299** (a) P. Yang, C. M. Lieber, *J. Mater. Res.* **1997**, *12*, 2981–2996; (b) P. Yang, C. M. Lieber, US Pat., 5897945, 1999.

**300** P. Yang, C. M. Lieber, *Science* **1996**, *273*, 1836–1840.

**301** G. Gundiah, G. V. Madhav, A. Govindaraj et al., *J. Mater. Chem.* **2002**, *12*, 1606–1611.

**302** (a) G. Gundiah, A. Govindaraj, C. N. R. Rao, *Chem. Phys. Lett.* **2002**, *351*, 189–194; (b) F. L. Deepak, A. Govindaraj, C. N. R. Rao, *J. Nanosci. Nanotechnol.* **2001**, *1*, 303–308.

**303** (a) Y. Chen, D. A. A. Ohlberg, R. S. Williams, *Mater. Sci. Eng. B* **2001**, *87*, 222–226; (b) Y. Chen, D. A. A. Ohlberg, G. Medeiros-Ribeiro et al., *Appl. Phys. Lett.* **2000**, *76*, 4004–4006.

**304** L. J. Lauhon, M. S. Gudiksen, D. Wang et al., *Nature* **2002**, *420*, 57–59.

**305** (a) J. Stejny, R. W. Trinder, J. Dlugosz, *J. Mater. Sci.* **1981**, *16*, 3161–3170; (b) J. Stejny, J. Dlugosz, A. Keller, *J. Mater. Sci.* **1979**, *14*, 1291–1300; (c) A. K. Cheetham, P. Day, Solid State Chemistry (Compounds), Clarendon Press, Oxford 1992, p. 31; (d) Extended Linear Chain Compounds, ed. J. S. Miller, Plenum Press, New York 1982; (e) Chemistry and Physics of One-Dimensional Metals, ed. H. J. Keller, Plenum Press, New York 1977.

**306** (a) H. R. van Kruyt, A. E. van Arkel, Kolloid-Z. **1923**, *32*, 29–36; (b) A. Gutbier, *Z. Anorg. Chem.* **1902**, *32*, 106–107.

**307** (a) B. Gates, Y. Yin, Y. Xia, *J. Am. Chem. Soc.* **2000**, *122*, 12582–12583; (b) B. Gates, B. Mayers, B. Cattle et al., *Adv. Funct. Mater.* **2002**, *12*, 219–227.

**308** B. Mayers, Y. Xia, *J. Mater. Chem.* **2002**, *12*, 1875–1881.

**309** (a) B. Messer, J. H. Song, M. Huang et al., *Adv. Mater.* **2000**, *12*, 1526–1528; (b) J. Song, B. Messer, Y. Wu et al., *J. Am. Chem. Soc.* **2001**, *123*, 9714–9715; (c) C. Wang, K. Tang, Q. Yang et al., *Inorg. Chem. Commun.* **2001**, *4*, 339–341; (d) E. I. Gerzanich, V. A. Lyakhovitskaya, V. M. Fridkin et al., *Curr. Top. Mater. Sci.* **1982**, *10*, 55–190; (e) J. S. Miller, A. J. Epstein, *Prog. Inorg. Chem.* **1976**, *20*, 1–151.

**310** (a) Huczko, *Appl. Phys. A: Mater. Sci. Process.* **2000**, *70*, 365–376; (b) M. Zheng, L. Zhang, X. Zhang et al., *Chem. Phys. Lett.* **2001**, *334*, 298–302.

**311** C. G. Wu, T. Bein, *Science* **1994**, *266*, 1013–1014.

**312** M. H. Huang, A. Choudrey, P. Yang, *Chem. Commun.* **2000**, 1063–1064.

**313** R. Ulrich, A. Du Chesne, M. Templin et al., *Adv. Mater.* **1999**, *11*, 141–146.

**314** Z. Liu, Y. Sakamoto, T. Ohsuna et al., *Angew. Chem. Int. Ed. Engl.* **2000**, *39*, 3107–3110.

**315** (a) R. Leon, D. Margolese, G. Stucky et al., *Phys. Rev. B* **1995**, *52*, R2285–R2288; (b) N. R. B. Coleman, K. M. Ryan, T. R. Spalding et al., *Chem. Phys. Lett.* **2001**, *343*, 1–6; (c) H. Ringsdorf, B. Schlarb, J. Venzmer, *Angew. Chem. Int. Ed. Engl.* **1988**, *27*, 113–158.

**316** Y. Yin, Y. Lu, Y. Sun et al., *Nano Lett.* **2002**, *2*, 427–430.

**317** (a) B. Gates, Y. Wu, Y. Yin et al., *J. Am. Chem. Soc.* **2001**, *123*, 11500–11501; (b) B. Gates, B. Mayers, Y. Wu et al., *Adv. Funct. Mater.* **2002**, *12*, 679–686.

**318** Y. Zhang, K. Suenaga, C. Colliex et al., *Science* **1998**, *281*, 973–975.

**319** N. Demoncy, O. Stephan, N. Brun et al., *Eur. Phys. J. B* **1998**, *4*, 147–157.

**320** W. K. Hsu, S. Trasobares, H. Terrones et al., *Chem. Mater.* **1999**, *11*, 1747–1751.

**321** J. Sloan, D. M. Wright, H. G. Woo et al., *Chem. Commun.* **1999**, 699–700.

**322** S. Link, M. B. Mohammed, M. A. El-Sayed, *J. Phys. Chem. B* **1999**, *103*, 3073–3077.

**323** C. H. Kiang, J. S. Choi, T. T. Tran et al., *J. Phys. Chem. B* **1999**, *103*, 7449–7451.

**324** B. K. Pradhan, T. Kyotani, A. Tomita, *Chem. Commun.* **1999**, 1317–1318.

**325** W. K. Hsu, W. Z. Li, Y. Q. Zhu et al., *Chem. Phys. Lett.* **2000**, *317*, 77–82.

**326** S. D. Dingman, N. P. Rath, P. D. Markowitz et al., *Angew. Chem. Int. Ed. Engl.* **2000**, *39*, 1470–1472.

**327** (a) T. J. Trentler, S. C. Goel, K. M. Hickman et al., *J. Am. Chem. Soc.* **1997**, *119*, 2172–2181; (b) P. D. Markowitz, M. P. Zach, P. C. Gibbons et al., *J. Am. Chem. Soc.* **2001**, *123*, 4502–4511; (c) O. R. Lourie, C. R. Jones, B. M. Bartlett et al., *Chem. Mater.* **2000**, *12*, 1808–1810.

**328** (a) J. D. Holmes, K. P. Johnston, R. C. Doty et al., *Science* **2000**, *287*, 1471–1473; (b) T. Henrath, B. A. Korgel, *Adv. Mater.* **2003**, *15*, 437–440.

**329** L. Manna, E. C. Scher, A. P. Alivisatos, *J. Am. Chem. Soc.* **2000**, *122*, 12700–12706.

**330** (a) C. Chao, C. Chen, Z. Lang, *Chem. Mater.* **2000**, *12*, 1516–1518; (b) Y. Sun, B. Gates, B. Mayers et al., *Nano Lett.* **2002**, *2*, 165–168; (c) Y. Sun, Y. Xia, *Adv. Mater.* **2002**, *14*, 833–837; (d) Y. Sun, Y. Yin, B. T. May et al., *Chem. Mater.* **2002**, *14*, 4736–4745.

**331** Y. F. Liu, J. H. Zeng, W. X. Zhang et al., *J. Mater. Res.* **2001**, *16*, 3361–3365.

**332** X. Jiang, Y. Xie, J. Lu et al., *Chem. Mater.* **2001**, *13*, 1213–1218.

**333** Y. Wu, H. Yan, M. Huang et al., *Chem. Euro. J.* **2002**, *8*, 1260–1268.

**334** Y. Wu, H. Yan, P. Yang, *Top. Catal.* **2002**, *19*, 197–202.

**335** M. Huang, S. Mao, H. Feick et al., *Science* **2001**, *292*, 1897–1899.

**336** C. L. Cheung, J. H. Hafner, T. W. Odom et al., *Appl. Phys. Lett.* **2000**, *76*, 3136–3318.

**337** B. Messer, J. H. Song, P. Yang, *J. Am. Chem. Soc.* **2000**, *122*, 10232–10233.

**338** Y. Huang, X. Duan, Q. Q. Wei et al., *Science* **2001**, *291*, 630–633.

**339** X. Duan, Y. Huang, Y. Cui et al., *Nature* **2001**, *409*, 66–69.

**340** F. Kim, S. Kwan, J. Arkana et al., *J. Am. Chem. Soc.* **2001**, *123*, 4360–4361.

**341** L. Brus, *J. Phys. Chem.* **1994**, *98*, 3575–3581.

**342** J. F. Wang, M. S. Gudiksen, X. F. Duan et al., *Science* **2001**, *293*, 1455–1457.

**343** J. C. Johnson, H. Yan, R. D. Schaller et al., *J. Phys. Chem. B* **2001**, *105*, 11387–11390.

**344** S. W. Chung, J. Y. Yu, J. R. Heath, *Appl. Phys. Lett.* **2000**, *76*, 2068–2070.

**345** Y. Cui, C. M. Lieber, *Science* **2001**, *291*, 851–853.

**346** Y. Huang, X. Duan, Y. Cui et al., *Science* **2001**, *294*, 1313–1317.

**347** J. Y. Yu, S. W. Chung, J. R. Heath, *J. Phys. Chem. B* **2000**, *104*, 11864–11870.

**348** Y. Cui, X. F. Duan, J. T. Hu et al., *J. Phys. Chem. B* **2000**, *104*, 5213–5216.

**349** X. T. Zhou, H. L. Lai, H. Y. Peng et al., *Chem. Phys. Lett.* **2000**, *318*, 58–62.

# 9
# Synthesis, Assembly and Reactivity of Metallic Nanorods

*C. J. Murphy, N. R. Jana, L. A. Gearheart, S. O. Obare, K. K. Caswell, S. Mann, C. J. Johnson, S. A. Davis, E. Dujardin, and K. J. Edler*

## 9.1
## Introduction

Metal particles with dimensions on the nanometer scale are of great current interest for their unusual properties [1–3]. Fundamentally, the mean free path of an electron in a metal at room temperature is ~10–100 nm, and one would predict that as the metallic particle shrinks to this dimension, unusual effects might be observed [3]. Indeed, gold nanoparticles of diameter ~100 nm or less appear red (not gold) when suspended in transparent media [1–3]; and gold nanoparticles of diameter ~3 nm are no longer "noble" and unreactive, but can catalyze chemical reactions [4].

The optical properties of silver and gold nanoparticles in the visible region of the spectrum, specifically, absorption and scattering, are exquisitely sensitive to nanoparticle size, shape, aggregation state, and local environment [2, 5–10]. Additionally, molecules adsorbed to the surface of gold and silver nanoparticles undergo surface-enhanced Raman scattering (SERS) effects, due to the coupling of the plasmon band of the irradiated metal (i.e., the collective oscillation of the conduction band electrons upon absorption in the visible for these particular metals, due to their dielectric constant) with the molecules' electronic states [11, 12]. Thus, one emerging application of metallic nanoparticles is optical sensors, and single-molecule detection via SERS has been reported [8, 9, 11–19].

The aspect ratio of a solid is defined as its length divided by its width; therefore, spheres have an aspect ratio of 1. We define, somewhat arbitrarily, a "nanorod" to be an object with an aspect ratio between 1 and 20, with the short dimension on the 10–100 nm scale, and a "nanowire" to be an object with an aspect ratio greater than 20 (with the short dimension on the 10–100 nm scale) [20]. The extinction spectra (the combination of visible absorption and scattering) of silver and gold nanoparticles are tunable throughout the visible, depending on the aspect ratio [5–10; Figure 9.1].

For nanorods and nanowires, the plasmon band of the metal is split in two: the

*The Chemistry of Nanomaterials: Synthesis, Properties and Applications, Volume 1.* Edited by C. N. R. Rao, A. Müller, A. K. Cheetham

ISBN: 3-527-30686-2

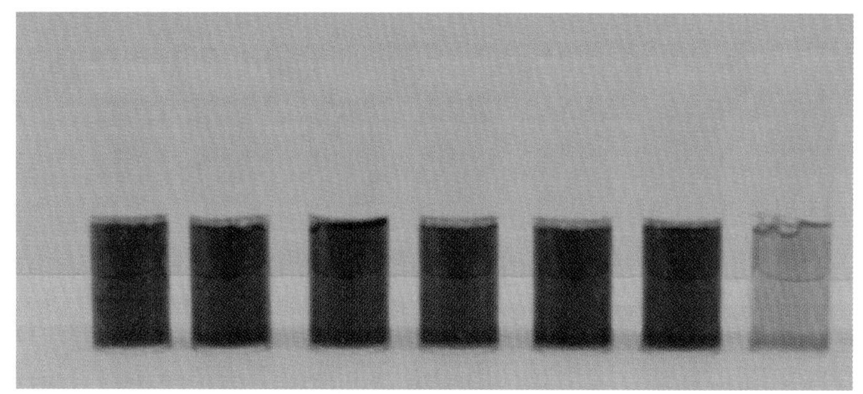

**Fig. 9.1.** Photograph of aqueous solutions of gold nanoparticles of aspect ratio 1 (far left) up to 18 (far right). The short axes of the nanoparticles are 15–30 nm.

longitudinal plasmon band, corresponding to light absorption and scattering along the long axis of the particle, and the transverse plasmon band, corresponding to light absorption and scattering along the short axis of the particle [2, 5–7, 10]. Other nanoparticle shapes are predicted to have other, more complicated extinction spectra [10, 21]. Thus, metallic nanorods and nanowires have tunable optical properties throughout the visible and into the infrared portions of the electromagnetic spectrum, and they are predicted to have enhanced SERS activity compared to spheres [9].

Information technology and storage is another arena for which metallic nanoparticles are of interest. As the size of integrated circuit elements shrinks to the ~100 nm scale and below, metallic and semiconducting nanowires of controllable size and ability to be positioned need to be developed [22]. The magnetic domain size in magnetic nanoparticles is of the order of 10 nm, and control of crystal structure and size of the relevant metals and metal alloys (e.g., Fe, Co) on this scale are key to advances in magnetic data storage [23, 24]. Hybrid data storage schemes that rely on melting metallic nanorods to nanospheres in polymer matrices, with subsequent detection of the altered optical properties for "reading" and "writing", are also being developed [25].

All of the promise of the future technology based on nanometer-scale inorganic solids relies on the production of nanoparticles of controlled size, shape and crystal structure, and further ultimately requires that these nanoparticles be rationally linked to make a working device. Enormous progress has been made in the synthesis of inorganic nanospheres; routinely, at present, control over the diameter of nanoparticles leads to particle size distributions that are within 10% of the mean diameter, and frequently within 5%. Only since the mid-1990s and later have there been good synthetic methods to make nanoparticles of controllable size *and* shape (other than spheres) [3, 20, 22].

## 9.2
## Seed-Mediated Growth Approach to the Synthesis of Inorganic Nanorods and Nanowires

We have developed a seed-mediated growth approach to make silver and gold nanorods and nanowires of controllable aspect ratio in aqueous solution, at room temperature [20, 26–28]. Our method (Figure 9.2) starts with the reduction of a metal salt in aqueous solution, by a strong reducing agent, to make 3–4 nm metallic "seeds" which are generally spherical. The seed reaction is done in the presence of sodium citrate to prevent the seeds from aggregating and precipitating. The seeds are then added to a "growth" solution that consists of fresh metal salt, and a surfactant, cetyltrimethylammonium bromide, CTAB, that directs the growth of nanoparticles into nanorods and nanowires. Growth is initiated by the addition

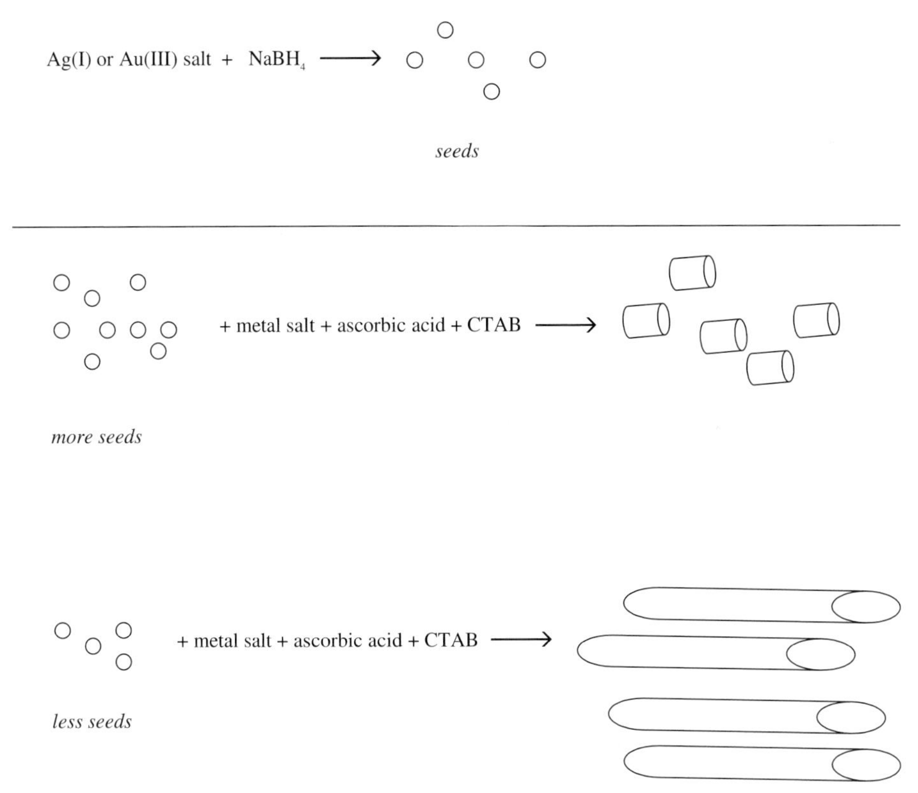


**Fig. 9.2.** Cartoon showing the seed-mediated growth approach to the synthesis of metallic nanorods of controllable aspect ratio. In the first step, metal salts are reduced with sodium borohydride, a strong reducing agent, to metal nanospheres ("seeds") that are 3–4 nm in diameter. In the subsequent growth steps, seeds are added to fresh solutions of metal salt in the presence of CTAB, and are reduced with the weak reducing agent ascorbic acid. The ratio of metal seed to salt controls the aspect ratio, fewer seeds leading to longer rod growth.

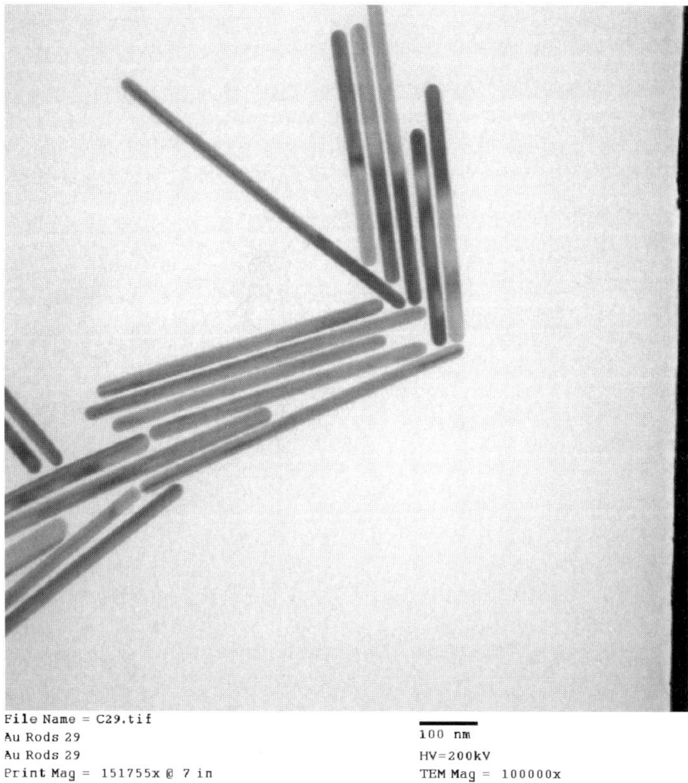


**Fig. 9.3.** Transmission electron micrograph of gold nanorods, prepared by the seed-mediated growth method in water, in the presence of CTAB. Scale bar = 100 nm.

of a weak reducing agent, usually ascorbic acid, that cannot reduce the metal salt on its own (at room temperature). Varying the seed to metal salt ratio controls the aspect ratio of the resulting nanorods. Additionally, one can use short nanorods as "seeds" on which to grow longer nanorods.

Our seed-mediated growth approach has been successful for the synthesis of gold nanorods, with diameters of ca. 20 nm and aspect ratios from 2 to 20, in a controllable fashion (Figure 9.3). For silver, we have been able to make both short nanorods (aspect ratio ~4) and long nanowires (aspect ratios 50–350); in the case of nanowires the level of control we have is more limited.

Despite the overall simplicity of the seed-mediated growth approach in aqueous solution, there are many complicating factors during the synthetic reactions that can alter the outcome of the reaction. The percent yield of nanorods, compared to spheres, is typically only ~20%, although with slight changes in reaction conditions, 90% yields have been obtained [29]. Glassware must be rigorously cleaned with aqua regia. Relatively small changes in the pH of the solution, for example, in the case of silver nanorods, leads to nanowires [26]. Gold, we find, is difficult to make into nanowires, but is relatively easy to make into nanorods, for reasons that are unclear. A few degrees difference in room temperature, as well as the timing of the growth steps, also affects the final product's size and shape. Changing relative concentrations of seed, salt, and surfactant can eliminate nanorods entirely, but produce nanospheres with well-controlled dimensions from 5–40 nm [30]. Also, depending on reaction conditions, it is possible to have seeds promote the formation of more seeds instead of growth [31]!

There is an ever-increasing list of other methods to make inorganic nanorods and nanowires. For example, porous alumina membranes with well-defined nanochannels can be used as hard templates in which to electrochemically deposit metals; the metal forms nanorods dictated by the channel dimensions, and the membrane must of course be attached to the electrode for this to work [5, 6, 19]. Metal nanowires can also be grown via electrodeposition along the step edges of highly-ordered pyrolytic graphite from metal salt in solution, using a scanning tunneling microscope tip to pulse the voltage in "activation", "nucleation" and "growth" steps [32].

Many methods of producing nonspherical, well-defined nanoparticles of various shapes and sizes use a "soft template" (e.g., microemulsions, polymers, or surfactants) that directs nanoparticle growth, in the absence of preformed seed [24, 33–40]. Some of these preparations are performed at higher temperatures (up to several hundred °C) in organic solvents, with organometallic precursors, while others are simple reductions of metals, or arrested precipitation reactions, in water. Metals, semiconductors, and metal oxides have been made by these soft solution routes [24, 33–40]. Heterogeneous seeded methods, in which the seed is not the same element or compound as the nanostructure grown from it, are also being developed [41, 42], with great success in the case of semiconductor nanowires [41].

Intuitively, one might imagine that the mechanism of anisotropic growth of a nanoparticle in the "soft template" methods would involve physical constraint on the part of the template; for example, that the size and shape of rodlike micellar templates would be mirrored in the size and shape of the resulting nanoparticles made in those matrices. However, it is now more accepted that preferential adsorption of molecules and ions to different crystal faces of the growing nanoparticle leads to different nanoparticle shapes [1, 40, 43–46]. The underlying mechanisms of anisotropic nanoparticle growth are of fundamental interest, but are also of practical interest; for example, poisoning of platinum nanoparticle catalysts by adsorbed sulfur compounds is a problem of great commercial importance, and it has been suggested that the poisoning is due to adsorbate-induced crystal facet changes of the Pt nanoparticle catalyst [47].

In the case of gold nanorods prepared in aqueous solution by the seed-mediated growth method, we have performed high-resolution transmission electron microscopy (HRTEM) and electron diffraction experiments as a function of growth time to crystallographically characterize the nanorods [45]. It is well-known that for the face-centered cubic (fcc) structure of gold, the (111) face is the most stable, close-packed face, and one might have expected that the surface of the nanorods would consist mostly of this face (e.g., the long sides) and that the dimensions of the nanorods would be limited to the rod-like micelle size of CTAB. But this is not what we have observed [45].

The crystallographic structure and 3-D crystal morphology of individual gold nanorods prepared by seed-mediated sequential growth in the presence of CTAB were determined by selected area electron diffraction (SAED) in combination with HRTEM. At zero degree tilt, not all the rods imaged on the TEM grid showed Bragg diffraction. Of those that did, SAED gave two patterns in equal proportions. Neither pattern could be indexed as a single zone, indicating that the gold nanorods were not single domain crystals. Instead, both patterns consisted of a superposition of two specific crystallographic zones of general form, $\langle 112 \rangle$ and $\langle 100 \rangle$ and $\langle 110 \rangle$ and $\langle 111 \rangle$, which were consistent with multiple twinning of a face-centered cubic structure. Both types of composite electron diffraction patterns can be rationalized on the basis that the gold nanorods consist of an elongated variant of a cyclic penta–tetrahedral twin crystal in which five $\{111\}$ twin boundaries are arranged radially to the direction of elongation. This type of twinning is common in isotropic gold nanoparticles with decahedral ($D_{5h}$) morphology because the interfacial angle between $\{111\}$ planes (70.53°) is close to $2\pi/5$ (72°). In the case of the nanorods, the shape anisotropy originates from a specific elongation along the common [110] five-fold axis to produce an idealized 3-D morphology based on a pentagonally twinned prism with five $\{100\}$ side faces and capped at both ends by five $\{111\}$ faces (Figure 9.4).

The absence of any preferred zone combination in the electron diffraction patterns suggests that the side faces of the nanorods are either not well-developed or consist of two forms, e.g. $\{100\}$ and $\{110\}$ with approximately equivalent surface area. HRTEM images of individual gold nanorods (Figure 9.5) show stripe patterns characteristic of the superposition of two diffraction patterns, i.e. a twinned defect structure, consistent with the SAED data.

In our experiments, growth of the isometric twinned "seeds" (diameter 4 nm) in the presence of CTAB results in the initial transformation of ca. 4% of the seeds into short nanorods, while the remaining crystals increase in size to around 17 nm [45]. Once formed, the nanorods grow almost unidirectionally in length when immersed in a fresh reaction solution to produce cylindrical penta-twinned particles with high aspect ratios and variable crystal lengths between 100 and 300 nm. Because the increase in width is minimal, the elongated crystals have a uniform thickness that is determined by the width of the short nanorods formed in the previous stage of the reaction sequence. At the same time, ca. 6% of the 17 nm-sized isometric twins are transformed into a new population of short rod-shaped nanoparticles, while the remaining crystals continue to grow isometrically. Subsequent transfer of the products into fresh reaction solutions reiterates the combination of isometric growth, nanorod elongation and nanorod formation to produce a trimodal distribution in rod widths [45]. The distribution corresponds to three types of nanorods with mean widths of 34, 40 and 58 nm and decreasing aspect ratios with values between 17–20, 8–11, and 2–3, respectively. Each type can be

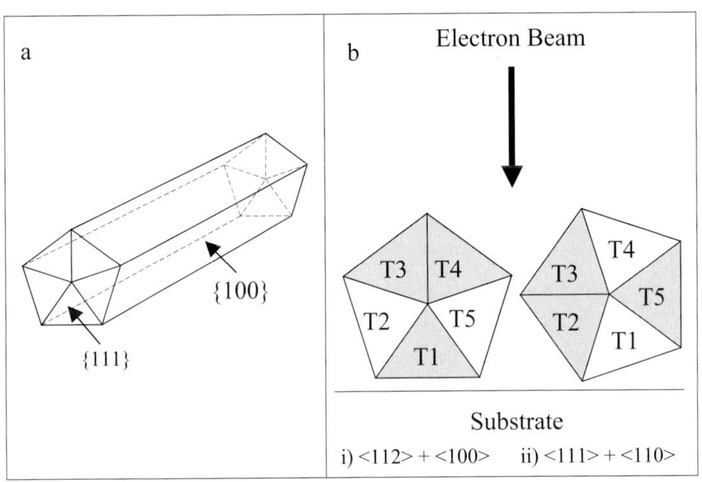


**Fig. 9.4.** Elongated cyclic penta–tetrahedral twin model of gold nanorods, taken from [45]. (a) Idealized 3-D morphology showing {111} end faces and {100} side faces. The common five-fold axis of elongation is [110]. (b) Cross-section of nanorod structure showing arrangement of twins T1 to T5, and possible orientations of domains with respect to the electron beam. These give rise to superimposed zone combinations of (i) ⟨112⟩ and ⟨100⟩, and (ii) ⟨110⟩ and ⟨111⟩ that are related in the diagram by an anticlockwise rotation of 18° around the five-fold ⟨110⟩ central axis. Twin domains in diffraction alignment are highlighted. Reproduced by permission of The Royal Society of Chemistry.

correlated with the widths of the transferred cylindrical, short rod-like or isometric nanoparticles, respectively, which in turn are related to the dimensions and shapes of crystals formed in earlier stages of the sequential process. Isolation of the higher-aspect ratio nanoparticles by centrifugation and precipitation works relatively well [27].

In general, the formation of mixed populations of gold nanoparticles and their associated morphologies and sizes can be rationalized by differences in the delay in the onset of shape anisotropy in the reaction sequence. Clearly, the mechanism responsible for the transformation from isotropic to anisotropic growth of the isometric penta-twinned nanoparticles is not highly competitive, although once achieved, the crystals rapidly elongate along the common [110] axis, suggesting that the process is essentially auto-catalytic. As the unidirectional growth rate is high and the onset of the shape transformation process occurs over an extended reaction period, the nanorods originate at different times to produce a marked variation in the particle lengths. In contrast, the width of each particle increases only slowly, which indicates that the {100} side edges are effectively blocked from further growth compared with the {111} end faces. Previous studies have indicated that the yield of nanorods rises with an increased concentration of CTAB [27], suggesting that CTAB molecules bind more strongly to the side edges than the {111} end faces, with the consequence that the crystal grows preferentially along the [110] direction as the side edges/faces become stabilized. The HRTEM data

**Fig. 9.5.** High-resolution transmission electron micrograph of a gold nanorod showing the double diffraction pattern, indicating twinning, taken from [45]. Scale bar = 5 nm. Reproduced by permission of The Royal Society of Chemistry.

suggest that this process is not specific enough to produce well-defined {100} side faces, particularly in the initial stages of nanorod formation.

It seems likely that steric and chemical factors play an important role in determining the preferential interactions between the cationic quaternary ammonium headgroups and growth sites on the side edges and faces. One possibility is that surfactant-containing complexes such as $[AuBrCTA]^+$ are specifically incorporated into the {100} side edges, whereas non-complexed ion-pairs or Au(0) atoms/clusters are added to the {111} end faces. The discrimination between sites could be due to the increased stability of the close-packed {111} surfaces compared with the edge sites, which will contain numerous defects. Moreover, binding of the large $[NMe_3]^+$ headgroup of CTAB (diameter = 0.814 nm, area = 0.521 $nm^2$) and associated long alkyl chain can be more readily accommodated at the {100} side edges than within the plane of individual close-packed {111} faces, where the Au–Au spacings are too small to facilitate epitaxy. As the nanorods grow in length, the area of the side faces increases, and this could facilitate the assembly of a bilayer of CTAB molecules at the crystal surface, in which the alkylammonium headgroup points down toward the gold surface, with the long hydrocarbon chain facing the solvent, interdigitating with a second hydrocarbon chain from a second CTAB layer, with the second CTAB's headgroup facing the solvent [46]. In turn, the bilayer would provide additional stabilization and growth inhibition, and this could explain why elongation of the nanorods is rapid once the shape anisotropy has been established, in a "zipping" type of mechanism.

Finally, we note that our proposed mechanism follows a classical description of crystal growth inhibition that involves the preferential attachment of individual molecules to the different crystal faces of the growing nanoparticle. The mechanism does not implicate the involvement of surfactant micelles per se in controlling the shape anisotropy of fcc metallic nanoparticles, as has been previously postulated. Instead, the data indicate that symmetry breaking in fcc metallic structures is an intrinsic structural mechanism (twinning) that is subsequently modulated extrinsically during growth in solution by edge-specific surfactant adsorption. This general conclusion is also being reached by many other research groups [1, 24, 40, 43–45].

## 9.3
## Assembly of Metallic Nanorods: Self-Assembly vs. Designed Chemical Linkages

While great progress in the synthesis and mechanism of the growth of metallic nanorods has been made, a real challenge in the field is the assembly of nanorods into a functional structure. The researcher faces two general choices in the matter: self-assembly vs. designed assembly. By self-assembly, we mean that the nanorods can be "left to themselves" to order. For example, there is experimental evidence that as one puts lateral pressure on a random "raft" of rods at the air–water interface, nanorods can assemble into ordered rafts [48]. Onsager long ago predicted [49] that increasing the concentration of "hard colloidal" rods in solution would lead to one-dimensional or two-dimensional liquid crystalline ordering, which has been experimentally and theoretically verified [50–55].

By "designed" assembly, we mean that the different crystal faces of the nanorods could be specifically reacted with some reagent to link them in a rational way. This last approach is made difficult by the usual lack of information about the crystal face identity on the ends and edges of nanorods, but progress is being made [1, 24, 40, 45].

In the case of liquid crystalline ordering as a function of nanorod concentration in the self-assembly approach, we have performed TEM, polarizing microscopy, and small angle X-ray scattering experiments to show that gold nanorods of high aspect ratio (13–18) do indeed stack in regular arrays as the concentration increases [55]. The system is not as ideal as the Onsager case, the CTAB outer bilayer

[56] makes our rods "softer" and more highly charged, but nonetheless we do observe concentration-dependent ordering.

In our experiments [55], gold nanorods of aspect ratios 13–18 were concentrated and separated from spherical nanoparticles by centrifugation. Thermogravimetric analysis showed that ~20% of the total mass of the nanorods was associated with the CTAB surfactant. This is significantly larger than that calculated for monolayer coverage (3 wt%) on the basis of particle surface area, and suggests that the nanorods are covered with multiple CTAB layers; indeed, El-Sayed has spectroscopic evidence for CTAB bilayers on gold nanorods [56]. We found that the presence of the surfactant coating was of key importance, not only for hydrophilic stabilization of the nanorods in water but also for controlling long-range self-assembly in concentrated dispersions; we found that the optimum conditions required for in situ liquid crystalline ordering involved redispersing the nanorods in 1–100 mM CTAB. Above this surfactant concentration, the nanorods precipitated instantaneously and were unable to be redispersed to image in the electron microscope. Below this CTAB concentration, nanorods were also unable to be redispersed in water. Although not fully explored, we believe these effects are not simply due to the ionic nature of CTAB, as ~1 mM NaCl precipitates the gold nanorods.

In general, the 18 aspect ratio nanorod solutions were dark brown in color (when concentrated) and had a weak absorbance maximum in the visible at ~530 nm, in addition to a near-infrared absorbance at ~1700 nm [27]. Thin films of the concentrated dispersions supported on glass slides showed iridescent droplets ~0.1 mm in diameter under polarizing light microscopy (Figure 9.6). The observed textures were indicative of localized regions of liquid crystalline ordering and are similar to nematic droplets observed in boehmite nanoneedle solutions [52]. Significantly, the liquid crystalline droplets were stable up to 200 °C in air, after which the surfactant began to degrade, although the nanorods remained unchanged in size and shape. Similar experiments with concentrated surfactant solution alone showed much smaller "speckles" in the polarizing microscope and no liquid crystalline textures.

Small-angle X-ray scattering (SAXS) experiments were undertaken to determine the extent of long-range ordering in concentrated (~5–10 wt% of solids) and diluted (by a factor of ~1000) dispersions of the gold nanorods [55]. The scattering curves show ripples in the scattered X-ray intensity due to particle shape and interparticle interactions. The data were fitted to a model consisting of core–shell cylinders stacked with a Gaussian distribution of interparticle distances using a method of non-linear least-squares fitting. The fitting parameters included the radius of the nanorods, the number of particles in a stack, the width of the Gaussian distribution of interparticle distances in the stack, the surfactant layer thickness and the major radius of the elliptical impurities. Attempts to fit the SAXS data to isolated rods failed; rod stacks were required. The fits suggest that the concentrated solutions contained self-assembled stacks of ca. 200 nanorods, each of which had a surfactant coating 3.9 nm in thickness, consistent with a CTAB bilayer [56]. In contrast, smaller clusters of ~30 rods were present in the more dilute sample. The

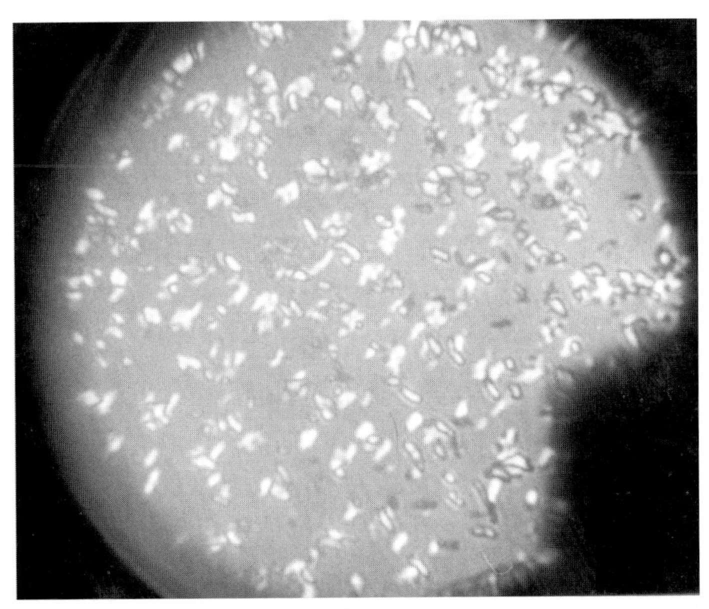

**Fig. 9.6.** Polarizing optical micrograph of a concentrated solution (~5–10% w/w) of aspect ratio 18 gold nanorods in water, taken from [55]. Reproduced by permission of The Royal Society of Chemistry.

width of the Gaussian distribution of spacings between the particle centers was an order of magnitude narrower in the concentrated dispersions compared to the dilute dispersions, consistent with more dense arrays in the concentrated dispersions and a high degree of disorder in the dilute dispersions. No evidence for hexagonal phases was found, under our conditions.

TEM images of air-dried dispersions prepared at low nanorod concentrations (<1% by weight, including surfactant) showed mainly discrete nanoparticles (Figure 9.7). Some short-range order involving side-on, end-to-side or end-to-end aggregation was observed, presumably due to capillary forces associated with the drying process [57]. At high nanorod concentrations (~5–10 wt%), in contrast, microscopic smectic-like arrays of closely packed nanorods were observed (Figure 9.7). The arrays consisted of nanorods that were aligned parallel to each other in micron-length rows, which in turn were stacked laterally to produce the higher-order superstructure. Such structures were observed predominantly at the edges of dried droplets (apparent by visual inspection of the TEM grid as brown rings), suggesting that capillary forces were responsible for the smectic-like organization [57]. In contrast, other areas of the TEM grid showed a predominance of micrometre-long rows of ordered nanorods (Figure 9.7), which probably correspond more closely to the in situ organization of the nanorods within the concen-

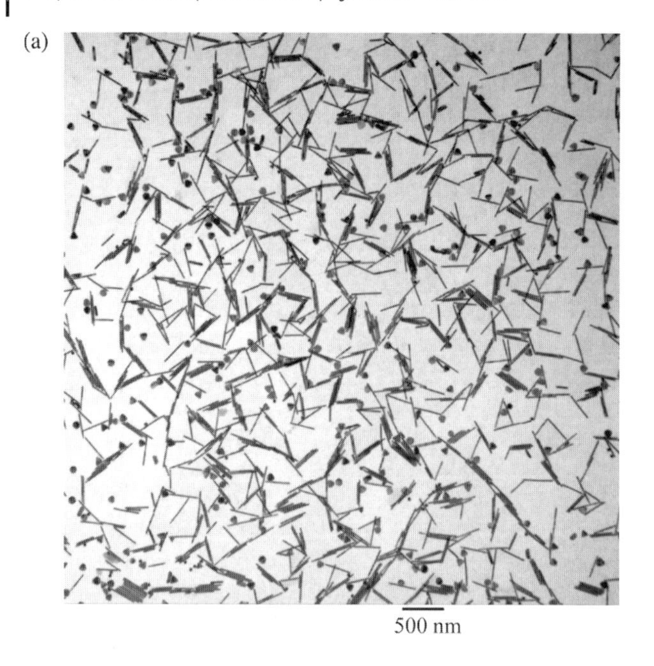


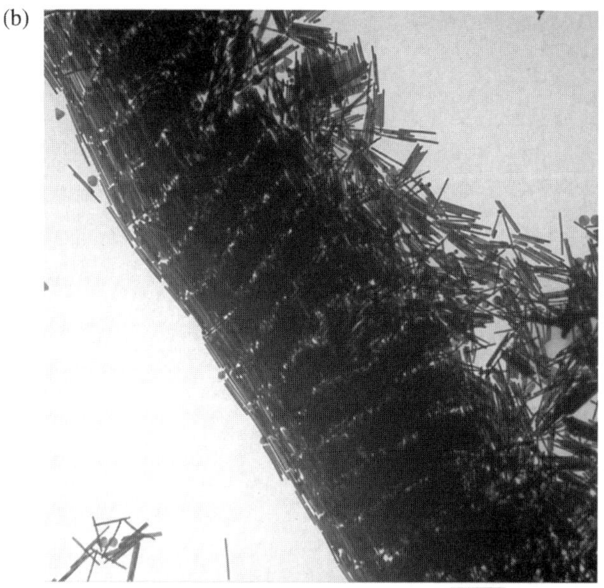


**Fig. 9.7.** Transmission electron micrographs showing concentration dependent ordering of gold nanorods, taken from [55]. (a) <1 wt% showing isolated nanorods; some spherical particles are still present even after centrifugation. (b) ~5–10 wt% showing smectic-like arrays, and

(c)

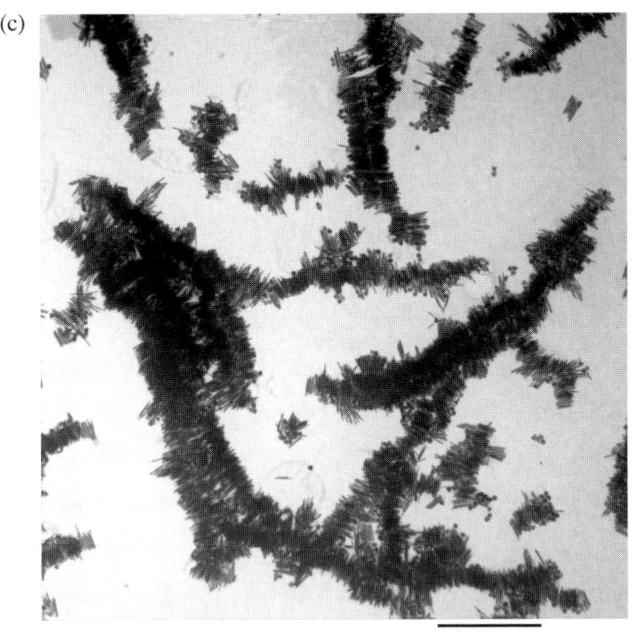


**Fig. 9.7.** (c) ~5–10 wt% showing linear stacks of gold nanorods. Reproduced by permission of The Royal Society of Chemistry.

trated dispersion. We estimate tens of rods assemble in vertical stacks, while ~100 stacks assemble into rows, in reasonable agreement with the SAXS data given the large uncertainties in this fitted parameter. The smectic-like phase presumably arises from secondary ordering of the preformed linear stacks on the TEM grid during drying.

El-Sayed has reported that gold nanorods of aspect ratio 4.6, coated with two different cationic surfactants, assemble into higher-order structures upon concentration from aqueous solution [53]. Our results with higher-aspect ratio nanorods are consistent with this and indicate that surfactant-mediated interactions between gold nanorods of uniform shape and size can give rise to ordered liquid crystalline arrays in concentrated suspensions. Multilayers of surface-adsorbed cationic surfactants such as CTAB can induce a remarkable degree of self-ordering of spherical gold nanoparticles due to a balance between short-range electrostatic repulsion and interchain attraction [58]. Moreover, interdigitation of surfactant chains on specific faces of prismatic nanocrystals can give rise to ordered single chains of other ($BaCrO_4$) nanoparticles [35] or self-assembled nanocubes of $Cu_2O$ [59]. For the gold nanorods described here, liquid crystalline arrays were only routinely observed in aqueous solutions containing the proper concentration range of CTAB, suggesting that interactions between surfactant molecules in solution with surface-adsorbed amphiphiles were important aspects of the assembly process. The hydrophilic nature of the gold nanoparticles prior to assembly indicates that the surfactant molecules in the outer layer of the surface coating are oriented with their cationic headgroups exposed to the solvent. However, as the surfactant-coated nanorods approach each other in solution, expulsion of the outermost "cationic head out" CTAB molecules and their associated counterions could result in the formation of hydrophobic nanorods in which the remaining CTAB hydrophobic tails face the solvent; thus, the resulting nanorods spontaneously self-assembly in a side-on fashion to minimize the unfavorable hydrophilic–hydrophobic interactions with water and promote interdigitation of the surfactant tails.

"Designed" assembly of metallic nanorods has been demonstrated recently [60, 61]. In one case [60], metallic nanorods were cross-linked with DNA in a manner analogous to Mirkin's approach for gold nanospheres [15]. Two different batches of gold nanorods were derivatized with two different DNA sequences, then a third linking DNA strand is added that can hydrogen-bond to both nanorod batches [60]. In this way, the spacing between gold nanorods can be controlled by the length of the DNA linkers. In the other case, gold nanorods derivatized with DNA were bound to a flat gold surface that was patterned DNA, and the assembly was directed by the specific hydrogen-bonding of the DNA [61]. Figure 9.8 shows TEM images of the linked gold nanorods from [60].

Functionalizing the ends only of gold nanorods is possible, if the sides are physically protected from reaction, which could then result in designed end-to-end

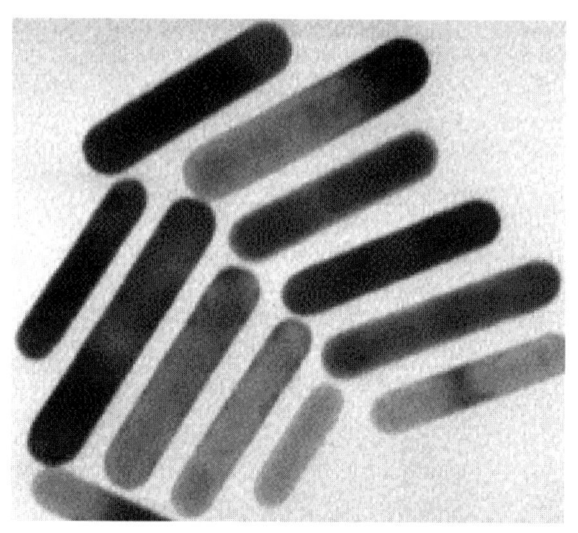

**Fig. 9.8.** Transmission electron micrograph of DNA-linked gold nanorods, taken from [60]. Reproduced by permission of The Royal Society of Chemistry.

alignment of the rods [61]. It is also possible, however, that the different crystal faces of gold nanorods might react differently with reagents, and also form the basis for rational assembly strategies of gold nanorods.

## 9.4 Reactivity of Metallic Nanoparticles Depends on Aspect Ratio

Understanding the chemical reactivity of nanoparticles as a function of size and shape is a research area which is likely to lead to many useful and perhaps surprising results. We have already mentioned that gold nanoparticles ~3 nm in diameter are no longer "noble" and can catalyze chemical reactions [4]. Poisoning of platinum nanoparticle catalysts by sulfur compounds results in different crystal facets of Pt becoming exposed to the environment, and the subsequent reactivity of the Pt nanoparticles is compromised [47]. We have found that the reactivity of gold with cyanide and persulfate depends on the size and shape of the gold nanoparticles [62].

The reaction of cyanide with gold, in the presence of air and water, produces the gold–cyanide complex ion [63]:

$$4\ Au + 8\ CN^- + O_2 + 2\ H_2O \rightarrow 4\ [Au(CN)_2]^- + 4\ OH^- \tag{1}$$

This reaction is the industrially important cyanide process for recovering gold from rocks [63]; in that case, elemental gold is precipitated out by adding Zn dust to reduce the Au(I) back to Au(0).

Gold spheroids have a transverse plasmon band, at ~530 nm and a longitudinal plasmon band at 600–1000 nm, depending on the aspect ratio. Various aspect ratio spheroids (aspect ratio 2–5) were used for cyanide dissolution experiments, and all gave similar results; typical results for spheroids of aspect ratio $2.5 \pm 0.5$ will be described here [62]. Figure 9.9 shows the ultraviolet–visible absorption spectra of the $2.5 \pm 0.5$ aspect ratio gold spheroids upon reaction with cyanide. In the absence of cyanide, the solution appears violet and has two absorption peaks at approximately 526 nm and 685 nm. Upon addition of cyanide solution to the nanoparticle solution, the absorption spectra changed, depending on the cyanide concentration (Figure 9.9b–g). The violet color gradually turned red from a to d as the cyanide concentration increased; in set e ($5 \times 10^{-4}$ M cyanide) the solution appeared deep red; and in f and g, it was colorless. For cyanide concentrations below $2.5 \times 10^{-4}$ M, both of the band positions decreased in absorbance, with the band maximum at 526 nm remaining unchanged and the 685 nm band blue shifting by ~20 nm. At $5 \times 10^{-4}$ M $CN^-$ the long wavelength band completely disappeared within 1 min, and the intensity of the 526 nm band decreased by ~30% within 2 min. At $10^{-3}$ M $CN^-$ only a weak band at 526 nm was observed. As

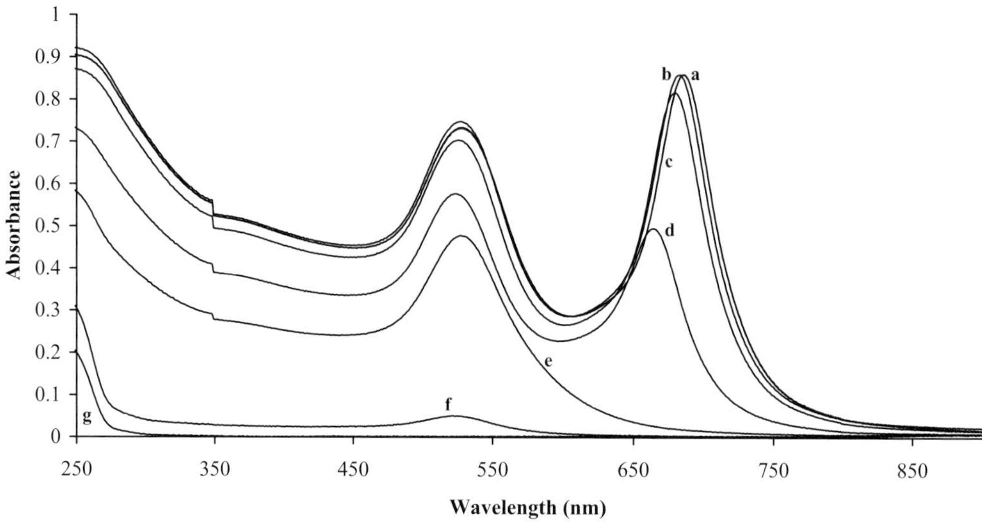


**Fig. 9.9.** UV–visible spectra of 2.5 ± 0.5 aspect ratio gold spheroids with 17 nm short axis ($2.5 \times 10^{-4}$ M in terms of gold) after 1 h of the addition of various amounts of cyanide solution. a: 0.0 M, b: $10^{-5}$M, c: $10^{-4}$ M, d: $2.5 \times 10^{-4}$ M, e: $5 \times 10^{-4}$ M, f: $10^{-3}$ M, g: $2.5 \times 10^{-3}$ M sodium cyanide. Reprinted with permission from [62]. Copyright (2002) American Chemical Society.

the cyanide concentration was further increased to $2.5 \times 10^{-3}$ M, this band completely disappeared within 8 min.

Figure 9.10(a) shows the TEM of 2.5 ± 0.5 aspect ratio spheroids (with 17 nm short axis) before $CN^-$ addition. The solution also contained a significant amount (~40%) of spheres of similar size (~20–25 nm). Those spheres contributed only a single absorbance at 526 nm, as they do not have a longitudinal plasmon band. Figure 9.10(b) shows the TEM corresponding to sets e of Figure 9.9. In set e only spheres in the size range 10–25 nm were present in the TEM. In set d (TEM not shown) the percentage of spheroids decreased from 60% to 40%, and size measurement of the spheroids showed that the length decreased but the width remained unchanged. This leads to a reduced average aspect ratio of spheroids from 2.5 ± 0.5 to 2.0 ± 0.5. In contrast, the spheres present in the solution did not change in size significantly. In sets f only spheres of diameter 2–5 nm (~90%) and 5–15 nm (10%) were observed. In set g no particles were found in the TEM.

Absorbance vs. time (Figure 9.11) (at both the longitudinal and transverse wavelength bands for the spheroids) and successive UV–visible spectra (Figure 9.12) were measured during the dissolution process of the gold spheroids. Figures 9.11 and 9.12 corresponds to set f. The rate of decrease in absorbance is faster for the longitudinal band than the transverse band (Figure 9.11). In the intermediate stage, the longitudinal plasmon band gradually blue shifts with a simultaneous decrease in intensity and finally mixes into the transverse band (Figure 9.12).

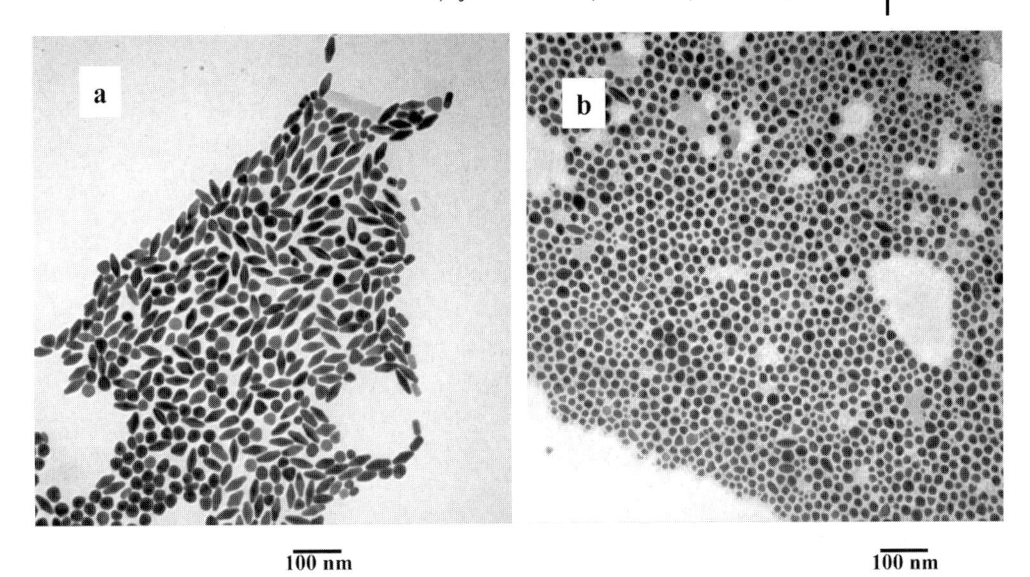


**Fig. 9.10.** Transmission electron micrograph of gold spheroids before and 1 h after cyanide treatment. a: no cyanide, b: $5 \times 10^{-4}$ M $CN^-$. Reprinted with permission from [62]. Copyright (2002) American Chemical Society.

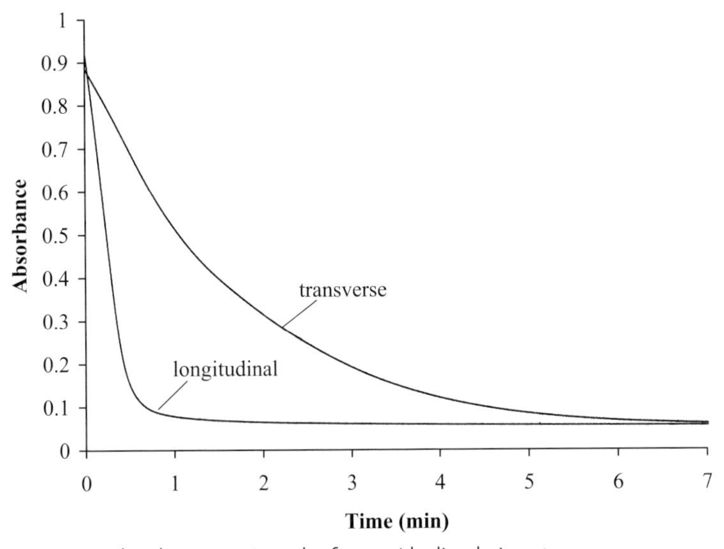


**Fig. 9.11.** Absorbance vs. time plot for cyanide dissolution at wavelengths corresponding to the transverse and longitudinal plasmon bands. Conditions were similar to set 'f' in Figure 9.9. Reprinted with permission from [62]. Copyright (2002) American Chemical Society.

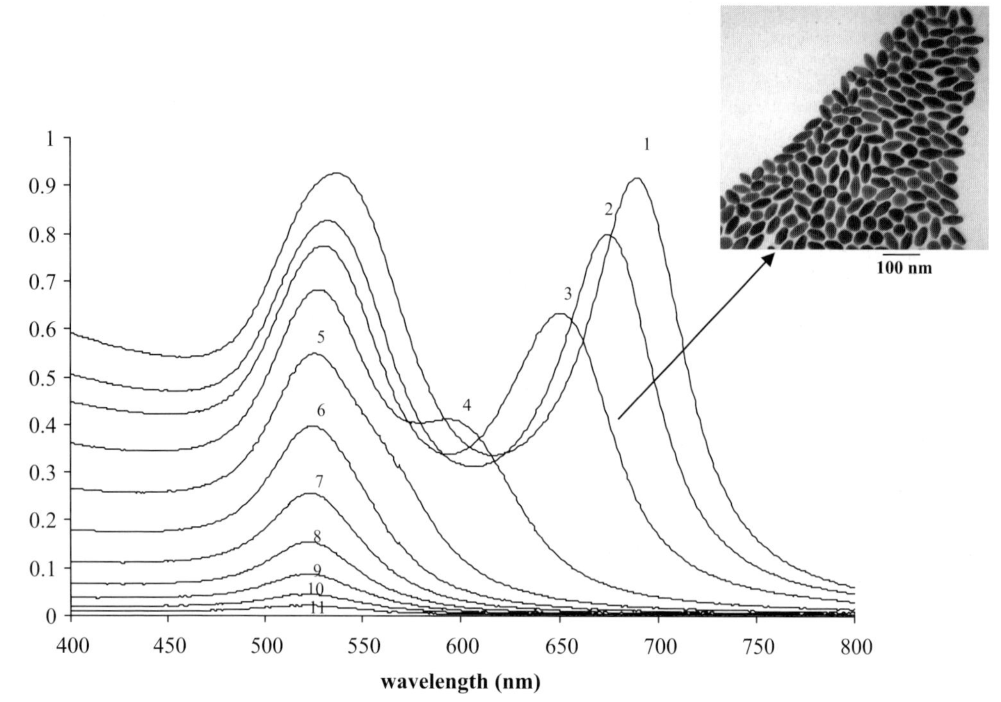


**Fig. 9.12.** Successive UV–visible spectra of gold plasmon bands during cyanide dissolution of set 'f'. (1) before cyanide addition, (2) after 20 s of cyanide addition and (3 to 10) after 50 s intervals from 2. The inset shows the TEM of intermediate short spheroids corresponding to spectrum 3 isolated by rapidly centrifuging the particles from reaction mixture. Reprinted with permission from [62]. Copyright (2002) American Chemical Society.

Thus, the apparent transformation of the spheroids to spheres in the presence of cyanide, in air, may be depicted by the following cartoon:

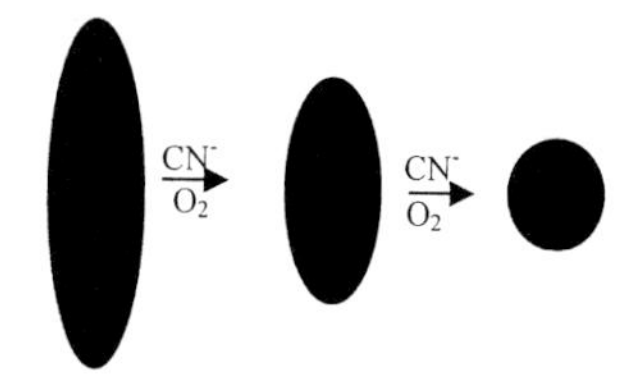


In this cartoon, the width of the nanoparticles does not change, only its length.

The appearance of an intermediate blue shifted longitudinal band also indicates that spheroid dissolution starts at the two ends. The longitudinal band disappears at a faster rate due to the faster dissolution rate of spheroids than spheres. We also isolated the particles at the intermediate stage of cyanide dissolution by rapidly centrifuging the particles and separating them from excess cyanide present in solution before the dissolution reaction could complete. This was performed within 1 min (compared to 5–10 min for the complete reaction). Both TEM and UV–vis

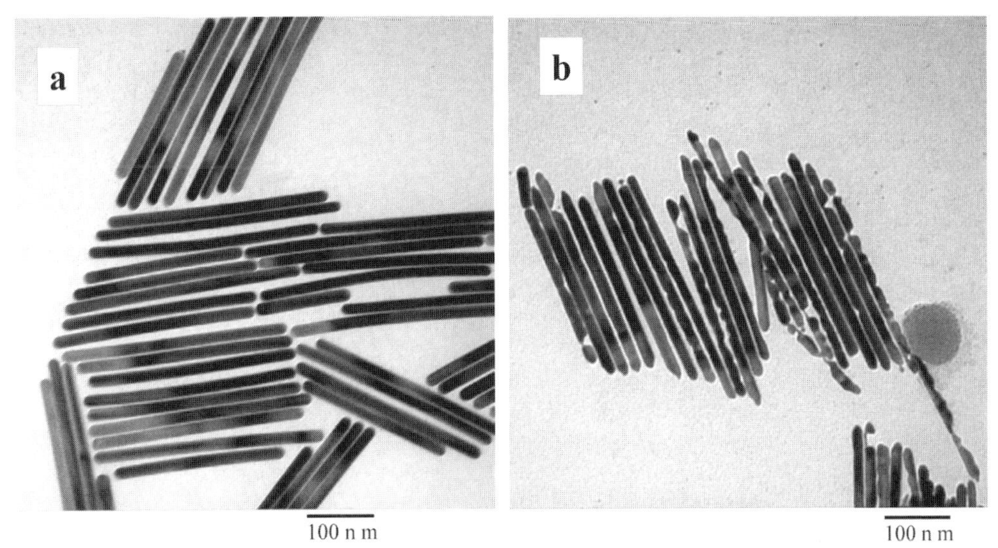


**Fig. 9.13.** Transmission electron micrographs of gold nanorods, aspect ratio 18, before (a) and 24 h after (b) cyanide treatment. Reprinted with permission from [62]. Copyright (2002) American Chemical Society.

spectra indicate that the intermediate particles were smaller 1.8 aspect ratio spheroids (inset of Figure 9.12).

Gold nanorods with 17 nm short axis and aspect ratio of 18 have a weak plasmon band at 503 nm and (presumably) a longitudinal plasmon band beyond ~2000 nm. With increasing cyanide concentration the 503 nm band gradually decreased in intensity and finally vanished in ~60 min (data not shown). No plasmon band in the 600–1850 nm region was detected at intermediate cyanide concentrations, suggesting that short rods were not intermediates. TEMs of the nanorods showed that they were uniform in shape (Figure 9.13) before reaction with cyanide. At intermediate cyanide concentrations, these higher-aspect ratio nanorods did not decrease in length, but dissolution occurred at many sites along the length of the rod (Figure 9.13), leading to pitted nanorods. At lower cyanide concentrations, many rods remained unchanged and, as the cyanide concentration increased, defective rods increased in number with the increased extent of corrosion. Our results indicate that gold spheroids (aspect ratio 2–5 with 12–30 nm short axis) are more reactive than spheres (20–30 nm) and nanorods (aspect ratio 18 with 16 nm short axis) for reaction with cyanide [62].

For the spheroids, we were able to obtain SERS signals of cyanide, even at concentrations of cyanide too low to show changes in the visible spectra [62]. Thus, it appears that cyanide can adsorb to the surface and yet not immediately react. This is consistent with a purported mechanism of cyanide reaction with gold, in which the adsorbed cyanide ions react to form a protective AuCN layer [62].

The high aspect ratio gold nanorods were not "eaten" from the ends by cyanide, as were their spheroidal counterparts. The topology and chemical potential of edge atoms of spheroids may not favor the formation of stable films of protective AuCN. This may be the reason for enhanced reactivity and anisotropic dissolution of spheroids.

Non-uniform cyanide dissolution of cylindrical nanorods starts at many sites along the length of the nanorod. This may be related to local crystal structures within the rods, which is a twinned defect structure [55]. It is also possible that the CTAB bilayer that is likely to be most strongly bound on the long sides of the nanorods may compete with cyanide for binding and subsequent reaction; thus, the rods are pitted along their long axis due to CTAB pinholes [62]. Statistically, there are many more gold atoms along the long axis of the nanorod than on the end faces, so one might expect dissolution along the long axis of the nanorods to appear there first. Clearly, however, this statistical argument is incorrect for the spheroids of aspect ratios 2–5, which do react from the ends preferentially.

We also examined the dissolution reaction of gold nanospheres, spheroids, and nanorods (aspect ratios 1, 2–5, and 18 respectively) with persulfate, which is thermodynamically a favorable reaction [62]:

$$2\ Au + S_2O_8{}^{2-} \rightarrow 2\ Au^+ + 2\ SO_4{}^{2-} \tag{2}$$

Our results, based on UV–vis spectroscopy and TEM [62], indicate that spheroids (aspect ratio 2–5 with 12–30 nm short axis) are more reactive than spheres (20–30 nm) and nanorods (aspect ratio 18 with 16 nm short axis). In the presence of persulfate, spheroids convert to spheres (Figure 9.14), but similar size spheres and rods do not react at all, on the timescale of days. We presume that during the shape transition from spheroid to sphere, a fraction of gold atoms at the spheroid edges oxidize; thus the particle diameter shrinks (~5–10%) after persulfate treatment.

Overall, then, we can conclude that the nanoparticle shape affects its reactivity [62].

## 9.5 Conclusions and Future Prospects

Metallic nanorods are highly interesting materials from many points of view: as elements in future nanoscale electronic circuits; as sensors; as catalysts; as optical elements in future nanoscale optical devices. Gold and silver nanorods have distinct visible absorption and scattering spectra that are tunable with aspect ratio. Many workers have developed wet synthetic routes to these nanomaterials, with control of aspect ratio a key improvement compared to the synthesis of simple nanospheres. Another key area for which improvements need to be made is the understanding of the atomic arrangements of the different faces of crystalline

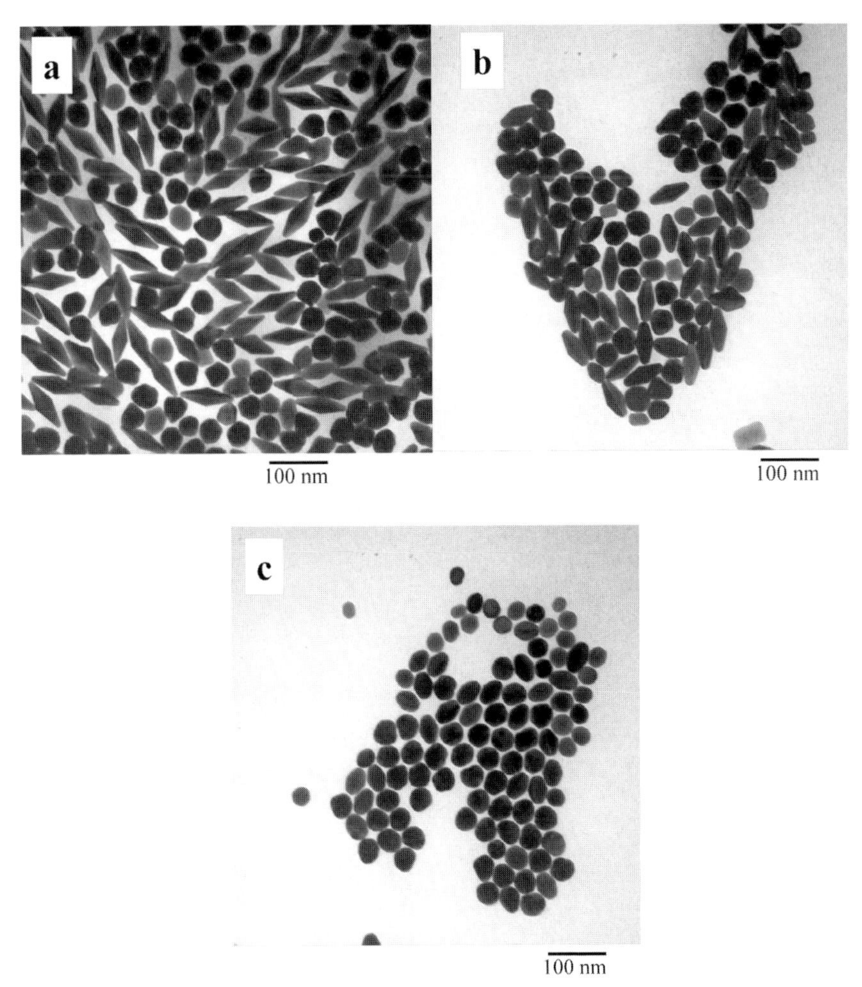


**Fig. 9.14.** Transmission electron micrographs of gold spheroids before (a) and 24 h after (b, c) persulfate treatment: a, no persulfate; b, $10^{-6}$ M persulfate; c, $10^{-5}$ M persulfate. Reprinted with permission from [62]. Copyright (2002) American Chemical Society.

nanorods. Assembly of nanorods into organized structures is at a very early stage, but self-assembly into liquid crystalline ordered arrays has been demonstrated. Linkage chemistries to "wire up" the nanorods are being developed, and that endeavor will be greatly aided by better understanding of the chemistry of different crystal faces of the nanorods. Evidence of the different reactivity of the different crystal faces of metallic nanorods, as a function of aspect ratio, is emerging.

## Acknowledgements

We thank the U.S. National Science Foundation (C.J.M.) and the University of South Carolina for support of sabbatical leave for C.J.M. to visit S.M. in Bristol, UK. We also thank the EPSRC, UK for support of a postgraduate studentship to C.J., Dr Igor Dolbyna and the other staff on BM26B at the ESRF for assistance in collecting the SAXS data, and the Royal Society for a Dorothy Hodgkin Research Fellowship (K. J. E.).

## References

1 Murphy, C. J. *Science* **2002**, *298*, 2139–2141.
2 *Metal Nanoparticles: Synthesis, Characterization and Applications*, eds. Feldheim, D. L., Foss, C. A., Jr., Marcel Dekker, New York 2002.
3 El-Sayed, M. A. *Acc. Chem. Res.* **2001**, *34*, 257–264.
4 Valden, M., Lai, X., Goodman, D. W. *Science* **1998**, *281*, 1647–1650.
5 Foss, C. A., Jr., Hornyak, G. L., Stockert, J. A. et al. *J. Phys. Chem.* **1994**, *98*, 2963–2971.
6 Hornyak, G. L., Patrissi, C. J., Martin, C. R. *J. Phys. Chem. B* **1997**, *101*, 1548–1555.
7 Link, S., El-Sayed, M. A. *J. Phys. Chem. B* **1999**, *103*, 8410–8426.
8 Schultz, S., Smith, D. R., Mock, J. J. et al. *Proc. Natl. Acad. Sci. USA* **2000**, *97*, 996–1001.
9 Kottman, J. P., Martin, O. J. F., Smith, D. R. et al. *Chem. Phys. Lett.* **2001**, *341*, 1–6.
10 Kelly, K. L., Coronado, E., Zhao, L. L. et al. *J. Phys. Chem. B* **2003**, *107*, 668–677.
11 Nie, S. M., Emory, S. R. *Science* **1997**, *275*, 1102–1106.
12 Kneipp, K., Kneipp, H., Itzkan, I. et al. *Chem. Rev.* **1999**, *99*, 2957–2976.
13 Obare, S. O., Hollowell, R. E., Murphy, C. J. *Langmuir* **2002**, *18*, 10407–10410.
14 Lin, S.-Y., Liu, S.-W., Lin, C.-M. et al. *Anal. Chem.* **2002**, *74*, 330–335.
15 Elghanian, R., Storhoff, J. J., Mucic, R. C. et al. *Science* **1997**, *277*, 1078–1080.
16 Taton, T. A., Mirkin, C. A., Letsinger, R. L. *Science* **2000**, *289*, 1757–1760.
17 Kim, Y., Johnson, R. C., Hupp, J. T. *NanoLett.* **2001**, *1*, 165–167.
18 Thanh, N. T. K., Rosenzweig, Z. *Anal. Chem.* **2002**, *74*, 1624–1628.
19 Nicewarmer-Pena, S. R., Freeman, R. G., Reiss, B. D. et al. *Science* **2001**, *294*, 137–141.
20 Murphy, C. J., Jana, N. R. *Adv. Mater.* **2002**, *14*, 80–82.
21 Jin, R., Cao, Y. W., Mirkin, C. A. et al. *Science* **2001**, *294*, 1901–1903.
22 Kovtyukhova, N. I., Mallouk, T. E. *Chem. Eur. J.* **2002**, *8*, 4355–4363.
23 Sun, S., Murray, C. B., Wells, D. et al. *Science* **2000**, *287*, 1989–1992.
24 Puntes, V. F., Krishnan, K. M., Alivisatos, A. P. *Science* **2001**, *291*, 2115–2117.
25 Wilson, O., Wilson, G. J., Mulvaney, P. *Adv. Mater.* **2002**, *14*, 1000–1004.
26 Jana, N. R., Gearheart, L., Murphy, C. J. *Chem. Commun.* **2001**, 617–618.
27 Jana, N. R., Gearheart, L., Murphy, C. J. *J. Phys. Chem. B* **2001**, *105*, 4065–4067.
28 Jana, N. R., Gearheart, L., Murphy, C. J. *Adv. Mater.* **2001**, *13*, 1389–1393.
29 Busbee, B. D., Obare, S. O., Murphy, C. J. *Adv. Mater.* **2003**, *15*, 414–416.
30 Jana, N., Gearheart, L., Murphy, C. J. *Langmuir* **2001**, *17*, 6782–6786.
31 Jana, N. R., Gearheart, L., Murphy, C. J. *Chem. Mater.* **2001**, *13*, 2313–2322.
32 Walter, E. C., Murray, B. J., Favier,

F. et al. *J. Phys. Chem. B* **2002**, *106*, 11407–11411.
**33** Ahmadi, T. S., Wang, Z. L., Green, T. C. et al. *Science* **1996**, *272*, 1924–1925.
**34** Tanori, J., Pileni, M. P. *Langmuir* **1997**, *13*, 639–646.
**35** Li, M., Schnablegger, H., Mann, S. *Nature* **1999**, *402*, 393–395.
**36** Peng, X. G., Manna, L., Yang, W. D. et al. *Nature* **2000**, *404*, 59–61.
**37** Kwan, S., Kim, F., Akana, J. et al. *Chem. Commun.* **2001**, 447–448.
**38** Peng, Z. A., Peng, X. G. *J. Am. Chem. Soc.* **2002**, *124*, 3343–3353.
**39** Sun, Y., Xia, Y. *Adv. Mater.* **2002**, *14*, 833–837.
**40** Sun, Y., Xia, Y. *Science* **2002**, *298*, 2176–2179.
**41** Hu, J., Odom, T. W., Lieber, C. M. *Acc. Chem. Res.* **1999**, *32*, 435–445.
**42** Yu, H., Gibbons, P. C., Kelton, K. F. et al. *J. Am. Chem. Soc.* **2001**, *123*, 9198–9199.
**43** Filankembo, A., Pileni, M. P. *J. Phys. Chem. B* **2000**, *104*, 5865–5868.
**44** Manna, L., Scher, E. C., Alivisatos, A. P. *J. Cluster Sci.* **2002**, *13*, 521–532.
**45** Johnson, C. J., Dujardin, E., Davis, S. A. et al. *J. Mater. Chem.* **2002**, *12*, 1765–1770.
**46** Yu, S.-H., Colfen, H., Antonietti, M. *Chem. Eur. J.* **2002**, *8*, 2937–2945.
**47** Harris, P. J. F. *Nature* **1986**, *323*, 792–794.
**48** Kim, F., Kwan, S., Akana, J. et al. *J. Am. Chem. Soc.* **2001**, *123*, 4360–4361.
**49** Onsager, L. *Ann. N. Y. Acad. Sci.* **1949**, *51*, 627–659.
**50** Frenkel, D., Lekkerkerker, H. N. W., Stroobants, A. *Nature* **1988**, *332*, 822–823.
**51** van Bruggen, M. P. B., Lekkerkerker, H. N. W. *Langmuir* **2002**, *18*, 7141–7145.
**52** Gabriel, J.-C. P., Davidson, P. *Adv. Mater.* **2000**, *12*, 9–20.
**53** Nikoobakht, B., Wang, Z. L., El-Sayed, M. A. *J. Phys. Chem. B* **2000**, *104*, 8635–8640.
**54** Li, L.-S., Walda, J., Manna, L. et al. *NanoLetters* **2002**, *2*, 557–560.
**55** Jana, N. R., Gearheart, L. A., Obare, S. O. et al. *J. Mater. Chem.* **2002**, *12*, 2909–2912.
**56** Nikoobakht, B., El-Sayed, M. A. *Langmuir* **2001**, *17*, 6368–6374.
**57** Oliver, S. R. J., Bowden, N., Whitesides, G. M. *J. Colloid Interface Sci.* **2000**, *224*, 425–428.
**58** Fink, J., Kiely, C. J., Bethell, D. et al. *Chem. Mater.* **1998**, *10*, 922–926.
**59** Gou, L., Murphy, C. J. *NanoLetters* **2003**, *3*, 231–234.
**60** Dujardin, E., Hsin, L.-B., Wang, C. R. C. et al. *Chem. Commun.* **2001**, 1264–1265.
**61** Mbindyo, J. K. N., Reiss, B. D., Martin, B. R. et al. *Adv. Mater.* **2001**, *13*, 249–254.
**62** Jana, N. R., Gearheart, L., Obare, S. O. et al. *Langmuir* **2002**, *18*, 922–927.
**63** Greenwood, N. N., Earnshaw, A. *Chemistry of the Elements*, 2nd edition, Butterworth-Heinemann, Oxford 1997, p. 1175.

# 10
# Oxide-Assisted Growth of Silicon and Related Nanowires: Growth Mechanism, Structure and Properties

*S. T. Lee, R. Q. Zhang, and Y. Lifshitz*


## Abstract

This chapter focuses on the oxide-assisted growth (OAG) of silicon-based nanowires. OAG, invented by the research team of City University of Hong Kong, is different and distinguishable from the conventional metal catalyst vapor–liquid–solid (VLS) growth. The 1D growth is initiated through suboxide droplets which are very reactive to $Si_xO_y$ in the gas phase, and no metal catalyst droplets are needed. The further 1D growth occurs via precipitation of a silicon core encapsulated by a $SiO_2$ sheath, which restricts the lateral growth. OAG was found to be a generic method capable of bulk production of a variety of different semiconducting nanowires. We summarize the research efforts at City University of Hong Kong during the past several years. We first describe the discovery of the OAG process, and its distinction from the metal catalyst VLS process as evident from the growth conditions and the structure of the resulting wires. Then we discuss the OAG nucleation and growth process. We follow by showing how we can modify the morphology and size (affecting the properties) of the nanowires by varying growth parameters: (1) morphology control by temperature, (2) diameter control by carrier gas, (3) large-area, aligned, long silicon nanowires (SiNWs) by flow control. Two-dimensional nanostructures, i.e. nanoribbons, have also been fabricated. Hybrid structures, such as nanocables, metalized SiNWs and SiC, were grown by applying multi-step processes, ion implantation and reduction in liquid solutions. The generic nature of the OAG approach was realized in a host of different semiconducting nanowires such as Ge, SiC, GaN, GaAs, and GaP nanowires, as well as ZnO whiskers. The variety of nanowires produced was characterized by different methods including electron microscopy, Raman scattering, photoluminescence, FTIR, field emission, electrical measurements, and scanning tunneling microscopy. The morphology, microstructure, optical, electrical and chemical properties of Si and related nanowires were systematically characterized. The work was supported by modeling efforts which gave additional insight into different aspects of the oxide-assisted nucleation and growth and the resulting properties.


*The Chemistry of Nanomaterials: Synthesis, Properties and Applications, Volume 1.* Edited by C. N. R. Rao, A. Müller, A. K. Cheetham

ISBN: 3-527-30686-2

## 10.1 Introduction

The discovery of carbon nanotubes (CNTs) by Iijima [1] in 1991, initiated an intensive study of one-dimensional (1D) nanomaterials including tubes, wires, cables and ribbons in general, and of the fundamental properties and potential applications of carbon nanotubes [2–6] in particular. The interest in CNTs stems from their small diameter (smallest, 4 Å) enabling unprecedented and exciting opportunities for the study of size- and dimensionality-dependent chemical and physical phenomena [7–13]. It is believed that these size effects open the door for many potential applications in nanotechnology, such as high-strength materials [14], electronic components [15], sensors [16, 17], field emitters [18, 19] and hydrogen storage materials [20].

The limitations of carbon nanotubes, such as the selective growth of metallic or semiconducting tubes and the difficulty of achieving controlled doping, motivated the alternative study of conventional one-dimensional (1D) semiconducting materials. These nanowires do not seem to face these problems, which make them much more adaptable for volume fabrication of nanodevices. Silicon nanowires (SiNWs) are of special interest since silicon is the most widely used and studied semiconducting material. In 1998, Lieber et al. [21] and the CityU team [22] independently reported the bulk-quantity synthesis of SiNWs. At CityU, we proposed an oxide-assisted growth (OAG) model to explain the growth of SiNWs [23–28], while Lieber et al. advocated the laser-assisted metal catalyst vapor–liquid–solid (VLS) growth [21]. In contrast to the conventional metal catalyst VLS growth [21], the OAG does not require a catalytic metal nanoparticle tip, thus providing a much "cleaner" method for the 1D material fabrication.

With this OAG approach, highly pure, ultra-long and uniform-sized SiNWs in bulk-quantity could be synthesized by either laser ablation or thermal evaporation of silicon powders mixed with silicon oxide or silicon monoxide only [23–28]. Section 10.2 discusses the physical chemistry aspects of the OAG. Transmission electron microscopic data and theoretical calculations are used to describe the nucleation and the growth of SiNWs via the OAG process.

In further efforts to achieve controlled growth, SiNWs of varying diameter, phase purity, morphology, defect density and doping have been obtained. This was achieved by varying the deposition parameters including growth temperature, carrier gas composition, carrier gas flow, and target composition. Different SiNW diameters were obtained by varying the carrier gas [29]. In contrast to the work of Lieber et al. [30] and Yang et al. [31] (who used metal nanoparticles of uniform size to control the diameter of SiNWs via the laser-assisted catalytic VLS growth) we found that the SiNW diameters had a wide distribution. The OAG method enabled not only fabrication of Si 1D nanostructures of different morphologies [32, 33], but also 2D nanostructures, i.e. silicon nanoribbon [34]. Section 10.3 reviews our work on the control of SiNW structure and size.

Future applications of SiNWs require the production of hybrid structures made of SiNWs integrated with nanostructures of other materials. We treated this issue by using multistep processes. One example was to grow nanocables. Metallization of the SiNW surfaces is another example of a hybrid configuration. We addressed it by ion implantation of SiNWs with metal ions. An additional route for incorporation of SiNWs with other materials/structures is the transformation of one nanowire to another by chemical reactions. We demonstrate this approach by our work on SiC nanowires. Our works on compound structures are detailed in Section 10.4.

The OAG is a generic method capable of producing different nanostructures from a variety of materials. We have extended the oxide-assisted approach to successfully synthesize a host of semiconducting materials, including Ge [35], GaN [36, 37], GaAs [38, 39], SiC [40], GaP [41], and ZnO (whiskers) [42]. Section 10.5 reviews these works.

The motivation for studying nanoscience and nanotechnology stems from the exciting properties predicted for nanomaterials due to size effects. Section 10.6 details our work on the chemical properties of SiNWs. We first discuss the stability of the hydrogen-terminated SiNWs produced by HF dipping which removes the $SiO_2$ sheath surrounding the crystalline Si core. This issue is central to the silicon wafer technology. A second study was dedicated to the reduction properties of the SiNWs in liquid solutions containing metal ions. Surface interactions of SiNWs with gases enable their use in chemical sensing. This was demonstrated for ammonia and water vapors in air or nitrogen. The reactivity of SiNWs in a liquid solution can be exploited for their use as templates to grow carbon nanostructures, which is the final topic of this section.

The optical and electrical properties of the nanowires (Section 10.7) have been characterized systematically by Raman scattering, photoluminescence and field emission [43, 44]. Understanding the atomic structure and electronic properties of SiNWs, including the dopant-induced conductivity, is an essential step towards the application of the nanowires. Although the structures and electronic properties of boron-doped silicon wafers have been investigated extensively, the corresponding study for SiNWs is relatively lacking, due to the insulating nature of the oxide sheath on most semiconductor nanowires and the difficulty in dispersing them. We have succeeded in removing the oxide layer of the SiNWs, obtaining atomically resolved STM images of H-terminated surfaces of SiNWs with diameters ranging from 1 to 7 nm. This enabled reliable scanning tunneling spectroscopy (STS) measurements of these wires, from which the electronic density of states and energy band gaps could be derived. The energy band gaps indeed increase from 1.1 eV for a 7 nm diameter SiNW to 3.5 eV for a 1.3 nm diameter SiNW, in accord with theoretical predictions, demonstrating the quantum size effect in SiNWs. In Section 10.7 we review our scanning tunneling microscopy (STM)/STS study on boron-doped and undoped SiNWs [45] and on the quantum size effect in SiNWs as well as our characterization work on other electrical and optical properties of SiNWs.

Modeling of SiNW structures, nucleation and growth processes and properties was done in parallel with the experimental work. The modeling work is most valuable in providing additional insight into the nature of the OAG and in explaining our experimental results. Our modeling efforts are described in Section 10.8.

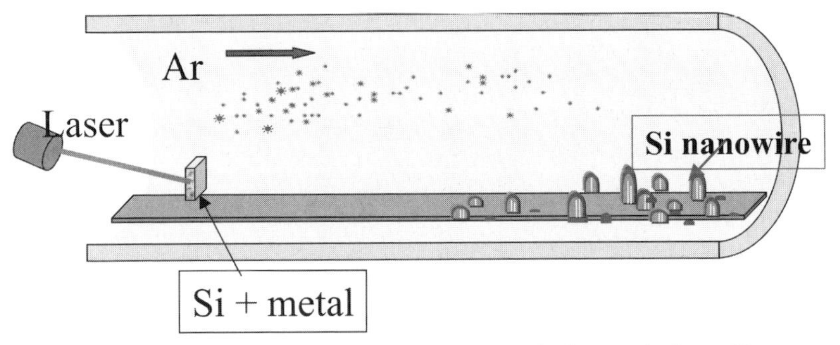


Fig. 10.1. Schematic diagram of metal catalyst VLS growth of SiNWs by laser ablation.

## 10.2 Oxide-Assisted Nanowire Growth

### 10.2.1 Discovery of Oxide-Assisted Growth [23–25]

Si nanowires were first produced using the classical metal catalyst VLS approach [21, 22, 46]. Laser ablation of a metal-containing Si target produces metal/metal silicide nanoparticles that act as the critical catalyst needed for the nucleation of SiNWs. The wires grow further by dissolution of silicon in the metallic nano-cap and concurrent Si segregation from the cap. In a typical experiment, an excimer laser is used to ablate the target placed in an evacuated quartz tube filled with an inert gas, e.g. argon [22].

We followed this idea by ablating a solid composite target (Figure 10.1) of highly pure Si powder mixed with metals (Fe, Ni, or Co). The target temperature was 1100–1400 °C and the nanowire growth temperature was selected as 900–1100 °C. Si wires with a typical diameter of 100 nm were formed (see Figure 10.2(a)) in the high furnace temperature zone (around 1100 °C). SiNWs were millimeters long and straight with metallic (Fe) spheres at the wire tip, indicating growth via a metal-catalyst VLS reaction. High-resolution transmission electron microscopy (HRTEM) observations showed that the growth direction of these Si wires was predominantly ⟨111⟩ (Figure 10.2(b)). The formation of such Si wires only at the relatively higher temperature was clearly due to the high melting temperature of Fe-silicides, e.g. $FeSi_2$.

Entirely different SiNWs grew in the lower temperature zone (~900 °C) (Figure 10.2(c)). TEM investigations showed that SiNWs obtained in this region were extremely long and highly curved with a typical smaller diameter of ~20 nm (see Figure 10.2©). Each wire consisted of a crystalline Si core in a sheath of Si oxide. The crystalline Si core had a high density of defects, such as stacking faults and micro-twins. HRTEM and electron diffraction showed that the most frequent axis of the SiNWs was along the ⟨112⟩ direction with the {111} surfaces of Si crystalline cores parallel to the nanowire axis [47]. This is in contrast to the ⟨111⟩ growth

(a)

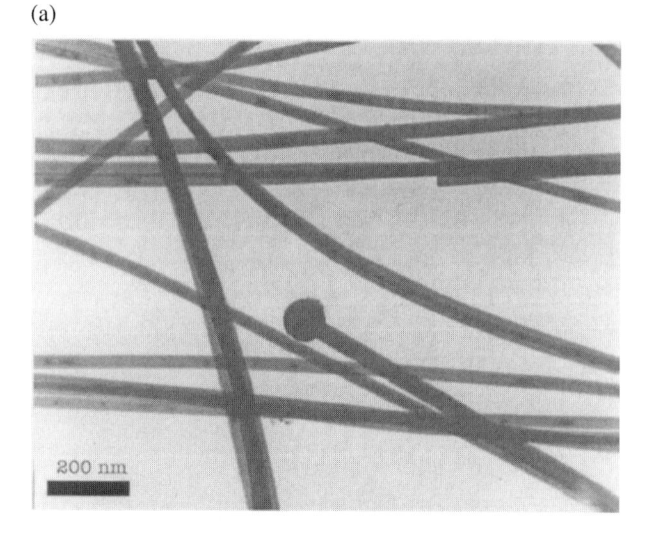


(b)

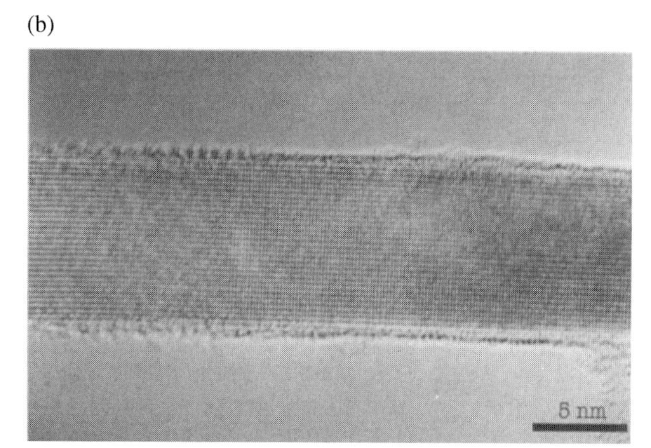


**Fig. 10.2.** TEM images of SiNWs from laser ablation [24]: (a) At high temperature, (1100 °C) note the metallic tip typical for the metal catalyst growth; (b) a typical HRTEM image of SiNWs formed at high temperature (1100 °C); and

direction common for the metal catalyst VLS growth. Most surprisingly, no evidence for metal particles was found either on the SiNWs tips or in the wires themselves, regardless of the metal used in the target (Fe, Ni or Co), in sharp contrast to those SiNWs grown in the high-temperature region. The SiNW tips were generally round and covered by a relatively thick Si oxide layer (2–3 nm) and no other component other than Si or O was detected by electron energy dispersive spectroscopy (EDS). The Si crystal core near the tip contained a high density of stacking faults and micro-twins [27], generally along the nanowire axis in the $\langle 112 \rangle$ direction.

(c)

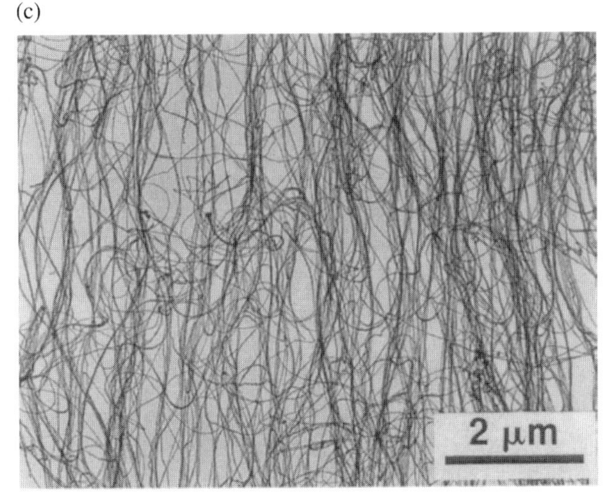


**Fig. 10.2.** (c) TEM micrograph of Si nanowires formed in the low temperature zone (900 °C).

Similar metal-free SiNWs were obtained for other metal catalysts using laser ablation or even by thermal evaporation (Figure 10.3) of either mixed powders of silicon dioxide and silicon or a pure silicon monoxide powder [23, 24, 27]. The morphology and structure of SiNWs obtained from thermal evaporation of SiO were similar to those grown from a ($Si + SiO_2$) solid source. The yield of SiNWs increased with increasing thermal evaporation temperature and pressure (see Figure 10.4). Using highly pure SiO powders, we obtained a high yield of SiNWs at temperatures ranging from 1130 to 1400 °C. This provided the direct evidence for the OAG process. These observations led us to propose that the SiNW growth at lower temperatures was induced by the oxide and not by the metal catalyst. This proposition was further substantiated by the observation that: (1) A limited quantity of SiNWs was obtained by laser ablation of pure Si powders (99.995%) or a high-purity Si wafer. (2) The growth rate of Si nanowires was greatly enhanced when $SiO_2$ was added to the Si powder targets. The yield of SiNWs produced from $SiO_2$-containing Si targets (at 50 wt% $SiO_2$) was up to 30 times higher than that from a

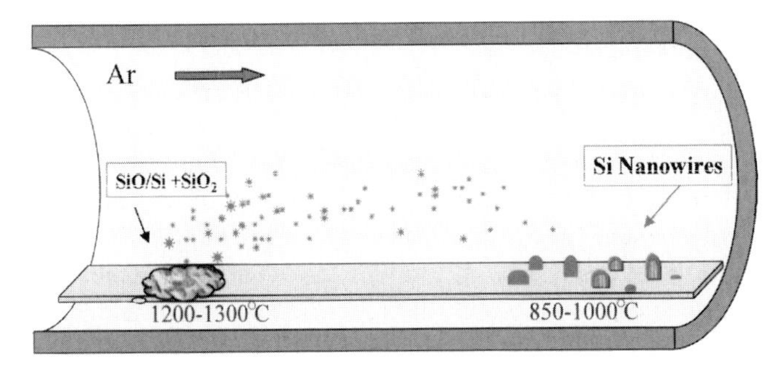


**Fig. 10.3.** Schematic diagram of oxide-assisted growth of SiNWs by thermal evaporation.

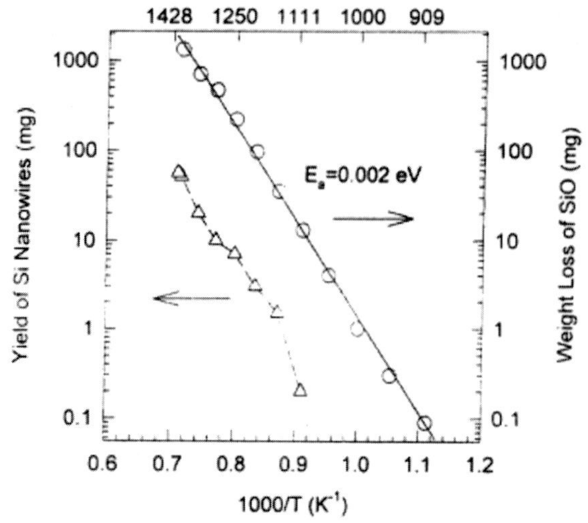


**Fig. 10.4.** The yield of the Si nanowire product increases with the weight loss of the SiO target temperature when increasing evaporation temperature [24].

Fe-containing Si target. No nanowires could be obtained using a pure $SiO_2$ target. We performed several experiments to further understand the nature of this OAG of nanowires. SiNWs could not be formed by ablation of a pure Si target in the absence of a pure metal catalyst. A two-stage experiment was carried out to explore the role of $SiO_2$ in the nucleation and growth of SiNWs. First, a $SiO_2$-containing Si target was laser ablated to form SiNW nuclei. Second, ablation of a pure Si target was attempted for further growth. SiNW growth could be observed only when a $SiO_2$-containing Si target was ablated in the second stage. The experiment showed that in the OAG nucleation of a SiNW a pure Si target was not sufficient for the further growth and the oxide was continuously needed throughout the entire SiNW nucleation and growth process. This is in contrast to the metal catalyst VLS mechanism in which the metal catalyst sustains the growth as long as the pure Si supply is maintained.

### 10.2.2
### Oxide-Assisted Nucleation Mechanism

We conducted experiments to reveal the nature of the Si core precipitation in the $SiO_2$ sheath in this new growth process. In these experiments, the vapor phase generated from the mixture of Si and $SiO_2$ at 1200 °C mainly consisted of Si monoxide [$Si(s) + SiO_2(s) \rightarrow 2SiO(g)$, where (s) and (g) represent solid and gas, respectively]. This was proven by the EDS observation that the material collected on the water-cooled Cu finger was $Si_mO_n$ ($m = 0.51, n = 0.49$). Si monoxide (SiO) is an amorphous semiconductor of high electrical resistivity, which can be readily generated from the powder mixture (especially in a 1:1 ratio) of Si and $SiO_2$ by heating [22, 48]. By heating the SiO sample, Si precipitation was observed. Such precipitation of Si nanoparticles from annealed SiO is quite well known [49]. The precipitation, nucleation and growth of SiNWs always occurred at the area near the cold finger, which suggests that the temperature gradient provided the external driving force for nanowire formation and growth. The nucleation of nanoparticles is assumed to occur on the substrate by decomposition of Si oxide as shown in Eqs. (1) and (2).

$$Si_xO(s) \rightarrow Si_{x-1}(s) + SiO(s) \quad (x > 1) \tag{1}$$

$$2SiO(s) \rightarrow Si(s) + SiO_2(s) \tag{2}$$

Our TEM data suggested that this decomposition results in the precipitation of Si nanoparticles, which are the nuclei of SiNWs, clothed with shells of silicon oxide.

We further initiated theoretical studies to explore the role of the oxide species in the OAG process. The gas-phase composition of silicon oxide clusters evaporated by laser ablation or thermal treatment should be considered to be important in the SiNW synthesis. We first used density functional theory (DFT) calculations to study the nature of the $Si_nO_m$ $(n, m = 1\text{–}8)$ clusters formed in the gas phase during OAG [50]. Our calculations show that silicon suboxide clusters are the most probable constituent of the vapor, and they have an unsaturated nature and are highly reactive towards bonding with other clusters. Moreover, a silicon suboxide cluster prefers to form a Si–Si bond with other silicon oxide clusters as shown in Figure 10.5 [51], while an oxygen-rich silicon oxide cluster prefers to form a Si–O bond with

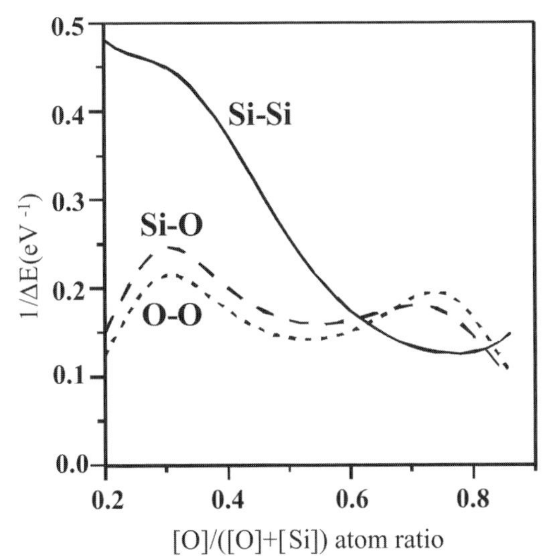


**Fig. 10.5.** The inverse of the energy difference $\Delta E$ = LUMO (electron acceptor) – HOMO (electron donor) and thus the reactivity (proportional to the inverse of the energy difference) for the formation of a Si–Si bond, a Si–O bond, or an O–O bond between two silicon oxide clusters as a function of Si:O ratio [51].

other clusters. Based on these calculations we proposed the following SiNW nucleation scheme [52]: First, a silicon suboxide cluster is deposited on the substrate and some of its highly reactive silicon atoms are strongly bonded to the substrate (silicon) atoms, limiting the cluster motion on the substrate. Non-bonded reactive silicon atoms in the same cluster are now exposed to the vapor with their available dangling bonds directed outward from the surface. They act as nuclei that absorb additional reactive silicon oxide clusters and facilitate the formation of SiNWs with a certain crystalline orientation. The subsequent growth of the silicon domain after nucleation may be crystallographic dependent. Oxygen atoms in the silicon suboxide clusters might be expelled by the silicon atoms during the growth of SiNWs and diffuse to the edge forming a chemically inert silicon oxide sheath [51]. In a certain orientation, e.g. [112], the diffusion might be lower and the high reactive silicon oxide phase can still be exposed to the outside and facilitate the continuous growth of the wire in such a direction. The oxygen-rich sheath formed in other directions may however possess lower reactivity and thus does not favor further stacking of silicon oxide clusters from the gas-phase, leading to growth suppression in such directions. The reactivity of silicon atoms in oxygen-rich clusters becomes very low at a Si:O ratio of 1:2 [51], while the reactivity of oxygen atoms changes to a lesser extent. The overall reactivity for Si:O = 1:2 is low. 1D growth in a specific direction is thus facilitated. In summary, the highly reactive $SiO_x$ layer ($x > 1$) at the tip of nanowires acts as a collector for the vaporized silicon oxide, while the outer $SiO_2$ layer of the SiNWs stops the diameter growth of the nanowires.

### 10.2.3
### Oxide-Assisted Growth Mechanism

The Si nanowire growth is determined by four factors: (1) The high reactivity of the $Si_xO$ ($x > 1$) layer on nanowire tips. (2) The $SiO_2$ component in the shell, which is formed from the decomposition of SiO and retards the lateral growth of nanowires. (3) defects (e.g. dislocations) in the Si nanowire core. (4) The formation of {111} surfaces, which have the lowest energy among the Si surfaces, parallel to the axis of the growth direction. The first two factors were discussed earlier. As far as the first factor is concerned we would like to add to the previous discussion that the melting temperature of nanoparticles can be much lower than that of their bulk materials. For example, the difference between the melting temperatures of 2 nm Au nanoparticles and Au bulk material is over 400 °C [46, 53]. The materials in the SiNW tips (similar to the case of nanoparticles) may be in or near their molten states, thus enhancing atomic absorption, diffusion, and deposition.

We suggest that the defects of SiNWs are one of the driving forces for the 1D growth. The main defects in Si nanowires are stacking faults along the nanowire growth direction of $\langle 112 \rangle$, which normally contain easy-moving 1:6 [112] and non-moving 1:3 [111] partial dislocations, and micro-twins. The presence of these defects at the tip areas should result in the fast growth of Si nanowires, since dislocations are known to play an important role in crystal growth. The SiNW growth

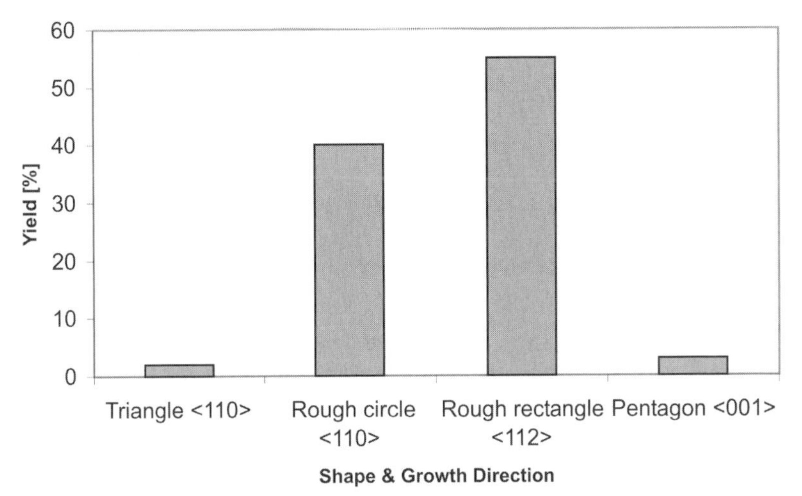


**Fig. 10.6.** The distribution of shapes, sizes, and growth directions of SiNWs [54].

rate in certain crystallographic directions is enhanced not only by existing dislocations in the growth direction but also by the formation of facets with a low surface energy (Si $\{111\}$ facets have the lowest surface energy). Figure 10.6 presents the statistical data of SiNW growth directions and shapes derived from the cross-sectional TEM images, showing that SiNWs grown by the OAG technique are primarily oriented in the $\langle 112 \rangle$ and $\langle 110 \rangle$ directions, and rarely in the $\langle 100 \rangle$ or $\langle 111 \rangle$ directions [54]. The cores of SiNWs are bounded by well-defined low-index crystallographic facets with a variety of shapes that can be circular, rectangular and triangular. We found a correlation between the cross-sectional shape and the growth direction, and proposed a model to explain these findings [55]. We suggest that the SiNW growth direction and cross-section are determined by four factors: (i) the stability of a Si atom occupying a surface site; (ii) the Si $\{111\}$ surface stability in the presence of oxygen; (iii) the stepped Si $\{111\}$ surface layer lateral growth process; and (iv) the effect of dislocations in providing perpetuating $\{111\}$ steps to facilitate SiNW growth. Theoretical evaluation of the SiNW growth along these criteria shows that indeed $\langle 112 \rangle$ and $\langle 110 \rangle$ are the preferred wire growth directions, and $\langle 111 \rangle$ and $\langle 100 \rangle$ are less likely, in accord with our experimental observations.

### 10.2.4
### Comparison between Metal Catalyst VLS Growth and OAG

To end this section we note that the OAG is vastly different from the metal-catalyst VLS growth. The two vary in the growth mechanism, in the growth conditions, in the yield of the grown wires in their abundant growth direction, in their diameters and in the chemical purity. Figure 10.7 compares schematically the two processes

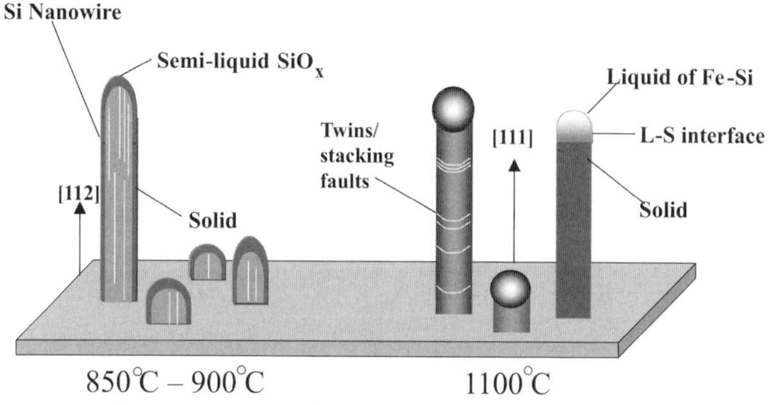


Fig. 10.7. Schematic comparison of the laser ablation of SiNWs via metal catalyst (VLS method) and the oxide assisted growth of SiNWs by thermal evaporation.

and Table 10.1 summarizes these differences between metal-free OAG and Fe or Ni catalyst VLS growth under the same conditions. Understandably, in the metal catalyst VLS growth, the characteristics of the grown nanowires depend on the nature of the metal catalyst used. When silane is used with Au nanoparticles, the growth temperature of SiNWs can be as low as 370 °C or close to the eutectic point of Au and Si alloy [56]. Furthermore, our recent work [57] shows that, like silane, SiO or other Si-containing vapor in the presence of Au nanoparticles or films could also decompose and lead to the growth of SiNWs, and the growth temperature was as low as 700 °C. We anticipate that nanoparticles of other kinds of metals can also induce the deposition of SiNWs from SiO vapor, providing the metal can induce SiO decomposition and form a eutectic alloy with Si.

Tab. 10.1. OAG versus metal catalyst VLS growth in a 3 in diameter tube.

| ***Property*** | ***Oxide-Assisted Growth*** | ***Metal Catalyst Growth*** |
|---|---|---|
| Source | SiO; Si + $SiO_2$ | (Fe or Ni) + Si |
| Growth temperature | 850–900 °C | >1100 °C |
| Pressure | 10–800 Torr | 10–800 Torr |
| Yield | 3 mg $h^{-1}$ | <0.1 mg $h^{-1}$ |
| Impurity | None | Metal |
| Tip composition | $SiO_x$ | Metal |
| Wire diameter | Typically 2–20 nm | Typically > 10 nm |
| Growth direction | Mostly ⟨112⟩ & ⟨110⟩ | ⟨111⟩ |
| Morphologies | Nanowires, nanoribbons, nanochains | nanowires |

## 10.3
## Control of SiNW Nanostructures in OAG

### 10.3.1
### Morphology Control by Substrate Temperature

One of the most important issues of nanomaterial growth is the control of the morphology. This can be done by varying different process parameters. The effect of the growth temperature on the structure of the Si nanowires has been studied systematically [32, 33].

The substrate temperature substantially affects the SiNW growth in several ways: (1) determination of the growth process (metal catalyst VLS growth or OAG), (2) determination of the SiNW shape in the growth process itself (e.g. the SiNW diameter), (3) annealing effects that change the structure and morphology of the SiNW after its formation. Examples of these effects were given in our study of SiNWs deposited by laser ablation of a mixture of Si, $SiO_x$ and metals (the metallic constituent introduced either intentionally or as impurities or contamination of the system). The study was focused on the nature of the nanostructures produced on the substrate at temperatures ranging from 850 to 1200 °C.

The Si nanostructures produced in this temperature region can be divided to two groups: (1) Region I (1200∼1100 °C) forming nanostructures by the metal catalyst VLS process (indicated by the presence ofmetallic caps), (2) Region II (1100∼850 °C) where the OAG process is dominant (the nanostructures do not have metallic caps). HRTEM analysis of samples prepared in region I (temperatures of 1190, 1160, and 1130 °C denoted I1, I2 and I3) and region II (temperatures 1050, 950, and 900 °C denoted II1, II2 and II3) revealed the structure evolution at different temperatures.

In region I the diameter of the SiNWs (all single-crystalline SiNWs) decreases with temperature (200, 80 and 50 nm for 1190, 1160 and 1130 °C respectively) as shown in Figure 10.8. While the first two types of nanowires are straight, continuous ones, the third one has a tadpole-like shape and appears to be broken into short Si rods. The head of the tadpoles is crystalline Si, while the tail of the tadpoles is amorphous $SiO_x$.

In region II the diameter of the Si nanostructures is constant (∼20 nm), but their shape changes from tadpole-like through chain-like to wire-like as the temperature decreased from 1050, through 950, to 900 °C respectively. The wires have $SiO_x$ caps rather than metallic caps, and are encapsulated in a $SiO_2$ sheath, all indicative of an OAG process.

We will now explain these results as a combination of: (1) the dominant growth process (metal-assisted VLS or OAG) and (2) extended annealing of Si nanowires at high temperatures.

Region I is characterized by a metal catalyst VLS growth, as indicated by the metal caps on top of the nanostructures. The diameter of the nanowires in this growth process is determined by the diameter of the liquid alloy droplet at their tips. Metal silicide clusters of different sizes are present in the flowing gas above

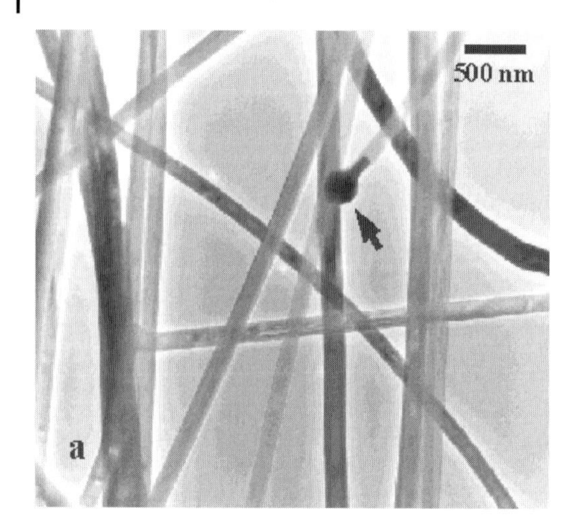


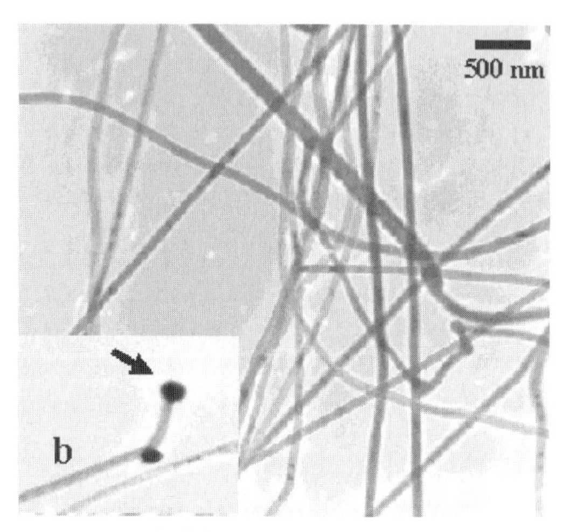


**Fig. 10.8.** Bright-field TEM images showing the typical morphology of Si nanowires grown at: (a) 1190, (b) 1160, and

the substrate, leading to condensation of droplets when the substrate temperature is lower than the melting point of the metal silicide clusters. The melting point of the nanoclusters decreases with decreasing size in the nanometer region. Large droplets will melt and serve as SiNW nucleation sites at higher temperatures than small droplets, explaining the decrease in the SiNW diameter with decreasing temperature Figure 10.8(d) gives a schematic diagram of the SiNW evolution at different temperatures. The size of the molten droplets decreases with temperature (d1). The droplet absorbs Si-containing clusters from the vapor and becomes supersaturated with Si. The excessive Si precipitates out, resulting in the 1D growth of crystalline SiNWs shown in Figure 10.8(d2) the diameter of which follows that

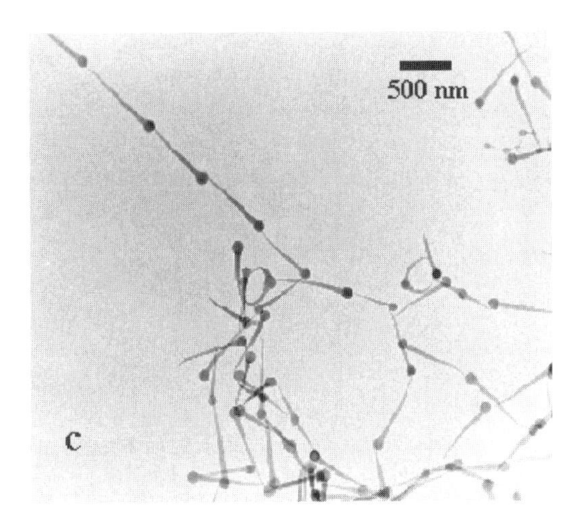


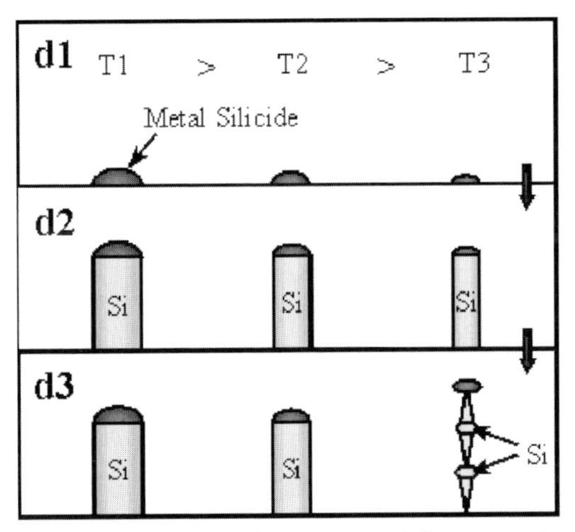


**Fig. 10.8.** (c) 1130 °C. The arrows reveal the metal catalyst present at the tip of the nanowires. The diameter of the Si nanowires can be seen to decrease with decreasing growth temperature. (d) Diagram showing the morphology evolution of Si nanowires with time in region I: (1) nucleation, (2) growth, and (3) annealing [33].

of the initial droplet. The formation of the SiNWs is restricted to the temperature region for which the temperature is high enough so that the solid particle melts and forms a liquid droplet (lower temperature of region I) on the one hand, and the temperature is low enough to melt and condense from the vapor (upper temperature region) on the other hand. Previous work on Si whiskers revealed that there is a critical whisker diameter at which growth stops completely, due to the Gibbs–Thomson effect [58]. This may be the reason why the smallest nanowires obtained in region I have diameters larger than 50 nm (Figure 10.8). Figure

10.8(d3) shows a spheroidization effect for the smaller diameter nanowires, which is attributed to annealing of the nanowires, as will be described later.

In region II, the OAG region, the diameters of the Si nanowires are quite uniform, irrespective of the substrate temperature. This may be explained by a vapor–solid (VS) process governing the initial nucleation of the OAG rather than a VLS process in which the size of the liquid droplet decreases with decreasing energy, leading to a respective decrease in the SiNW diameter as discussed above (for the metal catalyst growth). This would mean that the small nuclei of crystalline Si nanowires were directly solidified from SiO in the vapor phase. This explanation is, however, not in accord with the formation of a $SiO_x$ cap on the top of the SiNW and the alternative proposition that the oxide-assisted nucleation and growth is occurring due to: (1) the lower melting point of $SiO_x$ compared to that of $SiO_2$, (2) the high reactivity of the molten $SiO_x$ cap to Si-containing clusters in the vapor, and (3) the decomposition of Si suboxide to Si and $SiO_2$. We proposed that the $Si_nO_m$ clusters react with the $SiO_x$ cap, the crystalline Si core precipitates below and the excess oxygen diffuses to the sides forming a $SiO_2$ amorphous outer layer which solidifies, due to its higher melting point, and limits the further lateral growth of the nucleus. The lateral growth results from the energetically favorable adsorption of vapor clusters by the highly curved $SiO_x$ molten tip on the one hand and the lateral restriction imposed by the solid $SiO_2$ sheath on the other hand (as shown in Figure 10.9(d2)). The SiNW diameter may be determined not only by the diameter of the initial $SiO_x$ droplet, but also by the equilibrium between the condensation and the disproportionation of SiO to Si and $SiO_2$ and by diffusion of the excess O to the sides. It is still not completely understood why this equilibrium is not temperature-dependent under our experimental conditions. It could be that the dependence is weak in this limited temperature region and will be revealed if we enlarge this region by using different experimental conditions.

The SiO vapor phase is stable at a high temperature, so that the condensation and disproportionation of the SiO vapor into Si + $SiO_2$ occurs only below a certain substrate temperature (the upper limit of region II). On the other hand, below the lower limit in region II the SiO vapor condenses directly to form SiO solid [59], with no preferential adsorption nor disproportionation, so that the 1D growth is suppressed. This explains why the OAG of SiNWs was restricted to the temperature range 1100–850 °C (region II).

Finally, we discuss the formation of tadpole-like and chain-like Si nanostructures from the metal-catalyst VLS and the OAG (Figures 10.8 and 10.9) processes. Both can be described in terms of a spheroidization mechanism. One-dimensional SiNWs are less stable than the three-dimensional bulk Si, since the wire has a much larger surface area and thus higher surface energy. Annealing of SiNWs for a sufficient time results in spheroidization, as shown schematically in Figure 10.10. The chemical potential of the SiNWs varies with the local curvature so that small variations in their diameter generate a driving force for diffusive transport between different chemical potentials. The Si nanowire would convert into a nanospherical chain first (as shown schematically in Figure 10.10 and experimentally in Figure 10.9(b)). Later, with further diffusion, the inner crystalline Si core

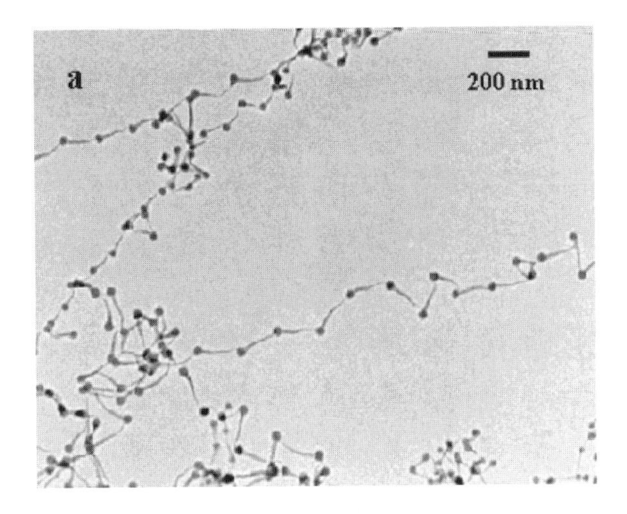


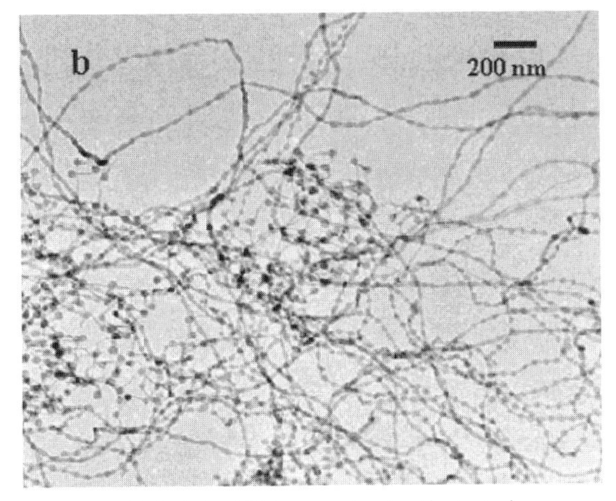


**Fig. 10.9.** Bright-field TEM images showing the typical morphology of Si nanowires grown at: (a) 1050, (b) 950, and

would break up and the spheroidization would become faster due to the larger variations increasing the driving force. The amorphous $SiO_x$ nanorods connecting the Si nanospheres then become thinner, and eventually break up. This is why the Si nanosphere chains convert into tadpole-like Si nanorods (as shown schematically in Figure 10.10 and experimentally in Figures 10.9(a) and 10.8(c)). Eventually, the amorphous $SiO_x$ tails would disappear and perfect Si/Si oxide spheres might also be formed. Note now the difference between Region I, in which the spheroidization occurs only at the lower temperature (Figure 10.8), and region II, where spherodization is more significant at higher temperatures (Figure 10.9). In region I the first two SiNWs grown at 1190 and 1160 °C are too thick (200 and 80 nm re-

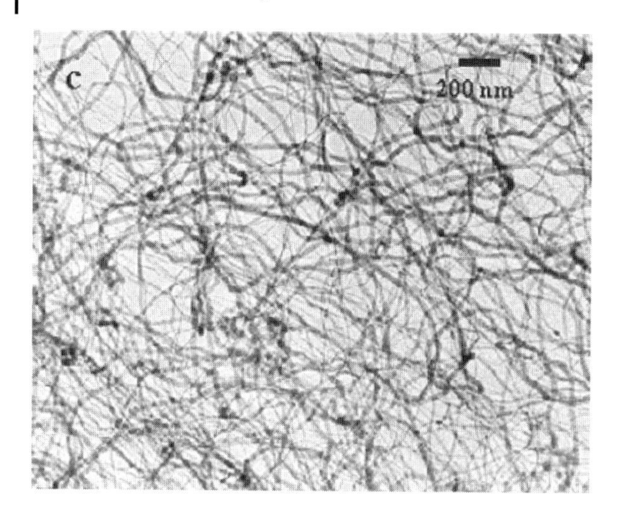


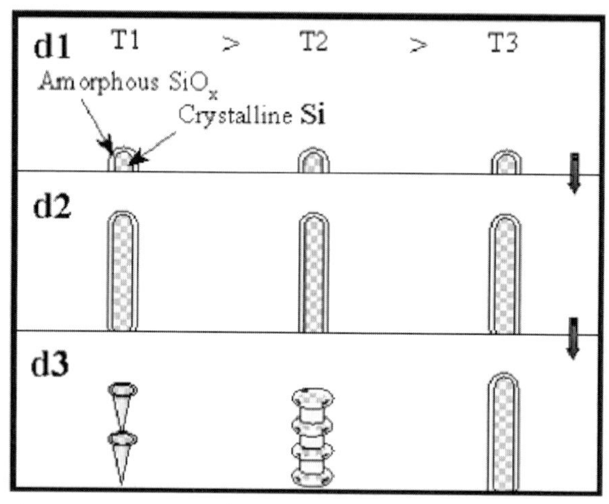


**Fig. 10.9.** (c) 900 °C. The SiNW diameter is independent of the growth temperature. (d) Diagram showing the morphology evolution of Si nanowires with time in region II: (1) nucleation, (2) growth, and (3) annealing [33].

spectively) for spheroidization, which is observed only at the lower temperature of 1130 °C (Si nanowire diameter 50 nm), in spite of the higher diffusion rate at higher temperatures. In region II, however, the SiNW diameter is relatively constant with temperature, so the diffusion rate increases with temperature and the spheroidization becomes more significant at higher temperatures.

We have, until now, discussed the spheroidization process in the context of the formation of different Si nanostructures by varying the substrate temperature at which these structures evolve. One of the structures reported was nanochains of Si

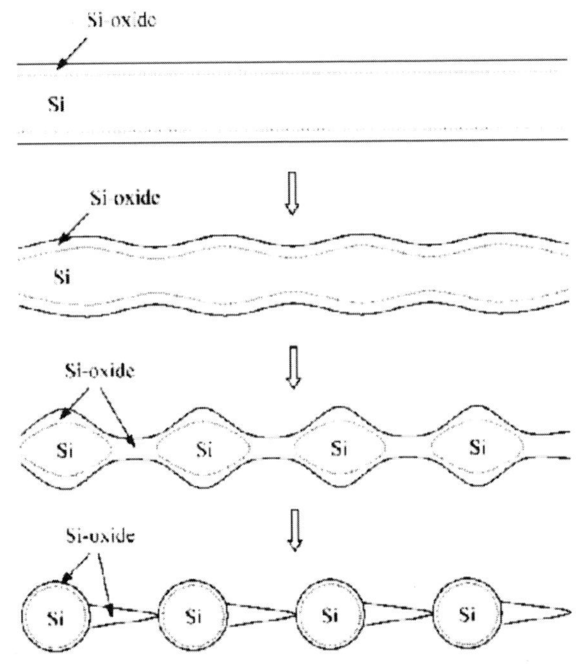


Fig. 10.10. Diagram showing the spheroidization process of one-dimensional Si nanowires [33].

nanospheres enclosed in and linked together to form a chain by $SiO_2$. Similar silicon nanochains (SiNC), this time of boron-doped Si were grown [60, 61] by laser ablation of a compressed target of a mixed powder of SiO and $B_2O_3$ at 1200 °C. TEM analysis including imaging (Figure 10.11), diffraction and EDS indeed verifies a structure of nanochains with uniform diameters of the Si nanospheres con-

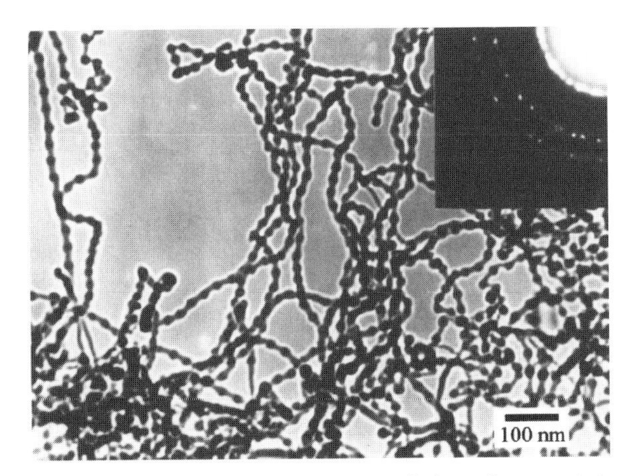


Fig. 10.11. TEM image showing the morphology of Si nanochains.

nected by amorphous $SiO_2$, like a chain of pearls. The SiNC consists of knots and necks with equal distances between them. The average diameters of the knots and necks are 15 and 4 nm, respectively and the thickness of the $SiO_2$ sheath surrounding the Si spheres is about 2 nm. The product is uniform with no isolated particles and consisting of 95% SiNC and 5% SiNWs.

To summarize, both the growth process and the morphology of SiNWs can be controlled by the growth temperature. The temperature required for a metal catalyst growth (using iron as the catalyst, for gold the temperature would be much lower) is higher than that required for the OAG. Post-growth annealing results in spheroidization and structural changes which occur faster at higher temperatures. The diameter control by temperature is readily possible for the metal catalyst VLS method but appears to be difficult for the OAG.

### 10.3.2
### Diameter Control of Nanowires [29]

The motivation for the study of nanomaterials stems from the expected size effects; SiNWs are no exception. This makes the issue of diameter control very important. In a previous section we have shown that the growth temperature affects the diameter of the SiNWs grown by the metal catalyst VLS method but not the diameter of SiNWs grown by the OAG method. Here we show that the SiNW diameter can be affected by the carrier gas used in the growth process.

The SiNWs were synthesized by laser ablation of a target made by compaction of a mixed powder of 90% Si and 10% $SiO_2$. Different carrier gases were used including He, $N_2$, and a mixture of Ar with 5% $H_2$. During growth the carrier gas pressure was about 300 Torr and the flow 50 sccm. The substrate growth temperature was $\sim$930 °C. No deposition was observed at places where the temperature was higher than $\sim$950 °C, while some nanoclusters and amorphous mixtures of Si and oxygen were deposited at lower temperature regions (for $N_2$). The SiNWs synthesized in He and in Ar (5% $H_2$) consisted almost entirely of nanowires. Some spherical particles, with diameters ranging from $\sim$9 nm to several hundred nm, composed of a mixture of crystalline Si and amorphous Si oxide, were found to coexist with the nanowires grown in a $N_2$ atmosphere, and the quantity of the spherical particles was a little less than that of the SiNWs. Most of the SiNWs were smoothly curved with some short straight sections, a few possessed bends and kinks. The SiNWs synthesized in a He atmosphere possessed many more bends and curves.

The diameter distributions of the SiNWs were measured from the TEM micrographs as given in Figure 10.12. The SiNWs (up to several mm long) had a distribution of diameters (Si core plus $SiO_2$ sheath) peaked at 13, 9.5 and 6 nm for carrier gas mixtures of He, (5% $H_2$ in Ar) and $N_2$ respectively. The smallest wires were mixed with spherical particles with diameters ranging from $\sim$9 nm to several hundreds of nm. High-resolution TEM images of several SiNWs produced in He, Ar (5% $H_2$), and $N_2$ atmospheres shows that every nanowire consists of a crystalline Si core and an amorphous $SiO_2$ sheath. The crystalline Si core has many lat-

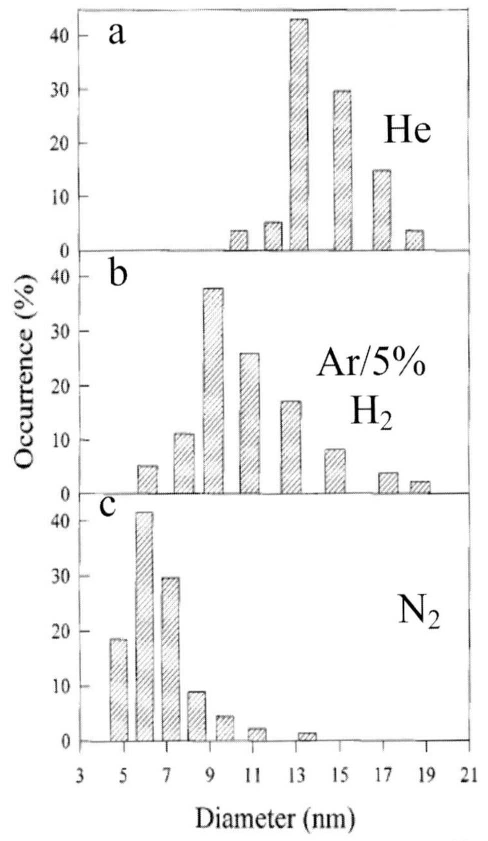


**Fig. 10.12.** Size distribution of SiNWs synthesized by laser ablation using different carryier gases [29]: (a) He, (b) Ar (5% $H_2$), and (c) $N_2$.

tice defects. The nanowires consisted of only Si and oxygen with no ambient gas atoms, as determined by EDS and XPS.

The mechanism by which the carrier gas affects the growth and diameter of the SiNWs is not clear. We propose that the ambient atoms affect the formation and transport of nanoclusters and the phase separation process at the growth front of SiNWs. We suggest that the dominant effect of the ambient is on the phase separation. First, the thermal conductivity of the ambient affects the cooling rate of the nanoclusters and thus the rate of phase separation. Second, the inert gas atoms are very likely incorporated in the deposited matrix at the growing tips of SiNWs, and their presence would influence the phase separation process, despite their eventual outdiffusion. Helium is smaller, faster moving, and more thermally conducting than $N_2$. The collective effect of these processes leads to a slower supply of heat by $N_2$ and thus faster cooling of nanoclusters. This, in turn, results in faster phase separation and the formation of thinner SiNWs. The fast cooling and elemental incorporation in the nanoclusters formed in an $N_2$ environment probably leads to incomplete phase separation and to the formation of the remnant spherical particles. The Ar (5% $H_2$) mixture is suspected to exert an intermediate effect between $N_2$ and He, thus giving rise to nanowires of intermediate diameters.

### 10.3.3
### Large-Area Aligned and Long SiNWs via Flow Control [62]

Until now we have discussed the effect of two parameters on the SiNW growth: temperature and carrier gas. Now we show that the carrier gas flow can be exploited as well. In this particular example we use it to grow millimeter-area arrays of highly oriented, crystalline silicon nanowires of millimeter length.

The growth in this particular case was performed by thermal evaporation of SiO powder. A carrier gas of argon mixed with 5% $H_2$ was used with a flow of 50 sccm at 400 Torr. The furnace temperature was 1300 °C, while the growth temperature was about 930 °C.

Large area (about 2 mm × 3 mm) of highly oriented, long (up to 1.5–2 cm) nanowires were grown on the surface of the silicon substrate under these conditions, as detected by SEM (Figure 10.13). The thickness of the oriented nanowire product was about 10 μm, as estimated from the cross-sectional image (Figure 10.13(b)) of the sample prepared by focused ion beam cutting. EDX shows that the nanowires are composed of silicon and oxygen with no metal impurities, suggesting an oxide assisted growth. Figure 10.13(c) shows the typical morphology of SiNWs. TEM shows that the SiNWs are quite clean, with very few particles attached to their surfaces, and are relatively homogeneous. The SiNW diameters vary from 18 to 46 nm, and the mean value is about 30 nm.

HRTEM shows the typical SiNW Si core encapsulated by a $SiO_2$ sheath and the {111} planes of crystalline silicon. The diameters of the crystalline silicon core varied from 13 to 30 nm, and the mean value was about 20 nm. The thickness of the amorphous silicon oxide shell varied from 2 to 10 nm, and the mean value was about 5 nm.

The growth of the oriented SiNWs may be related to the flow of the carrier gas, because it was found that the orientation direction of the SiNWs is parallel to the direction of flow of the carrier gas in the alumina tube. A mechanism for the growth of the oriented silicon nanowires is illustrated below. First, the nucleation of silicon nanowires from silicon oxide ($SiO_x$) vapor started at the proper position on the substrates. Because there was a temperature gradient along the alumina tube, and the planes with the same temperature were perpendicular to the axial direction of the tube, only some particular positions with the appropriate temperature may be suitable for the nucleation of SiNWs [21, 27]. These positions should be located on a line of equal temperature on the substrate and will also be perpendicular to the flow direction of the carrier gas. Once initial nucleation is established, nanowire growth will tend to continue on the substrate. Secondly, the strength of the flow of the carrier gas will force the growing nanowires to grow in

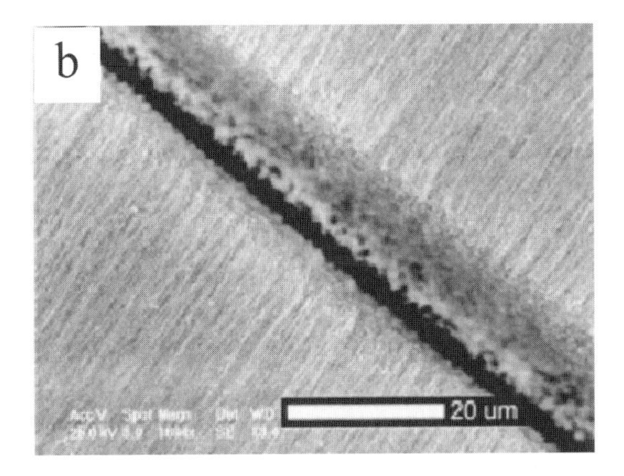


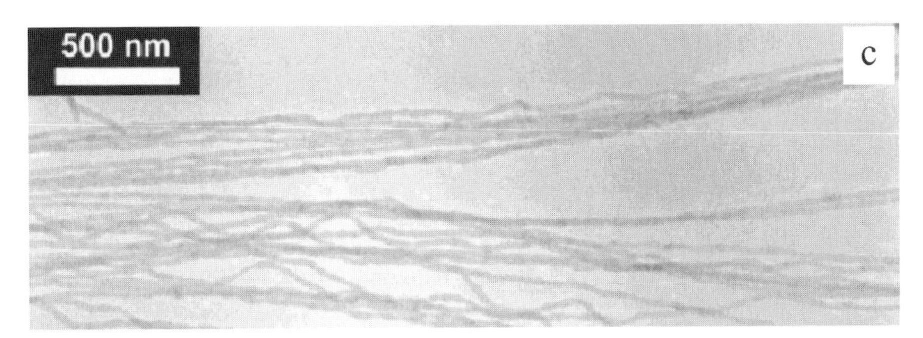


**Fig. 10.13.** (a) SEM image of oriented SiNWs at low magnification; (b) The cross-section of the SiNW in (a) cut by a focused ion beam; (c) TEM image of the oriented silicon nanowires [62].

the direction of the flow. At the same time, overcrowding of the nanowires will limit the possibility of nanowire propagation in other directions [63]. In addition, it was found that a smooth plane substrate is very helpful for oriented growth.

### 10.3.4
### Si Nanoribbons

SiNWs are a 1D nanostructure. A distinct feature of the OAG process revealed in our studies is the variety of different configurations it can form. Some were discussed in the previous sections. Now we present an exciting and unexpected 2D configuration, nanoribbons, discovered in the course of studying OAG. A 2D configuration is not expected to exhibit the same magnitude of size effects as a 1D structure, but it may be advantageous in processing and in obtaining signals with more measurable intensities in single object characterization.

The single-crystal silicon nanoribbons were grown by simple thermal evaporation of silicon monoxide (SiO) heated to 1150 °C. No templates or catalysts were used. The nanoribbons have a thickness of only about 10–20 nm (average 15 nm), widths of several hundreds of nanometers (50–450 nm), and lengths of many micrometers [34]. Most of the ribbons have rippling edges (Figure 10.14(a)), and a small portion of the ribbons has smooth edges (Figure 10.14(b)). Due to their small thickness the ribbons seem transparent in TEM imaging (Figure 10.14(b)) using 200 keV electrons. Nanoribbons of different width and morphology have a similar thickness which is constant throughout each individual ribbon. In the typical nanoribbon shown in (Figure 10.14(b)) the thickness is about 14 nm and the width is about 370 nm. The thickness to width ratio of the SiNWs varies from 4 to 22. The rippling and curling features at the edge of most ribbons also confirm that the nanoribbons are quasi-2D structures distinctly different in shape from the 1D SiNWs.

HRTEM imaging of a single nanoribbon revealed that the ribbon has a crystal core nipped by amorphous layers with atomically sharp interfaces. The in-plane layers of the nanoribbon were determined to be silicon (110) facet with a perfect atomic, defect-free, single-crystal structure grown along the ⟨111⟩ direction. This direction is different from the predominant ⟨112⟩ and ⟨110⟩ direction of SiNWs synthesized by the OAG method. It is the same as the most abundant growth direction of Si nanowires synthesized by the metal catalyzed VLS method. The wide part of the ribbon is along the ⟨112⟩ direction. The amorphous edges of the ribbon consist of silicon oxide ($SiO_x$), as determined by EDS and EELS attached to the TEM. The width of the amorphous edge is about 10 nm. Analysis of a number of nanoribbons with different widths shows that the width of the oxide edges varies from 3 to 25 nm, similar to the thickness of the amorphous silicon oxide shell of the nanowires synthesized by OAG. The thickness of the oxide layer covering the flat surfaces of the ribbons is much less than the width of the oxide edges; that is, the thickness of the oxide layers is anisotropic. This result may be understood in terms of the OAG process, in which the silicon oxide shell was formed by the re-

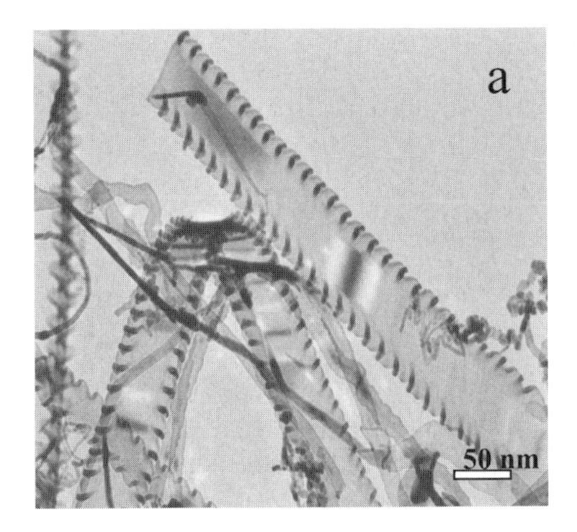

**Fig. 10.14.** TEM images (TEM, Philips CM 20 TEM at 200 kV) of (a) rippling-edge nanoribbons, and (b) a smooth-edge nanoribbon. The thickness of the ribbons is about 14 nm [34].

action of $SiO + SiO = Si + SiO_2$, and subsequently separated from the silicon core during growth. According to this reaction, the amount of segregated silicon oxide is proportional to the amount of silicon at the same place.

The growth of Si nanoribbons cannot be via the twin-plane growth mechanism, (i.e. controlled by a twin plane parallel to the flat surface of the nanoribbons) as suggested for microribbons [64]. A twin-plane mechanism is impossible from crystallographic considerations taking into account the observed structure of the nanoribbons and, indeed, no twins were observed. The precise growth mechanism of the nanoribbons is not yet clear, but it is likely that it is governed by anisotropic growth kinetics along different crystallographic directions (i.e. two fast growing directions).

## 10.4
## Nanowires of Si Compounds by Multistep Oxide-Assisted Synthesis

### 10.4.1
### Nanocables [65]

Coaxial three-layer cables offer a potentially simple way of producing nanojunctions. The idea of applying the nanocable configuration was indeed recently reported by Iijima et al. [66]. Such nanocables can be produced by a multistep process. We give as an example a coaxial three-layer nanocable synthesized by combining high-temperature laser ablation of SiC as the first step, and thermal evaporation of SiO at a higher temperature as the second step. Figure 10.15 is a TEM image showing the structure of the nanocables synthesized. Uniform, tens of micrometer long nanocables with diameters smaller than 150 nm were formed. The nanocables were made of: (1) a crystalline Si core with a diameter ranging from 30 to 50 nm, (2) an amorphous $SiO_2$ interlayer (second layer) 12–23 nm thick, (3) an amorphous carbon sheath (external third layer) 17–31 nm thick. The average dimensions of the nanocable are: core 43 nm in diameter, second layer 16 nm thick, and third layer 24 nm thick. The interfaces between the layers are sharp. Most of the products were nanocables but small amounts of Si and SiC nanowires were also detected.

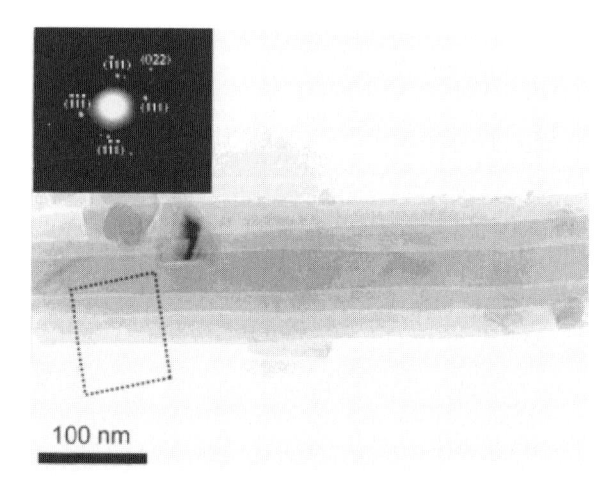


**Fig. 10.15.** A magnified image of the coaxial nanostructure, showing a crystalline core and two additional amorphous layers (a-$SiO_2$ and a-C). The inset shows the selected area diffraction pattern [65].

### 10.4.2
### Metal Silicide (MS)/SiNWs from Metal Vapor Vacuum Arc Implantation [67]

A conventional method to produce contacts in the semiconductor industry is ion implantation. The possible advantages of ion implantation for SiNWs are in the control in adding precise (and small) amounts of metal atoms to SiNWs, which might be difficult for bulkier techniques. Here we describe ion implantation of Ni and Co into 20 nm diameter SiNWs produced by thermal decomposition of SiO. The SiNWs were mounted on copper folding grids and directly implanted by metal vapor vacuum arc (MEVVA) implantation with a 5 keV $Ni^+$ or $Co^+$ dose of $1 \times 10^{17}$ $cm^{-2}$ at room temperature. The implanted samples were later annealed in argon.

Ni implantation results in the formation a Ni silicide layer on the implanted SiNW surface. The layer contains lots of defects. Rapid thermal annealing (RTA) at 500 °C smoothed the surface of the Ni-implanted SiNWs, which was transformed to a continuous outer layer with a typical thickness of about 8 nm, as shown in Figure 10.16.

The Co-implanted SiNWs surface is much rougher than that of the Ni-implanted surface with isolated $CoSi_2$ particles 2–40 nm in diameter (Figure 10.17).

The generation of $NiSi_2$ and $CoSi_2$ is schematically described in Figure 10.18. Room temperature $Ni^+$ or $Co^+$ implantation of the as-grown SiNWs (Figure 10.18(a)) results in the formation of a metal/Si mixture (Figure 10.18(b)). The energy of the ion beam should be optimized (5 keV in the present experiment) to avoid excessive damage of the SiNWs. Post-implantation annealing was found to be efficient in reducing the ion implantation damage. The metal silicides are expected to give an improved electrical conductivity of the SiNWs and provide electrical contacts to the SiNWs. The structure of the MS/SiNWs layer is sensitive to

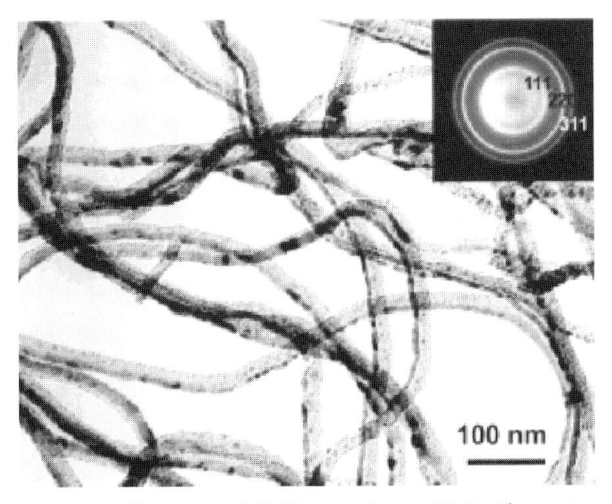


**Fig. 10.16.** Ni implanted SiNWs annealed at 500 °C. The inset shows the TED pattern of the Ni layer [67].

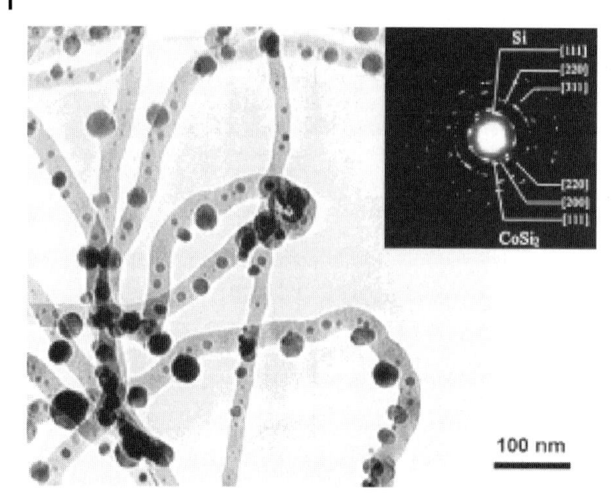


**Fig. 10.17.** Co implanted SiNWs annealed at 900 °C. The inset shows the TED of the Co polycrystals [67].

annealing treatment. Under proper annealing conditions, the MS layer can exhibit a highly oriented relationship to the SiNW core.

### 10.4.3
### Synthesis of Oriented SiC Nanowires [68]

One of the fascinating options opened by the OAG method is the possibility of transforming from one type of nanowires to the other. A notable example is the

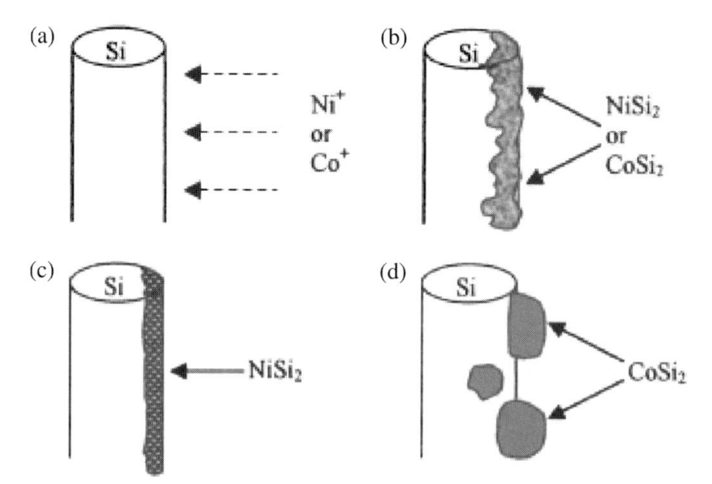


**Fig. 10.18.** The formation mechanism of $NiSi_2/Si$ and $CoSi_2/Si$ on the surface of bare SiNWs [67]. (a) Bare SiNWs implanted with metal ions. (b) Formation of the metal/Si mixture layer on one side of SiNW. (c) The $NiSi_2/Si$ nanowire after low temperature annealing. (d) $CoSi_2$ nanoparticles formed by coarsening at high temperature annealing.

**Fig. 10.19.** SEM images of oriented SiC nanowire array showing high density of well-separated, oriented nanowire tips [68].

synthesis of oriented SiC nanowires by reacting SiO with aligned carbon nanotubes prepared via the established method of pyrolysis of acetylene over film-like iron/silica substrates [69, 70, 71]. Solid SiO powders (purity 99.9%) were placed in a graphite crucible and covered with a molybdenum grid. The highly aligned carbon nanotubes were placed on the molybdenum grid. The crucible was covered with a graphite lid, placed in the hot zone inside the alumina tube, and held in a flowing argon atmosphere (50 sccm) at 1400 °C for 2 h. After reaction, the aligned carbon nanotube arrays were converted to oriented SiC nanowire arrays. These highly oriented SiC nanowires were similar in appearance to the original aligned carbon nanotubes. The bottom end of the nanowire array is composed of a high density of well-separated and highly oriented nanowire tips (Figure 10.19). TEM (Figure 10.20(a)) imaging and diffraction showed that the transformed wires are single crystalline $\beta$-SiC with the wire axes along the (111) direction and a high density of stacking faults perpendicular to the wire axis. In contrast to the carbon nanotubes (Figure 10.20(b)) the SiC were full wires and not hollow tubes.

## 10.5 Implementation of OAG to Different Semiconducting Materials

The OAG method has a general nature and can be applied to a variety of materials other than Si. Based on the OAG method, we have synthesized nanowires of a wide range of semiconducting materials including Ge [35], GaN [36, 37], GaAs [38, 39], GaP [41], SiC [40], and ZnO (whiskers) [42]. The actual OAG process was activated by laser ablation, hot-filament chemical vapor deposition (HFCVD) or thermal evaporation.

(a)

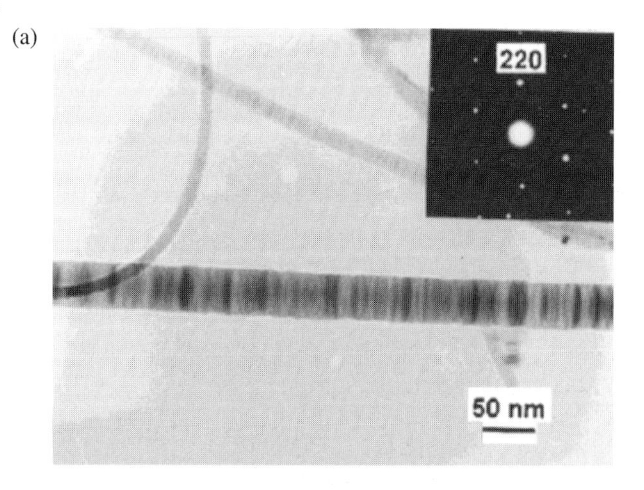


(b)

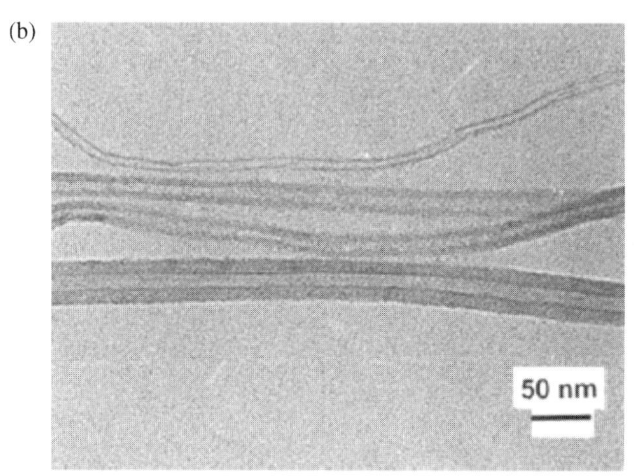


**Fig. 10.20.** (a) TEM image of $\beta$-SiC nanowires. The SiC nanowires exhibit a high density of stacking faults perpendicular to the wire axes. The inset shows a selected area electron diffraction pattern of the $\beta$-SiC nanowires. (b) TEM image of the initial carbon nanotubes. Note the transition from tubes to filled SiC wires [68].

Similar to the production of Si nanowires, we used laser ablation of a mixed $GeO_2$/Ge target to synthesize Ge nanowires at 830 °C. In comparison, Morales and Lieber [21] used metal-containing targets for the same purpose. We show that Ge nanowires are obtained by the OAG method in an analogous way to SiNWs. TEM studies (Figure 10.21) show that the structure of the Ge nanowires is similar to that of SiNWs, with a crystalline Ge core and a thick amorphous oxide shell. The diameters of the Ge nanowires in this particular experimental set-up had a larger size distribution than that of SiNWs (ranging from 16 to 370 nm).

Gallium arsenide nanowires with a zinc-blende structure were fabricated by laser ablation of GaAs powders mixed with $Ga_2O_3$ (no metal catalyst used). SEM obser-

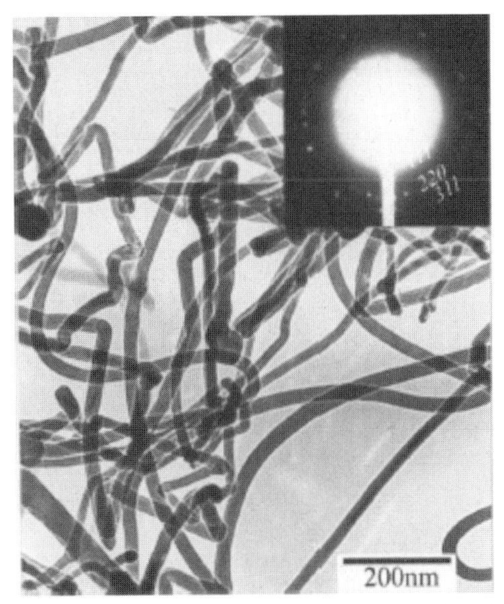


**Fig. 10.21.** A TEM image of Ge nanowires and a selected-area electron-diffraction pattern (inset) [35].

vation (Figure 10.22(a)) shows that the product consists of wire-like structures with length up to 10 μm and diameters of the order of 50 nm. The EDX spectrum shown in the inset of this figure demonstrates that the nanowires consist only of gallium, arsenic and oxygen. The silicon signal originates from the silicon substrate. HRTEM, selected-area electron diffraction (SAED) and electron energy-loss spectrometry (EELS) of individual nanowires (Figure 10.22(b)) revealed a zinc-blende GaAs core enclosed in a gallium oxide ($GaO_x$) sheath. The [111] growth direction of the present nanowires is the same as that of GaAs nanowires grown by a metal-catalyzed VLS process [72]. The diameter of the crystalline GaAs cores range from 10 to 120 nm with the thickness of the outer sheath ranging from 2 to 10 nm. The average diameter of the core was about 60 nm, and the average thickness of the outer sheath was 5 nm. As expected from an OAG process, the crystalline GaAs tip was coated with a thin amorphous layer of $GaO_x$, similar to the $SiO_x$ tip of SiNWs, and different from the GaAs nanowires synthesized by the metal catalyzed VLS growth, in which the tips were terminated at metal-alloy nanoparticles [72].

We can thus suggest that the oxide-assisted nucleation and growth of GaAs nanowires advances through the following reactions: (1) laser-induced decomposition of GaAs into Ga and As, (2) reaction of $4Ga + Ga_2O_3 = 3Ga_2O$ in the high-temperature zone, (3) transport of volatile $Ga_2O$ and As to the low-temperature-deposition zone, (4) reaction of $3Ga_2O + 4As = 4GaAs + Ga_2O_3$ in the low-temperature-deposition zone leading to the nucleation and growth of the GaAs nanowires.

The above model similarly applies to the successful synthesis of GaN and GaP and other binary compounds. Figures 10.23 and 10.24 show the typical HRTEM

(a)

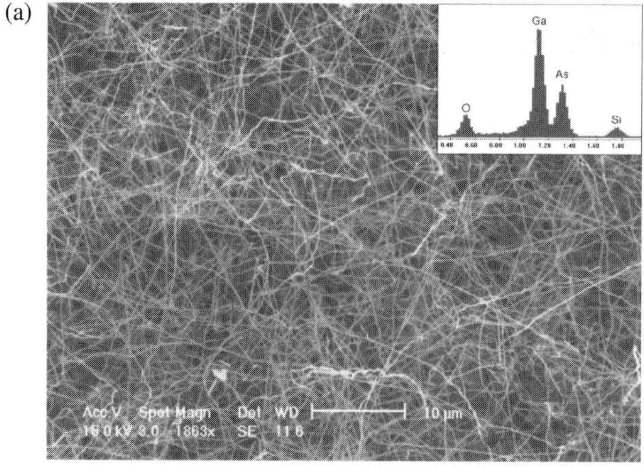


(b)

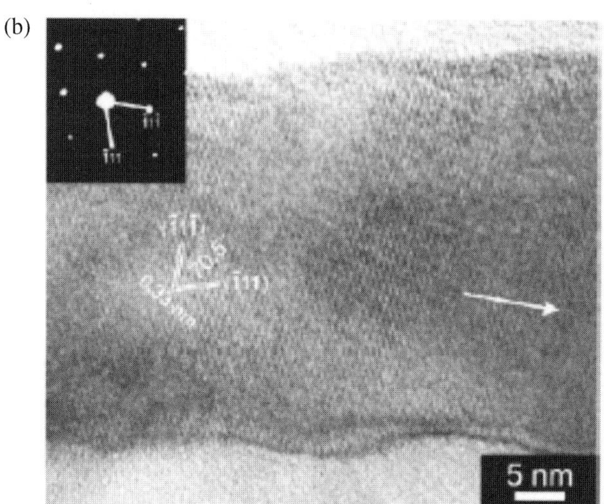


(c)

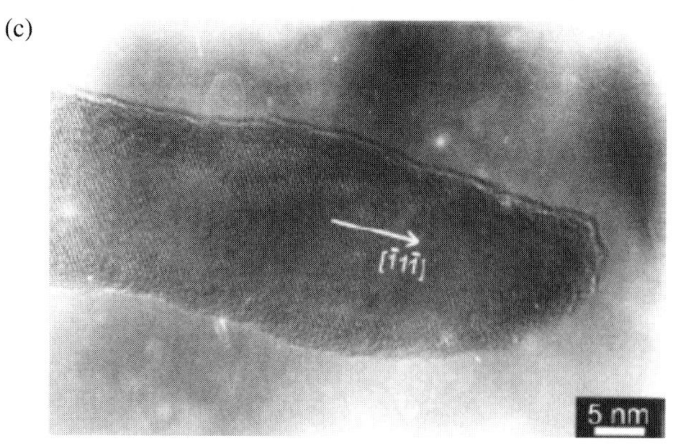


**Fig. 10.22.** (a) A typical SEM image of the GaAs nanowires synthesized by the oxide-assisted method. The EDS in the inset indicates Ga, As, O and Si; (b) a HRTEM image of a GaAs nanowire. The growth axis is close to the [$\bar{1}1\bar{1}$] direction (white arrow). The inset is the corresponding ED pattern recorded along the [110] zone axis perpendicular to the nanowire growth axis; (c) A HRTEM image of the tip of a GaAs nanowire. The growth direction is close to the [$\bar{1}1\bar{1}$] direction (white arrow) [38].

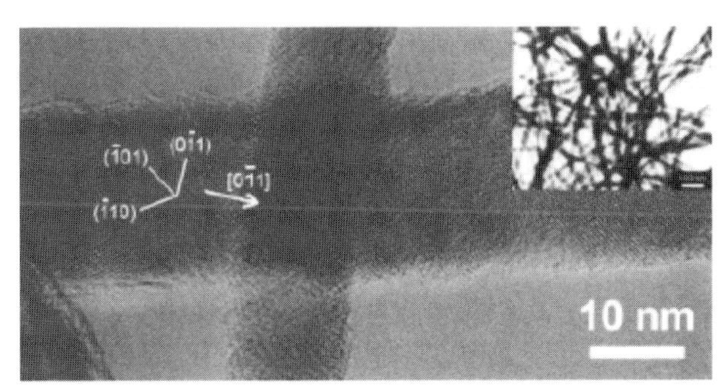


**Fig. 10.23.** A typical HRTEM image of GaP nanowires grown at about 750 °C on a silicon (100) substrate. The growth direction is close to the [0$\bar{1}$1] direction. The inset is a TEM image of GaP nanowires [40].

images of GaP and GaN nanowires, respectively. A TEM image of GaP nanowires and a SEM image of GaN nanowires are shown as insets in Figures 10.23 and 10.24, respectively. Again, both kinds of nanowires had a core of crystalline GaP or GaN wrapped in a thin layer of amorphous gallium oxide ($GaO_x$). The formation of GaP and GaN nanowires is similar to that of GaAs nanowires, replacing As by P and N, respectively. The critical reactions responsible for the formation of GaP and GaN nanowires are, $3Ga_2O + 4P = 4GaP + Ga_2O_3$ and $3Ga_2O + 4N = 4GaN + Ga_2O_3$, respectively.

It is thus obvious that for all the cases described above the oxide reacted with the element to form a volatile oxide. A chemical reaction (oxidation–reduction reaction

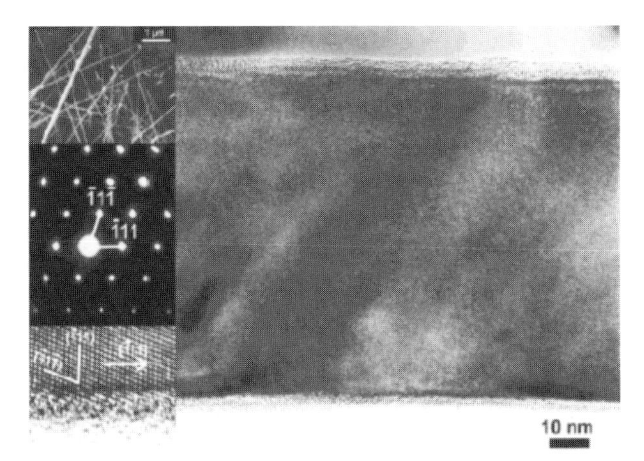


**Fig. 10.24.** A HRTEM image of a GaN nanowire grown at about 900 °C on a silicon (100) substrate. The insets are a SEM image (top), the corresponding ED pattern (middle) recorded along the [110] zone axis perpendicular to the nanowire growth axis, and the local enlarged HRTEM image (bottom). The growth direction is along the [−111] direction (white arrow in bottom inset) [36].

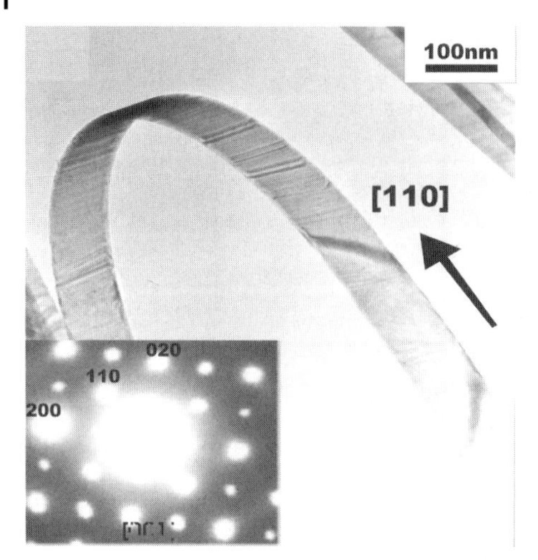


**Fig. 10.25.** TEM image of a single $SnO_2$ nanoribbon with [110] growth direction, inset showing the SAED pattern along the [001] axis [74].

where the oxide serves as the reactant) induced nucleation and growth of a nanowire embedded in an oxide sheath follow [22–27, 73]. The chemical reaction induced nucleation and growth of nanowires differentiates the OAG from the conventional metal catalyst VLS growth. The oxide-assisted nanowire growth process is free of metal catalysts thus enabling the formation of pure nanowires.

Similar to the growth of silicon nanoribbons, 2D nanostructures of $SnO_2$ with a ribbonlike morphology were also prepared on a large scale via rapid oxidation of elemental tin at 1080 °C [74]. As shown in Figure 10.25, the as-synthesized $SnO_2$ nanoribbons were single crystals and had preferred [110] and [203] growth directions. The lengths of the nanoribbons were up to several hundreds of micrometers, and the typical width and thickness were in the range 30–150 nm and 10–30 nm, respectively. Similarly, ZnS nanoribbons were grown (Figure 10.26) with a perfect 2H structure and a [120] growth direction [75].

## 10.6 Chemical Properties of SiNWs

### 10.6.1 Stability of H-Terminated SiNW Surfaces [76]

Silicon-based technology requires the removal of the surface oxide layer, and the termination and stabilization of the Si surfaces. This is conventionally performed by dipping in HF, which not only removes the oxide layer, but provides H-terminated Si surfaces. Examples of the significance of such a treatment are

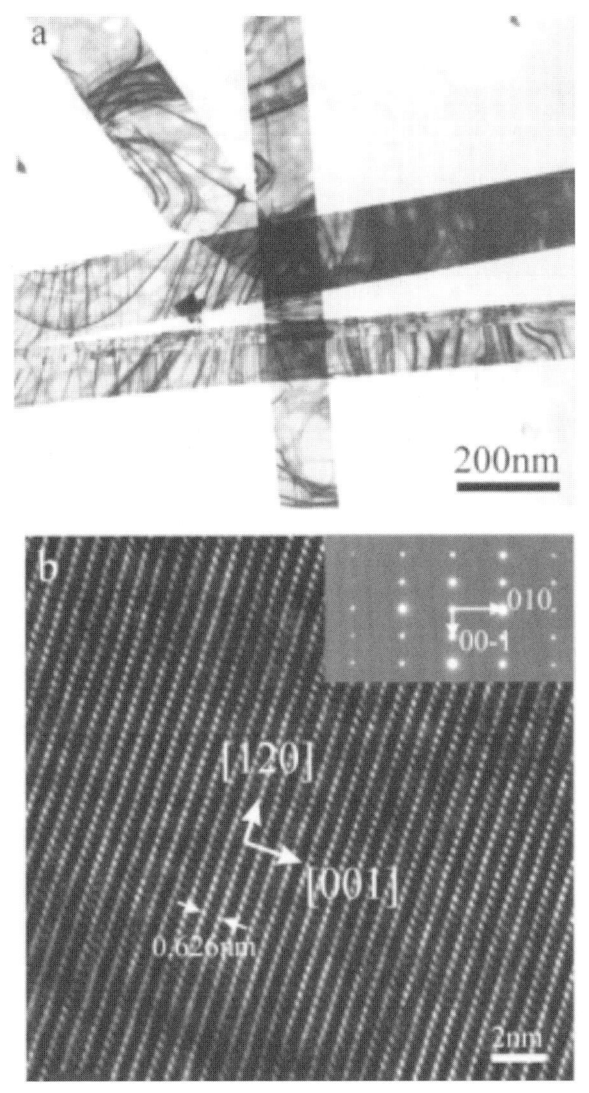


**Fig. 10.26.** (a) Low magnification TEM image and (b) high resolution TEM image and SAED of ZnS nanoribbons [75]. Note the high crystalline quality of the ZnO nanoribbons.

given in the previous and following sections. The stability of the oxide-removed H-terminated Si wafer surfaces has been extensively investigated, and many techniques were developed to suppress its re-oxidation upon exposure to humidity.

These considerations initiated our FTIR study of the nature of the HF-dipped SiNWs and their stability upon exposure to air and to water. The SiNWs had a distribution of diameters from a few nm to tens of nm. The thickness of the oxide layer was about 1/4 to 1/3 of the nominal diameter. Micro ATR-FTIR was used to monitor the wires (1) as grown, (2) after 5 min immersion in a 5% HF (or DF) solution, (3) upon exposure to air or water for different times, (4) after annealing to different temperatures. We note that the FTIR signals originate mostly from the large-diameter SiNWs (the surfaces of which are larger) and only a small fraction of the signal represents the small-diameter SiNWs. This means that if the stability of SiNW surfaces is size-dependent (as indicated by theoretical calculations, see Section 10.8), this would not be revealed in the experiment.

The as-grown SiNWs clearly show (Figure 10.27) only Si–O vibrations at 1050 and 800 $cm^{-1}$. Immersion in HF (Figure 10.27) removes the Si–O related lines and Si–$H_x$ ($x = 1, 2, 3$) absorption modes appear. These modes can be attributed to mono-hydrides and tri-hydrides on Si(111) and di-hydrides on Si(100) as already

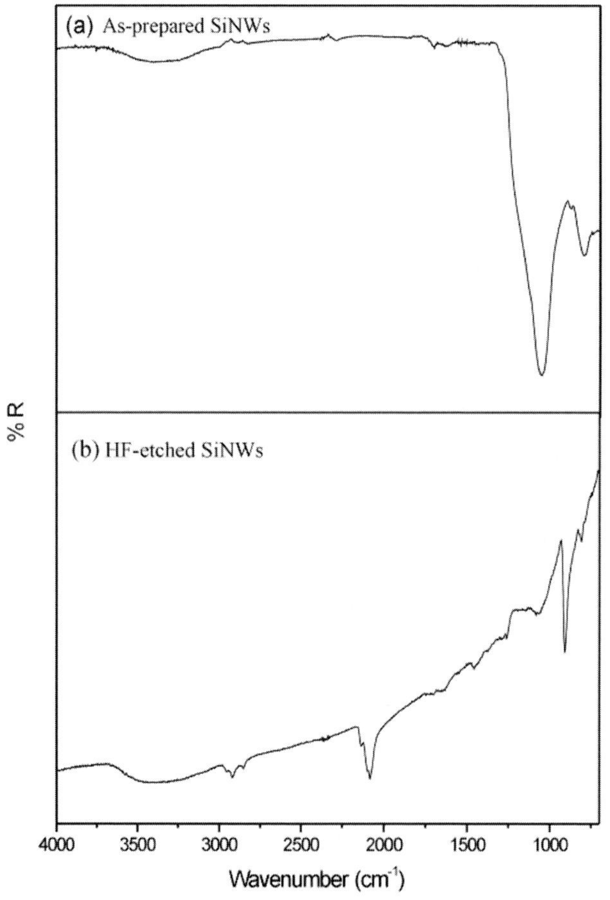


**Fig. 10.27.** ATR-FTIR spectra of (a) as-prepared SiNWs and (b) HF-etched SiNWs in the range 700–4000 $cm^{-1}$. Note the removal of the oxide absorption line and the formation of Si–$H_x$ absorption lines in the etched SiNW [76].

identified for Si wafers. This identification was substantiated by the isotope shift introduced by substitution of hydrogen with deuterium. These results are in accord with atomically resolved STM images of SiNWs described in a later section. Annealing of the SiNWs results in the weakening and disappearance of the trihydride, whereas the monohydride peaks remain strong. The hydrogen is completely removed only at 850 K. Hydrogen is observed on a SiNW surface even 26 days after exposure to air. Si–O bands are however detected 17 h after exposure to air and increase with time while the SiH bands decrease. It is likely that this is a superposition of the incorporation of O in the large-diameter SiNWs while the smaller-diameter (and probably more stable) SiNWs maintain the Si–H signal. It is obvious that the stability of H-terminated SiNW surfaces in water is much lower than in air, and Si–O bands are apparent after immersion in water for 15 min. These results indicate that it is possible to remove the oxide layer from the SiNW surfaces and terminate them by H by immersion in HF, similar to Si wafers. The H-terminated surfaces seem to be stable in air for at least a day, whereas the stability of small-diameter SiNWs, as determined by single-wire STM measurements, seems to be substantially better than that of Si wafers.

### 10.6.2 Reduction of Metals in Liquid Solutions

The reductive deposition of silver and copper ions on (oxygen-removed and hydrogen-terminated) SiNW surfaces in a solution was investigated [76] as an alternative method to ion implantation. The SiNWs surface is indeed capable of reducing silver and cooper ions to metal aggregates of various morphologies at room temperature.

Laser ablation was used [22] to produce SiNWs ~ 20 nm in diameter with a polycrystalline silicon core in a thin silicon oxide sheath with 1/4–1/3 of the nominal diameter and 1/3 of the weight of the SiNW. The oxide layer (which makes the SiNWs surfaces inert) was removed by a 5% HF dip for 5 min resulting in smooth, stable, H-terminated SiNW surfaces [77]. The etched SiNWs were immersed into solutions of silver nitrate and copper sulfate of different concentrations. Silver and copper ions were reduced to metallic aggregates deposited onto the surface of SiNWs. The TEM image of the sample treated with a $10^{-4}$ M silver nitrate solution (Figure 10.28) shows dark, round silver particles 5–50 nm in diameter. The HF-etched SiNWs treated with $1.0 \times 10^{-3}$ M copper sulfate show much smaller (a few nm) particles (Figure 10.29) identified by EELS as Cu particles.

The silver metal deposition on SiNWs in different concentrations of silver solution ($10^{-6}$ to 0.1 M) was studied in detail with SEM, EDS, and XPS. The SEM images show that the morphology of the deposited silver depends on the concentration of silver nitrate. At high concentrations the redox reaction is controlled by the concentration of Ag ions (mass action), so that large quantities of Ag ions in the vicinity of SiNWs are reduced and aggregated as dendrites. At low concentrations, the conditions required for dendrite formation are not reached and silver is de-

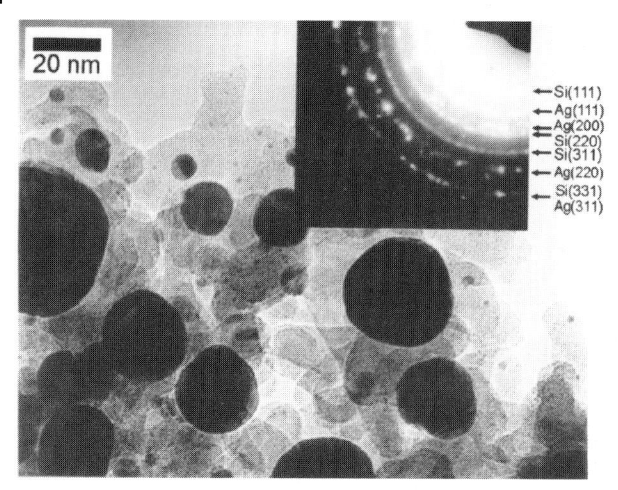


**Fig. 10.28.** A TEM image and the corresponding electron diffraction pattern (inset) of the SiNWs treated by a $1.0 \times 10^{-4}$ M silver nitrate solution [76].

posited as clusters or small aggregates on the SiNW surfaces. XPS analysis verifies the redox reaction (Figure 10.30). Most of the Si in the initial SiNWs is oxidized (see the Si 2s and Si $2p_{3/2}$ lines in Figure 10.30) and becomes elemental Si after the HF dip that removed the oxygen. The immersion in the Au solution results in the oxidation of Si, which increases with increasing Au concentration, as expected for a redox reaction of the Au ions. The metal concentration in the solution thus controls the size and the morphology of the metal deposited by the SiNWs, from small clusters to dendrites. This can be used for controlled deposition of metal nanoparticles on SiNWs on the one hand and larger self-supported metal configurations on the other hand.

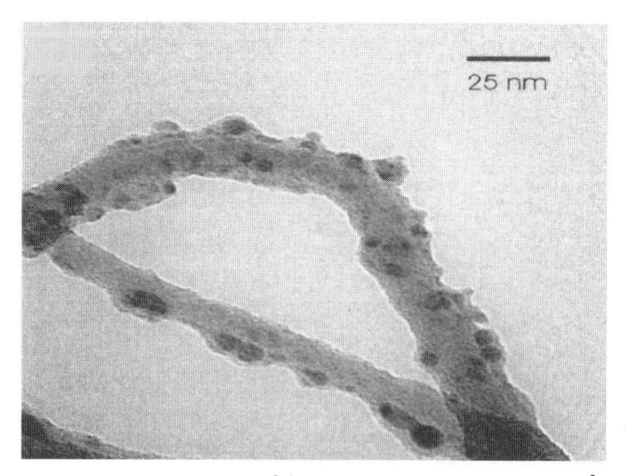


**Fig. 10.29.** A TEM image of the SiNWs treated by a $1.0 \times 10^{-3}$ M copper sulfate solution [76].

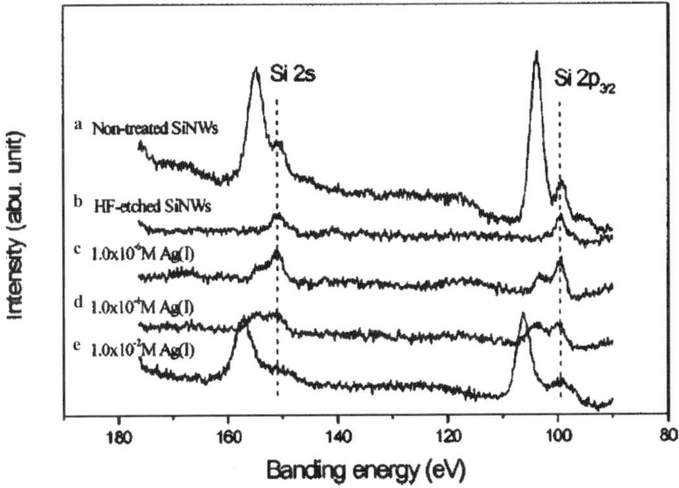


**Fig. 10.30.** Si 2p and 2s core level XPS spectra of (a) untreated SiNWs, (b) HF etched SiNWs, (c) $1.0 \times 10^{-6}$ M silver nitrate treated SiNWs, (d) $1.0 \times 10^{-4}$ M silver nitrate treated SiNWs, and (e) $1.0 \times 10^{-2}$ M silver nitrate treated SiNWs. Note the oxidation of the Si associated with the reduction of the silver ions [76].

### 10.6.3
### Chemical Sensing of SiNWs [78]

*I–V* measurements performed on an ensemble of SiNWs with a variety of diameters, growth directions, defect densities etc. are expected to yield only averaged behavior, which is dominated by those wires with the lowest resistivity [as in a parallel configuration of wires (resistors)]. While such ensemble measurements cannot be used to study the electrical conduction properties and mechanisms of nanowires, they can however give rough indications to check a variety of possible applications, one being gas sensing.

We have fabricated bundles of SiNWs of two types: (1) as-grown $SiO_2$ sheathed wires, (2) SiNWs dipped in HF to remove the $SiO_2$. Silver contacts were glued to the edges of the bundles and their resistivity was measured at different ambient conditions (vacuum ($2 \times 10^{-2}$ Torr, air with ~60% humidity, dry $N_2$, $NH_3$:$N_2$ 1:1000). The resistivity of the oxide-removed bundles was strongly reduced (by more than three orders of magnitude) upon exposure to humid air and to ammonia, but was hardly changed by exposure to dry nitrogen (Figure 10.31(a)). The process was found to be reversible, i.e. the resistivity increased to the initial value after pumping (Figure 10.31(b)). In contrast, the resistivity of the SiNWs embedded in the $SiO_2$ sheath did not change when exposed to different ambient environments.

The gas molecules may affect either the contact resistance across two nanowires or the surface resistance along individual wires, e.g. through charge exchange similar to polycrystalline semiconductor $SnO_2$ sensors. This would not happen for $SiO_2$ sheathed wires (having a high resistivity) for which gas incorporation has no effect. The chemical sensitivity of HF-etched SiNWs to $NH_3$ and water vapor exposure indicate their possible use in gas sensing applications.

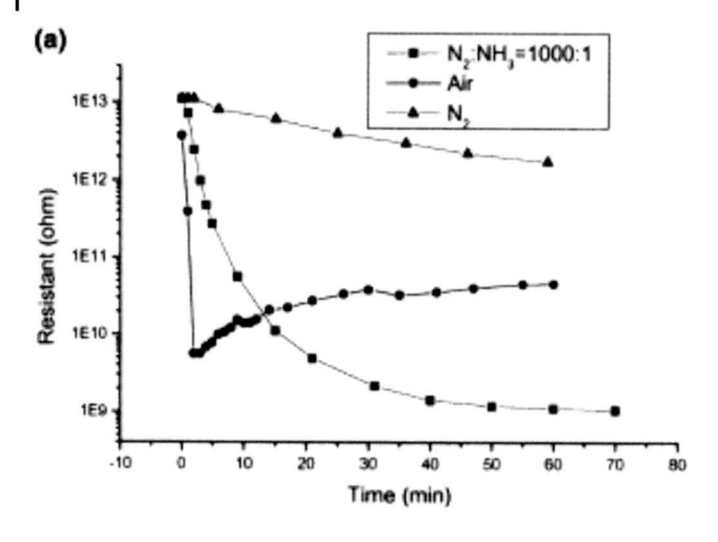


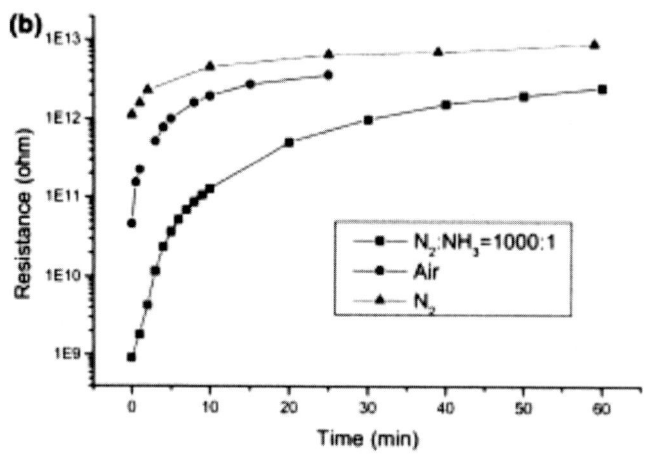


**Fig. 10.31.** Electrical responses with time of the Si nanowire bundle to $N_2$, a mixture of $N_2$ and $NH_3$ ($NH_3$ concentration: 1000 ppm), and air with a relative humidity of 60%; (a) when the gases were introduced into the chamber and (b) when the gases were pumped away [78].

### 10.6.4
### Use of SiNWs as Templates for Nanomaterial Growth [79]

A novel template effect of SiNWs was discovered accidentally trying to disperse SiNWs in common solvents such as $CHCl_3$, $CH_2Cl_2$ and $CH_3I$. A 15 min bath sonication resulted in a colloidal solution, the products of which were characterized by HRTEM, EELS and Raman. The analysis revealed that under sonication the SiNWs acted as templates on which carbon nanotubes and carbon nano-onions formed (Figure 10.32). Moreover, in addition to these known carbon structures,

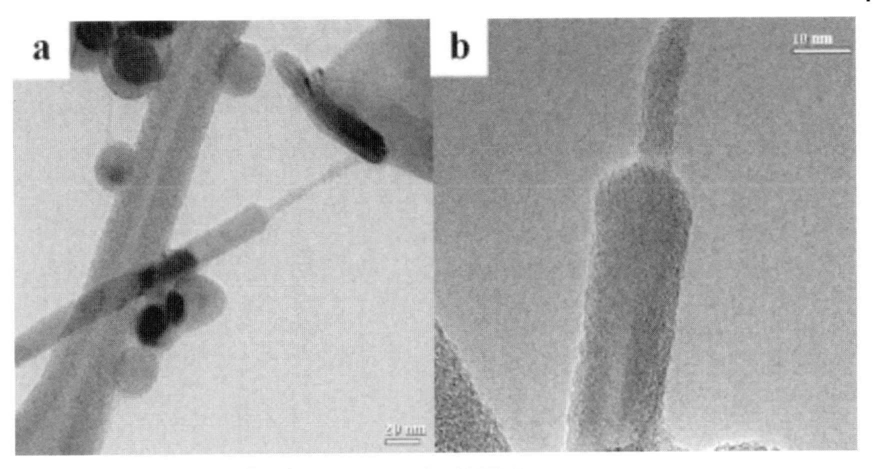


**Fig. 10.32.** Carbon nanotube (thin) grown on the SiNW tips (thick) which serves as a template [79].

nanotubes and nano-onions with plane spacings larger than those of graphite (3.5–5.8 Å) were also formed. The latter were interpreted as hydrogenated carbon nanotubes in which hydrogen atoms are bonded between graphitic layers forming $sp^3$ bonding.

The TEM data indeed verifies that some carbon nanotubes (NTs) and nano-onions (NOs) were attached to the SiNWs (Figure 10.33). We believe that all the NTs and NOs were formed on SiNW templates, since no NT/NO formation was detected in the absence of SiNWs. Moreover, the NT and NO formation occurred only when the oxide layer of the SiNWs was removed by HF dipping (H-terminated SiNWs) but not on as-grown SiNWs with a $SiO_2$ sheath.

The template mechanism of the SiNWs is still unclear. We nevertheless believe that the carbon nanostructures result from reactions between the $SiH_x$ species on the SiNW surfaces. The substituents of the solvent material are eliminated by the local heating caused by the sonication, giving rise to either C or CH units that wrap around the SiNWs (templating effect). Further sonication causes the SiNWs to shed off the NTs, refreshing the SiNW surfaces for additional templating of new NTs. Prolonged sonication transforms all hydrogenated carbon NTs and NOs to regular carbon structures: hydrogen free CNTs and CNOs.

## 10.7 Optical and Electrical Properties of SiNWs

### 10.7.1 Raman and PL of SiNWs [24]

The Raman spectrum of Si nanowires (Figure 10.34(a)) shows a broad and symmetric peak at 521 $cm^{-1}$ compared to that of a bulk single crystal Si. The peak

(a)

CNT

0.34nm

(b)

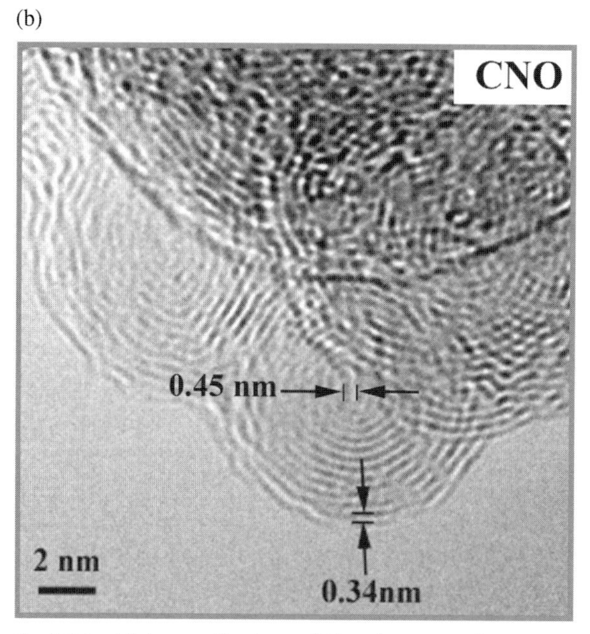


**Fig. 10.33.** High magnifications of (a) carbon nanotube and (b) carbon nano-onion grown on the SiNW template [79].

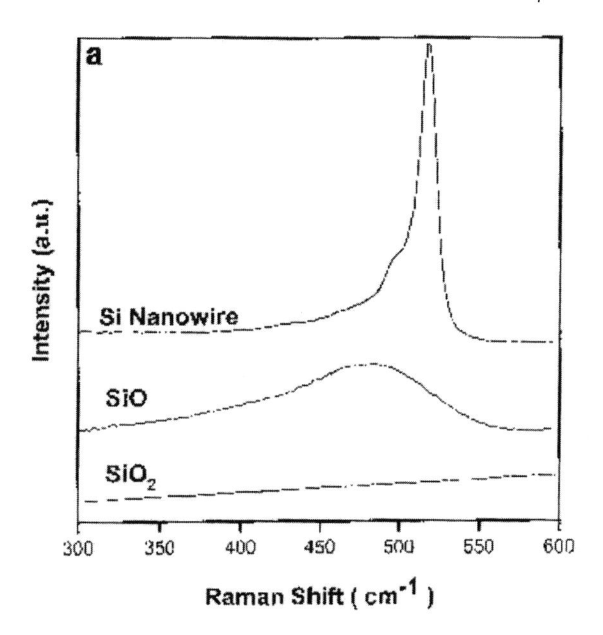


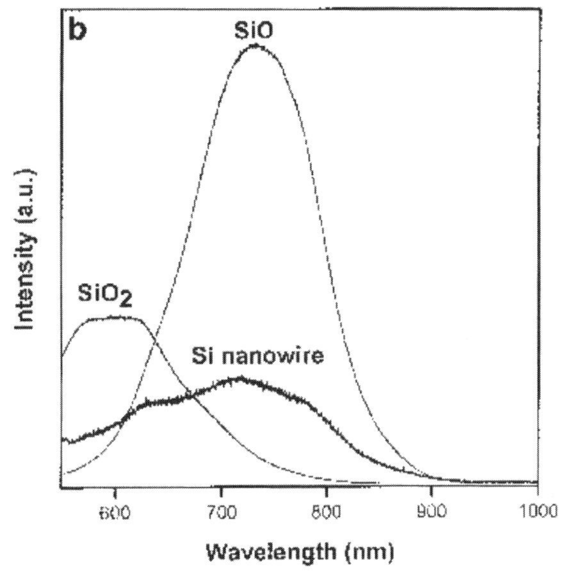


**Fig. 10.34.** (a) Raman spectra taken from the as-grown Si nanowires, Si monoxide and fully oxidized Si nanowires; (b) PL spectra taken from the as-grown Si nanowires, Si monoxide and fully oxidized Si nanowires [24].

profile may be associated with the effect of the small size of Si nanocrystals or defects. The presence of nonstoichiometric Si suboxide may also contribute to the peak asymmetry. For comparison, the spectrum from a SiO film contains a broad peak at 480 $cm^{-1}$, whereas the fully-oxidized SiNWs (prepared by annealing in air) show no Raman scattering (Figure 10.34(a)).

Si monoxide has a strong photoluminescence (PL) at 740 nm, while the oxidized nanowire gives a weak PL peak at 600 nm (Figure 10.34(b)). The PL from the SiNW product is weak and complicated. A typical PL spectrum from SiNWs covers the range 600–800 nm. Clearly, the SiO and Si suboxide components in the nanowires are the main contributors to this spectrum. The SiO generated by thermal evaporation is indeed a mixture of various oxides of Si. Si nanoparticles also coexist with the SiO generated.

### 10.7.2
### Field Emission from Different Si-Based Nanostructures

It is well known that nanotubes and nanowires with sharp tips are promising materials for application as cold cathode field emission devices. We have investigated the field emission of different nanowire structures. The first is from SiNWS. SiNWs exhibit well-behaved and robust field emission fitting a Fowler–Nordheim (FN) plot. The turn-on field for SiNWs, which is needed to achieve a current density of 0.01 mA $cm^{-2}$, was 15 V $\mu m^{-1}$ [26]. The field emission characteristics may be improved by further optimization, such as oriented growth or reducing the oxide shell, and may be promising for applications.

The second example of field emission of Si-based nanowires is that of B-doped Si nanochains [80]. The SiNCs were attached onto a Mo substrate by a conductive carbon film. The anode-sample separation ranges from 120 to 220 μm. The turn-on field was 6 V $\mu m^{-1}$, and smaller than that (15 V $\mu m^{-1}$) for the SiNWs. The field-emission characteristics of the SiNCs were analyzed according to the FN theory [81]. All the FN curves with different anode-sample separations fall in nearly the same region and have similar "Y" intercepts, showing that the SiNCs are uniformly distributed. A stability test showed no obvious degradation of current density and the fluctuation was within $\pm 15\%$, indicating that the B-doped SiNCs are a promising material for field emission applications.

The third example of field emission from Si-based nanowires is from the aligned SiC nanowires. The field emission measurements [68] were carried out in a vacuum chamber at a pressure of $\sim 5 \times 10^{-7}$ Torr at room temperature. An oriented SiC nanowire array, which was used as the cathode, was stuck to a stainless steel substrate by silver paste with the bottom end of the nanowires facing upward. A copper plate with a diameter of 1 cm, mounted on a precision linear feedthrough, was used as the anode. Field emission current densities of 10 μA $cm^{-2}$ were observed at applied fields of 0.7–1.5 V $\mu m^{-1}$, and current densities of 10 mA $cm^{-2}$ were realized at applied fields as low as 2.5–3.5 V $\mu m^{-1}$, as shown in Figure 10.35. These results represent one of the lowest fields ever reported for any field-emitting materials at technologically useful current densities. We attributed this emission

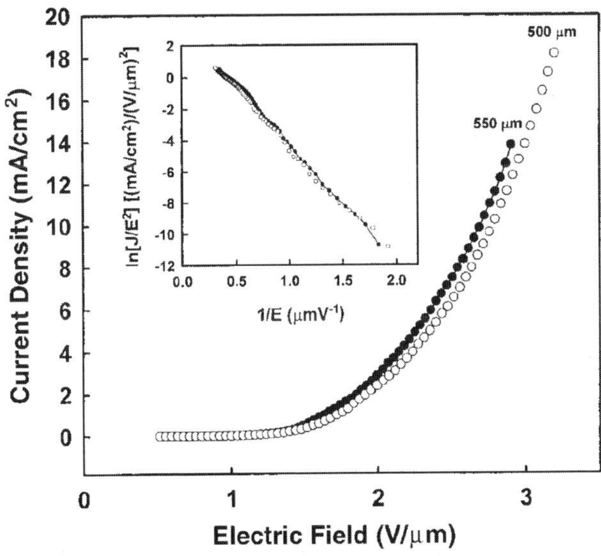


**Fig. 10.35.** Emission *J–E* curves from an oriented SiC nanowire emitter (emitting area 3.65 $mm^2$). The average turn-on field and threshold field for this sample are about 0.9 V $mm^{-1}$ and 2.7 V $mm^{-1}$, respectively. Inset: Fowler–Nordheim plot. The linearity of these curves indicates that the emission of the oriented SiC nanowires agrees with the properties expected for field emission [68].

behavior to the very high density of emitting tips with a small ratio of curvature at the emitting surface. The fact that when the oriented SiC nanowire array was pressed flat the current density was an order of magnitude lower than for the initial sample under the same electric field strongly supports this point.

### 10.7.3 STM and STS Measurements of SiNWs and B-Doped SiNWs

Almost all the characterizations performed by us until now are ensemble characterizations (i.e. probing many nanostructures simultaneously). HRTEM and HRSEM do probe the structure (and elemental composition) of individual nanostructures, but they do not correlate this structure with a specific property. STM and STS measurements are real single-object measurements that reveal the size, shape, and surface atomic structure, as well as the electronic density of states (deduced the *I–V* characteristics). The STM/STS measurements offer a way to correlate the electronic properties of SiNWs with the nanostructure size.

STM imaging of SiNWs with atomic resolution requires the complete removal of the oxide layer and the termination of the exposed SiNW surface by hydrogen. This was achieved by HF etching of the SiNWs. Oxide removal and H-termination were confirmed by FTIR measurements and indeed, atomically resolved STM images of SiNWs oriented along the two abundant growth directions ([112] and [110]) were

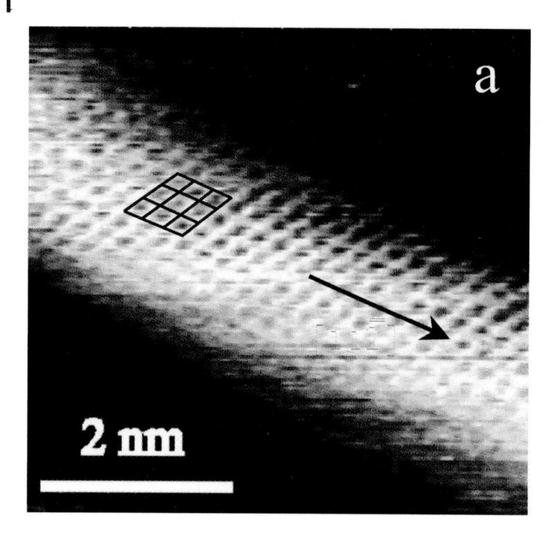


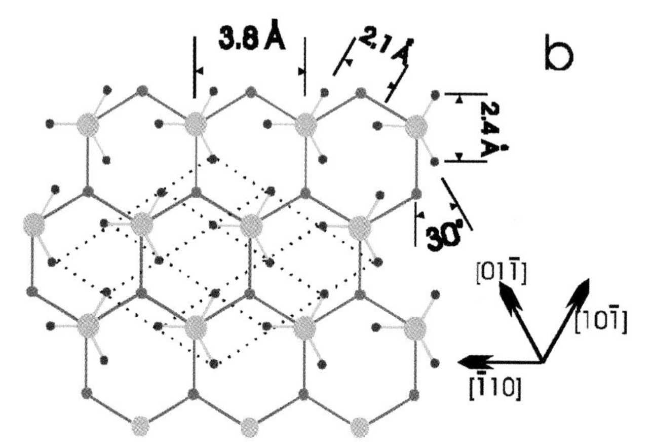


**Fig. 10.36.** STM image and schematic view of a SiNW with Si (111) facet. (a) Constant current STM image of a SiNW on a HOPG substrate. The wire's axis is along the [112] direction, and (b) schematic view of $SiH_3$ on Si (111) viewed along the [111] direction. Red and large blue circles represent the H atoms and Si atoms in the $SiH_3$ radical, respectively. Small blue circles represent Si (111) atoms in the layer below. The crystallographic directions are shown in the inset [82].

obtained (Figures 10.36 and 10.37). The SiNW grown along the [112] direction (Figure 10.36) can be interpreted as a Si(111)-H-terminated surface (tri-hydride). The SiNW grown along the [110] direction (Figure 10.37) is a Si(100)-H-terminated surface (bi-hydride). The later has a hexagonal faceted structure. The removal of the oxide and the formation of the H-terminated facets enabled the performance of reliable STS measurements. $I-V$ curves and their normalized derivative $(dI/dV)/(I/V)$ (representing the electronic density of states) of SiNWs with different diameters ranging from 7 nm to 1.3 nm were measured (Figure 10.38(a) and (b)). The

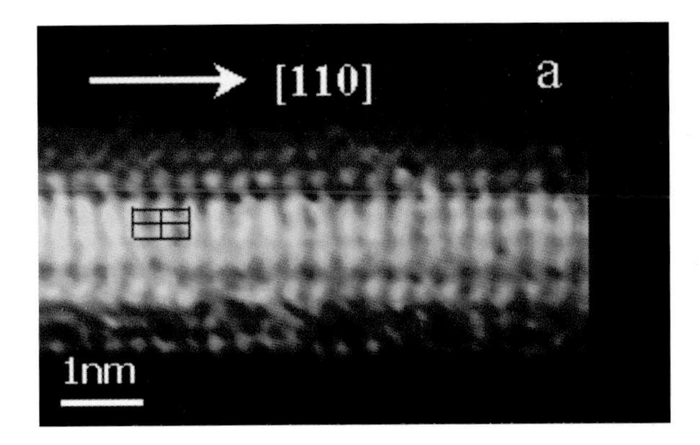


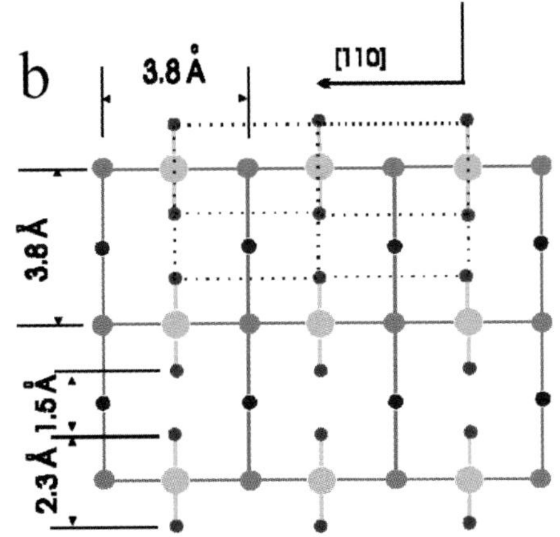


**Fig. 10.37.** STM image and schematic view of a SiNW with Si (001) facet. (a) Constant current STM image of a SiNW on a HOPG substrate. The wire's axis is along the [110] direction. (b) Schematic view of the dihydride phase on Si (001). Red and large blue circles represent H and Si atoms in the dihydride phase, respectively. Small blue circles correspond to Si atoms on the layers below. The crystallographic directions are shown in the inset [82].

electronic energy bandgaps increase with decreasing SiNW diameter from 1.1 eV for 7 nm to 3.5 eV for 1.3 nm in accord with previous theoretical predictions (Figure 10.38(c)), demonstrating a quantum size effect [82].

STM and STS measurements have been also performed on B-doped and undoped SiNWS [45] produced by OAG [23, 80]. The as-grown sample consisted primarily of SiNWs and nanoparticle chains coated with an oxide sheath. Samples for STM and STS measurements were prepared by dispersing the SiNWs into a suspension, which was then spin-coated onto highly oriented pyrolytic graphite (HOPG) substrates. The presence of nanoparticle chains and nanowires in the B-doped SiNWs sample was observed. Clear and regular nanoscale domains were observed on the SiNW surface, which were attributed to B-induced surface recon-

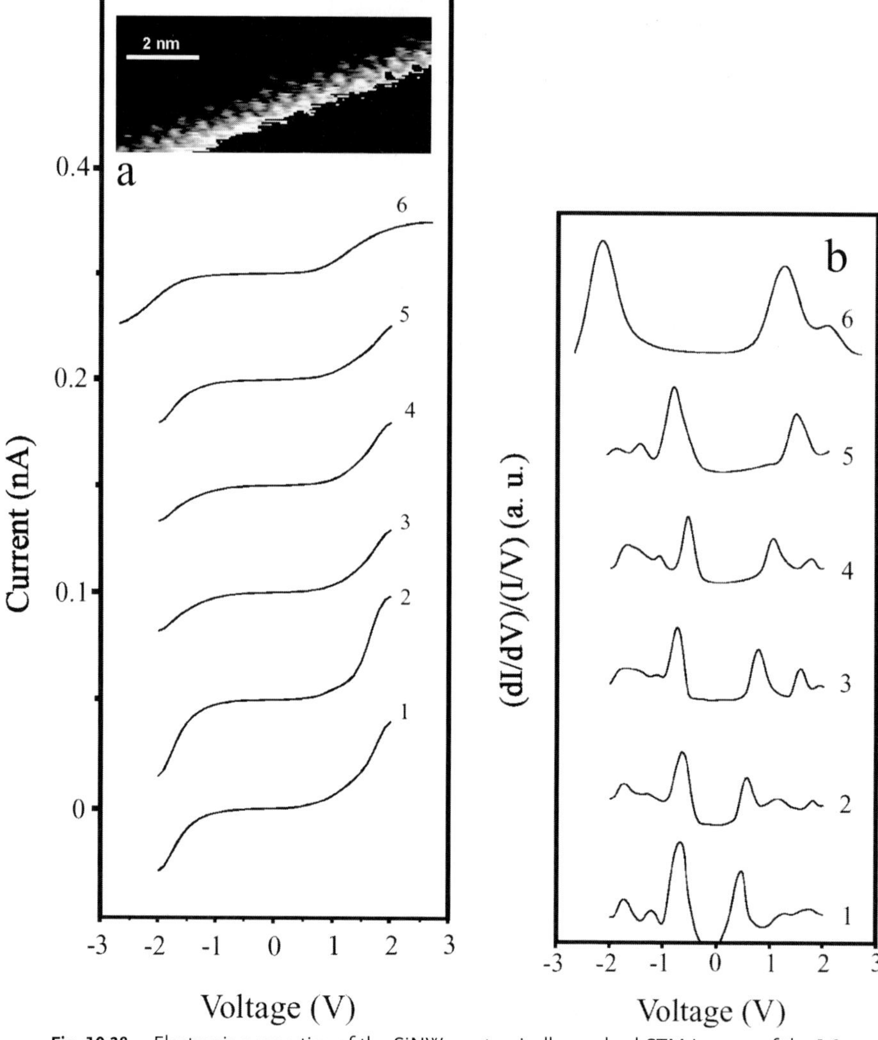


**Fig. 10.38.** Electronic properties of the SiNWs surfaces: (a) Current ($I$)–voltage ($V$) curves obtained by STS on six individual SiNWs; the diameter of wires 1 to 6 being 7, 5, 3, 2.5, 2, and 1.3 nm respectively. The inset shows the atomically resolved STM images of the 1.3 nm wire (6) which is the smallest ever reported. (b) The corresponding normalized tunneling conductances, $(dI/dV)/I/V$; the curves are offset vertically for clarity [82].

struction. STS measurements have provided current–voltage curves for SiNWs, which showed enhancement in electrical conductivity by boron doping.

STM images of several typical SiNWs are shown in Figure 10.39. Figure 10.39(a)–(c) show the images of B-doped SiNWs with different morphologies. Figure 10.39(a) shows a nanoparticle chain with a diameter of 30 nm. The particles in

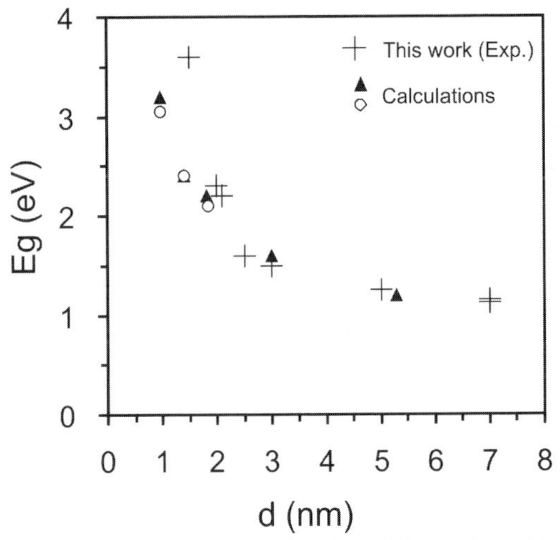


**Fig. 10.38.** (c) Experimental bandgap deduced from 38b versus the diameter of wires 1–6 plus additional three wires not shown in (a) and (b). The references of the calculated bandgaps are found in [82].

the chain are clearly revealed as evenly spaced bright dots of uniform size. Figure 10.39(b) shows a typical straight nanowire, the diameter of which is about 35 nm. Occasionally, sharp images of SiNW of ca. 40 nm in diameter, as shown in Figure 10.39(c), were obtained via air STM. The image reveals clearly resolved rectangular domains of several nanometers in size. These domains are associated with B-induced reconstruction of the silicon surface, and the clear images were made possible by B-enhanced conductivity.

STS measurements have also been performed on the undoped and B-doped SiNWs shown in Figure 10.39. The $I$–$V$ and the corresponding differentiated $\mathrm{d}I/\mathrm{d}V$ curves of the nanowires reveal several features. First, while the curves for the two B-doped nanowires with different morphologies are quite similar, they are distinctly different from that for the undoped wire. This is because, as far as tunneling from a STM probe is concerned, the particle in the nanochain is the same as any point on a nanowire with a similar doping concentration and oxide sheath. Second, the steeper rise in the $I$–$V$ curves and the higher values of $\mathrm{d}I/\mathrm{d}V$ for the B-doped SiNWs are consistent with the expected B-induced conductivity enhancement. As $\mathrm{d}I/\mathrm{d}V$ values can be regarded as a measure of local density of states (LDOS), the low value between $-1$ to $+1$ V in the $\mathrm{d}I/\mathrm{d}V$ curve of the undoped wire indicates relatively little LDOS within the gap, while the higher $\mathrm{d}I/\mathrm{d}V$ in the same region of the doped wires is in accord with the presence of the B dopants. The minima of the curves are at 0.3 V, indicating that the Fermi level in the B-doped SiNWs lies 0.3 eV closer to the valence band relative to the undoped wire. The position of the Fermi level corresponds to a hole carrier concentration of $1.5 \times 10^{15}$ cm$^{-3}$ in the boron-doped nanowires.

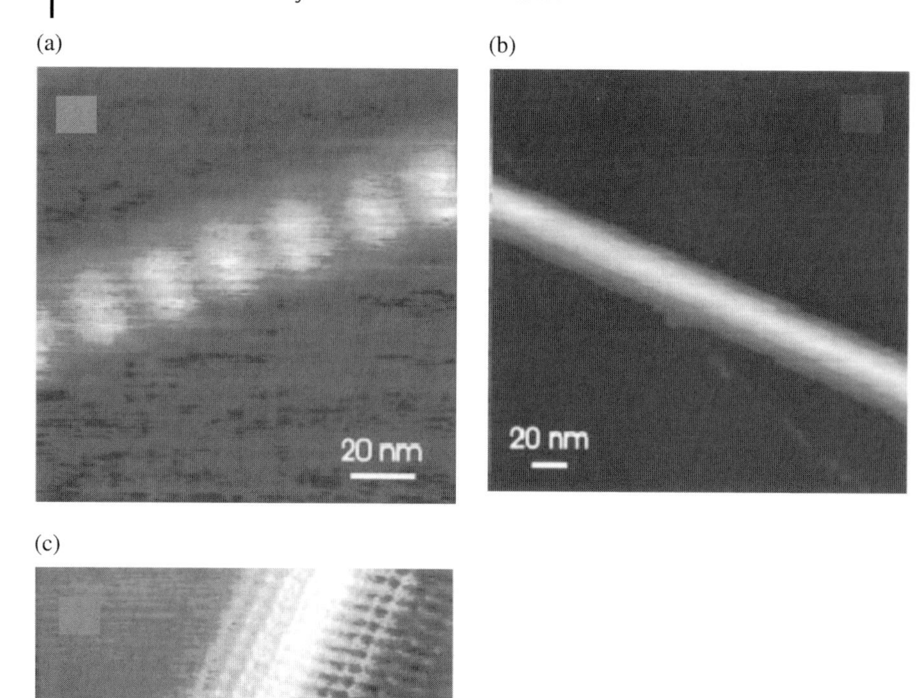


**Fig. 10.39.** STM images of: (a) a B-doped nanoparticle chain, (b) a B-doped straight nanowire, and (c) boron-induced reconstruction of SiNW [45].

## 10.7.4 Periodic Array of SiNW Heterojunctions [83]

The OAG of SiNWs using thermal evaporation of SiO at 1200 °C was studied in detail as previously discussed and the growth at ~900 °C leads to straight, uniform (in diameter) nanowires with a crystalline silicon core embedded in a $SiO_2$ sheath. STM analysis was performed on such SiNWs that were dipped into HF to remove the oxide sheath and then dispersed on a HOPG substrate and loaded into the

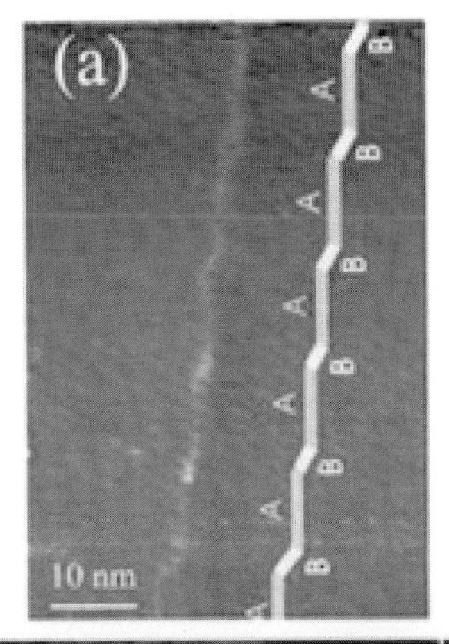


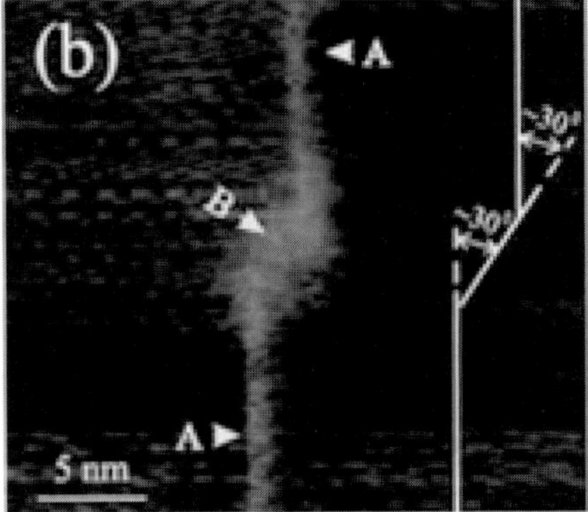


**Fig. 10.40.** STM images of a SiNW showing the periodic array of Si segments: (a) a medium magnification showing six sequences as further illustrated by the schematic sketch denoting segments A (~10 nm long) and B (~5 nm long) and (b) a higher magnification showing an image of a pair of junctions (ABA), also denoting the angle between the segments at ~30° [83].

UHV STM system. Among the less abundant forms of SiNWs, ~1% have a zigzag shape, composed of many bends or junctions.

Figure 10.40 shows the structure of such a typical zigzag-shaped SiNW with a diameter of 3 nm and a length of several microns. The wire is composed of a periodic array of long (~10 nm) and short (~5 nm) segments, denoted A and B respectively. The angle between the segments is ~30°. The junctions repeat themselves regularly so that the length of the different segments is fairly constant along the entire wire. We have discussed in a previous section the growth directions of SiNWs in the OAG, and have reported that the two most abundant directions are

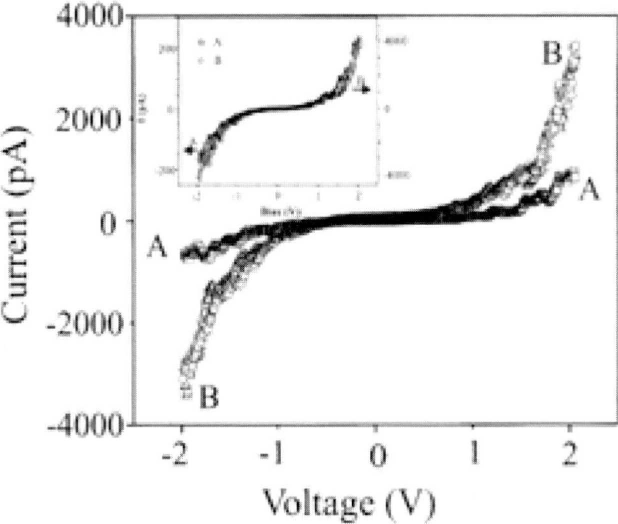


**Fig. 10.41.** *I–V* STM curves of segments A and B. Note the significant difference between the two *I–V* curves indicating different electronic properties. The inset shows the same *I–V* curves for A (squares) and B (circles and triangles) in which the right-hand side scale (for segment B) is the same as in Figure 10.32 but the left-hand side scale (for segment A) was increased (50 pA per bar instead of 1000 pA per bar) until the *I–V* curves of the two segments overlapped. It is evident that the shape of the *I–V* curves of the two different segments is similar (the electronic energy gap is almost the same) but the *I* values differ by a factor of ~20. The *I–V* curve is the same for all A segments or B segments and along each segment. The change from one *I–V* curve to the other along the junction (going from A to B or from B to A) is however very sharp [83].

[112] and [110]. We thus speculate that the zigzag shape originates from a periodic transition between two growth directions (which is 30 between [112] and [110] and 35 between [110] and [111]).

STS of these wires performed on several of each segment indicates that the *I–V* curves are almost the same along identical segments (i.e. along all As or along all Bs), but very different for the two types of segments as indicated in Figure 10.41. The transition between one type of *I–V* curve and the other along the segment interface (junction) is very sharp. We can conclude that periodic arrays of Si intramolecular junctions were grown in a single SiNW growth process (segmented growth was previously reported by a periodic change in the growth conditions). The difference in the electronic properties of the different segments is unclear at present. Among the possible origins we include: (1) different surface electronic structures, (2) different diameters of the different segments, (3) defect and impurity induced variations, (4) stress effects. The observation of arrays of 66 pairs of junctions per micron indicate that self assembly of SiNWs in the OAG process may be manipulated to grow highly dense devices ($1.1 \times 10^4$ ABA "transistors" per micron $\times$ micron).

## 10.8 Modeling

### 10.8.1 High Reactivity of Silicon Suboxide Vapor

Silicon oxide is a critical source material in the oxide-assisted growth as described above. It also plays important roles, as is well known, in many fields such as electronics, optical communications, and thin-film technology. Our recent finding of silicon oxide in the synthesis of silicon nanowires, as we reviewed in the previous part of this chapter, would extend further the important new application of silicon oxide.

We have studied the silicon oxide-assisted formation of Si nanostructures based on quantum-mechanical calculations of $Si_nO_m$ ($n, m = 1$–8) clusters [50, 51, 84]. We found that most of the structures contain planar or buckled ring units. Pendent silicon atoms bonded only to a single oxygen atom are found in silicon-rich clusters. Oxygen-rich clusters have perpendicular planar rings, while silicon monoxide-like clusters usually form a large buckled ring. Structures made up of tetrahedrally bonded units are found only in two clusters. Furthermore, the energy gap and net charge distribution for clusters with different Si:O ratios have been calculated. We further found that: (i) the most energetically favorable small silicon oxide clusters have O to Si atomic ratios at around 0.6 (see Figure 10.42); and (ii) remarkably

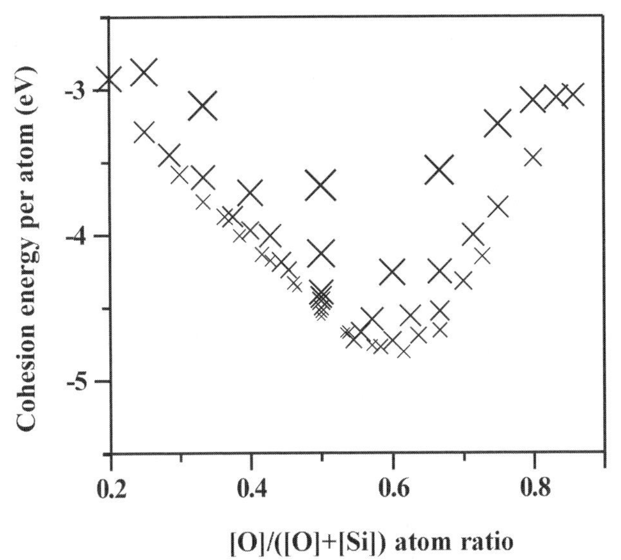


**Fig. 10.42.** Cohesion energy per atom of $Si_nO_m$ ($n, m = 1$–8) clusters as a function of O ratio based on total energy calculations with B3LYP/3-21G:Si; 6-31G*:O. The decreasing size of the symbol × is related to increasing cluster size ($n + m$) [51].

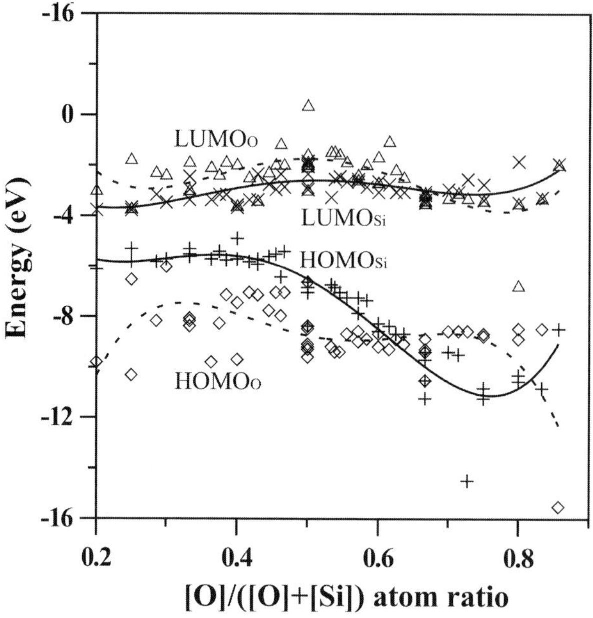


**Fig. 10.43.** $LUMO_{Si}$ ("×"), $LUMO_O$ ("Δ"), $HOMO_{Si}$ ("+") and $HOMO_O$ ("◇") of $Si_nO_m$ clusters determined based on the calculations using B3LYP/3-21G:Si; 6-31G*:O. Their fittings using four-order polynomials are shown with solid curves for $LUMO_{Si}$ (upper) and $HOMO_{Si}$ (lower) while dashed curves represent $LUMO_O$ (upper) and $HOMO_O$ (lower).

high reactivity at the Si atoms exists in silicon suboxide $Si_nO_m$ clusters with $2n > m$ (see Figure 10.43). The results show that the formation of a Si–Si bond is preferred and thus facilitates the nucleation of Si nanostructures when silicon suboxide clusters come together or stack to a substrate. Based on these findings, the mechanism of oxide-assisted nucleation of silicon nanowires has been drawn clearly [52] as we reviewed in Section 10.2.2 of this chapter.

### 10.8.2
### Thermal and Chemical Stabilities of Pure Silicon Nanostructured Materials

For application of the SiNWs in advanced areas, their oxide sheath has to be removed. If the silicon core were not saturated, the stability of the structure would be very poor. Demonstrations of their stability would be useful for understanding the related problems. Other issues of pure silicon nanostructures including the structure and property of the thinnest nanowire and the stability of silicon nanotubes are also interesting. Summarized below are our computational efforts regarding these issues.

#### 10.8.2.1 Structural Transition in Silicon Nanostructures

As is well known, small silicon clusters do not have any structural feature similar to that of bulk silicon (tetrahedral). Between the small silicon clusters and bulk silicon, there may be structural transition from amorphous to ordered tetrahedral structure. The structural transition to bulk diamond structure in nanosized silicon clusters has been studied by the tight-binding molecular dynamics method combined with a simulated annealing technique [85]. For intermediate-sized clusters ($<200$ atoms), the energetically favorable structures obtained consist of small subunits like $Si_{10}$ and $Si_{12}$ (Figure 10.44), qualitatively consistent with the experimental fragmentation behavior of these clusters. For spherical silicon nanocrystals, the surface atoms reconstruct to minimize the number of dangling bonds, forming a continuum surface (Figure 10.45). The large curvature of the continuum surface causes lattice contraction in the nanocrystals. Present calculations predict the lattice contraction versus the particle radius as $\Delta a = 0.38/R$, with $\Delta a$ and $R$ in Å. By comparing the cohesive energies of the two sorts of structures with the same number of atoms, the structural transition is estimated to occur at about 400 atoms, or 2.5 nm in diameter.

#### 10.8.2.2 **Thinnest Stable Short Silicon Nanowires**

Using a full-potential linear-muffin-tin-orbital molecular-dynamics method, we have studied the geometric and electronic structures of thin short silicon nanowires consisting of tri-capped trigonal prism $Si_9$ sub-units and uncapped trigonal prisms, respectively [86]. Comparing to other possible structures, these structures are found to be the thinnest stable silicon nanowires, being in particular much more stable than the silicon nanotubes built analogously to small carbon nanotubes (Figure 10.46). As for their electronic structures, these silicon wires show very small gaps of only a few tenths of an eV between the lowest unoccupied energy level and the highest occupied energy level, and the gaps decrease as the stacked layers increase. The results provide guidance to experimental efforts for assembling and growing silicon nanowires.

#### 10.8.2.3 **Silicon Nanotubes**

In contrast to the synthesis of large quantity of SiNWs, no Si nanotube has ever been observed experimentally, indicating that silicon is an element very different from carbon in forming nanotubular structures although they are in the same group of the Periodic Table. The difficulty in the synthesis of silicon nanotubes is widely attributed to the property of $sp^3$ hybridization in silicon. How and to what extent such hybridization affects the tubular structural formation still needs further clarification. To understand the reason(s) for the hitherto unsuccessful synthesis of silicon nanotubes we have studied [87] the differences in the structures and bonding between cubic (diamond-like) and tubular nanostructures of carbon and silicon, and their relative stabilities in terms of their characteristic electronic structures. Our calculated results indicate that when the dangling bonds at the open ends of the tubular structure are properly terminated, Si nanotubes with a severely puckered structure can, in principle, be formed. Such computationally stable, energetically minimized, and geometrically optimized Si nanotube structures may serve as models for the design and synthesis of silicon nanotubes.

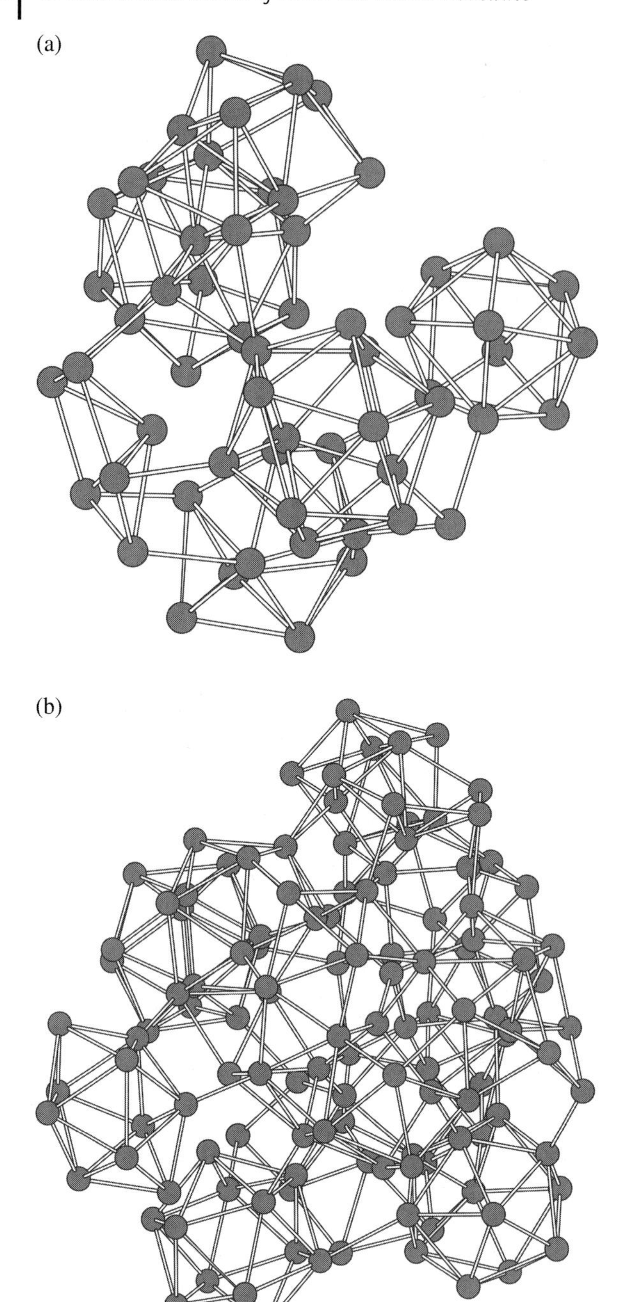


**Fig. 10.44.** Structures of (a) $Si_{60}$ and (b) $Si_{123}$ after annealing. All bonds below 2.8 Å are drawn out. The structures are fully relaxed, with the root-mean-square force to be 0.015 eV $Å^{-1}$ [85].

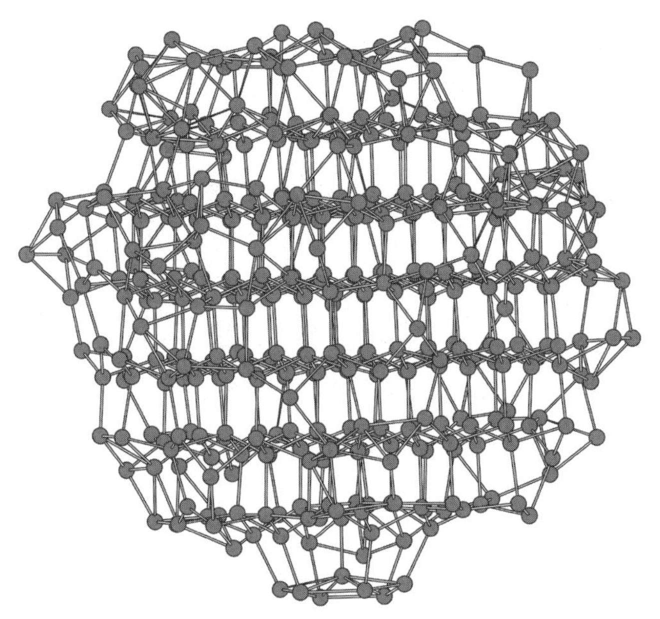

**Fig. 10.45.** Structure of the 417-atom Si nanocrystal with reconstructed surface. All bonds below 2.8 Å are drawn out. The structure is fully relaxed, with the root-mean-square force to be 0.015 eV $Å^{-1}$ [85].

### 10.8.3 Thermal and Chemical Stabilities of Hydrogenated Silicon Nanostructures

As reviewed in Section 10.6, the SiNWs could be etched using HF solution so that the surface oxide sheath is removed and the exposed silicon surface is saturated with hydrogen. The stability of the oxide-removed H-terminated Si wafer surfaces has been investigated and many techniques were invented to suppress its re-oxidation upon exposure to humidity. For the case of nanostructured hydrogenated silicon, its chemical stability would be different from the case of the wafer, as has been demonstrated in our recent computational works. One of our predictions, that the hydrogenated SiNWs possess better stability than that of a silicon wafer, could find a number of supportive evidences from our experiments as described above.

#### 10.8.3.1 Structural Properties of Hydrogenated Silicon Nanocrystals and Nanoclusters

The structures of hydrogenated Si nanocrystals and nanoclusters were studied using the empirical tight-binding optimizations and molecular dynamics simulations [88]. It was shown that the structural properties of the hydrogen-saturated Si nanocrystals have little size effect, contrary to their electronic properties. The surface relaxation is quite small in the hydrogen-saturated Si nanocrystals, with a lat-

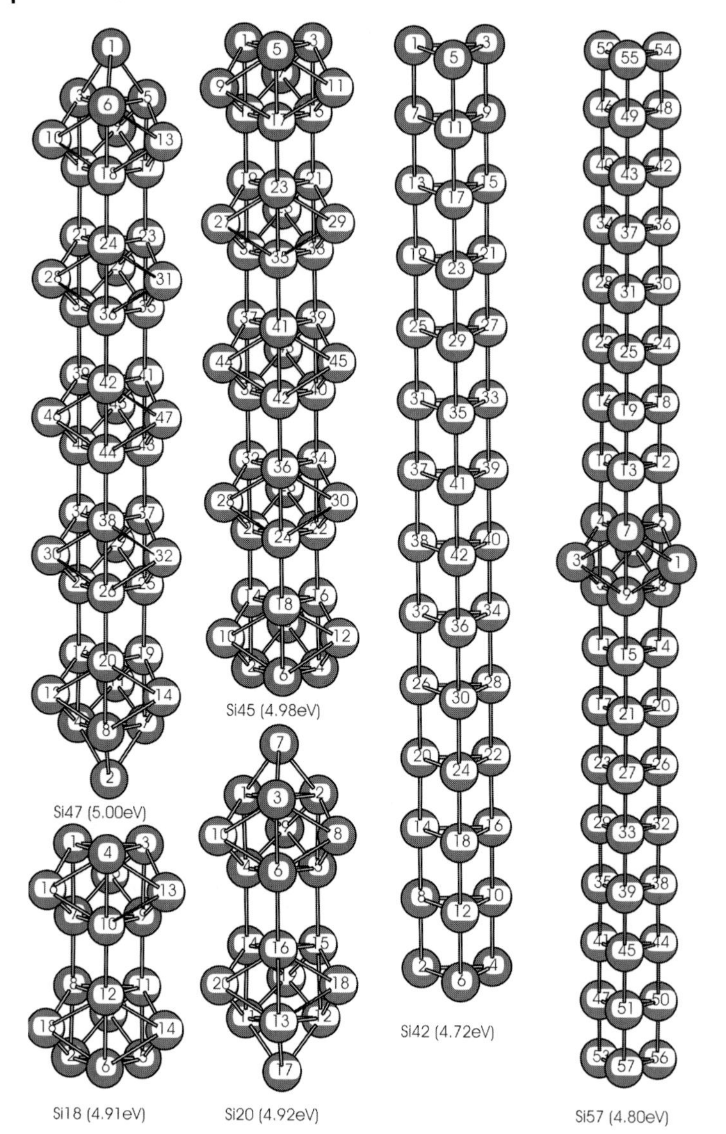


**Fig. 10.46.** Stable structures of some selected Si clusters. $Si_{18}$, $Si_{20}$, $Si_{45}$, and $Si_{47}$ correspond to the stacked structures from the tricapped trigonal prisms. $Si_{42}$ consists of the trigonal prisms. $Si_{57}$ refers to the stacked trigonal prisms inserted among trigonal prisms by one tricapped trigonal prism. The binding energy per atom is listed below the corresponding structure [86].

tice contraction of 0.01–0.02 Å within the outermost two to three layers. Inside the hydrogenated Si nanocrystals, there is only a very small strain (lattice expansion) of the order of $10^{-4}$–$10^{-3}$, in good agreement with the X-ray diffraction measurement. The fully hydrogenated Si nanocrystals are the most stable structures compared to partially hydrogenated ones. Removal of up to 50% of the total terminating H atoms only causes distortion to the crystal structure, while the tetrahedral structures are retained. By removing more than 70–80% of the total terminating H atoms, the clusters evolve to more compact structures (Figure 10.47).

#### 10.8.3.2 Size-Dependent Oxidation of Hydrogenated Silicon Clusters

We have shown explicitly the size-dependent chemical reactivity of hydrogenated silicon clusters towards water [89]. A unique trend of decreasing reactivity with decreasing cluster size has been deduced from reaction energetics, frontier orbital analysis, and chemical reaction rates determined by the transition state theory in conjunction with *ab initio* calculations at Hartree–Fock and Møller–Plesset perturbation levels for water reaction with both dihydride and trihydride silicon configurations, as shown in Figure 10.48. This study indicates the possibility of fabricating stable hydrogenated silicon structures with sizes close to nanometers. We predict that, with the nanosized hydrogenated silicon structures, it is possible to fabricate nonreactive, stable nanodevices.

## 10.9 Summary

A new method based on OAG has been developed that is capable of producing high-quality and bulk-quantity of various semiconductor nanowires. The presence of oxides in the target is a common and essential ingredient for the synthesis using laser ablation or thermal evaporation, so that the targets are capable of generating semiconductor oxides in the vapor phase. Subsequent decomposition of the vapor phase oxides at high temperature and defect structures play crucial roles in the nucleation and growth of high-quality nanowires. The developed OAG approach has been applied to grow a host of semiconducting nanowires such as Ge, GaN, GaAs, GaP, SiC, ZnO (whiskers), and carbon nanowires, by either laser-ablation, HFCVD or thermal evaporation. Large-area, aligned, and long SiNWs via flow control, and diameter and morphology control by temperature have been achieved. High-quality SiNW-based nanocables and Si nanoribbons using oxide-assisted methods have been additionally grown. Oxide-stripping and hydrogen-termination along with nanowire dispersion enabled the study of SiNW surfaces by STM and STS. The quantum size effect of the SiNW diameter on the electronic bandgap was demonstrated for the first time. The nanochemistry of SiNWs has been investigated by performing surface reactions with Ag and Cu ions on the SiNWs. Nanosized ligated metal clusters on a SiNW surface were realized to achieve good metal–nanowire contact. The surface reactivity of SiNWs was exploited for growing carbon-based nanostructures on a SiNW template (liquid solution) and for chemical sensing in the gas phase. Our synthesized highly-oriented SiC nanowires were shown to be excellent field emitters with large field emission current densities at very low electric turn-on and threshold fields. Modeling of SiNW structures, nu-

(a)

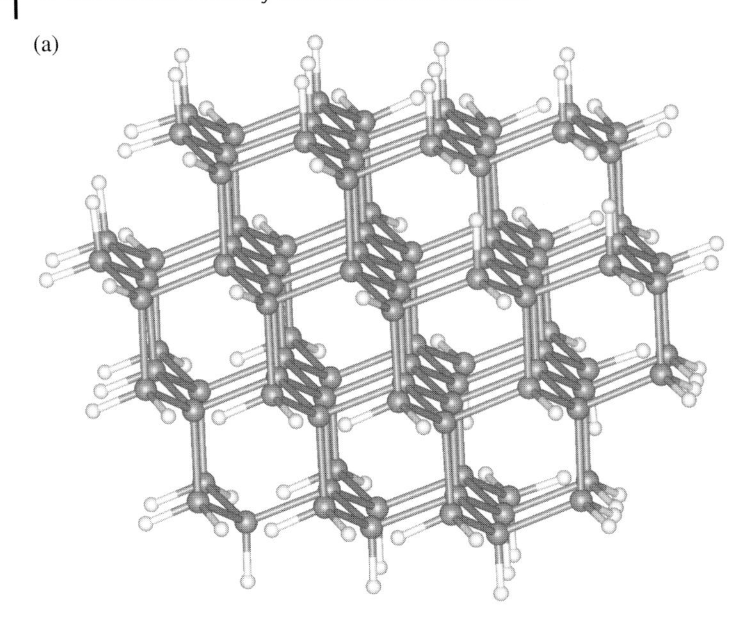

(b)

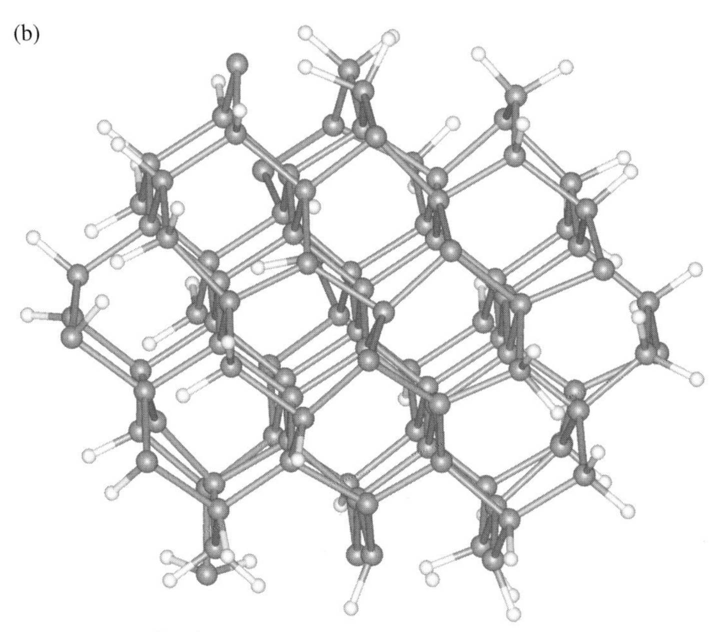

**Fig. 10.47.** Ball-and-stick diagrams of the structures of $Si_{100}H_x$ clusters obtained from simulated annealing. (a) Fully H-saturated $Si_{100}H_{86}$; (b) $Si_{100}H_{60}$;

(c)

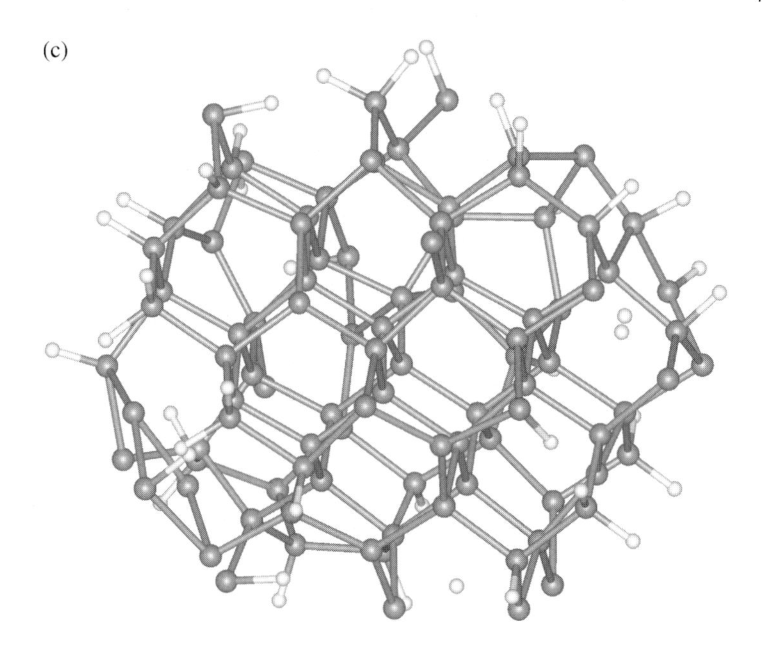

(d)

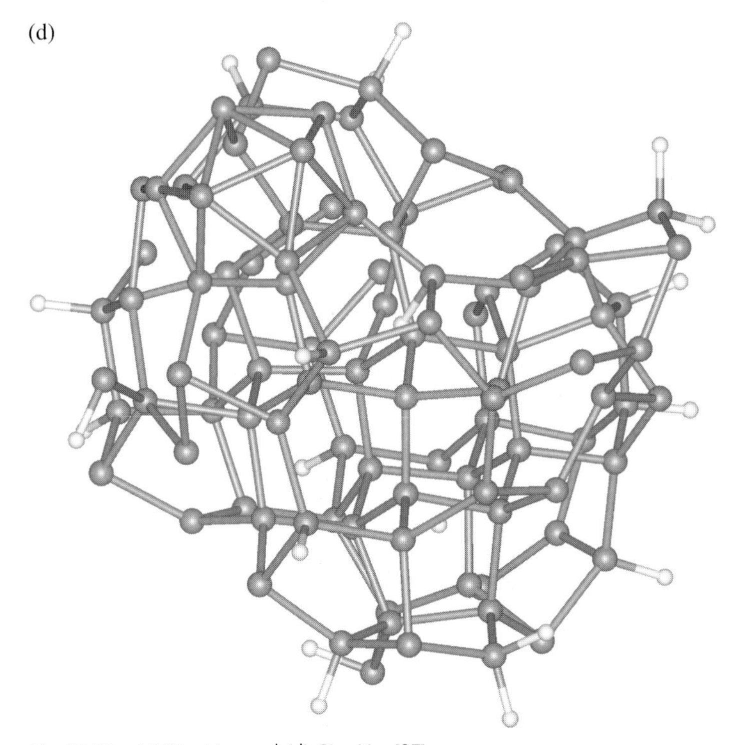

**Fig. 10.47.** (c) $Si_{100}H_{40}$; and (d) $Si_{100}H_{20}$ [87].

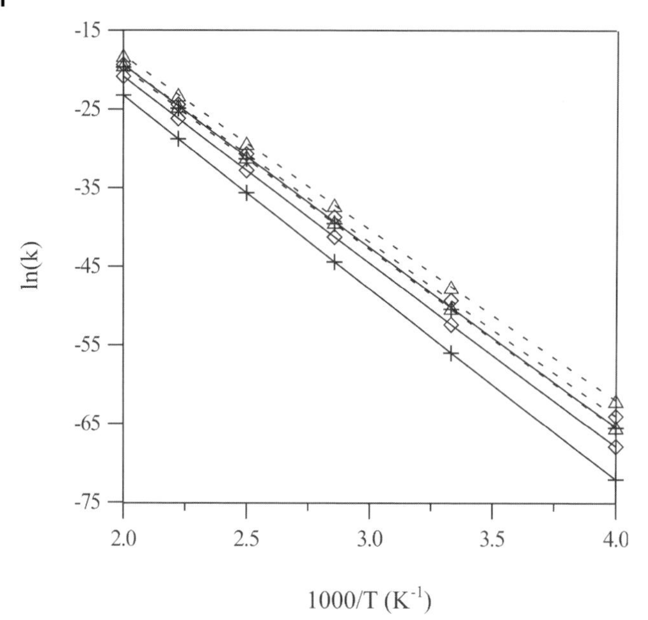


**Fig. 10.48.** Calculated total rate constants ($cm^3\ mol^{-1}\ s^{-1}$) at a pressure of 1 atm in a temperature range between 250 and 500 K. Solid lines are those of water reactions on the dihydride configurations: $Si_2H_6$–$SiH_2$ ("+"); $Si_5H_{10}$–$SiH_2$ ("◇"); and $Si_9H_{14}$–$SiH_2$ ("△"). Dashed lines are those of water reactions on the trihydride configurations: $SiH_3$–$SiH_3$ ("+"); $Si_4H_9$–$SiH_3$ ("◇"); and $Si_{10}H_{15}$–$SiH_3$ ("△"). The inset illustrates the potential energy profile for a representative $H_2O$ reaction with the hydrogenated silicon cluster [89].

cleation and growth processes and properties carried out in parallel with the experimental work is highly valuable in giving additional insight into the nature of the oxide-assisted growth and in explaining our experimental results.

## Acknowledgement

We are grateful to our colleagues (I. Bello, C. S. Lee, D. S. Y. Tong, N. Wang, N. B. Wong), post-doctoral fellows (J. Q. Hu, Y. Jiang, B. X. Li, D. D. D. Ma, Z. W. Pan, H. Y. Peng, W. S. Shi, D. K. Yu), graduate students (F. Au, C. P. Li, X. M. Meng, X. H. Sun, Y. H. Tang), and visiting professors (T. K. Sham, B. K. Teo, D. P. Yu, Y. F. Zhang) to COSDAF for their contributions to the work described in this chapter. The work described herein was supported financially by the Research Grants Council of Hong Kong SAR (Project No. 8730016 (e.g. CityU 3/01C); 9040459; 9040533 (e.g. CityU 1033/00P); 9040633 (e.g. CityU 1011/01P); and 9040746 (e.g. CityU 1138/02P)) and City University of Hong Kong.

## References


1 S. Iijima, *Nature, 354,* **1991**, 56.
2 T. W. Ebbesen, *Carbon Nanotubes, Preparation and Properties*, CRC Press, Boca Raton 1997.
3 M. Dresselhaus, G. Dresselhaus, P. Eklund et al., *Phys. World, 33,* **1998.**
4 J. W. G. Wildoer et al., *Nature, 391,* **1998**, 59.
5 P. M. Ajayan, *Chem. Rev., 99,* **1999**, 1787.
6 C. Dekker, *Phys. Today, 22,* **1999**.
7 J. W. G. Wildoer, L. C. Venema, A. G. Rinzler et al., *Nature, 391,* **1998**, 59.
8 T. Odom, J. Huang, P. Kim et al., *Nature, 391,* **1998**, 62.
9 L. Langer, V. Bayot, E. Grivei et al., *Phys. Rev. Lett., 76,* **1996**, 479.
10 T. W. Ebbesen, H. Lezec, H. Hiura et al., *Nature, 382,* **1996**, 54.
11 A. Kasumov, I. I. Khodos, P. M. Ajayan et al., *Europhys. Lett., 34,* **1996**, 429.
12 S. Frank, P. Poncharal, Z. L. Wang et al., *Science, 280,* **1998**, 1744.
13 A. Batchtold, C. Strunk, J. P. Salvetat et al., *Nature, 397,* **1999**, 673.
14 M. M. J. Treacy, T. W. Ebbesen, J. M. Gibson, *Nature, 381,* **1996**, 678.
15 S. J. Tans, R. M. Verschueren, C. Dekker, *Nature, 393,* **1998**, 40.
16 J. Kong et al., *Science, 287,* **2000**, 622.
17 P. G. Collins, K. Bradley, M. Ishigami et al., *Science, 287,* **2000**, 1801.
18 Q. H. Wang et al., *Appl. Phys. Lett., 72,* **1998**, 2912.
19 S. Fan et al., *Science, 283,* **1999**, 512.
20 A. C. Dillon et al., *Nature, 386,* **1997**, 377.
21 A. M. Morales, C. M. Lieber, *Science, 279,* **1998**, 208.
22 Y. F. Zhang, Y. H. Tang, N. Wang et al., *Appl. Phys. Lett., 72,* **1998**, 1835.
23 S. T. Lee, Y. F. Zhang, N. Wang et al., *J. Mater. Res., 14,* **1999**, 4503.
24 S. T. Lee, N. Wang, C. S. Lee, *Mater. Sci. Eng. A, 286,* **2000**, 16.
25 S. T. Lee, N. Wang, Y. F. Zhang et al., *MRS Bull., 24,* **1999**, 36.
26 N. Wang, Y. F. Zhang, Y. H. Tang et al., *Appl. Phys. Lett., 73,* **1998**, 3902.
27 N. Wang, Y. F. Zhang, Y. H. Tang et al., *Phys. Rev. B, 58,* **1998**, R16024.
28 Y. H. Tang, Y. F. Zhang, H. Y. Peng et al., *Chem. Phys. Lett., 314,* **1999**, 16.
29 Y. F. Zhang, Y. H. Tang, H. Y. Peng et al., *Appl. Phys. Lett., 75,* **1999**, 1842.
30 Y. Cui, L. J. Lauhon, M. S. Gudiksen et al., *Appl Phys. Lett., 78,* **2001**, 2214.
31 Y. Y. Wu, P. D. Yang, *J. Am. Chem. Soc., 123,* **2001**, 3165.
32 H. Y. Peng, N. Wang, W. S. Shi et al., *J. Appl. Phys., 89,* **2001**, 727.
33 H. Y. Peng, Z. W. Pan, L. Xu et al., *Adv. Mater., 13,* **2001**, 317.
34 W. S. Shi, H. Y. Peng, N. Wang et al., *J. Am. Chem. Soc., 123,* **2001**, 11095.
35 Y. F. Zhang, Y. H. Tang, N. Wang et al., *Phys. Rev. B, 61*(7), **2000**, 4518.
36 W. S. Shi, Y. F. Zheng, N. Wang et al., *Chem. Phys. Lett., 345,* **2001**, 377.
37 H. Y. Peng, X. T. Zhou, N. Wang et al., *Chem. Phys. Lett., 327,* **2000**, 263.
38 W. S. Shi, Y. F. Zheng, N. Wang et al., *Adv. Mater., 13,* **2001**, 591.
39 W. S. Shi, Y. F. Zheng, N. Wang et al., *Appl. Phys. Lett., 78,* **2001**, 3304.
40 W. S. Shi, Y. F. Zheng, H. Y. Peng et al., *J. Am. Ceram. Soc., 83,* **2001**, 3228.
41 W. S. Shi, Y. F. Zheng, N. Wang et al., *J. Vac. Sci. Technol. B, 19,* **2001**, 1115.
42 J. Q. Hu, X. L. Ma, Z. Y. Xie et al., *Chem. Phys. Lett., 344,* **2001**, 97.
43 K. W. Wong, X. F. Zhou, F. C. K. Au et al., *Appl. Phys. Lett., 75,* **1999**, 2918.
44 F. C. K. Au, K. W. Wong, Y. H. Tang et al., *Appl. Phys. Lett., 75,* **1999**, 1700.
45 D. D. D. Ma, C. S. Lee, S. T. Lee, *Appl. Phys. Lett., 79,* **2001**, 2468.
46 D. P. Yu, Z. G. Bai, Y. Ding et al., *Appl. Phys. Lett., 72,* **1998**, 3458.
47 N. Wang, Y. H. Tang, Y. F. Zhang et al., *Chem. Phys. Lett., 283,* **1998**, 368.
48 U. Setiowati, S. Kimura, *J. Am. Ceram. Soc., 80,* **1997**, 757; G. Hass,

C. D. Salzberg, *J. Opt. Soc. Am.*, *44*, **1954**, 18.
49 G. Hass, C. D. Salzberg, *J. Opt. Soc. Am.*, *44*, **1954**, 18.
50 T. S. Chu, R. Q. Zhang, H. F. Cheung, *J. Phys. Chem. B*, *105*, **2001**, 1705.
51 R. Q. Zhang, T. S. Chu, H. F. Cheung et al., *Phys. Rev. B*, *64*, **2001**, 113304.
52 R. Q. Zhang, T. S. Chu, H. F. Cheung et al., *Mater. Sci. Eng. C*, *16*, **2001**, 31.
53 J. P. Borel, *Surf. Sci.*, *106*, **1981**, 1.
54 C. P. Li, C. S. Lee, X. L. Ma et al., *Adv. Mater.*, *15*, **2003**, 607.
55 T. Y. Tan, S. T. Lee, U. Gösele, *Appl. Phys. A: Mater. Sci. Process.*, *74*, **2002**, 423.
56 J. T. Hu, T. W. Odom, C. M. Lieber, *Acc. Chem. Res.*, *32*, **1999**, 435–445.
57 S. T. Lee et al., unpublished results.
58 E. I. Givargizov, *J. Cryst. Growth*, *20*, **1973**, 217.
59 G. Hass, *J. Am. Ceram. Soc.*, *33*, **1950**, 353.
60 Y. Cui, X. F. Duan, J. T. Hu et al., *J. Phys. Chem. B*, *104*, **2000**, 5213.
61 Y. H. Tang, Y. F. Zhang, N. Wang et al., *J. Appl. Phys.*, *85*, **1999**, 7981.
62 W. S. Shi, H. Y. Peng, Y. F. Zheng et al., *Adv. Mater.*, *12*, **2000**, 1343.
63 W. Z. Li, S. S. Xie, L. X. Qian et al., *Science*, *274*, **1996**, 1701.
64 R. S. Wagner, R. G. Treuting, *J. Appl. Phys.*, *32*, **1961**, 2490.
65 W. S. Shi, H. Y. Peng, L. Xu et al., *Adv. Mater.*, *12*, **2000**, 1927.
66 Y. Zhang, K. Suenaga, C. Colliex et al., *Science*, *281*, **1998**, 973.
67 C. P. Li, N. Wang, S. P. Wong et al., *Adv. Mater.*, *14*, **2002**, 218.
68 Z. W. Pan, H. L. Lai, F. C. K. Au et al., *Adv. Mater.*, *12*(16), **2000**, 1186.
69 W. Z. Li, S. S. Xie, L. X. Qian et al., *Science*, *274*, **1996**, 1701.
70 Z. W. Pan, S. S. Xie, B. H. Chang et al., *Nature*, *394*, **1998**, 631.
71 Z. W. Pan, S. S. Xie, B. H. Chang et al., *Chem. Phys. Lett.*, *299*, **1999**, 9.
72 X. Duan, J. Wang, C. M. Lieber, *Appl. Phys. Lett.*, *76*, **2000**, 1116.
73 N. Wang, Y. H. Tang, Y. F. Zhang et al., *Chem. Phys. Lett.*, *299*, **1999**, 237.
74 J. Q. Hu, X. L. Ma, N. G. Shang et al., *J. Phys. Chem. B*, *106*, **2002**, 3823.
75 Y. Jiang, X. M. Meng, J. Liu et al., *Adv. Mater.*, *15*, **2003**, 323.
76 X. H. Sun, H. Y. Peng, Y. H. Tang et al., *J. Appl. Phys.*, *89*, **2001**, 6396–6399.
77 Y. F. Zhang, L. S. Liao, W. H. Chan et al., *Phys. Rev. B*, *61*, **2000**, 8298.
78 X. T. Zhou, J. Q. Hu, C. P. Li et al., *Chem. Phys. Lett.*, *369*, **2003**, 220.
79 X. H. Sun, C. P. Li, N. B. Wong et al., *J. Am. Chem. Soc.*, *124*, **2002**, 14856.
80 R. H. Fowler, L. W. Nordheim, *Proc. R. Soc. London, Ser. A*, *119*, **1928**, 173.
81 D. D. D. Ma, C. S. Lee, F. C. K. Au et al., *Science*, *299*, **2003**, 1797.
82 Y. H. Tang, X. H. Sun, F. C. K. Au et al., *Appl. Phys. Lett.*, *79*, **2001**, 1673.
83 D. D. D. Ma, C. S. Lee, Y. Lifshitz et al., *Appl. Phys. Lett.*, *81*, **2002**, 3233.
84 R. Q. Zhang, T. S. Chu, S. T. Lee, *J. Chem. Phys.*, *114*, **2001**, 5531.
85 D. K. Yu, R. Q. Zhang, S. T. Lee, *Phys. Rev. B*, *65*, **2002**, 245417.
86 B. X. Li, R. Q. Zhang, P. L. Cao et al., *Phys. Rev. B*, *65*, **2002**, 125305.
87 R. Q. Zhang, S. T. Lee, C. K. Law et al., *Chem. Phys. Lett.*, *364*, **2002**, 251.
88 D. K. Yu, R. Q. Zhang, S. T. Lee, *J. Appl. Phys.*, *92*, **2002**, 7453.
89 R. Q. Zhang, W. C. Lu, S. T. Lee, *Appl. Phys. Lett.*, *80*, **2002**, 4223.